PRAISE FOR *THE NATURAL BUILDING COMPANION*

"As the title clearly states, this book is a true 'companion' to take along on one's natural building journey. A well-thought-out account of Jacob Deva Racusin's and Ace McArleton's experiences and the methods they have evolved over the years. Details that are useful in just about any climate, but particularly in cold climates. One of the best natural building books published in recent years."

—Bill Steen, author, *The Straw Bale House*

"*The Natural Building Companion* is a joy to read. The approach is holistic, the style is generous, and the authors gracefully balance technical details, beautiful spaces, and big ecological questions. This book empowers the reader to make choices that matter—for their own home and for the health of our planet."

—Paul Lacinski, coauthor, *Serious Straw Bale*

"This excellent and thoroughly researched book reflects the progression of natural building. It eloquently expresses the beautiful marriage between experience and experimentation, fact and anecdote, science and soul. Jacob Deva Racusin and Ace McArleton have built us a much-needed bridge between natural building and green building. *The Natural Building Companion* will transform how we build!"

—Adam Weismann and Katy Bryce, authors,
Using Natural Finishes and *Building with Cob*

"A thorough treatment of both the context for and specifics of natural building, this book is a wealth of resources all in one place. Though geared to the Atlantic Northeast, there is still more than enough here to make it a must-read for natural builders everywhere."

—Bruce King, author, *Design of Straw Bale Buildings*
and *Buildings of Earth and Straw*

"*The Natural Building Companion* brings age-old building techniques and materials into the modern world. It is a comprehensive evaluation of natural building methods that shares results from up-to-date testing and monitoring, explains how these results apply to building codes, and covers new tools and resources that make installation more user friendly and durable. By combining a wealth of information in a single volume, this book is indispensable to anyone wanting to build a healthy home using natural materials."

—Will Beemer, director, The Heartwood School for the Homebuilding Crafts

"'What is a high-performance building?' Racusin and McArleton answer that question by taking a values-based approach that integrates social and ecological good with health, resource efficiency, and durability. Fusing the knowledge of the building-science community with the wisdom and experience of the best natural building practitioners, *The Natural Building Companion* provides plenty of detailed, how-to information to help readers create structures and communities that are, as the authors put it, 'worthy of our highest aspirations.'"

—Marc Rosenbaum, founder, Energysmiths

the Natural Building COMPANION

A COMPREHENSIVE GUIDE TO INTEGRATIVE DESIGN AND CONSTRUCTION

Jacob Deva Racusin and Ace McArleton

CHELSEA GREEN PUBLISHING
WHITE RIVER JUNCTION, VERMONT

Project Manager: Patricia Stone
Editorial Contact: Joni Praded
Developmental Editors: Dan Eckstein and Tara Hamilton
Copy Editor: Cannon Labrie
Proofreader: Nancy W. Ringer
Indexer: Peggy Holloway
Designer: Melissa Jacobson
Illustrators: Ben Graham of Natural Design/Build and Aaron Westgate/FreeFlow Studios

Printed in the United States of America
First printing March 2012
10 9 8 7 6 5 4 3 2 1 12 13 14 15 16

Photographs featured in this book portray a variety of natural buildings from around the country and the buildings are the products of a wide variety of designers, builders, students, and craftspeople.

Chelsea Green Publishing is committed to preserving ancient forests and natural resources. We elected to print this title on paper containing at least 10% postconsumer recycled paper, processed chlorine-free. As a result, for this printing, we have saved:

27 Trees (40' tall and 6-8" diameter)
12,358 Gallons of Wastewater
11 million BTUs Total Energy
784 Pounds of Solid Waste
2,740 Pounds of Greenhouse Gases

Chelsea Green Publishing made this paper choice because we are a member of the Green Press Initiative, a nonprofit program dedicated to supporting authors, publishers, and suppliers in their efforts to reduce their use of fiber obtained from endangered forests. For more information, visit www.greenpressinitiative.org.

Environmental impact estimates were made using the Environmental Defense Paper Calculator. For more information visit: www.papercalculator.org.

Our Commitment to Green Publishing
Chelsea Green sees publishing as a tool for cultural change and ecological stewardship. We strive to align our book manufacturing practices with our editorial mission and to reduce the impact of our business enterprise in the environment. We print our books and catalogs on chlorine-free recycled paper, using vegetable-based inks whenever possible. This book may cost slightly more because it was printed on paper that contains recycled fiber, and we hope you'll agree that it's worth it. Chelsea Green is a member of the Green Press Initiative (www.greenpressinitiative.org), a nonprofit coalition of publishers, manufacturers, and authors working to protect the world's endangered forests and conserve natural resources. *The Natural Building Companion* was printed on FSC®-certified paper supplied by RR Donnelley that contains at least 10% postconsumer recycled fiber.

Library of Congress Cataloging-in-Publication Data

Racusin, Jacob Deva.
The natural building companion : a comprehensive guide to integrative design and construction / Jacob Deva Racusin and Ace McArleton.
pages cm
Yestermorrow Design/Build Library.
Includes bibliographical references and index.
ISBN 978-1-60358-339-8 (pbk.) — ISBN 978-1-60358-340-4 (ebook)
1. Ecological houses—Design and construction. I. McArleton, Ace. II. Title.
TH4860.R33 2012
690'.8047—dc23

2011047471

Chelsea Green Publishing
85 North Main Street, Suite 120
White River Junction, VT 05001
(802) 295-6300
www.chelseagreen.com

MIX
Paper from responsible sources
FSC® C101537

CONTENTS

ACKNOWLEDGMENTS

This book is the result of many years' work and dedication by many wonderful people. Those who contributed to its creation are too numerous to mention. The authors, however, would like to acknowledge the following individuals and organizations for their support:

The creators of the natural building and traditional building movements in both North America and across the world, whose courage and innovation enabled our own progress: Bill and Athena Swentzell Steen, Judy Knox (may she rest in peace) and Matts Myhrmann, David Eisenberg, David Bainbridge, Steve MacDonald, Ianto Evans, Bruce King, Catherine Wanek, Frank Anderson, Paul Lacinski, Clark Sanders, Michelle and Will Beemer, Jack Sobon, Stafford Holmes and Michael Wingate, and many others, including the thousands of owner-builders and professional builders who have taken great risks and made great efforts to create the path toward a just and sustainable future by walking it. Without their failures and successes both we would not know what we know today. Also the traditional craftspeople and indigenous builders whose techniques we use to this day, in large part for whom all building was natural building.

Our incredible clients, who have shown us unstinting generosity both during our working relationship and far beyond, allowing us to photograph and performance-test their homes, giving us feedback on our work, and hosting innumerable house tours: Maggie Cahoon and Joe Marks, Dylan Ford, Bobby Farlice-Rubio and Indigo, Jango, and Niobe, Mary Ellen Blakey, Will and Theresa White, Linda Ides and Brad Viatje, Lynne Hadley, Peter and Helen Forbes and Taz Squire of the Center for Whole Communities at Knoll Farm, Michael Brodeur, and Stephen Le Blanc and Kate Abbott.

Our brilliant colleagues and friends in Natural Builders NorthEast (NBNE), from whom we have learned and shared so much and without whom the advancement of this work would not be possible: Ben Graham, Josh Jackson, Stephen Dube, Mark Krawczyk, Mark Piepkorn, Sarah Machtay, Clark Sanders, Daniel Fullmer, Bryan Felice, Micah Whitman, Aaron Dennis and ej George, Sarah Highland, Brent and Diana Katzmann, Dave Lanfear, Kevin Connors, Dave Vail, Steve Paisley, Ben Simpson, Jim Luckner, Jonah Vitale-Wolff, Andy Mueller, Megan McNally, and Liz Johndrow. Also our beloved friends and colleagues beyond the NBNE fold: Tim Rieth, José Galarza, Sasha Rabin, Andy Burt, Danny Viescas, Adam Weismann, Katy Bryce, Skip Dewhirst and Lizabeth Moniz, Ryan Chivers, Thea Alvin, Patti Garbeck, Ben Falk, Keith Morris, Missa Aloisi, and William Davenport.

Our teachers and mentors, who have shared with us both directly and indirectly so much about designing, building, and much, much more: Paul Lacinski, Chris Magwood, John Straube, Robert Riversong, Bill and Athena Swentzell Steen, Brad Cook, Josh Jackson, Buzz Ferver, John Unger Murphy, and Bill Hulstrunk.

Our New Frameworks Natural Building family, past and present: Chloe Jhangiani, David Ludt, Nick Jackson, Nick Salmons, Julie Krouse, Kelly Cutchin, and Joanna Ware.

The amazing staff of Yestermorrow, who have supported us fully and tirelessly in our work from the beginning, and given us the opportunity to work in a community of some of the finest educators, designers, and craftspeople alive: Kate Stephenson, Dan Eckstein, Monica DiGiovanni, José Galarza, Heidi Benjamin, and the many fabulous interns, staff, and directors with whom we've worked over the years.

Our students, from whom we have learned so much, and with whom we have worked together to co-create a built environment as a place we all want to live in, humans and nonhumans alike. Special thanks to everyone involved in the Natural Building Intensive Program in 2009, where much of the DVD footage was taken—the initial inspiration for this book.

All of the people who gave so generously of themselves to help bring this book to life: the good people of Chelsea Green, including Joni Praded, Patricia Stone, and Margo Baldwin, with whom it was a great pleasure to work; Ed Dooley of Mad River Media, for taking the risk on filming many hours of footage on spec and bringing the DVD to life; Dan Eckstein and Tara Hamilton, who shared with us the journey of developmental editing with grace and patience; Robert Riversong, for his guidance, instruction, contributions, and many hours of editing the most technical chapters of this book; Aaron Westgate and Ben Graham, for their friendship and artistic brilliance; Kelly Griffith of Closed Circle Photography, Jen Smith, and the other credited photographers, for their stunning photography; Alicia Fraser, for epidemiological expertise on dioxin; Mike Beluso and Melody Emily Martin, for their terrific research assistance; Maureen Raymo for geology consultancy; and Carey Clouse, Jeff Schoellkopf, Mark Krawczyk, John Straube, City Repair Portland, and ReUse Action Western New York, for their contributions to this project.

Jacob Deva Racusin would personally like to thank Mary Niles, Elijah, Naomi, and Micah, for their incredible support, patience, and humor during many years of writing and editing; Will Moses, who taught me how to build; Remi Gratton, my friend and mentor, who leads by his example; Ace McArleton, who helps me daily to achieve more than I could ever achieve alone; John Ringel, Kathy Meyer, and Lisa Williams, the first instructors I had on my path toward learning how to design and build; and Cherry, Robert, and Jessica Racusin, for believing in me and helping me to believe I can create anything I choose.

Ace McArleton would personally like to thank my life partner, Diana González, for strong love and faith in me; my work partner Jacob Deva Racusin, for daily joy and being my co-conspirator in taking over the world; my queer family: Chloe Jhangiani, Llu Mulvaney-Stanak and Kate Van Wagner, Alicia Fraser, Megan Nealis, Joanna Ware, and Marina Weisz, for support along every step of this book process; Bricklayers and Allied Craftsworkers Local 3 and my tile-setter Stack Daddy, for teaching me much-needed trades lessons in my apprenticeship; Amber Wiggett, for being bold and fiercely creating space for women and trans people in construction; Cindy Milstein, Chaia Heller, Ian Grimmer, Joshua Stephens, Arthur Foelsche, Walter Hergt, and the good people surrounding the Institute for Social Ecology, who shared brilliant, important theory and practice with me that has structured my worldview; Jean McPhail, Dan Carleton, and Jordan McArleton, who made writing and words a part of my life always; and AndreA Neumann-Mascis, for always holding the light.

INTRODUCTION

The building of structures—for living, for working, for storage, for play and gathering—is a practice that has been with us for as long as humans have been on this planet. As the authors of this book and as builders, our goal is to improve and enhance the built environment. In our view, the act of building should promote social and ecological health and well-being while creating structures that perform well in all weather conditions and are comfortable, beautiful, and long lasting. The flourishing green-building movement is a step in the right direction as more members of the architecture and building communities are clearly starting to take notice that design solutions and building practices focused on reduced energy consumption and improved indoor air quality need to be part of their standard operating procedures.

Green-building practitioners generally hold performance as their approach's highest metric: How much less energy will a given material or building method use than traditional choices? How well will the building perform under temperature, pressure, and moisture stresses? Green building has become more mainstream because the language of performance and energy efficiency is an easily understood and welcome benefit of this type of building strategy. These are undeniably important performance criteria. We need to consume less energy in both building and using our structures to ensure a healthy future on our planet. We need our buildings to maintain the correct levels of temperature and humidity with the minimum amount of resource use to accomplish these goals. We need to know our buildings will perform well when confronted with fire, high winds, excessive rain, or other extreme weather events. We need to know our systems will resist organic threats: mold, mildew, and insect or rodent infestations. We need to understand appropriate materials and systems that work really well for us as builders (and dwellers), and that create not just adequate but excellent shelter.

But what if the performance of our structures is evaluated by other metrics as well? From our standpoint, green-building practitioners' focus on performance that is assessed solely by the structure's level of energy efficiency and durability has resulted in blinders that impede the peripheral vision necessary for the creation of truly ethical and ecologically sound structures. (Note: In referring to "green building" we are speaking of the industry-scale community of design and construction professionals whose standards are largely defined by certifying agencies, such as the United States Green Building Council [USGBC] and its Leadership in Energy and Environmental Design [LEED] program.) While we support the goals imparted by the green-building movement, they are goals that do not dig deeply enough nor span broadly enough to address humanity's ongoing and increasing challenges. For instance, green building has promoted, among other things, the profligate use of foams, which meet performance and efficiency criteria but are toxic, bio-persistent compounds that will be on this planet for an unknown amount of time, harming the ecological systems in which they remain. Already foams, plastics, and other trash items have aggregated in a massive gyre in the North Pacific Ocean and other oceans across the planet, causing serious harm to pelagic ecosystems. These compounds are not biodegradable, and we must ask ourselves whether using these materials is the best option we have to meet our efficiency and performance goals.

Natural building is broader in its approach and scope than green building. It holds us to a standard that redefines the concept of "high performance" to include consideration of the materials used throughout their entire life stream, including how the materials—and the methodologies with which they are applied—affect the people and communities involved, as well as the broader ecology over time. In addition, natural building is about relationships: we choose to work with natural materials not only because they are "natural" but also because their use helps us develop and sustain as many relationships and connections as possible within the context of the development of a building. People often feel disconnected from their shelter, and natural building—through the use of local materials that are more familiar, a part of the natural world, and less toxic in their manufacture or method of installation—offers a way for more people to connect to the process and practice of creating and maintaining shelter. Natural building is about encouraging access, opening up pathways for involvement, while simultaneously striving for levels of built excellence. Rediscovering how to build high-performance structures with natural materials will take education, research, and investment, but we prefer to strive for this higher level of quality by encouraging more education and by continuing to promote open access to this work, as open-source information and democratic involvement are fundamental principles of natural building.

Natural builders ask such questions as:

- *Is a given material toxic in origin, in production, during use, or at the end of its life?* For example, the human and nonhuman communities in the Kanawha Valley (nicknamed "Chemical Valley") in West Virginia suffer health risks due to exposure to toxins from from plastic and foam manufacturers in their towns (Ware et al. 1993). When we do not use a building material such as latex paint made from ingredients synthesized in Chemical Valley and choose instead a natural paint made from local clays and wheat paste, the entirety of these life-cycle conditions and effects are considered. Ideally, building materials should not pose a threat to the health and well-being of the builder and the ecologies of which they are a part for the duration of their life cycles.

- *What is the aesthetic and sensory experience for people who live or work with the material?* Often those who use natural paints for the first time tell us how surprisingly enjoyable it is not to smell the fumes that accompany the use of industrial paints. Using natural paints can have practical implications as well: the lack of off-gassing of these paints opens up the possibility of painting in the middle of winter, with minimal ventilation—something one could never do with latex or conventional oil-based paint and reside safely and comfortably in the space at the same time.

- *Does the material fit with the climatic demands, the local ecology, and the local vernacular building style?* Sometimes straw is not available nearby in baled form, and the cost and carbon footprint of transporting it make it not the best choice, so other, more local options may be the more "natural" choice. Additionally, the design detailing relevant to that region—practically and culturally—must at least be considered, with the understanding that the technologies available to us as natural builders can be selected and adapted to support the priorities of regional design.

- *Does the material or process connect us to our planet—such as watching the straw that will become the walls of your house grow in your neighboring farmer's fields?* And does it connect us to our communities and to each other, such as by acquiring stone or locally grown lumber from a neighbor, or through work parties for a straw bale wall or timber frame raising?

> With intention and action natural builders create structures that enhance social and ecological well-being, incorporate what we have learned from the past, address current best practices, and push us toward a brighter future.

Natural buildings have often performed well historically, but just as often they have fallen well short of meeting performance expectations owing to a lack of cultural knowledge about how to work effectively with natural materials in the climate in which the building is located—or because of a general deficiency in design or building skills and techniques. We therefore acknowledge that natural building practitioners can learn a lot from green builders about durability, moisture management, energy efficiency, and testing. The effective integration of these skills and techniques, which have been fine-tuned by green-building practitioners, into the practice of natural building is one goal of this book. We are cognizant that building healthy and ethical structures is complicated work, sometimes requiring thoughtful and intelligent assimilation of a variety of building philosophies and methodologies. Natural building and green building do not have to stand in opposition to one another, but rather can complement each other. Our ambition is to draw from multiple philosophies of construction to address building challenges while still being guided by the standard that is based on the consideration of the full life-cycle performance of a material or process, including social as well as ecological considerations. We feel this integrative approach to the design/build process is the most effective, taking the best from all building disciplines to achieve shared goals of improvement in our built environment.

Such a complex approach moves the field of natural building away from the fixation on "purity" of materials and form as these relate to architectural building and design. For example, cob is not always the best choice in the Northeast given mass-based walls' low R-value, and the consequent high heating needs that would need to be designed into such a building. When choosing between cob and a cellulose-insulated wall for a structure in the Northeast, cellulose may easily be the "natural building" choice, despite its more industrial origins and machine-intensive installation, and cob's status as the ultimate do-it-yourself, low-impact natural building material.

The natural building approach is more than just a materials choice. It is an approach that emphasizes interconnectedness and multiple metrics for establishing what is appropriate to build, how best to go about it, and what materials to use to achieve the desired outcome. Decisions on materials and methods are arrived at after answering questions that take different ethical and practical considerations into account, rather than accepting a predetermined decision based on its designation as "preferable" in the abstract, outside of the context of its application. This allows us to be creative, to innovate in our locations, and to build high-performance buildings in a multitude of different ways that suit our local ecologies, economies, and cultures.

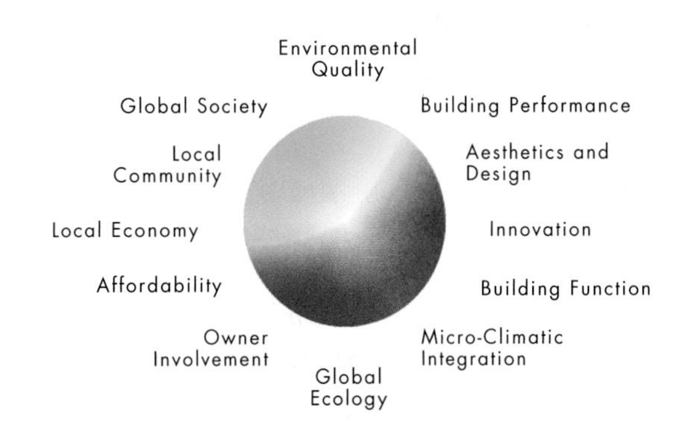

Good design and construction involve negotiating many dynamic values throughout the process, and the structure reflects these values. IMAGE BY JACOB DEVA RACUSIN.

If we ask the question, "What is a high-performance building?" and use a set of metrics grounded in ecology, environment, economics, empowerment, performance, and enjoyment to determine the answer, we then get away from predetermined ideas of how to build. We are consequently able to evaluate straw, clay, lime, cement, or foam simply as a material or method we could choose to use, each of which might meet different goals with different costs and benefits.

This strategy asks more of us. It asks us to think in complex ways, to comprehend the multiple aspects of any material or process, and to problem-solve, just as natural materials require us to. Just because a material is unprocessed, from the earth, or readily available doesn't make it easy to work with. We become the manufacturers, which is more challenging than utilizing pre-processed materials but which also provides abundant opportunity for connection to our human past, the earth, and each other—as well as opportunities for creativity, empowerment, and innovation. The more we invest our energies, the more we gain from the results.

Unfortunately, much of the knowledge of how to work with natural materials has been largely obscured or lost. It is also segmented into different trades and academic studies, and assembling it all into a coherent strategy is a challenging undertaking. One of the reasons we have written this book is to bring together what might be seen as seemingly unrelated fields such as soil science, agriculture, architecture, permaculture, and building science into a useful framework for building structures that perform exceptionally well while also enhancing or otherwise contributing to their ecological and social context.

Through our work in the field over the last decade, we have become increasingly aware of the genuine need to propel natural building toward the next phase in its evolution so that it is accepted as a serious, legitimate approach to designing and building structures. Natural building methods using materials such as straw bales, straw-clay (a mixture of loose straw and clay slip placed in formwork), and natural plasters are authentic, viable, successful building strategies that address ecology and social health as well as energy and durability performance challenges. We make the case in this book that the materials and methods examined within these pages must be considered when the goal is to build high-performance buildings as defined by a wide-ranging set of criteria.

There are many who remain resistant to accepting or even considering natural building principles and methods as realistic options. We maintain that it is a lack of information that causes people to scoff at these materials and building styles, and to think them inherently flawed. It is our lost knowledge of our building heritage that causes us to think them "weird" or obscure, when it is actually such current-day common building materials like polyisocyanurate (foam) and everyday 4 × 8 foot plywood that are the new kids on the block with questionable long-term impacts on our environment and our health. By evaluating natural building practices in a comprehensive context and through the lens of the principles of building science, we will attempt to communicate to skeptics the reasoning behind our assertions of validity, and challenge them to hold their accepted modern, conventional technologies to the same standards of criticism.

Our aspiration in writing this book is also a practical one. We have been building natural structures for over a decade and teaching how to build them for nearly as long at Yestermorrow Design/Build School in Warren, Vermont—a unique school that teaches laypeople and professionals cutting-edge design and building techniques. Through our work as builders we have experimented and tested and explored countless strategies and methodologies and have developed techniques and skills that are in need of greater dissemination. Clients who have come to us for design or construction assistance have often asked for published resources they could consult to gain a better understanding of how to detail their buildings in our cold, wet northeastern climate. Both within our own company and collectively with our colleagues in Natural Builders NorthEast (a group of builders, designers, consultants, educators, and professionals located throughout the northeastern

United States and eastern Canada who are practicing the art of natural building and design), we were not seeing the methods that we were utilizing and evolving in the field reflected in the literature. Through designing and continually improving the curriculum for the natural building courses we offer regularly at Yestermorrow we found ourselves compiling information and presenting it in a particular framework that we did not see elsewhere. The reading lists we create for students are filled with excellent articles that touched on parts of the picture, but nothing exists that brings the seemingly disparate elements together into a coherent whole. Our curriculum is unique in its depth, breadth, and complexity, and we seek to provide for our students an appropriate resource to match the content of the courses.

Ace McArleton started off in the mainstream trades as an apprentice in Bricklayers and Allied Craftsworkers Local 3 in the Bay Area in California in 2000, working on commercial jobs and learning the culture of the traditional masonry trades. Since then, propelled by a desire for craftspersonship that is not driven solely by profit, builder culture that is not steeped in interpersonal and social violence, and structures that are responsive to the surrounding ecology, Ace moved to Vermont, worked on a natural-building crew for two years, and then started New Frameworks Natural Building, LLC in 2006. Jacob Deva Racusin joined New Frameworks Natural Building as a partner in 2009, having built professionally since 2004 and having begun his path as a designer and builder with the construction of his own straw bale and green wood home in 2000. NFNB offers services in green remodeling, new construction, consultation, and education featuring natural building technologies.

When the opportunity arose to document Yestermorrow's 2009 Natural Building Intensive program on film, and to create a book and accompanying DVD package exploring updated natural building methods, we immediately saw the utility of this project in elucidating key methods and concepts in our field. In this book you will find up-to-date information, instructions, and guidance on natural materials, building methods, and design principles. This book is divided into three parts: "The Context for Natural Building," "Building Science and Performance," and "Natural Building Practices." Part 1 provides background on the historical, ecological, and geologic underpinnings of natural building practices, as well as insight into the common materials utilized in the field. Part 2 explores the fundamentals of building science and performance and specifies effective strategies for the integration of natural materials and methodologies within these metrics. Part 3 details the specific methods and skills needed for the practical and effective utilization of natural materials. Along the way, you'll find callouts directing you to the accompanying DVD, so you can see firsthand the methods and processes being described in action.

You'll also find more. We make the case in this book that natural building is both a process and an approach: it is a set of decisions that are guided by ethical criteria with multiple—perhaps infinite—good answers. Natural building does not have a set of rules for the practitioner to follow. Therefore, rather than a prescription or an imperative, we share in these pages a way of thinking—a framework—and information that can serve as tools for making decisions applicable to building shelter in any climate or situation.

Building can be more than knowing how to put things together with tools. We human beings are unique in that we have the capacity for self-reflection and can weigh ethical concerns to inform the creation of our habitat. Building is an opportunity to (literally) construct the world as we would like to see it: not just through choice of materials performance, but through a process that creates a stronger social fabric and a healthier ecology. The very human, basic ethical question of "What is good?" is central to all realms of social activity. This book advocates making this question the guide for our choices and our work as designers and builders of structures. Natural building is the process and practice of designing and creating shelter that is supportive of ecological well-being and thriving social networks throughout all phases of creation, maintenance, and end-of life breakdown.

This book hails from the northeastern United States, a relatively cold and wet climate. In it, we walk you through the most up-to-date strategies we know to meet building challenges with the use of natural materials. Because these challenges will differ based on the climate, the project, the people involved, and local materials, among other factors, ultimately this work is best seen as a case study that showcases how you can approach designing and building with these materials in any climate or situation. The emphasis is placed on how to go about determining a good design or construction detail, or the most relevant material, or the soundest economic practice in the specific place in which you find yourself as a builder.

Ultimately, it is this ability to think through building problems given the surrounding context, finding local solutions to global problems guided by the shared human and ecological goals of health, growth, and happiness, that will help us create structures and communities worthy of our highest aspirations.

PART 1

The Context for Natural Building

CHAPTER 1

A Brief History of Natural Building in the Northeast

In this chapter we will explore building practices that trace back to pre-Columbian times to discover the origins of many of the materials we use today and to provide context for their use in natural building processes and techniques. The past provides insight into design and construction practices that we can choose to draw on as a part of a natural building heritage: indigenous people, European settlers, and others have used site-specific design and have innovated with straw, clay, lime, timber, and stone. These traditions have survived despite the growth of post–World War II standardized, industrialized construction methods, which are now ubiquitous.

One of our favorite things to do in our classes and presentations is to ask people to look at the room we are in and notice what the walls, floors, and ceilings are made of. We do this because most of us spend so little time reflecting on our built environment, relative to the amount of influence it has on our lives. The materials—their feeling, their performance, their ecological impact, their aesthetic, as well as the social context of the development patterns and architecture itself—are all around us every day.

Today's built environment has evolved as a result of numerous historical human-influenced factors, including culture, economics, power and decision-making structures, technologies of the era, and which natural resources were available, among others. Buildings, as well as the development patterns they are part of, are the result of actions taken based on a given era's priorities. These priorities derive in part from what current mores dictate is the appropriate answer to the question, "What is good?" (although this question unfortunately has often been watered down to, "What is the best we can do given the circumstances?").

Intentionality, or acting to manifest one's priorities, has always been a component of the building process. Looking ahead, there is tremendous opportunity to steer this intentionality toward the deliberate creation of buildings that reflect priorities focused on increasing the health and well-being of the social and ecological life of our societies.

Learning from History

Historical connections are easier to make when history is visible in our environment. European natural builders, for instance, have around them structures built with clay, sand, stone, timber, straw, and lime that have endured the passing of time. In the United Kingdom, Katy Bryce and Adam Weismann, authors of *Using Natural Finishes*, find themselves surrounded by lime plasters, cob, timber framing, thatch roofing, and structures made using other natural building methods as well—some of which go back 500 years or more. These existing structures can be evaluated for longevity and performance characteristics. There are vibrant trades organizations and apprenticeships—such as Les Compagnons (the French timber apprentice program)—whose missions are to develop professionals devoted to increasing living knowledge of how to work with these time-tested systems. There is still much work to be done to develop natural building into an accepted, high-performance building strategy, even in places like Europe where such historical links are alive, although the historical connections that are present make the prospect of integrating natural building seem like less of a foreign or brand-new idea.

In the southwestern United States visible examples of ancient building traditions also remain, in this case the adobe pueblos and cliff dwellings of pre-Columbian people. The combination of clay, manure, and sand used to create blocks or sculpted walls, the use of pre-existing stone cliffs, and very dry conditions have together left a legacy of buildings on that landscape that allows us now to see and be inspired by the massive earth vernacular traditions that exist in that region.

Few structures remain from this comparable period of building in the Northeast. Why is this? The biodegradability of the materials that indigenous people in the Northeast were using is often stated as the main reason, combined with the lifestyle pattern of moving every year that led to the creation of structures that could be put up and taken down with relative ease. Animal skins, bark, and woven mats of reeds stretched over a light frame of saplings worked for a migratory life, which is how we understand people lived at that time. For many tribes, winter was a time to retreat to the deep woods, where they could hunt and fish in streams, and as the thaw came they moved to places where they could cultivate the ground.

More permanent wigwams and longhouses did also exist, however, but these examples are difficult to find and see today. This is likely related to the systematic cultural and physical genocide and ongoing racism that indigenous people have experienced in this region since Europeans arrived.

Had this eradication not occurred, we might see in the built landscape how such methods evolved among native peoples and how perhaps those traditions have today blended with those of the new European immigrants. What would it be like to have interwoven historical traditions and social connections between indigenous people and immigrants—then, and today? Colonialism, racism, and genocide have robbed us of these opportunities, and continue to do so (Wiseman 2001, 147–49).

Another factor, we hypothesize, is the climate. In chapter 4 we tell a story of the Egyptian obelisk taken by an American explorer from the very dry land of its creation, where it had stood for a thousand years in almost perfect condition, and brought to New York City's Central Park. The powerful forces of weathering and erosion here in the northeastern United States have all but worn away the carved lettering in fewer than 200 years. The role of climate, specifically precipitation and its role in greatly reducing the longevity of structures, is a theme that will arise often in this book. It is a unique challenge in our region—for all building styles and materials—that requires innovation and creativity to overcome.

Regardless of the longevity of the materials and designs that indigenous people and European immigrants employed, how long could structures have withstood the elements without the important cultural intervention of maintenance and commitment to the structures? Many of the "rude" cottages the settlers built with timbers infilled with wattle and daub, or straw and clay, and plastered, were taken down in favor of the new wooden buildings. The settlers wanted progress, to move "up" from the peasant roots that the old cottages represented. One of the reasons that these structures remain standing in Europe, which is less cold than the northeastern United States but similarly wet, is constant human interaction with these structures, and the passing from generation to generation of the knowledge necessary to maintain them.

So, is all lost? Between climate challenges, colonialism, racism, and a culture that chose to eliminate its past in favor of "progress," can we learn nothing of the history of the built environment, nothing from the people who were building here in the Northeast before the United States was founded? Must we conclude that unlike Europe, and unlike the southwestern United States, natural building does not belong here since no legacy buildings exist? Thankfully, the answer to these questions is no.

WABANAKI AND HAUDENOSAUNEE BUILDING PRACTICES: 500 BCE–1600 CE

The eastern woodlands, thick with pine and hardwood, stretched from the maritime provinces of eastern Canada to the Great Lakes. The region was bounded to the north by the subarctic taiga and to the south by the rise of the Appalachian highlands. The first people came to this region crossing the Pacific Ocean via the Bering Strait land bridge 12,000 years ago. By 1600, two main groupings of people resided in the northeastern region of the United States: the Haudenosaunee Confederacy or League, comprising the Mohawk, Oneida, Onadaga, Cayuga, and Seneca tribes, which was located mainly in present-day New York State, and the Wabanaki (meaning "Dawnland" or "East Land") Confederacy, formed of the eastern tribes of the Abenaki, Penobscot, Maliseet, Passamoquodee, and Mi'kmaq, encompassing present-day Maine, New Hampshire, Vermont, New Brunswick, and Nova Scotia. (We have used the names the people gave to themselves, to the best of our knowledge, rather than the names given to them by European immigrants. Iroquois, for instance, though more commonly known, is a derogatory term for Haudenosaunee people, and Wabanaki is the name for the confederacy of eastern tribes that encompassed five tribes in what we now know as New England.)

Wigwams of the Wabanaki people included both temporary and more permanent dwellings. They both followed the basic building system that prevailed in the Northeast, which was a structural frame made of saplings covered with bark sheets or sewn reed mats.

The framing work of sapling harvesting and installation seems to historically have been a job for Wabanaki men, while the stories tell us that Wabanaki women took over to cover and furnish the skeleton. The reed mats were carefully collected and dried out so they would not mold, then woven into an air-filled loose web that was effectively insulative. It was a developed skill to know the best bulrushes for wall mats and floor mats, and what size cattails should go on outer roofing mats. Stalks were harvested and stacked so as to dry properly, woven together into flexible, lightweight, portable effective insulation.

A wigwam built in the Vermont Historical Society's museum. PHOTO COURTESY OF THE VERMONT HISTORICAL SOCIETY.

The utilization of bark sheets as another wall-covering option offered different performance attributes: they were heavier and were reserved for winter wigwams. Bark was harvested from trees in the spring and sewn together into sections that were stiffened on each end with cedar battens. These bark sheets were softened near the fire before application to the exterior of the structure. Longhouses were the chosen structure for the Haudenosaunee to the west. The name Haudenosaunee means "People of the Longhouse," or more accurately, "They Are Building a Longhouse." The longhouse was both a physical dwelling—where anywhere from twenty families, related matrilineally, would live together under one roof with separate hearths along the main path for each family—and a metaphor for the political structure of the confederacy, in which each tribe, in accordance with their geographic relationship, was envisioned as part of a huge longhouse, with the Mohawk guarding the eastern door, the Seneca the west, and the Onadaga keeping the hearth or central flame. The longhouse was built of sharpened saplings driven into the ground as the main framework, with woven bark for walls and grasses and leaves for roofing.

What can we learn from what we know of indigenous people's building practices? How does this help us as natural builders today? We see the use of materials close at hand used effectively in relationship to the lifestyle and culture of the people. Most notably, the use of natural materials such as reed and straw mats with air pockets for insulative wall coverings is of particular interest, given that this is what makes straw wall systems also thermally effective. The use of wood saplings and timber, for framing, also provides a tie across the centuries for our paired system of today of timber frame with straw bale wrap, or stud frame with bale infill.

From what we know from historical records, neither the longhouse nor the wigwam used woven branches pressed with coblike material, a longtime building method in the United Kingdom, where it is known as wattle and daub. Nor did the people use adobe block or stone, as did their distant neighbors in the Southwest. The choice of the Wabanaki and Haudenosaunee to use animal skins, bark mats, and woven reeds and straw is an intriguing one. As mentioned above, we can speculate that they found these materials more insulative due to intelligent use of air-pocket-filled mats. Daniel Gookin in Massachusetts in 1673 claimed that wigwams were as warm as the best English house. It is also true that the structures were less permanent than woven branches pressed with cob and would have fit better with a nomadic lifestyle.

It is possible to conclude that these indigenous people were not using earth-building techniques in the Northeast either, likely for the same reason that cob and adobe block are not more emphasized as natural building methods in this region today—those massive wall systems are simply not as effective in the relatively cold and wet climate of the Northeast as a well-detailed and protected insulative-straw wall. And possibly rather than try to make straw resistant to elements and build legacy structures, they worked with the material's tendency to biodegrade in this harsh climate by making maintenance and re-creation a part of their lives.

EUROPEAN SETTLERS: 1600–1900, EVOLVING BUILDING TECHNOLOGIES

The colonial architecture of New England has sometimes been viewed as a new achievement in a new world. It has been assumed that the colonists adapted the traditions of their homeland to the new environment, evolved new forms, and achieved an "American" architectural style. In fact, as far as can be determined, no single new building technique was invented, and no new architectural form evolved in the English colonies in the seventeenth century. The homes of the English colonists were in large part derived from both the manor house and the humble cottage of their mother country, which meant a predominant use of post-and-beam and timber framing, wattle and daub, clay, and lime plaster for walls, and clay and lime mortar for masonry work.

Wattle and Daub

Europeans came across the Atlantic Ocean in ships, and their first concern upon arrival was shelter. Strong evidence exists that the early shelters of these immigrants "reveal the employment of many primitive modes of building . . . [such as the use of] branches, rushes, and turf, of palings and hurdles, of wattle, clay and mud . . . that represent neither invention of necessity nor borrowing from the Indians, but transplantation and perpetuation of types current in England" (Kimball 1922, 3–4). Lieutenant Governor Dudley of Massachusetts Bay spoke in 1631 of "some English wigwams which have taken fire in the roofs covered with thatch or boughs" (Kimball 1922, 4). Shepherds and agricultural laborers were still using similar conical-shaped huts in England in the early 1900s during harvest, as well as charcoal burners (Kimball 1922). Saplings were driven into the ground to form the structure, and branches were woven throughout. This framework was plastered with mud to create wattle and daub. The wattle and daub that exists in the Northeast (and there is very little of it left) comes from European immigrants, who brought the traditional building styles of their

Wattle and daub was frequently used by early European settlers. PHOTO BY ACE MCARLETON.

homeland, most often as infill for timber structures, but also as wall systems in their own right. "Many of the early cottages were of wattle, with or without a daubing of clay" (Kimball 1922, 6).

Wood

Timber framing was the method most often used by colonists to frame their houses. "The frames of houses built before the 1830s or 1840s, whether large or small, generally exhibit many features in common throughout most of northern New England. Such frames are based on English precedent and follow the general principle that all stresses in the frame are transmitted to the ground through relatively few principal posts" (Garvin 2001, 9).

The frame of the New England house long retained its identity as a hewn frame covered and floored with mill-sawn boards. During the 1790s, possibly because of the introduction of mills with longer carriages, sawn rafters and other major framing elements began to make their appearance in some areas.

As we move forward into the 1830s, the traditional craft of the building framer began to move toward greater standardization. We see the evolution of a uniquely United States style of timber framing: square rule.

Square rule is the name for a system of laying out and designing all the joints in a frame as theoretically perfect joints. It then standardizes a way to compensate for the inevitable variations that exist in timber dimensions by introducing a "factor of uncertainty"—in the form of a "½-inch" housing for the joinery. That "½ inch" is anywhere from ½ inch to nothing, depending on the variation in the timber. In this system, all joints are expected to fit perfectly without having ever been anywhere near each other in layout. This runs in complete opposition to the methods of scribe rule, the traditional method, where each timber is physically laid on top of the other to exactly transfer the shape during layout. Square rule was made possible by the fact that timbers were mill-sawn rather than hewn and thus were perfectly regular in cross section. Square rule did not do away with traditional mortise-and-tenon joinery; it just standardized it so each joint was not a unique feature.

Other standardizations came about in the 1820s and 1830s, including an ever-increasing use of sawn timbers, joists, rafters, and other components. Houses built after 1830 frequently have wholly sawn frames. Circular sawmills resulted in faster sawing and came into use in the early nineteenth century. "Many advances in the design and metallurgy of circular saw blades, as well as the introduction of steam to power very large sawmills, made cheap, circular-sawn scantling easily available in all areas and encouraged the adoption of the balloon frame. The further abandonment of older framing methods was encouraged by the introduction of the inexpensive wire nail during the 1880s" (Garvin, 27).

Balloon framing, the very first style of "light" or "stud" framing, where the long, sawn 2-inch studs extended from the sill to the roof, with the joists for the second floor framing attached to them, soon gave way to platform framing. Also called the western frame, the platform frame is built of light sawn studs, joists, and rafters all nailed together and is constructed one story at a time. Platform framing uses shorter lumber, and hence is more standard mill stock, and creates a stiffer structure than the long members of the balloon structure. People needed to be able to frame structures with a minimum of skill and traditional craft. Milled studs, of set dimensions, that were attached using mechanical fasteners rather than tenon-and-mortise joinery provided a way to standardize and simplify framing.

Infill and Plaster Systems

"We are so accustomed to thinking of the old New England houses as structures covered with clapboards that we are in danger of forgetting what is underneath. . . . [I]f we rip it off many of the oldest buildings, we shall find behind it nothing more nor less than an old English half-timber house, built precisely as were the half-timber houses in the reigns of the Tudors or Stuarts. . . . [T]he spaces between the studs are 'pugged' with rough bricks or stones and coarse clay stiffened with chopped straw, also in the time-honored English manner" (Eberlein 1915, 150–51). Fairly quickly, these new immigrants learned that this infill method from the old country, when left exposed to the elements, resulted in the wind whistling through the gap that opened up along the posts. This was due to the greater temperature range (0–100 degree swing) that the northeastern United States experiences than in the milder European climate (0–60 degree swing). Some timber frame historians have argued that this was an additional reason why the shift from timber framing to stud framing occurred—the difficulty of finding good "infill" or a "wrap" system to make walls in the wide open spaces between the posts.

J. F. Watson writes in *Annals of Philadelphia and Pennsylvania*: "Some old houses seem to be made with log frames and the interstices filled with wattles, river rushes, and clay intermixed." Watson provides several historic examples of these methods, including the back wall of the Corwin House in Salem, Massachusetts, finished in 1675, in which clay and hay were used as infill; the filling of the walls of the old Stoughton House in Windsor, Connecticut, destroyed in 1809, but described in 1802 as "built of mud and stones built in on the outside between the joists and timbers"; and in the Ward House in Salem, in which bricks were used as infill. In all these cases the filling of the frame

Half-timbered German cottages and other systems where infill materials went between timbers performed poorly in the colder northeastern United States. PHOTO BY PHILIP LANGE/BIGSTOCK.COM

was found covered on the exterior with some form of wooden boarding (Watson 1857, 2:19).

Interior plastering in the form of clay daub preceded even the building of timber frame houses, and historians postulate that it must have been visible in the inside of wattle filling in those earliest frame houses. Clay continued in use long after the adoption of laths and brick filling for the frame. Records of the New Haven Colony in 1641 mention clay and hay as well as lime and hair. In Massachusetts Bay, where lime was scarce, the town of Dedham voted in 1657 to "have the meeting house lathed upon the inside and so daubed and whited over, workmanlike," indicating the walls would be daubed with clay and then coated with whitewash, to use less lime (Kimball 1922, 30). As late as 1675, in the Corwin House in Salem, clay plaster was left exposed in the walls of the garret and was used as a first coat in all the rooms. In the German houses of Pennsylvania the use of clay as a wall finish persisted much later still (Kimball 1922, 30).

Clay and lime were frequently both used also as mortar. Many basements of stone and chimneys of clay brick were mortared with just clay mortar, and sometimes also with lime mortar (Kimball 1922, 66). Lime was initially hard to come by in the Northeast as it took a while for veins of limestone to be discovered and limekilns to be constructed (and even after that, distribution was a challenge).

Straw as Insulation: The First Straw-Bale Houses

Insulation did not receive as much consideration as did the infill between the posts, and from what research shows, materials such as bricks and clay blocks, which tend to be not so much "insulation" as "mass," were used as often as straw and clay.

People living on the U.S. plains before the European settlers came were using prairie grasses: there were structures called grass houses that predominated in the southern plains of the United States in Kansas and south to Texas. These structures were beehive-shaped sapling frames, with tiers of overlapping grasses, similar to thatch, providing wall covering. The looseness of the grasses did not provide very good insulation; it took the invention of the baler to provide the more modern equivalent to the carefully constructed straw mats of the northeastern Wabanaki in which lots of air space existed in a highly structured, contained form.

The utilization of straw by European settlers is also part of the plains states' building history. Settlers who traveled to the western grasslands found themselves surrounded by a lot of grass but not a lot of wood, unlike their East Coast counterparts. It was these settlers who (encountering the ecology around them and far away from their home country's influence, and also using the new technology of the hay baler) discovered that grasses bound into bales and stacked up into walls do a nice job of keeping out the weather. These early Nebraska structures were what is referred to as "load bearing," in that the roof loads bear right onto the bales themselves, without any additional wood framing. Here, the straw is used not in pairing with wood framing, as it was in Wabanaki building practices, or in today's popular Northeast strategy of timber frame or stud frame with straw "wrap," but as both insulation and support for the roof loads.

The Burke House, in Alliance, Nebraska, is the oldest bale structure in the United States. It was built in 1903, the year that a Nebraskan named Ummo F. Leubben updated a version of the baler based on the original design created in the 1850s. In 1990, nine surviving bale buildings from this time period were reported in Arthur and Logan counties.

AFTER 1900: AN INDUSTRIALIZED BUILDING LANDSCAPE

The Industrial Revolution in the second half of the 1800s brought with it sweeping changes on all fronts of social life. The invention or synthesis of new building materials and construction processes was one way that the Industrial Revolution exerted a great influence on architecture.

New England was chosen by the new industries "partly because of its plentiful water power, partly because of its ready supply of labor in the towns, and partly because of its well-established commerce and banking" (Gelernter 1999, 128). Many of the remnants of these old factories, mills, hydroelectric dams, and other industrial infrastructure in the Northeast can be seen as abandoned, crumbling relics of this time.

New construction technologies removed the traditional limitations on buildings, such as those that climate presented or those posed by engineering with existing materials. With the invention of reliable artificial lighting, for example, buildings were no longer limited in width by the need for natural light (Gelernter 1999).

The Suburbs

This apparent ability to overcome the limitations of time, space, or energy use that had informed construction practices leading up to the Industrial Revolution found fruition in the explosion of the suburban building style after World War II. Aided by the mass production of the automobile and the infrastructural transformations brought on by its ubiquity, houses no longer needed to be clustered within walking distance of the railway stations and could expand infinitely in any direction. This pattern of the individual home necessitating travel to everyday functions (work, grocery, etc.) is also the most energy-intensive development model.

The postwar United States' landscape was defined largely by low-density housing spreading over previously rural land. Methods of mass production dominated the building trade. Prefabricated building components and materials were shipped to the site and assembled as in a factory into rows of similar houses.

From a social standpoint, in the years following World War II, American houses largely abandoned their traditional front porches in favor of a rear patio. The front porch is an inheritance from African-American architectural influence, and it tends to articulate an orientation to public sociality that the builders of suburban homes rejected. The anti-urban, anti-public, family-centered mood of the postwar era is epitomized by these structures.

Responses

Beginning in the 1960s, some in the United States began to doubt, among other things, the unquestioned superiority of science and technology. The reasons for this doubt included the fact that the vast technological superiority of the United States in the Vietnam War seemed increasingly incapable of bringing the war to conclusion; people watched as Agent Orange was dropped in Vietnam, it was discovered that school children studied and played on toxic waste dumps in the working-class Love Canal neighborhood in New York State, and Rachel Carson published *Silent Spring* in 1962, exposing the effects of toxic chemicals on avian populations in the United States. The horrors propelled many into vigorous opposition to war, imperialism, and much that mainstream society represented. The emerging environmental and socio-ecological movements were claiming that not only had industrial processes polluted and defoliated the planet, but they were intimately connected to and in fact stemmed from hierarchical social relations (Bookchin 1997). For many, technology and the social "progress" associated with it was less something to be celebrated and more something to be feared, or at least questioned—as demonstrated by the strong anti-nuclear movement nationally and globally that was most active in the 1970s and 1980s during the final decades of the cold war. Social ecologists like Murray Bookchin argued that technology carries within it the goals of the dominant social logic that created it, and so therefore it is up to human beings to create orientations to society, and related technologies, that further the kind of society we would like to create. These types of changing perspectives caused many to look at their built environment afresh, and to seek to construct various alternatives (Gelernter 1999, 291; Bookchin 1997).

This re-imagining took many forms. Here in the rural Northeast there existed a strong back-to-the-land movement, which incorporated calls for a more socially and ecologically minded life. There were also many experiments with architecture, community, and construction, both urban and rural. Urban farms and building takeovers, like the one in Loisaida (a neighborhood on Lower East Side of Manhattan) in the 1970s, combined direct democracy with fish farming, and housing occupation and social justice with ecology. The Jersey Devils design/build movement, lesbian-feminist collective farms such as Red Bird in Hinesburg, Vermont, the birth of the Radical Faeries (lesbian, gay, and transgender people interested in rural, earth-connected communal celebration), and the rebirth of the timber framing craft in the Northeast region as led by Jack Sobon, Will Beemer, and Tedd Benson all invigorated new

Radical Faeries and other LGBTQ people participate in the calls for a new relationship between gender, sexuality, freedom, and ecology. PHOTO BY ACE MCARLETON.

forms of interaction between society and the built environment. Homesteaders and permaculturists built with cob and straw—mostly in the southwestern United States but also with increasing frequency in the northeastern states. Paul Lacinski, Clark Sanders, and Michel Bergeron from Québec were a prominent few of the early pioneering builders to initiate and refine the craft in our region, and their work has become a part of a growing movement of professional builders and designers, educators, and owner-builders bringing natural building into the current day.

Mainstream Building Materials Today

The standard construction method of the houses of today is platform stud framing, built with sawn studs from spruce, pine, or fir—mostly coming from Pacific Northwest tree farms. Trees are often grown and logged without regard for sustainable forestry practices and then are shipped across the continent to building-supply houses. This contributes to the perception, by some, that wood is not a sustainable building material, when in actuality it is the practices surrounding its growth, harvesting, and transport where the unsustainable aspects abound.

Plastics and resins predominate in sheet goods like plywood treated with formaldehyde, latex paints, silicone caulks, and PVC (polyvinyl chloride) plastics. According to Anne McGinn, currently about 60% of PVC production is destined for the construction industry, where it becomes everything from water pipes to siding (McGinn 2000, 47). PVC, plywood and OSB (oriented strand board), silicone caulk, butyl rubber sealant, and polystyrene foams are just a few examples of the chemical content of conventional building materials and methods (see chapter 2).

"Reconciling personal consumption with social awareness is perhaps one of the greatest challenges of our era. That effort might begin with the kind of question used to advance the precautionary principle within the chemical economy as a whole. What do we really need? The answer will obviously depend on individual tastes and circumstances. Do we 'need' to consume Persistent Organic Pollutants and similar chemicals?" (McGinn 2000, 27). Why use these chemicals when viable alternatives are available and simply require a shift in consciousness, educating ourselves and each other, and supporting these alternatives in our communities?

We also saw how technological innovations of the Industrial Revolution like artificial lighting removed natural constraints on form. Designers and builders in the twentieth century were able to put any building form in any climate and simply rely on the brute force of energy-intensive technology to keep it inhabitable. By the mid-1990s, it became abundantly clear that the economic, political, and environmental costs of this strategy were too high to continue. "The sustainable architects in this decade set out to relearn ecological lessons of their pre-industrial predecessors" (Gelernter 1999, 317).

As we look to the future and evaluate our housing stock in light of the many challenges of the early twenty-first century, including housing and financial crises, the rising costs of gas and electricity based on petroleum or nuclear dependence, federal budget cuts (to low-income heat assistance), and other such challenges to our existing building and energy infrastructure, along with natural and social disasters such as Hurricane Katrina in New Orleans and the devastating earthquake in Haiti, it is going to become more and more crucial to navigate the interplay between the social and the ecological.

It will be imperative that we learn to address in creative ways new construction practices, as exemplified by GreenSpace Collaborative's work in Haiti to build straw bale structures in the wake of the 2010 earthquake, and by Dignity Village, a straw-bale-structure village built by a sector of the homeless community in Portland, Oregon, in collaboration with the community-based art and re-construction organization City Repair. We must simultaneously focus on developing ethical strategies for renovation to existing housing and industrial stock.

Dignity Village brought together natural building and self-determinative housing for homeless folks in Portland, Oregon. PHOTO COURTESY OF CITYREPAIR.

Patterns in U.S. architecture have always reflected ideas about ourselves and our social life. The materials and shape of a building are only a part of the meaning and role that shelter plays in our lives as human beings. It is important for us to remember that even the noble aspirations of environmentally sound building are not enough to create a just world—natural building calls for a practice that integrates social *as well as* ecological vibrancy.

As we will discuss in chapter 10 on design, it is essential to look more broadly, to see patterns of development that support the local economy and local community life, as well as efficiency and joy, so that our social and community lives are reflected in our living structures, and vice versa. Permaculture is a practical design philosophy that embraces this vision and emphasizes caring for the earth and for people in an integrated manner. In terms of city and town planning, housing development, architecture, and construction, we are proposing to use this type of holistic compass to guide the ethical development of our built environment.

CHAPTER 2

Ecology

A critical consideration in gaining an awareness of the context of a building is appreciating the ecological impact of the structure as a whole and of its material components—from the extraction of the feedstock to the disposal of the building upon demolition—as well as the ecology of the proposed building site. There is far more on this subject than can be covered in this book; therefore, we will lay out the most significant ecological considerations and the role natural building plays in causation and/or remediation of these considerations, and we encourage you to engage in further research on the topics of greatest relevance to your project.

Global Warming and Climate Change

The largest, most complex, and most urgent social and ecological issue we face as a global community is that of the steady warming of our Earth's climate, which is the result of, at least in part, the release of anthropogenically sourced carbon compounds into the atmosphere. While there are still people who question the validity of climate change—the extent of its potential negative impact, or the human causation factors involved—the overwhelming majority of scientific research shows measurable evidence of rising global temperatures and shifts in climate, with direct human causation. Associated ecosystem degradation, biodiversity and habitat loss, phenological changes (how plant and animal life-cycle events are influenced by variations in climate), and myriad other changes to systems ecology are happening now, and are not just potential future outcomes. In accepting this reality, we then must also accept responsibility for changing these outcomes.

The primary cause of the current global warming trend is the release of greenhouse gases into the atmosphere; the gases prevent heat from radiating through the atmosphere into space, trapping the heat and raising global climate temperatures. As designers and builders we are part of an industry that contributes 35% of the greenhouse gas emitted in North America (Biello 2008), so we can play a big role in achieving significant greenhouse gas reductions. From material selection to energy efficiency detailing, much of what makes natural building a more ecologically favorable building practice relates to the fact that it contributes less to global warming than do other building practices.

The effectiveness of greenhouse gases in trapping heat in our atmosphere is measured as global warming potential (GWP); by association, the GWP of a material, product, or building is directly related to the emissions of these gases. Carbon dioxide (CO_2) is the most common and most heavily produced by anthropogenic sources. However, methane has 25 times, nitrous oxide 298 times, and hexafluoride 22,800 times the GWP (Malin 2008). Since CO_2 is used as a baseline, the value of these gases is referred to as CO_2 equivalent, or CO_2e. For example, 1 ton of methane is represented as 25 tons CO_2e.

To have an effective impact on reducing CO_2e, we have to identify the contributing factors of CO_2e in buildings. Given that energy consumption is a primary source of CO_2e production, a common tool

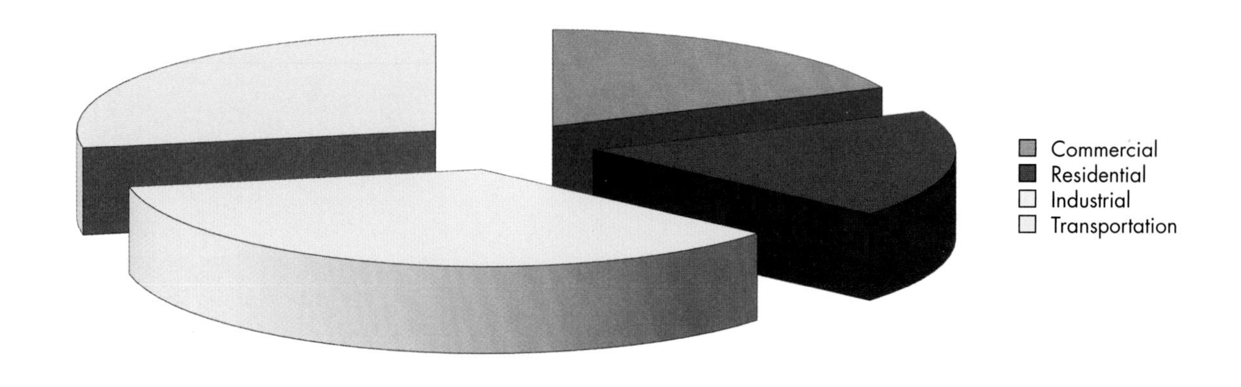

Greenhouse gas production in the United States in 2005, by sector. GRAPH BY JACOB DEVA RACUSIN; SOURCE EMRATH AND LIU 2007.

for analyzing the impact of a material is through its energy use. The amount of energy used to produce a material from raw feedstock extraction through production and manufacturing is called its *embodied energy*; this period of analysis, or boundary, is referred to as *cradle-to-gate*, and it is the common boundary used to calculate the embodied energy for materials. The *cradle-to-grave* boundary, by comparison, includes all energy used in resource extraction, manufacturing, production, transportation to site, inclusion within a building, and disposal (Hammond and Jones 2011). In analysis of a new construction

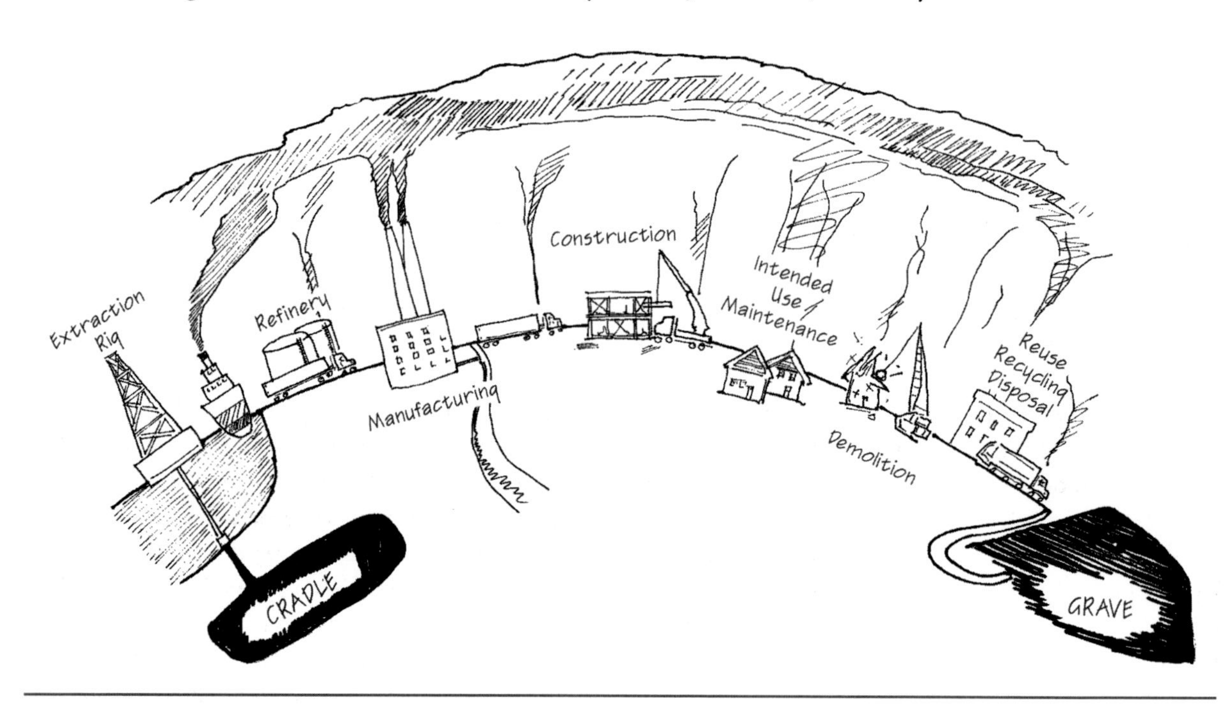

There are ecological impacts throughout a product's life cycle, from cradle to grave. ILLUSTRATION BY BEN GRAHAM.

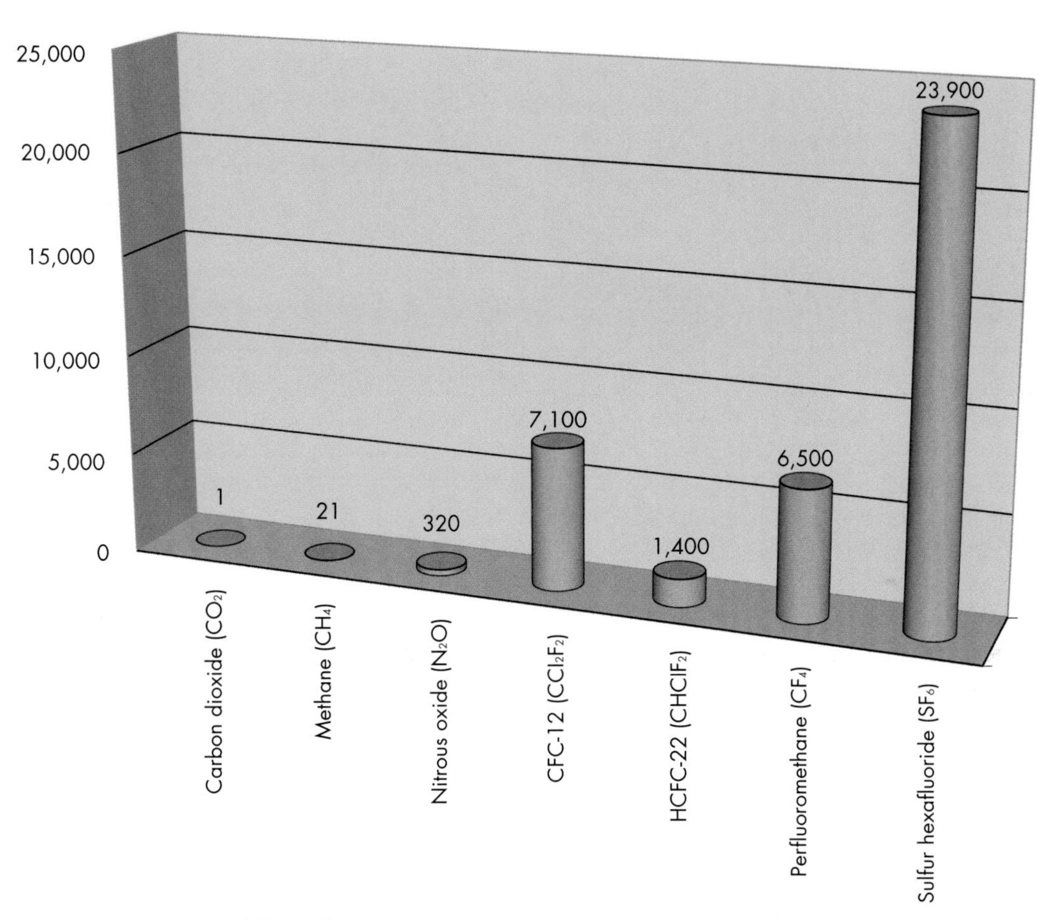

Greenhouse warming potential of different greenhouse gases. GRAPH BY JACOB DEVA RACUSIN; SOURCE: INTERGOVERNMENTAL PANEL ON CLIMATE CHANGE 1996, 22.

project, the calculation may also include carbon released from disturbing soil, or transportation of workers to and from the site, as well as operational energy use.

Understanding the factors used to determine the value of embodied energy and the applied boundaries is critical to having useful comparisons between different products. A good example is a comparison between bamboo flooring and concrete. In one analysis, the material embodied energy of bamboo flooring is 15 megajoules(MJ)/ton, whereas that of the concrete is 1452.3 MJ/ton (a joule is a measurement equivalent to the energy expended to produce one watt for one second, and a megajoule is one million joules). However, once transportation to Denver, Colorado, is considered, the embodied energy of the bamboo flooring, which has traveled from Hunan Province, China, inflates to 4943.1 MJ/ton—while the concrete, with in-state sourcing of cement production and quarried aggregate,

TABLE 2.1. IMPACT OF TRANSPORTATION ON EMBODIED ENERGY OF BUILDING MATERIALS

Material	Embodied energy	With transportation
Bamboo	15	4943.1
Concrete	1452.3	1537.3

Source: Symbiotic Engineering 2007, 6.

Note: This chart shows the influence of transportation on the embodied energy of different building materials.

grows only to 1537.3 MJ/ton. Careful sourcing and understanding of relevant data is critical to making informed decisions based on embodied energy values.

Embodied energy is a helpful guide, but it looks at only the energy footprint of a given product, not the overall CO_2e footprint. This metric is referred to as *embodied carbon*, and better data sets will include full CO_2e values, and not just CO_2 values (again, the data conditions must be understood when comparing values). The source of the embodied energy will significantly affect embodied CO_2e. For example, coal releases forty times the amount of CO_2 per kilowatt-hour (kWh) compared to wind. Most CO_2e production is sourced from energy consumption, but not all of it. The current methods of producing extruded polystyrene (XPS) foam board and closed-cell spray foam use blowing agents with very high GWP ratings, and these blowing agends are released over time as the material ages and during disposal. The *payback* period—the length of time the insulation would have to be in use (reducing energy consumption in operation of a building) in order to offset its embodied GWP—was calculated to be in the range of thirty to sixty years (depending on thickness, given assumed variables and a location of Boston, Massachusetts) as compared to less than one year for cellulose. This clearly highlights how embodied energy does not equate to the GWP of a material; it also highlights the advances needed in manufacturing to improve the embodied GWP of important materials such as insulation.

More sophisticated life-cycle analysis (LCA) tools are becoming available to help broaden the context and give as accurate a picture of the ecological impact of a product as possible. According to International Standard ISO 14040, LCA is "a compilation and evaluation of the inputs, outputs, and the potential environmental impacts of a product system throughout its life cycle." Considered factors include fossil fuel and nonrenewable-resource depletion, water use and contamination, GWP, and toxic release to air, water, and land. LCA and the tools used in its analysis are limited in their accuracy given the broad scope and plethora of variables affecting outcomes, and the potential to produce misleading results based upon calculations of differing boundaries. Additionally, LCA does not address the "precautionary principle" or the health risks associated with toxic release and exposure, both of which will be discussed later in this chapter.

Regardless of the nuances of the debates raging in the industry and government arenas surrounding targeted greenhouse gas production limits, the simple answer to the question "How much should we reduce greenhouse gas emissions?" is "As much as possible." With that in mind, in the context of the built environment, the following are opportunities and priorities for reduction strategies that we as natural builders can consider for creating positive change in this field.

Create Efficient Buildings

Depending on the life span of the building and the nature of its materials, typically 13% to 18% of the overall energy consumption of a building comes from the production of the materials used in its construction, while the vast majority is consumed in operating the building during the life of its use. Given that the largest amount of embodied carbon comes from energy consumption, the strong focus on energy-efficient construction by the green-building community is appropriate as a meaningful response to reducing structures' GWP. (See chapter 7 for detailed examples of designing and building energy-efficient natural buildings.)

Building energy-efficient structures is an essential part of natural design/build. PHOTO BY JACOB DEVA RACUSIN.

Natural building is a relevant and appropriate strategy for development in urban environments. PHOTO BY JAN TYLER ALLEN.

Renovate Instead of Demolishing, Rebuilding, or Building New

Whether to renovate is a significant consideration, not only because of the potential influence of reducing the GWP of a project but because of the implications beyond GWP impact. Certainly, improving existing housing stock has fewer ecological impacts than new construction with regard to material consumption. Providing that the improvements bring the building up to sufficient energy-performance standards, there will be little to no penalty incurred in operational impacts relative to new construction. Additionally, the land-use impacts of new construction can be very significant, including loss of agricultural or productive forestland, loss of habitat, extension of infrastructure such as roads, power, and water services, increased transportation impact (which is significant when considering GWP), and possible cultural fragmentation due to urban flight and the loss of vibrancy in village, town, and city centers. Considering the quantity (plentiful), age (increasing), and location (villages, towns, cities) of the existing housing stock in many parts of our country, choosing renovation of these houses will result in overwhelming ecological benefits as compared to new construction and site development.

Create Community

Design communities to reduce transportation impacts and foster a sense of community. Placing buildings and their occupants in a relevant context of place is critical to truly achieve goals of impact reduction—of which transportation is a major contributing factor. No building, and no occupant, is an island. See the conclusion of this book for more on this topic.

Build Small

Consider this: the material with the lowest embodied carbon is the material that is never used or produced. By building a smaller building, impacts of heat loss (in relation to total surface area of the shell), energy and resource consumption, material consumption, and site disruption are all decreased, just by virtue of a reduction of the building's size.

Select Materials Carefully

While operational energy consumption may be the largest contributor to a building's cradle-to-grave embodied carbon, the embodied carbon of the

Small buildings have lighter impacts on their local and global ecological support systems. PHOTO BY JAN TYLER ALLEN.

building's materials also matters greatly. Citing the earlier example of the GWP payback of XPS and closed-cell spray foam, there is a clear case for understanding how the embodied carbon of a material can drastically undermine its role as part of an overall carbon-reduction strategy for the building. Additionally, the payoffs of carbon reduction by way of energy efficiency are realized over a longer time span, as a result of reduced operational impact. Low-carbon materials such as those used in the practice of natural building, on the other hand, realize their carbon reduction benefits early on in the process, which addresses an immediate problem in an immediate time frame.

It is interesting to look at how different materials play out in their embodied carbon profiles. Table 2.2 compares embodied energy and CO_2e profiles for a host of common building materials. Natural materials are clearly advantageous in this regard: by this analysis, softwood has 32% less embodied CO_2e than a glue-laminated member; 5% cement-stabilized soil has 55% less embodied CO_2e than reinforced concrete; and straw has less than 0.75% of the embodied CO_2e as fiberglass insulation!

When comparing statistics like those in table 2.2, it is important to remember that there are many variables to consider in the true GWP of the material, including how the data is aggregated and the boundary condition of the data (the boundary for data in this chart is cradle-to-gate/site).

To that end, there are some important considerations in selecting and evaluating the use of natural materials. As we see from the chart, lime has a comparable embodied energy to cement—by some calculations, even greater—despite lower kiln temperatures in its firing (potential causes include longer firing times, fuel sources, and production efficiency). However, its embodied carbon cradle-to-gate is lower, and these metrics do not factor in carbon sequestration during the curing process, which is greater for lime than for cement. Softwood, by this analysis, has a higher embodied carbon value than that assigned by other analyses, showing the potential discrepancies that can exist in data depending on location (for example, the data in tables 2.2 and 2.3 are taken from a U.K.-based report), energy source, scale of production, and many more variables—a caveat that was clearly identified by the report's authors.

Straw falls into a similar category as timber in this regard: while it proves to have a significantly lower embodied carbon level than any other insulation material, the scale of the agriculture that produces the straw will have everything to do with the true GWP of the material. This point is made clear by comparing the GWP of site-baled grasses by horse- and human-power, or even small-scale regionally sourced farming operations, to the more conventional agri-industrial production process that is heavily reliant upon diesel-powered machinery and other intensive inputs. A similar dichotomy exists for sustainable silvicultural management practices compared to industrial commodity-scale timber harvesting. In comparing insulations, payback relative to a material's efficacy in reducing operational loads is also important, as discussed earlier. Carbon sequestration by use of cellulosic materials such as wood and straw is another factor not considered in this analysis.

TABLE 2.2.
EMBODIED ENERGY (EE), CARBON, AND CARBON-EQUIVALENT (EC) VALUES FOR COMMON BUILDING MATERIALS

Material	Embodied energy (MJ/kg)	Embodied CO_2 ($kgCO_2/kg$)	Embodied CO_2e ($kgCO_2e/kg$)
Aggregate (gravel or crushed rock)	0.083	0.0048	0.0052
Aluminum, virgin/recycled	218.00/29.00	11.46/1.69	12.79/1.81
Brick, common	3.00	0.23	0.24
Cement, CEM I Portland 94% clinker/20% fly ash	5.50/4.51	0.93/0.75	0.95/0.76
Mortar, 1:3 cement:sand mix/ 1:2:9 cement:lime:sand mix	1.33/1.03	0.208/0.145	0.221/0.155
Cement-stabilized soil, 5%	0.680	0.060	0.061
Concrete, reinforced (12% cement by mass)	1.894	0.179	0.192
Copper, 37% recycled	42.00	2.06	2.71
Glass	15.00	0.86	0.91
Gypsum wall board (plasterboard)	6.75	0.38	0.39
Insulation, cellulose	0.94–3.3	—	—
Insulation, fiberglass	28.00	1.35	—
Insulation, rockwool	16.80	1.05	1.12
Lime, hydrated	5.30	0.76	0.78
Paint, water-based	59.00	2.12	2.54
Plastic, high-density polyethylene (HDPE) pipe/PVC pipe	84.40/67.50	2.02/2.56	2.52/3.23
Plastic, polystyrene/ polyurethane	86.40/101.50	2.71/3.48	3.43/4.26
Sand	0.081	0.0048	0.0051
Soil, rammed	0.45	0.023	0.024
Steel, virgin/recycled	35.40/9.40	2.71/0.44	2.89/0.47
Stone, granite/limestone	11.00/1.50	0.64/0.087	0.70/0.09
Straw	0.24	0.01	—
Timber, sawn softwood/ glue-laminated	7.40/12.00	0.58/0.84	0.59/0.87
Timber, oriented strand board (OSB)	15.00	0.96	0.99

Source: Hammond and Jones 2011.

Note: All values taken from U.K. data for cradle-to-gate/site boundaries; cross-referencing with other EE/EC data produced comparable results, with the exception of elevated softwood EE/EC in the U.K. relative to U.S. analysis.

TABLE 2.3.
EMBODIED ENERGY OF INSULATION MATERIALS

Material	Embodied energy
Cellulose	2.12
Fiberglass	28
Polystyrene	86.4
Rockwool	16.8
Straw	0.24

Source: Hammond and Jones 2011.

Note: Not all insulation is created equally—straw bales have a fraction of the embodied energy of fiberglass batts.

Maximize Longevity

Design for durability and adaptability. The longer a building lasts, the longer the period of time over which the environmental impacts from building it can be amortized. Premature repair and maintenance increase the ecological footprint of a building, and the harder it is to maintain and adapt a building to changing needs, the earlier the building will reach the end of its useful life.

Efficient Consumption—Reuse/Recycle Materials

Reusing materials can save upwards of 95% of the embodied energy that would otherwise be used by procurement of a new product. As shown in table 2.2, dramatic improvements in embodied carbon are seen as the recycled content increases, especially of metals.

Minimize Waste

Considering end-of-life impacts is highly relevant, even if not always reflected in embodied energy/carbon-equivalent calculations. Designing for material reuse, waste reduction/optimized design, on-site recycling systems, and the use of recyclable and compostable materials all have direct positive impacts on a building's GWP.

Evaluate Payback

The use of a higher embodied-carbon material—such as high-mass materials or foam—to make performance improvements that will pay back the GWP of the material within a reasonable time can be justified, depending on the scale of analysis being used to compare different design options.

Utilize Modeling and Evaluation Tools

There are a wide variety of tools available to designers, builders, engineers, and clients to model projected GWP in design, evaluate during construction, and analyze real-world performance in operation. Design-modeling software ranges from free online embodied-carbon calculators to energy-performance and heat-loss modeling to more comprehensive and expensive LCA tools, with new and more powerful software hitting the marketplace every year. Increasingly, design software such as products developed by Autodesk and Google SketchUp feature plug-ins that will provide carbon estimates for modeled projects. Performance diagnostic equipment, including blower door tests and infrared thermography, are also readily available, and discussed in greater detail in chapter 7.

Green Up Your Business and Life

Minimize impact and maximize outreach beyond the scope of the building project. As professionals, practice what you preach throughout all your business practices. As owners, let the values reflected in your design program be reflected throughout your lives.

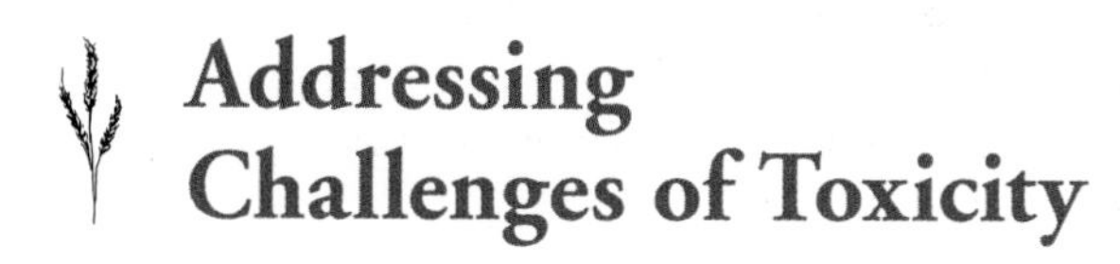

Addressing Challenges of Toxicity

Despite the massive challenge we face in addressing global warming, we cannot ignore other socio-ecological impacts of the built environment. As mentioned earlier, a comprehensive inventory of all ecological hazards is well beyond the scope of this book, let alone a chapter. Many materials' toxic impacts are well known—asbestos, lead, mercury, wood-treatment chemicals such as arsenic and creosote, ozone-depleting CFC-releasing propellants. We will therefore highlight a few of the issues that seem particularly relevant, in that they can be readily addressed by the adoption of natural building techniques and materials, or are both underreported and highly pervasive.

Just as every product has a cradle-to-grave energy and carbon footprint, so too does each product have a cradle-to-grave toxic footprint. Understanding where a material's primary impacts lie along this process from production to disposal is very important. In some cases, cradle-side impacts can be remediated through reuse or recycling, while in others, post-construction use and disposal issues may lead to myopic acceptance of a product that will leave an out-of-sight, out-of-mind legacy of problems for subsequent generations to face—a condition we have created for ourselves time and time again, from lead paint to asbestos insulation. Let's look at a handful of case studies spanning the cradle-to-grave life cycle to illustrate some of the primary considerations.

PORTLAND CEMENT

The manufacturing of Portland cement is responsible for 5% of anthropogenic CO_2 emissions worldwide, and 2% of total CO_2 emissions in the United States. This is due more to the sheer volume of Portland cement production (3 billion tons worldwide in 2009, or 900 pounds for every person on the planet) than to cement's embodied carbon, which is lower by weight than that of most common materials. In fact, the processing of iron and steel, as referenced above, is a larger contributor to annual CO_2 production in the United States than the manufacturing of Portland cement (Ehrlich 2010). Therefore, finding ways to reduce Portland cement consumption is a key strategy in reducing carbon loading. It is the other ecological impacts of Portland cement production that we focus on here, however. The processing of Portland cement releases significant amounts of numerous toxic materials, including chromium, arsenic, and mercury, as well as particulates and hydrochloric acid gases. Mercury is a particularly harmful substance, being a bioaccumulative neurotoxin that also attacks the kidneys; a single gram of mercury can kill off the life in an entire 25-acre pond. In many regions across the United States, there are restrictions on the consumption of fish caught from waterways as a result of mercury contamination, frequently sourced from industrial air pollution. Until August 2010, when the EPA announced a series of air pollution restrictions for cement production, the industry was largely unregulated.

While fuel-sourced contamination from coal combustion is a contributor to mercury release, further study has shown the limestone used in cement production to be contaminated with mercury. This indicates that the use of lime as a replacement for Portland cement may not achieve the objective of reducing mercury release, depending on the nature of the feedstock.

PLASTICS

As discussed in chapter 1, a sea change in building technology arrived in the 1950s with the "Age of Plastic." Industrial development of fossil fuels into a wide array of plastics changed formulations in everything from insulation to mechanicals to paint, and plastic is still a ubiquitous component of every building assembly. Unfortunately, the impacts of plastic production in its many forms are heavy in every phase of its life cycle. While there is a common general understanding that plastics have negative ecological associations, a closer understanding of what types of plastics create what types of impacts will empower us to improve the toxic footprint of our buildings; refer to table 2.2 for information concerning embodied energy and carbon-equivalent of different plastic products.

Plastics are not inherently bad, and they have many redeeming ecological features; in fact, many of the techniques we utilize in our designs involve targeted use of plastic products. Their durability and low maintenance reduce material replacement, their light weight reduces shipping energy, their formulation into glue products allows for the creation of engineered lumber and sheet products from recycled wood, and their formulation into superior insulation and sealant products improves the energy performance of our structures.

Plastics have become invaluable components of a modern building. PHOTO BY KELLY GRIFFITH.

The feedstock of plastic is primarily petroleum- or natural-gas-derived, although bio-plastics are making inroads in the overall market share of plastic products. Obvious issues emerge regarding the finite amount of available petroleum resources, as well as the pollution associated with oil extraction and refinement; the massive Gulf Coast oil spill of 2010 is only one of the more notorious of the many ecologically devastating accidents that are not frequently considered in addition to the standard pollution impacts of extraction and refinement, which are extensive.

Toxic release during manufacture is another significant source of impact. A whole host of carcinogenic, neurotoxic, and hormone-disruptive chemicals are standard ingredients and waste products of plastic production, and they inevitably find their way into our ecology through water, land, and air pollution. Some of the more familiar compounds include vinyl chloride (in PVC), dioxins (in PVC), benzene (in polystyrene), phthalates and other plasticizers (in PVC and others), formaldehyde, and bisphenol-A, or BPA (in polycarbonate). Many of these are persistent organic pollutants (POPs)—some of the most damaging toxins on the planet, owing to a combination of their persistence in the environment and their high levels of toxicity. These are discussed in greater detail later in this chapter as a consideration

The ecological impacts of petroleum extraction are devastating, as witnessed during the Gulf oil spill of 2010. PHOTO BY JEFF WARREN.

of human health; however, their unmitigated release into the environment affects all terrestrial and aquatic life with which they come into contact.

It is in the use phase that the benefits of plastics in durability and effectiveness are most evident. Though most plastics are benign in their intended use form, many release toxic gases in their in-place curing (such as spray foam) or by virtue of their formulation (as with PVC additives off-gassing during their use phase). Occupational exposure during installation, such as inhalation of dust while cutting plastic pipe or off-gassing vapors of curing products, is also a great concern.

The disposal of plastics—the "grave" phase, if you will—is one of the least-recognized and most highly problematic areas of plastic's ecological impact. Ironically, one of plastic's most desirable traits—its durability and resistance to decomposition—is also the source of one of its greatest liabilities. Natural organisms have a very difficult time breaking down the synthetic chemical bonds in plastic, creating the tremendous problem of the material's persistence. A very small amount of total plastic production (less than 10%) is effectively recycled; the remaining plastic is sent to landfills, where it is destined to remain entombed in limbo for hundreds of thousands of years, or to incinerators, where its toxic compounds are spewed throughout the atmosphere to be accumulated in biotic forms throughout the surrounding ecosystems.

Unfortunately, because of plastic's low density, it frequently migrates "downstream," blowing out of landfills and off garbage barges. For decades, marine biologists and researchers had been witnessing increasing amounts of plastic contamination in the ocean. Then, in 1997, as mentioned in the introduction, Captain Charles Moore discovered widespread plastic garbage contamination in an area larger than the state of Texas that had formed within a cyclonic region, called a gyre, in the North Pacific Ocean. By 2005, the estimated area of contamination expanded to 10 million square miles, nearly the size of Africa. Ninety percent of this garbage was determined to be plastic, and 80% was originally sourced from land, such as construction waste—so Captain Moore found where "downstream" goes. Early sampling determined approximately 3 million tons of plastic on the surface; the United Nations Environment Program reports that 70% of marine refuse sinks below the surface, which would suggest a staggering 100 million tons of plastic in this one area of the Pacific alone—with more entering every day. There are six similar gyres across the planet's oceans, each laden with plastic refuse (Weisman 2007).

The effects of this plastic on aquatic life are devastating, and accelerating. In addition to suffocation, ingestion, and other macro-particulate causes of death in larger birds, fish, and mammals, the plastic is ingested by smaller and smaller creatures (as it breaks down into smaller and smaller particles) and bioaccumulates in greater and greater

Plastic trash aggregating in bodies of water is a clear example of the problem of the bio-persistence of petroleum products. PHOTO BY LEUNG CHO PAN/BIGSTOCK.COM

concentrations up the food chain—with humans at the top. Exacerbating these problems of persistence and bioaccumulation is plastic's propensity to act as a magnet and sponge for persistent organic pollutants such as polychlorinated biphenyls (PCBs) and the pesticide DDT. So, in addition to ingesting the physically and chemically damaging plastic compounds, aquatic life is also ingesting concentrated quantities of highly bioaccumulative compounds that are some of the most potent toxins found on the planet. Again, this bioaccumulation increases in concentration as it works up the food chain and into our diets.

A final consideration of plastic disposal comes from the release of POPs and other toxic chemicals into the environment from the plastics themselves. These compounds present a host of ecological and human health issues and, like plastic, are also bioaccumulative. Polyvinyl chloride (PVC) is particularly noxious, owing to its formulated inclusion of halogenated compounds (those containing bromine or chlorine), and are particularly dangerous if burned, in which case dioxins are produced, some of which are among the most harmful of all human-made compounds. Consider, then, the terrific health liability of exposure through accidental or unwitting incineration or house fire. Halogens are also sourced from a class of flame retardants that are commonly formulated into a variety of plastic products found in the building industry, particularly polystyrene insulation (XPS, EPS); the effects of flame retardants are discussed in the next section. Collectively, these harmful chemicals are known to cause the following severe health problems: cancer, endometriosis, neurological damage, endocrine disruption, birth defects and child developmental disorders, reproductive damage, immune damage, asthma, and multiple organ damage.

While we recognize the need for plastic products in our homes, in light of the tremendous ecological impact throughout plastic's life cycle, we are compelled to select alternatives when possible. In many cases, we can elect to utilize a different material altogether; examples of alternatives include using straw or cellulose-based insulation in walls and roofs and mineral board insulation below basement walls instead of foam insulation, using wood or cement-board siding or plaster as an exterior finish instead of vinyl, and using clay, lime, or casein-based finishes instead of acrylic or latex paints. In other cases, our best option may be to replace a more toxic plastic, such as PVC, with a less toxic one, such as polyethylene, ABS, or metallocene polyolefin (a newly developed plastic of lesser environmental footprint) pipe instead of PVC pipe, fiberglass instead of PVC window profiles, polyethylene instead of PVC-jacketed wire, or polyester instead of PVC commercial wall coverings. The field of bio-plastics is also growing rapidly. These products have the benefits of being nonpetroleum in feedstock, supportive of the farm sector (although LCA must also evaluate industrialized farming practices), and, perhaps most importantly, biodegradable. Additionally, vegetable oils such as soy have been proven to effectively replace pthalates as plasticizers in PVC, reducing its POP load.

HUMAN HEALTH CONCERNS

While substances that are toxic to human health are generally toxic to other biotic forms, and vice versa, there are a few categories of toxins that are quite insidious in their ubiquity within our built environment that enact particularly harmful effects on occupant safety, which we feel bear specific mention.

Halogenated Flame Retardants (HFRs)

As mentioned in the discussion of plastics, halogenated compounds are chlorine- or bromine-based formulations that have the potential—when exposed to fire, for example—to create dioxins, a family of very highly carcinogenic, immune-, hormone-, reproductive-, and neurologic-damaging bioaccumulative POPs. These halogenated compounds are created in the formulation of PVC; they are also found as chlorinated or brominated fire retardants

(C or BFRs). BFRs are added to a host of common materials and products, including plastics like polystyrene foam insulation, hard plastic cases for electric and electrical products, and furniture and bedding products.

Avoiding BFRs—and HFRs in general, for that matter—is tricky, in part because it can be very difficult to determine the nature and quantity of flame retardants in products in absence of legislation requiring adequate labeling, and in part because of a lack of alternatives to products containing HFRs. Certainly, fire safety should not be compromised in an effort to move away from HFRs, nor should compromises in building performance and durability be made. That said, there are a series of steps that can be taken to reduce occupant and environmental exposure to HFRs in buildings, as recommended by the editors of *Environmental Building News*, including the use of inherently non-flammable products, avoiding the use of foam products and halogen-clad wiring, and pressuring manufacturers to develop safer alternatives.

Formaldehyde

Formaldehyde is a simple organic compound, built of oxygen, carbon, and hydrogen molecules; it is part of a large family of compounds called *volatile organic compounds* (VOCs), discussed later in this chapter. At atmospheric conditions and temperatures, it is a colorless gas with a pungent odor. Formaldehyde occurs naturally in trace levels; however, its ubiquity in industrial chemistry has increased our exposure to dramatically higher concentrations of formaldehyde, which prove harmful to human health. In 2011, the National Toxicology Program (a division of the United States Department of Health and Human Services) formally changed the status of formaldehyde to that of a "known human carcinogen." It is used within the building industry primarily as a binder and as a biocide.

From upholstery to carpets, laminated flooring products, particleboard and medium-density fiberboard (MDF), paints, and insulation, our potential exposure via degraded indoor air quality (IAQ) is very high. In fact, a U.S. governmental study on newly constructed, inhabited temporary relief housing provided by FEMA for survivors of Hurricane Katrina in the Gulf Coast region concluded that baseline levels of formaldehyde in the trailers were sufficient to cause acute health symptoms.

In 2007, the California Air Resource Board (CARB) released regulations that dramatically limit allowable concentrations of urea formaldehyde (UF) in products; while this does not fully outlaw UF, it certainly creates incentives for alternative binders for common products. Already in the marketplace are many no-added-UF and no-added-formaldehyde board products in wide distribution that are performance- and cost-competitive. While these alternatives all have occupational and production-phase exposure health hazards, they do not have deleterious effects on IAQ. There are formaldehyde-free fiberglass batt insulation products commonly available, as well as cotton batt insulation to replace not only the formaldehyde, but also potentially carcinogenic and highly irritating fiberglass. As natural builders, our approach—whenever practical and cost-feasible—is to use all-wood or no-added-formaldehyde casestock cabinet construction, 2 × 6 tongue-and-groove flooring (in lieu of a subfloor or phenol formaldehyde-based plywood for a subfloor), and straw or cellulose insulation.

If avoiding formaldehyde is impossible or impractical, it is recommended to seal the products with paint, hard sealer, or other effective barrier to reduce emissions, or isolate their exposure to the indoor environment through physical barriers. Additionally, in all cases and all houses, adequate ventilation rates and distribution should be designed to help control IAQ. New product developments, such as wallboards that absorb formaldehyde and other pollutants from the air and more effective and affordable air filtration systems, can assist us in airborne toxin remediation, and it can be expected that future developments will come online as the marketplace continues to respond to this issue.

Volatile Organic Compounds (VOCs)

VOCs are carbon-based (organic) compounds that vaporize (become volatile) at room temperature, and are therefore highly mobile and easy to inhale. Formaldehyde, discussed above, is the most common VOC in our built environment, but there are plenty of other compounds that find their way into our buildings through a host of products.

Further study over the last few decades has conclusively linked long-term exposure to petrochemical VOCs to a host of human disorders, including neurological damage, respiratory damage, nervous system impairment, multiple organ damage (kidney, liver, lung), multiple cancers (leukemia, lung, lymphoma), chemical sensitivity, and more. Acute symptoms include dizziness, nausea, headaches, blurred vision, and fatigue. Common VOCs include formaldehyde, toluene, isocyanates, and benzene; they can be found in nearly every component of a conventional building, as identified in our exploration of formaldehyde above. Some VOCs will volatize, or off-gas, very rapidly over a short period of time, such as epoxy binders and certain spray-foam products. Others, like paints and varnishes, can off-gas for years.

Leaving aside the ubiquitous formaldehyde, one of the greatest common sources of VOCs in buildings comes from interior flooring and finishes, especially carpeting and paint. The story of paint and its formulation relating to VOC content is told in chapter 20. In light of the health concerns listed previously, it is important to stress the risk of exposure from contact with conventional paint. The health impacts are so great that associated occupational hazards have been officially deemed "painter's syndrome" in Australia, or "painter's dementia" in Denmark. As far back as 1987, the World Health Organization stated that there was a 40% increased occupational risk for cancer among painters.

Many of these VOCs also create ozone, which may be helpful at stratospheric levels, but at ground levels this smog production causes respiratory disease and plant damage and contributes to global warming. U.S. reports have stated smog production from the creation of paint to be close to that of automobiles.

As mentioned above in discussing formaldehyde, the combination of VOCs' ubiquity, direct exposure in the interior of the building, highly concentrated and potent toxicity, and frequently inadequate ventilation create substantial IAQ problems. In fact, some studies have found the indoor built environment to be up to ten times more polluted than the outdoor environment, as highlighted by recent publicity surrounding "sick building syndrome," which plagues many structures that were tightened for energy efficiency and now have inadequate ventilation and a host of airborne contaminants. The U.S. Environmental Protection Agency estimates indoor pollution from VOCs to be responsible for more than 11,000 deaths a year in the United States. We believe these deaths are preventable, and that the use of VOCs in our buildings can be dramatically reduced simply by choosing nontoxic paints and finishes, furniture, carpeting, and flooring—simple changes that will make a big difference.

Moving Forward

In reading through this chapter, an appropriate response might be to feel overwhelmed, frustrated, or downright terrified by the variety and severity of ecological and human hazards created by our built environment in its contemporary form. While we provided proactive strategies for improving our design and construction practices with each topic we covered, and we devote much of the rest of this book—particularly in part 3—to describing positive solutions that can be taken in response to these issues, we now provide some fundamental tools and principles that we can use to help us take progressive steps in creating positive change. None of us alone can bring about the momentum needed to change the current tide, but by advancing and adopting a major philosophical shift that is responsive to these pressing concerns, we can address these problems the way most major problems in this world are successfully addressed—through a collective movement built one person, and one building, at a time.

THE PRECAUTIONARY PRINCIPLE

A cornerstone of this philosophical shift can be realized in changing our approach to the adoption of new technology. The burden of proof of the environmental or human safety of a product has, in the United States, largely fallen on the shoulders of consumers, consumer and environmental advocates, and regulators, who often can prove that a product is dangerous only long after it has been released into the marketplace and in many cases resulted in widespread damage. A handful of recent examples in the building industry include lead paint, asbestos, and POPs. This "catch-up" process is currently being played out with formaldehyde and BFRs, now that they have been shown through multiple studies to have already widely contaminated the planet and our bodies. In the aforementioned FEMA formaldehyde study, it was reported that "additional research is needed to better clarify the potential reproductive and developmental toxicity of formaldehyde," well after widespread contamination has been built into nearly all residences in the United States.

A proactive approach can be found in the *precautionary principle*, defined at the 1992 United Nations' Earth Summit: "Where there are threats of serious or irreversible damage, lack of full scientific certainty shall not be used as a reason for postponing cost-effective measures to prevent environmental degradation." This principle shifts the burden of proof from presuming a product or material is safe until proven hazardous to presuming a product or material may be hazardous until testing proves its safety before general dissemination into the world via the marketplace. We can already see effective adoption of the precautionary principle affecting market viability and increased scientific and regulatory scrutiny of new technologies, such as Europe and Japan's rejection of widespread adoption of genetic engineering (particularly in the food sector). New unproven technologies such as nanotechnology have the potential to reshape our world much in the same way plastic did in the last century. Learning from the past, we can see the potential danger we face should the precautionary principle not be applied to new innovations as they appear in the markets and in our daily lives.

The Living Building Challenge helps buildings such as the Omega Center for Sustainable Living create a paradigm for a new built environment. PHOTO COURTESY OF OMEGA INSTITUTE FOR HOLISTIC STUDIES, RHINEBOOK, NY.

DESIGN FOR CHANGE

We need useful metrics to assist us in making sound decisions in the design process. Tools such as life-cycle analysis (LCA) can go a long way toward helping us understand the impact of one material or system compared to another based upon a cradle-to-grave boundary. As identified earlier, there are limitations in LCA tools in their current form: the quality of an LCA report relies upon accurate, quantifiable, and directly comparable data, which is very difficult to produce to the required breadth and scope. We believe, however, that as research continues to be produced and LCA tools become more sophisticated, designers will be able to effectively use LCA as *one of a set of analytical tools* with which to make appropriate design decisions.

Progressive rating programs that set high bars for social and ecological quality will help challenge designers to make active, positive changes in the interface between the built and natural environments. Rather than simply "reducing impact," as is the goal for many rating and evaluation programs, programs such as Cascadia Green Building Council's Living Building Challenge (LBC) go far beyond that to set standards that define the future of how we want our buildings to perform.

In reviewing the many issues surrounding buildings and ecology, we find that natural building technologies offer solutions across the board, from reducing emissions that are contributing to global warming to decreasing toxins that are released into the environment and improving health conditions for builders and occupants alike.

CHAPTER 3

Beyond the Building: Siting and Landscape Design

To achieve the true potential of natural design, full consideration of the building's relationship with its site is critical. It is not enough to design only within the walls of the structure; we must design the structure in the context of its supporting environment. Our colleague and friend Ben Graham of Natural Design/Build in Plainfield, Vermont, has a lovely way of considering a building's relationship with its site. He uses the Buddhist principle of *shunyata*, "emptiness" or "nonbeing," as a way of identifying the mutability of the boundary between the building and its environs. The concept behind *shunyata* is that there cannot be a pure isolation of an individual entity, such as the self or a building, because all things are deeply interconnected and rely on active and dynamic exchanges that make boundaries of separation arbitrary and illusory. Take a tree, for example. We may at first glance define the edges of a tree as its trunk and branches. Then, looking below the soil, we add roots. But the roots themselves are hosts to vast colonies of mycorrhizal organisms spreading out through the soil, which enable the tree to readily uptake nutrients. Looking up to the leaves, sunlight is taken in for energy, and water is released into the air, having been drawn up the trunk from the roots, which pulled it from the soil. Birds, insects, and mammals take residence in the branches, further extending the connection to the surrounding ecology. The tree truly exists not by itself as an entity, but only as part of a greater whole—this is *shunyata*. Recognizing buildings' inherent interconnectedness with their surrounding human and natural ecologies allows us to help facilitate the quality and quantity of those relationships.

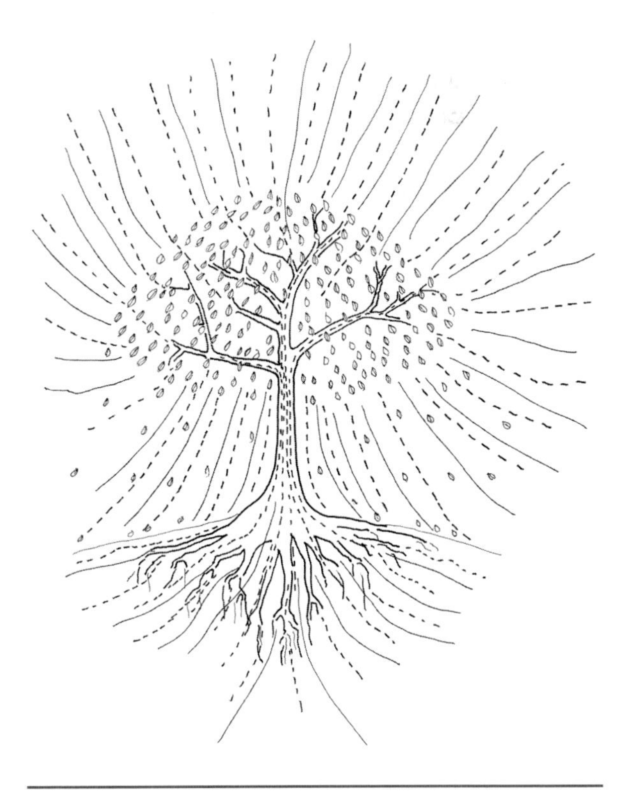

Just as a tree has a nebulous border defining it as being distinct from its environment, so too are the edges of our buildings difficult to define in the context of their environments. ILLUSTRATION BY BEN GRAHAM.

Fundamentals of Siting

We begin our exploration of the building's relationship with the world beyond its drip line with an identification of the fundamentals of siting and site considerations to ground our building in the context of place. While some of these considerations are more relevant to a rural context, where site options are more plentiful, others may be more important in a suburban or urban environment, where population density offers different considerations.

EXPOSURE AND ORIENTATION

The exposure of a building refers to its placement respective to wind vectors and wind shelters. The ideal is to strike a balance between allowing sufficient airflow to encourage drying and provide ventilation and protecting the building from wind-driven moisture and winter gales that will incite convective heat loss. Designing for solar orientation is also important and is based upon an understanding of where on the eastern and western horizons the sun rises and sets in both winter and summer (the azimuth), and the height off the horizon (the altitude) it will reach at either solstice.

VIEW

It is important to have a sense of being connected to the world beyond our buildings even when we are inside them. Our attention is captivated, and our feeling of connection enhanced, by a dynamic view such as one that takes in a bird feeder or a garden. Depths of perspective are important as well—foreground, midground, background—to give a greater sense of context of the surrounding landscape, even if this simply means a yard feature and a horizon line. Rather than a mammoth picture window in the living room that forces continued experience of a dramatic vista, a more powerful opportunity may come from a view experience that is created in select places in the building, including areas of motion such as stairwell landings, kitchen sinks, and bathrooms. One can see a view from a building; one can also see a view of a building. "Curb appeal," or the orientation of the building to the primary access route, may indeed be a relevant consideration, especially in a neighborhood environment or one where zoning ordinances apply to the design of the exterior of the structure. Privacy is certainly a consideration, particularly if the southern orientation of a house looks out on a busy street and offers passers-by a view into the living room. In more rural environs, visual impact on the landscape is a concern as well.

ACCESS

The primary access consideration involves transportation access to a site—whether by walking, biking, driving, or wheeling in a chair. Easy vehicular access is generally considered a high priority for year-round inhabited structures for many reasons. Construction vehicles need access while building, and emergency vehicles must be able to reach the building for the safety of the occupants, not to mention the logistics of moving people and belongings to and from the building. In urban areas, this is usually achieved fairly easily due to tight population density and small lots, although stairs are often challenges to universal access. In rural areas on larger sites, this access has liabilities, however. The logistics of plowing and maintenance of extended or difficult driveways can be intense, especially in snowy regions. Landscape impact for utility services such as electricity and heat are other serious concerns. On the other hand, a lack of vehicular access can make for a unique, cozy experience when using the structure after it is built. Providing vehicle access to a remote and beautiful part of the property will more likely than not strip much of the beauty from that place. The effects of the construction process, mixed with the impacts of a permanent settlement, are incompatible with the remote, natural experiences of those "magic places" on a property. Keep them magical and site the building elsewhere. While remote-access buildings offer significant privacy, they do so at a cost, and compromising access for high-use buildings is rarely a viable solution.

Having difficult vehicular access to a site can lead to challenges during construction, as well as a lifetime of higher maintenance costs. PHOTO BY KELLY GRIFFITH.

SLOPE (GRADE)

Slope—the angle at which the land is pitched—has a great deal of impact on many other siting considerations. A gentle slope (5%–12%) offers easy site development, great flexibility, adequate surface drainage, and good potential for wastewater drainage (given adequate soil conditions). A moderate slope (12%–20%) features more challenging site development, options for creativity in building into grade, good surface drainage, and potential wastewater (i.e., leachfield) limitations. A steep slope (greater than 20%) is very challenging for site development and can lead to difficulty in construction, potential danger to inhabitants, erosion concerns, difficult access, wastewater disposal problems, and the need for terraced agriculture, and building on such a slope may be prohibited by zoning ordinances. No slope (less than 5%) creates easy site access and flexibility and good wastewater drainage potential, while making bermed or earth-integrated building and foundation drainage difficult. Surface drainage can also be difficult.

WATER AND DRAINAGE

As can be seen from the various considerations that may need to be addressed given different slopes, water drainage in its many manifestations, such as surface and subsurface (foundations and footings) drainage, must be addressed to effectively integrate the building with the site. Freshwater supply and wastewater treatment must also be considered for successful site integration, as must existing natural water features.

SOILS

Soil quality is another factor that affects many other siting considerations. Soil conditions can be highly variable across any region, and even across a given site, and should be evaluated before an owner commits to a given site. To support the structural loads of a building, the soil must be stable; a very sandy soil type may require modifications in the foundation design. Having the soil tested by a civil engineer is advisable, especially if there is any doubt concerning its quality. A site with particularly good topsoil may best be reserved for agricultural use, rather than disturbed or removed for building purposes. Areas with protruding bedrock or ledge may best be avoided, as they could complicate the installation of the foundation unless the foundation design is flexible (such as a pier foundation), in which case anchoring into ledge will provide stable support for the building and reduce concrete use by eliminating the need to dig further below the frost line in cold climates. Finally, if soils are to be used for construction such as for earth walls or clay plasters, a careful evaluation should be made to ensure they are suitable for such use. If so, and excavation is to take place for grading or foundation work, topsoil should be removed carefully and set aside to later repair the site, while subsoil can similarly be set aside in an appropriate location for manufacturing of the building products (care should be taken to locate it such that it will not interfere with other parts of the construction and development process).

FIRE

Fire patterns in a landscape vary widely from region to region. Site-responsive fire-prevention strategies for buildings will vary accordingly, but there are some common strategies that are likely to be used wherever threat of fire in the landscape is expected. Creating firebreaks, or cleared areas around the settlement to remove fuel sources from the vicinity of the building, is one such approach. Providing good access for escape routes or emergency vehicular access is very important, as is providing access to water for firefighting such as a river, pond, or an approved yard hydrant, particularly in rural environments where access to municipal water is limited.

Natural Building and Permaculture

In most cases where landscaping is given more than peripheral attention, it is often designed to enhance or control the experience of entering, leaving, and being around a building, including fostering the leisure or hobby activities of the building's inhabitants. Bridging between the building and the site, the landscaping links the experience of being in the home with that of being on the property or in the neighborhood. Certainly this is a relevant and important goal of landscaping. There is however a subtle yet important distinction when we consider the greater context in which our landscapes and buildings exist when seen as being in relationship with each other. Let us revisit the example given earlier explaining *shunyata*. We can perceive a context as surrounding the tree, and see the sun as supplying the tree with energy, the soil as anchoring the tree and supplying it with water, and the mycorrhizal fungi as enabling nutrient uptake by the tree. In a more holistic and ultimately more relevant context, however, we see the tree, the sun, the soil, the water, the mycorrhizae as all supporting each other to create a healthy and functional ecological system, which in turn supports and is supported by neighboring ecosystems.

So too can we consider a building in relation with its surrounding landscape in this way. While indeed trees supply shade and windbreak, wind provides ventilation and drying, and the sun heats and lights our buildings, this is only one small part of a larger context, a vision in which we see the building, the sun, the wind, and the trees all working together with the occupants and their neighbors and neighbors' buildings to create a vibrant and successful ecological system. This is a core principle of regenerative design (discussed in more detail in chapter 10), which entails creating designs that, beyond "doing less harm," work toward creating positive social and ecological change.

Whether in rural, suburban, or urban environments, we have the potential to meet a significant portion of our basic needs from our own properties or from our own neighborhoods. How to address the need for safe, healthy, delicious food and clean, safe energy are two requirements among many that can be woven into the design for a property or a community. In doing so, we have the opportunity to create habitats for others in the plant, animal, insect, fungal, and bacterial communities. Permaculture is a design system that focuses on these goals. A portmanteau of "permanent culture" and "permanent agriculture," permaculture instructs through a series of core principles how humans can create permanent habitats in whatever region they occupy that are both sustainable (in the true sense of the word, meaning that they can be sustained over time) and harmoniously integrated into the ecological systems that support their habitat.

Buildings are most successful when closely integrated into the landscape, in both form and function. PHOTO BY KELLY GRIFFITH.

According to Mark Krawczyk, a permaculture designer and educator from Burlington, Vermont, and director of Keyline Vermont, "At its very essence, permaculture encourages us to work actively to meet our needs using the resources we have available. As such, food, water, shelter, waste management, livelihood, and community well-being all converge as we strive to design landscapes and lifestyles that accomplish these goals with an eye toward efficient use of resources and our sights set on regenerating community and landscape.

"More than a set of techniques and strategies, permaculture provides a holistic design process and philosophy that inform effective decision making. Founded on a basic set of ethical principles, permaculture design aims to care for the earth and care for people, and to reinvest surplus to further facilitate these two fundamental tasks."

Krawczyk believes that natural building shares a deep connection with the philosophy of permaculture, whose practitioners aim to create systems that make the best use of local and on-site resources; conserve, collect, and store available energy inputs; serve multiple functions while providing security in the form of backup systems for those essential functions (i.e., heat, water); and harmoniously integrate into the landscape and the community. Permaculture principles provide open-ended, solutions-based guidelines that can be applied to any situation regardless of scale, location, or context.

Thus, permaculture design assists us in planning for the design and development of not only the landscapes surrounding our structure but also the structure itself, its temporal installation and development, access and circulation patterns, microclimates and more, ultimately serving to create an integrated whole rather than a disparate set of parts.

"Many find that some permaculture concepts already play a role in their own decision-making processes," says Krawczyk. "That comes as little surprise seeing as how permaculture's co-originator Bill Mollison once referred to the interdisciplinary field as 'applied common sense.' But just as 'natural' building only becomes relevant in a world dominated by energy-intensive, synthetic products and processes, applied common sense becomes revolutionary in a world where common sense is no longer all that common."

There are many ways that buildings integrate into a permaculture design. Creating our structures directly in concert with our agricultural systems is a key way of supporting our own food production and ensuring agricultural systems that are healthier for our ecological systems (in contrast with industrial agricultural systems). Edible landscaping and agricultural zones surrounding the immediate vicinity of the home can take place in rural and urban environments alike (rooftop container gardens and backyard chicken coops being two great urban examples) and ensure active management of intensive sustainable agriculture practices. House designs that support agricultural activities such as processing and storage enable these activities to happen efficiently.

Buildings themselves can have positive relationships with our productive landscapes. We will discuss in later chapters the potential to use the collection area of a building's roof to serve as water catchment to support irrigation or aquatic landscape features while at the same time keeping erosive effects of water from damaging the structure. The heat energy stored by and emanated from a building can also be used to support ecological communities in the immediate vicinity of the building. Not only can attached greenhouses and solariums be used to extend the range of plant life that can survive in that region, but the building itself—particularly the south face—can be used as a heat source to create a distinctly warmer microclimate within the site, allowing propagation of plants that might not otherwise be cold-hardy enough to survive.

Systems that harvest energy from the site—including solar electricity and heat, wind power, micro-hydroelectric power, and wood heat—are all options for closing the loop of energy usage on the property. As this energy can also be used to maintain and enhance the quality and productivity of the landscape, we see this need extending out beyond the service of the building itself. In fact, sustainable silvicultural practices that promote high-density, multi-age, multi-species forest ecosystems can produce terrific quantities of fuelwood, as well as material for fiber, construction, medicine, food, and other uses.

The products of our buildings' operations can also be integrated into the surrounding ecology in a regenerative fashion. In addition to water catchment from the structure, treated graywater can be piped to irrigate agricultural zones or supply aquatic landscape features. Nutrients from food scraps and even human waste can readily be integrated into the landscape in the form of compost. When vegetables that are harvested from the garden are irrigated from water captured from the roof and filtered from the house and fed by compost generated from the home, we are living in line with the concept of *shunyata*.

CHAPTER 4

Soil and Stone: Geology and Mineralogy

Part of the enchantment of natural building is that it is not available at Home Depot in premade packages, nor is there a simple checklist of how to do it well. It entails developing an understanding of materials found on our planet Earth, born of relationship with those materials. This is related to the cultivation of a sense of place, of how these materials fit in our lives, and how we fit in theirs. Natural building is about relationships—and the intent of this chapter is to make visible the soil and stone that is all around us, quietly making life and work possible.

To make inquiries into the main inorganic materials we use in natural building—lime, clay, sand, mica, and stone—is to tell the story of the lithosphere, or the world of stone and rock and mineral. What is it, where does it come from, where does it go? What is lime, and how did it get placed in the earth, to be excavated by humans starting 4,000 years ago for building and other purposes? What is clay? How was it made, where do we find it, why does it exhibit the properties so useful for natural building systems?

These questions are partly about gaining understanding and knowledge in order to form a relationship with the lithosphere, not only to enrich our lives on this planet but also, as builders, to work with these materials more effectively. They are also partly a practical aspect of our information gathering on the source, toxicity, and ecological impact of these materials—which will then inform how they fit into our building plan. What is their genesis, what is their life span, and what does their end-of-life breakdown look like? What is their carbon footprint? We hope to provide you with enough information on the nature of lime, clay, sand, silt, stone, mica, and other materials to help you understand them so you can make your own decisions on their appropriate use in your context.

Geology

Fundamentally, the so-called inorganic parts of our planet—rocks, minerals, the mantle, the crust (all seemingly solid and unchanging materials)—come from a common source and are part of a constant cycling, wherein their form changes and they are reinvented. The inorganic, mineral-based materials that we are interested in as natural builders, such as clay, sand, lime, stone, and mica, all come from the same rock that makes up our mountains and bedrock, which is primarily granite in the Northeast. Weathering and erosion cause the eventual breakdown of mountains into boulders, then rocks, then pebbles, sand, silt, and eventually clay. The same mineral ions and compounds are found at all levels of the size continuum, just in different arrangements and sizes.

Lime is a little different in its genesis—it still comes from a concentrated deposit of inorganic minerals and is classified as a sedimentary rock, yet it comes directly from the ancient bones of marine life (which had extracted dissolved ions of broken-down granite from seawater to make their shells) laid down millions of years ago. The interconnections of biologic life in the geologic cycle, and vice versa, are many and varied.

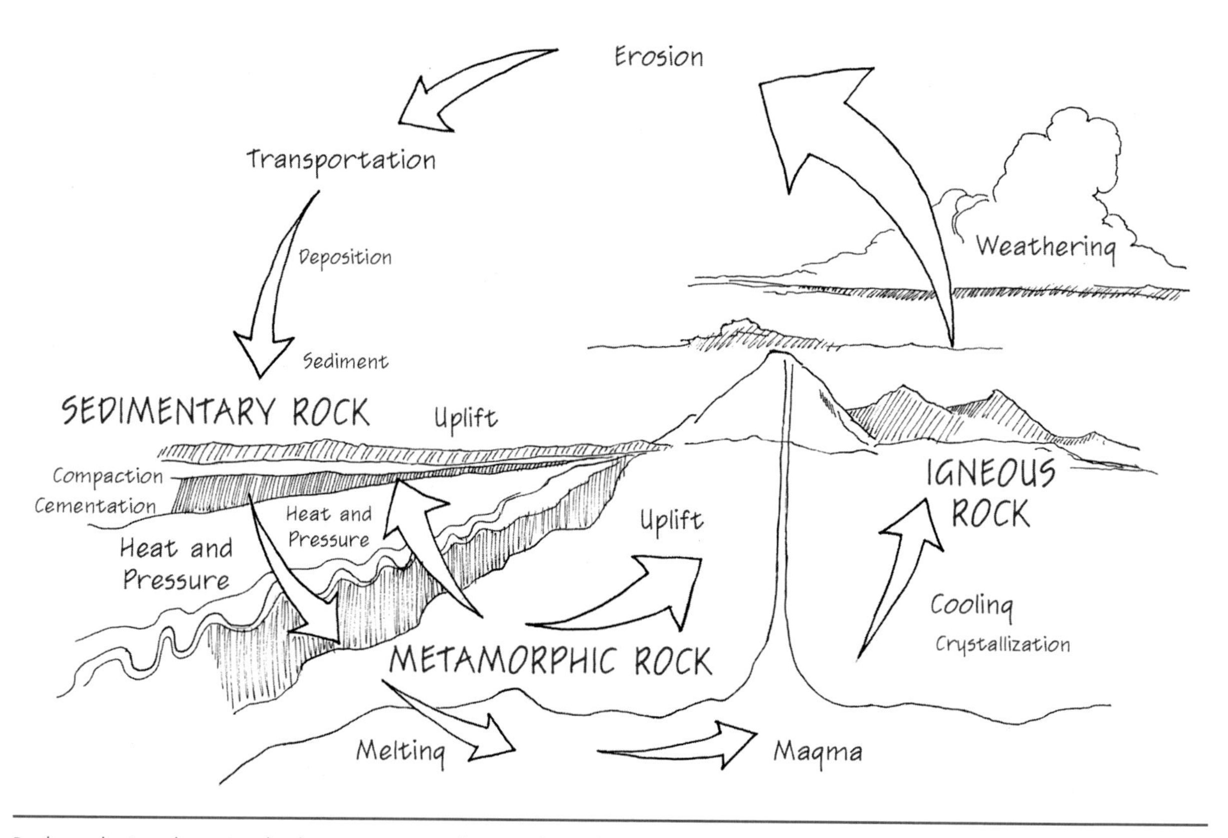

Rocks and minerals are involved in a constant cycling, similar to the water cycle. ILLUSTRATION BY BEN GRAHAM.

LAYERS OF EARTH'S CRUST

Earth is 4.6 billion years old. We have some idea of the most recent 3.6 billion years, but of the time before that, we have only the murkiest of ideas. The earth's core, analogous to the pit of a fruit, is thought to consist of a nickel-iron alloy. It is very hot at its center, and solid, owing to extreme pressure. The outer section of the core is liquid nickel-iron alloy, as the temperature is high enough to promote a liquid state, and pressure is low enough to allow for it.

The earth's mantle makes up the largest part of the substance of the earth, similar to the body of a fruit. It is also understood to have two layers. The layer closer to the core is made of soft, but not liquid, rocks. The upper layer consists of more cool and brittle rocks, which cause earthquakes as they are subjected to stresses and movement, during which they often break, unlike the lower section, which flows under similar pressures.

The crust of the earth is the smallest in volume—it's 1% of the total volume of the planet—and can be thought of as the thin skin of the fruit. There are two types of crust: the oceanic crust and the continental crust. The continental crust is primarily granite and related rocks, while the oceanic crust is basalt. The oceanic crust has been erupted out of volcanoes at the mid-ocean ridges—liquid rock that solidifies on the surface—while the continental crust comes from uplift at the faults, where it is made under the crust in the upper mantle.

THE ROCK/MINERAL/SOIL CYCLE

Our understanding of the cycling of the lithosphere comes from plate tectonics theory: the most game-changing scientific discovery in geology of the twentieth century. Plate tectonics explains the earth's crust as alive and moving, countering the longstanding

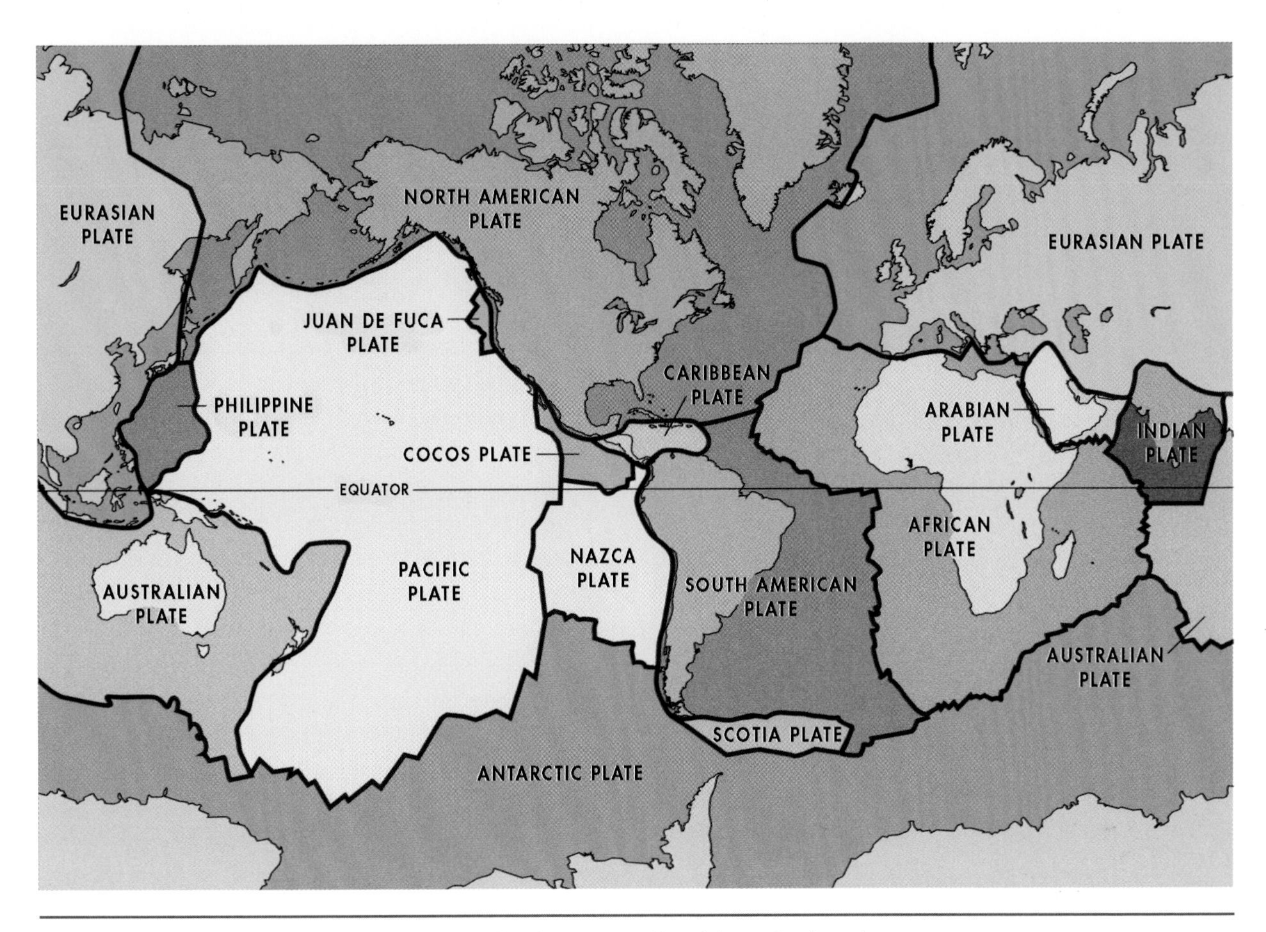

Plate tectonics theory transformed our understanding of the formation and breakdown of rocks and continents. MAP COURTESY OF THE U.S. GEOLOGICAL SURVEY.

notion that the earth is solid and unchanging, and that episodes like earthquakes are the exception rather than the norm. "We now recognize that continents are continually being fragmented and rearranged, chunk by chunk, and growing by accretion of new material from the mantle," write geologists Maureen Raymo and Chuck Raymo, in *Written in Stone: A Geological History of the Northeastern United States* (2001, 70). The theory of plate tectonics purports that new crust is constantly being created at mid-ocean ridges, at rates up to 120 mm/yr, while old crust is subducted at plate convergences.

Next at play are the forces of weathering and erosion. At the surface, exposure to ambient conditions that are much less hot (and have less pressure than the mantle) as well as the effects of water, carbon dioxide, plants, and animals immediately begins to erode and weather the rock. The average rate of erosion in North America is 2 inches every 1,000 years. No matter how strong the rock, how high the mountain, or how thick the stone, erosion will eventually level it. Raymo and Raymo (2001) point out that if it weren't for the crumpling and lifting of the earth's surface—through earthquakes, volcanoes, and other major earth-building events—on a regular basis, the land would be as flat as the sea.

The rate of erosion depends, of course, on the ambient climate and the composition of the rock. The classic example of this phenomenon (as mentioned earlier) is the Egyptian obelisk that was brought to New York City's Central Park in the 1800s. Popularly dubbed "Cleopatra's Needle," it had stood mainly unchanged for thousands of years in dry Egypt, whereas in rainy New York City, in a mere

This Egyptian obelisk taken from its home in an arid climate and brought to New York City's Central Park illustrates clearly the forces of weathering and erosion in the northeastern United States. PHOTO BY SEAN PAVONE/SHUTTERSTOCK.COM.

matter of 175 years the inscriptions on it became almost unreadable. Each day, the red granite that the obelisk is made of is exposed to water and the weak acids formed in water, such as carbonic acid, as well as temperature swings that cause weathering. The broken-down pieces of red granite are transported by wind and other forces (erosion) and make their way back into the water, the soil, and aggregate to become the clays or sands we excavate for building. Eventually, these fragments of the obelisk will be remade into rock: either in the crust, as a sedimentary and perhaps secondarily metamorphic rock; or carried on a subducting plate back into the mantle, to be reborn again in the heat and pressure. This is the force of erosion and weathering, breaking down rock into its constituent parts, smaller and smaller, to create the richness and diversity of soils, silts, sands, and clays that sustain life as we know it on this planet.

Soil particles—sand, silt, and clay—get picked up by surface water and carried downhill. As this water slows down, the sediments drop out in order according to size. The coarsest sands and pebbles come out of suspension first, then the finer sand, then the silt, and finally mud and clay. Clay minerals, owing to their microscopic size and affinity for water, will stay in suspension for much longer than the other macroscopic particles.

Looking to the sea, soil erosion parallels the mineral cycle on land. Sediment becomes sedimentary rock, which becomes metamorphic or igneous rock. Near shore, coarser particles accumulate, while the finest particles settle farther out. As thick layers are formed, water is squeezed out and the sediment is compacted to form new sedimentary rocks. These new layers then may be pushed up by tectonic upheaval, creating new bedrock, which becomes weathered and forms soil, and so on. When subjected to extreme pressure by being pushed down or by colliding forces, sedimentary rock will metamorphose, and if exposed, these metamorphic rocks will weather like the original igneous rocks to form sand, silt, and clay—the building blocks of soil.

TYPES OF ROCK

We recognize three types of rock: igneous, sedimentary, and metamorphic. We find materials we use in natural building in all three categories.

Igneous Rock

Igneous rock forms when molten rock cools from very high temperatures, when it is in a liquid state, to a cooler temperature, where it solidifies. The rate of cooling is indicated by grain size: smaller grains equal faster cooling.

Granite is an igneous rock that cools and

solidifies slowly within the body of the planet, in the mantle. Granite features mineral grains, easily visible, that give it a speckled appearance. Vast outcroppings exist in the Northeast, exposed after covering layers have been stripped away by weathering and erosion (Raymo and Raymo, 2001). We use granite for countertops, windowsills, and tiling. Granite is strong, local to us in the Northeast, and beautiful. But granite is also important as a parent material for other very well-known natural building materials: sand and clay.

Mica is igneous rock made of phyllosilicate compounds that solidify into translucent sheets. Mica is most prominently known in natural building as the "glitter and glow." In fine finishes, paints, and washes mica adds a luster, and sometimes sparkle, to the body of the finish that catches, reflects, and refracts light. There are micaceous clays found in North America (in the Southwest for example) in which mica is an intrinsic element and visible in the clay.

Mica lends sparkle and shimmer to many natural clay plasters and paints. PHOTO BY KELLY GRIFFITH.

Sedimentary Rock

Sedimentary rock forms when particles of broken-down rock—clays, silts, and sand—accumulate in layers. Eventually these layers are subjected to pressure and chemical cementation, which turns them into rock. Sedimentary rock weathers quickly, so the exposed parts of most mountains are igneous/metamorphic, but uplift periodically moves sedimentary rocks up to the surface.

Sand is most often made up of tiny quartz granules and is in actuality the pieces of mountains, broken apart by weathering. Eventually these leavings accumulate in great thicknesses, and pressure and chemical cementation turn them to sedimentary rocks. In the case of sand, it becomes sandstone.

Shale is a sedimentary rock formed from clay-particle depositions. Shale is soft and quite buttery and crumbly to the touch.

Limestone is also sedimentary, and it is an example of biologic life playing a role in the soil/rock/mineral cycle. What today is the northeastern region of the United States was 359 million years ago Laurentia—the ancient geological core of North America—located 25 degrees south of the equator. Laurentia was mostly submerged under a warm, shallow sea in the mid-Cambrian period. Organisms called crinoids lived on the shallow sea floor and extracted carbonate from seawater and secreted hard circular plates. Calcium carbonate deposits accumulated on the older sands, eventually forming a thick carbonate platform, lightly covered by sea, which stretched from Newfoundland to Alabama. This accumulated carbonate bank, which occurred during the mid-Cambrian through Ordovician periods, was 2 miles thick in many places. Time, pressure, and chemistry turned it into the limestone and dolomite that we harvest in the Northeast today.

Some of these deposits of limestone and dolomite were transformed during the Taconic orogeny—a metamorphic process—and became marbles.

It is interesting to note that while igneous rock is created along with tectonic activity that builds mountains up and pulls continents apart, sedimentary rock is associated with the long weathering, erosion, and buildup that tears mountains down (Raymo and Raymo, 2001).

Metamorphic Rock

Metamorphic rock is formed when any kind of rock is altered by high temperature and pressure without remelting. Metamorphic processes do not change the rock's bulk chemical composition but do produce new minerals and textures.

Slate is a metamorphosed variety of shale—a sedimentary rock made of clay sediments. Slate is a much softer stone than granite or marble, which makes sense when we compare the properties of their constituent materials. Slate is useful in natural building for countertops, floors, windowsills, walls, foundations, roofs, and shelves built into niches.

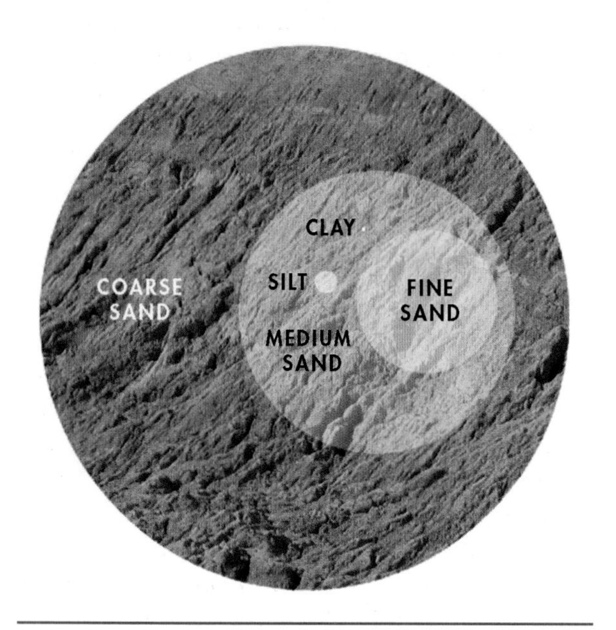

Soil particles are differentiated primarily by their size. Clay is microscopic, while sand and silt are macroscopic particles.

Marble is metamorphosed limestone. Marble can be used for countertops, floors, showers—any place stone tile or slab is desired.

Quartzite is metamorphosed sandstone.

Each of these three rock types—igneous, sedimentary, and metamorphic—can be turned into metamorphic (or other metamorphic) rocks. It is also true that change can happen in the other direction as well. All rocks can be weathered and eroded into sediments, which can eventually form sedimentary rock. Rocks can also be completely melted into magma and become reincarnated as igneous rock. The rock/mineral cycle is constantly on the move—although most often at rates difficult for humans to see.

Mineralogy: Minerals, Sand, Clay, and Soil

Mineralogy is the study of minerals: their chemical and physical properties, their uses and formation. Here, we will examine the relevant mineral categories for natural building.

SAND

Sand grains on Long Island and Cape Cod beaches are tiny crystals of quartz that were once part of inland New England granite rock (Raymo and Raymo 2001). Some minerals in granite dissolve in water, like feldspar, while more resistant quartz grains remain intact—which is why we find the quartz remaining in tiny, strong granules, while the feldspar has dissolved into water and been carried away to some other fate—perhaps to mineralize the soil, or to become clay particles.

Sands come in sizes from 0.0625 mm to 2 mm. In natural building applications, like plastering or earth-floor pours, we are most often looking for sharp, well-graded sand. The ideal sand for these purposes has particles with sharp edges, irregular shapes, and a mix of sizes, so the grains will lock together into a strong formation.

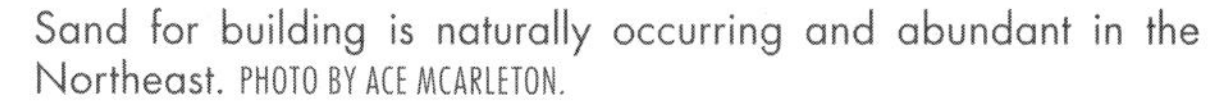

Sand for building is naturally occurring and abundant in the Northeast. PHOTO BY ACE MCARLETON.

The Northeast has abundant clay for building. PHOTO BY ACE MCARLETON.

CLAY

> "Every clay is simply an evolved expression of the earth's crust. It contains in corresponding proportions the important elements that go to make up crustal rock: silica, alumina, iron oxides, and in smaller quantities, oxides of manganese, calcium, potassium, and other elements. Long weathering has reduced these elements as far as water is able, making the particles very small and dissolving whatever compounds could be dissolved. The composition of clays determines their colors."
>
> —William Logan Bryant,
> *Dirt: The Ecstatic Skin of the Earth*

Clays are microscopic phyllosilicate particles. They are unique in our cast of characters so far in that they are microscopic—the weathering action of water has taken apart the parent rock material to sizes between 0.0002 mm and 0.004 mm. There are three main types of clay: the kandites (kaolinites), smectites, and illites. These classifications have to do with the type of crystalline structure the clay has, which is related to where it derived from geologically. The most salient piece of information for natural builders regarding these classifications is to know to stay away from the smectites, which are the expanding clays. The most common example of this is bentonite, which will expand to several times its original volume if it comes into contact with water.

The size and structure of the clay that we use for natural building purposes are the reasons it is useful to us: clay is small enough that it has lots of surface area to hold water in and among its molecules, it readily attracts water molecules with the ionic charges on its surfaces, and it also readily releases water into the atmosphere.

As William Logan Bryant says lyrically: "Clays, unlike their parent rocks, have no inaccessible interior, but instead a very large reactive surface. . . . Clays stack, wrap, pile, and exfoliate, like leaves or sheaves of paper. In fact, experiments have shown that a single gram of clay powder can have a total surface area larger than a football field" (Bryant 1995, 125). This immense water-holding capacity, paired with clay's hydrophilic (water-loving) nature, are two main reasons clay works so well for us in natural building.

Clay exhibits plasticity, which makes it joyful to work with. Clay is plastic owing to its composition of infinitesimal plates that slide across each other, loosely held in place by intervening layers of chemically combined water. "It is very hard to pull the plates

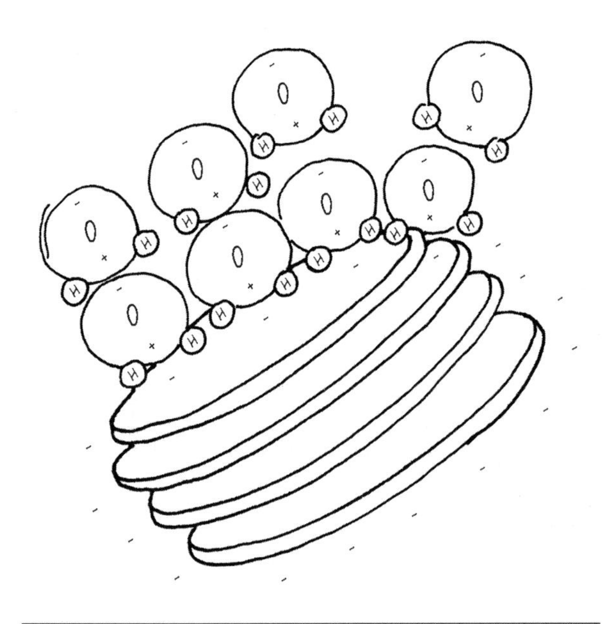

The ionic charges of clay molecules attract water. This hydrophilic property is what makes clay so useful for protecting straw in the wet, cold northeastern United States. ILLUSTRATION BY BEN GRAHAM.

apart, but comparatively easy to slide them one across the other—as anyone who has tried to walk in the sticky gumbo-till clays that sometimes cover the surface of the Earth will tell you" (Bryant 1995, 130). This "flexible" quality of clay is another useful attribute for us: when made into plaster on a wall, it is not a rigid finish, which makes it a good pairing for softer wall surfaces like straw. Pairing rigid finishes over softer finishes can cause failures such as cracking and shearing.

For natural building purposes—plasters, floors, bonding material for wall systems like cob or straw and clay, and paints—the kaolinites are the group we are most interested in. Also known as the stable clays, kaolins do not expand significantly when water is present. This means that when the water leaves and the plaster dries out, the body of clay will be less likely to crack as the volume retracts.

Kaolin

Kaolin is, in general, a white or light-colored soft mineral mainly composed of the clay mineral kaolinite with minor and variable amounts of illite, montmorillonite, quartz, and mica. There are two types of kaolin: in-situ kaolin, often called primary kaolin or china clay, is the direct result of the weathering of granite rocks, and so is found near the granite source; kaolinic clays, like ball clays, plastic kaolins, or sedimentary or secondary kaolins, are sedimentary soft minerals that formed as the primary kaolin was transported by wind and water, mixed with organic materials and other dissolved minerals, and deposited in low basins.

Ball Clays

Ball clays are secondary kaolin clays, formed from the mixture of kaolin with carbonaceous material, and ultimately settle in basins. When clay-rich site soil is unavailable or impractical, we turn to bagged clay called OM-4 ball clay for natural building projects. The "OM-4" refers to a specific mine, "Old Mine #4," that the clay is harvested from in Mayfield, Kentucky. Ball clay is actually a fairly rare type of clay, found only in specific places throughout the world. In the United States, the main deposit of ball clay is in the Mississippi Embayment, which extends over 45,000 square miles, and the clay is mined primarily in Missouri, Kentucky, and Mississippi. It was laid down during the early to middle Eocene epoch, which occurred approximately 40 to 45 million years ago and is the second geologic epoch of the Tertiary period, marked by the emergence of terrestrial mammals and present-day vegetation (Ferrario et al. 2000).

The composition of ball clays varies from 20% to 80% kaolinite, 10% to 25% mica, and 6% to 65% quartz. "In addition, there are other 'accessory' minerals and some carbonaceous material (derived from ancient plants) present" (Industrial Minerals Association 2007). The addition of this carbonaceous material gives ball clay its characteristic strength and flexibility, whereas the primary kaolins are finer and less dynamic without these organic additives.

Ball clays are used not just in natural building but also more widely in construction ceramics. Building materials such as bricks, clay pipes, and roof tiles all contain ball clay.

Dioxins in Ball Clays

Natural materials, like clay deposits, are safe and nontoxic—right? This is not always the case. Recently, testing of some harvested ball clays found them to have significant levels of dioxin.

Levels: How much dioxin and where is it from? Dioxin levels have been tested in ball clays originating from the Mississippi Embayment (Ferrario et al. 2007), and the average Polychlorinated-dibenzo-dioxins toxic equivalent (TEQ) concentrations in the tested ball clays was approximately 800 pg/g (picogram/gram). The Environmental Protection Agency's limit for dioxin is 0 pg/g—no level of exposure is deemed acceptable.

The catalyst for discovering this contamination was the finding of elevated dioxin levels in chickens. When their feed was tested, it was found that the ball clay used as an anti-caking agent in the feed was contaminated with dioxin.

Mechanism: Is it a natural source of occurrence or human made? Through human activities, dioxin is formed from the burning of chlorine-containing compounds. Dioxin in clay deposits so ancient that human activity is unlikely to have been involved leads to theories of natural sources. More research is needed, but some researchers are pointing to the burning or metabolism of halogen-containing compounds by bacteria, or fire, or other situations on pre-historic Earth that would have led to such compounds forming from chlorine—and other halogen-containing compounds.

Exposure: How does it affect natural builders or homeowners? More study is needed on this. Preliminarily, it seems that one major route of exposure is through the heating of these ball clays, either during drying and processing in the plant (resulting in worker exposure) or through the firing of the clays in kilns (resulting in craftsperson, artist, teacher, and student exposure). Dioxins volatilize at temperatures ranging from 421°C to 510°C, and it is hypothesized that the compounds are released into the air without being destroyed during kilning and/or drying (Ferrario et al. 2007).

How would builders be exposed? The potential routes of exposure for builders would be dermal (absorption through skin) or from dust inhalation, either from handling powder in mixing plaster or paint or from cleanup/vacuuming/dust on floor/in surroundings after plastering. Is it a concern for residents? For residents, it is hypothesized that only the dust from construction or later deconstruction would be a route of exposure. More research is needed.

There is some initial reason to hope that we will find the risks are not that great—the case study on the kiln exposure article is potentially helpful in this regard (Franzblau et al. 2008). In it, the individual with the highest measurable levels of dioxin had a kiln in her residence. She lived with both the kilning of clay and also the dust from sanding of the fired pieces. Her levels were much higher than those of either of her two lifelong potter friends, who also worked with ball clay in their studios for comparable amounts of time, but whose studios were not inside their homes. Ferrario and Byrne (2002) also compared dioxin levels in processed ball clay to those of fired ceramic products made from the same type of clay. They found lower and nondetectable amounts in the fired products, from which they speculate that the dioxin volatilized. This is good news for the natural builder on one front: inhalation of volatilized dioxin is an insignificant route of exposure because we do not fire the clays. However, it does mean that the dioxin that is present in the clay could be a risk through dermal absorption, or through the inhalation of the powder itself.

Safety: There is obviously going to be some level of exposure: dioxins are ubiquitous in our environment, due to both natural (prehuman) sources and human industrial sources. Every animal and human on this planet has been exposed to some level of dioxin. What we need to understand is how to limit our exposure. Commonsense precautions, such as using masks, gloves, and protective clothing, are advisable if using ball clay from sources identified to contain dioxin. Until more research is done, we won't conclusively know which activities either promote or minimize risk.

Clay and Organic Life

There is an intriguing theory about clay in the field of abiogenesis (the arising of life from nonlife). "Clay theory," as it is called, was first explored in the 1980s and 1990s. Clay theory postulates clay molecules as the partner necessary for the first amino acids and nucleotides to form and replicate. It theorizes that the way in which clay crystalline structures repeated themselves from parent crystal to daughter crystal provided a model for replication by the early proteins. Bryant, expounding on this clay theory as proposed by Graham Cairns-Smith, argues that clay's structure and its positive and negative charges (cations and anions) provided both the perfect container and the template for the first amino acids and the beginnings of organic life. This theory is one among some fifteen currently under debate about the nature of organic life arising from inorganic life on ancient Earth.

In terms of clay's function today, we know it plays a pivotal role in making biologic life possible by providing minerals in soil that react with plant roots. Soil is a mixture of sand, silt, and clay, in varying amounts. The charged nature of the surface of clay minerals in the soil attracts oppositely charged particles such as calcium, potassium, and nitrates, which are needed for organic reactions inside plants and at the roots. Again, the poetic approach of Bryant is appropriate here. He says, "From a locked closet of the mineral world, [clay] has become an open shelf, where the roots of plants may shop for what they need" (Bryant 1995, 135).

WHAT IS SOIL?

In natural building books you will often see references to "site soil," "subsoil," or "clay-rich soil" as terms for earth material in which clay can be found for plastering. But what is soil? How do we discern between soils that are useful for building and others that are better left for agriculture or some other purpose? As we have examined above, forces of weathering and erosion create soil by breaking down parent rock material into increasingly smaller pieces. So from larger to smaller, first we have boulders and stones, then sand and silt, then finally clay. All of what we call "soil" is a mix of sand, silt, and clay that together form what is called loam. Many combinations of the three components are possible.

Soil specialists use names for the various loams, such as silty loam, sandy clay loam, etc. A loam can be dried and pounded in the laboratory and passed through sieves to separate the mix by particle size.

Most usefully, for natural builders, the U.S. Department of Agriculture's Natural Resources Conservation Service (NRCS) has an online synopsis of its national soil survey data, which can be searched by area of interest. Anyone can search within a geographic region and find data classifying soils in that area. The data will indicate where the highest percentage of clay soils are to be found, for instance, which can be useful for locating promising sites to find locally sourced clay-rich soils for building.

In the flood plains of rivers, silt and sand are deposited along with mud (mixed clay and organic elements), a process that creates some of the most fertile and workable soils of the world; however, clays that have high silt content are undesireable and can be confusing to the natural builder to identify. Utilizing the ribbon test (see chapter 17) will help ensure good clay content.

Having come full circle in our description of the rock/soil/mineral cycle, and having identified the relevant minerals for natural building, we have set the stage for the next chapter, in which we discuss the roles of plants and animals in natural building processes.

Geology and the Northeast

One note before we leave this chapter on the specific geology of the Northeast: no other part of the continent has had a more violent history—geologically—than the Northeast, and much evidence of this claim is visible to

The Northeast has an abundant resource in stone, which is ripe for use in building. PHOTO BY KELLY GRIFFITH.

us today, revealing clues to the continent's physical and topographical past (Raymo and Raymo 2001).

One of the most formative episodes for our current northeastern landscape was the most recent ice age, when glaciers covered most of the Northeast. When these great continent-spanning ice sheets melted, they dropped the debris they were carrying frozen within the ice. This resulted in a landscape that is covered with till—"an intransigent, hodgepodge of boulders, rocks, gravel, sand and silt" (Raymo and Raymo 2001, 142).

A common joke in the Northeast is that we are farming rocks, and the residue from the ice age is why. Traversing the northeastern landscape, we see stone walls everywhere, and we can now recognize them not just as quaint, picturesque boundary markers, but as much-needed repositories for boulders removed from fields. To further connect the built environment to geologic events, the abundance of glacial till is why many Northeast farmers picked up and moved west to more fertile, less rock-infested lands with the construction of the Erie Canal and the laying of the railroads. The ones who stayed had to get creative to compete with the superior agricultural production potential of the Midwest—hence we see the emerging connected farmstead pattern of "big house–little house–back house–barn," where value-added functions were stacked for maximum profit and efficiency (Hubka 1987).

The Northeast is not great farming land for large-scale production; this, in addition to our more mountainous topography, is why most of our straw comes from the grain-growing plains of Québec just over the Canadian border, or from New York State. We do have a lot of great stone, some sand, some lime, and some clay deposits, and plenty of surface water and groundwater, yielding a richness of natural resources for green and natural builders to utilize.

CHAPTER 5

FLORA AND FAUNA

Following our look at soil and stone, the geologic elements important for natural building, we next turn the spotlight on the living things that play a similarly central role as building materials. Humans have utilized animals and plants in countless ways for centuries; however we rarely consider them as construction materials or methods. Natural building practice underscores that it is not only compounds synthesized in a lab that make useful, durable building materials, but resources that were once alive as well.

Natural building can engage a range of plants, both cultivated and wild, in its systems. This list includes yet is not limited to trees (from sawn or round timbers to framing lumber and boards, twigs and branches, and the bark from the cork oak tree); wood products like resins (turpentine) and paper that becomes cellulose insulation; and grasses. While the nutritive part of cereal grains like wheat, rice, and rye (as well as bamboo, thatch, and cattail) can be ground into flour for food, the flour can also be used to make building materials like wheat paste and rice paste, and the woody stems of these plants can be used as insulation. Seeds of trees and grasses can be pressed to yield oils—flax produces linseed oil; tung oil results from pressing nuts from the tung tree. Coir is a fiber made from coconut shells, and sisal is a plant fiber from the agave plant.

There is also an assortment of animals and animal products useful in natural building systems. From manure (cow, horse, elephant) to eggs, milks, urine, hair and wool, blood, beeswax, animal body remains known as "offal," and invertebrate secretions—a variety of animal products can be found in the spectrum of natural-building-material options. As we discussed in chapter 4, lime originates from the relics of animals. Much of the lime deposits we mine today are layers of the secretions of prehistoric plantlike organisms that filtered calcium out of seawater in the Laurentian era.

Plants

Plants are multicellular organisms that characteristically produce their own food from inorganic matter through the process of photosynthesis. It is the cellulose in the plants' walls, and sometimes the oils and resins the plants generate, that make various plants useful as building materials.

WOOD: TREES

The relevance of wood in natural building systems depends greatly on region and location. In the northeastern United States, the use of wood as a framing material, as an infill wall system (e.g., woodchip and clay; see chapter 15), or for siding, for example, is a fairly obvious choice that fits well with the ethics of natural building. However, in a place like the desert of Arizona where wood is less abundant and less a part of the vernacular building tradition—and would require more inputs in travel, ecological impact, and trades support—wood may not be the best choice. Thus, the materials we examine in these chapters should not be taken as prescriptive for natural building. We are overtly attempting to change the approach inherited from the postwar United States suburban boom that asserts that climate and region are irrelevant to building. Alternatively, we argue that understanding and operating within the opportunities and constraints of climate and materials is essential to natural building—or indeed, any form of intelligent and ethical building.

The landscape of the Northeast has been dramatically shaped by the interaction between trees and people. Accounts tell of indigenous people doing yearly burnings of the forests to rid them of understory and litter that got in the way of hunting and travel. The burning also created ripe conditions for berry plants like blueberries and huckleberries. "Over time, burning also selected for fire-tolerant, nut-producing trees, such as the American chestnut, white oak, and shagbark hickory" (Wessels 1997, 35). Large crops of nuts and berries also sustained large populations of game, and the increased sunlight caused by the elimination of subcanopy trees allowed for grasses to grow, which provided a grazing source for animals. "These pre-colonial, fire-managed woodlands looked dramatically different from New England's present forests. They were park-like, with massive hardwoods creating a canopy over forest floors carpeted with grasses and berry bushes" (Wessels 1997, 36).

All this changed with the arrival of the colonists, who did not continue the burning tradition but set to clearing the land for cultivation, and eventually sheep. "By 1840, approximately 75% of the region's landscape had been converted to open land for agricultural use, the bulk of it sheep pasture" (Wessels 1997, 58). A migration then quickly followed to richer farmland to the west. "By 1850, 100,000 Vermonters—almost half of the state's population—had moved west. . . . It wasn't until the 1980s that many rural central New England towns regained the population levels they had had in 1840" (Wessels 1997, 60). These settlers abandoned their farms and pastures, which then turned back into forest, ultimately resulting in the conditions of the woodlands of the Northeast today.

Currently, there exists much local woodlot management, felling, harvesting, and sawmills in the northeastern United States. Some small mills, such as Fontaine Sawmill in East Montpelier, Vermont, are starting to build their own kilns, to be able to provide locally sourced nominal 2× lumber.

Spruce, pine, and hemlock are common softwoods in the Northeast. PHOTO BY KELLY GRIFFITH.

Trees are large plants that produce the most abundant and concentrated source of "wood" for building purposes. "The main physiological function of wood is the mechanical one of giving strength to resist the increasing weight of the structure as it grows erect and branches. Submerged aquatic plants, buoyed up, as they are, by the water, do not form wood in their stems" (Boulger 1908, 2). Interestingly, wood comes from the xylem that exists in the stems of trees—gymnosperm and angiosperm stems. The straw of cereal grains used for straw construction, such as from grasses in the family Poaceae, are also stems. They are called monocot stems, as are bamboo. Similarly, the various other plant-based materials we use in natural building are from the woody, cellulosic part of the plants.

The lumberperson's terms "softwood" and "hardwood" bear little relation to actual hardness but instead refer to the two classes of trees, gymnosperms and angiosperms. The wood of angiosperms or deciduous trees (such as oak, mahogany, teak) is called hardwood or porous wood. Hardwoods have a more complex structure than softwoods. They have pores or vessels that conduct liquids, some of which are large enough to be seen with the naked eye. In gymnosperms (coniferous or softwood trees), water conduction occurs in tracheids, which are elongated cells in the xylem. Many softwoods also contain resin, either in very thin tubes (resin ducts) parallel to the tracheids or in short cells like those of the California redwood. It is important to know that there are some softwood trees that are harder than the softer hardwoods, and likewise, some hardwoods that are softer than the harder softwoods.

Looking at a cross section of a pine or spruce tree, we also see distinct annual rings. The softer, light-colored portion is the spring wood, and the darker, firmer portion is the summer wood. In general the central portion of the trunk of a tree is darker than the outer portion. The darker center is called heartwood; the lighter, outer portion is sapwood. Many of the sapwood cells are still in a sufficiently active state of vitality to store up starch in winter. Growth is confined to the outermost layer of all, the cambium. The inner rings of the heartwood, or duramen, are physiologically dead and serve the primary function of resisting the lateral strain of the wind.

Scents and grain patterns are characteristic and unique for each species and can be used as a means of identification. The various odors of different woods are a result of aromatic gums, resins, and essential oils that accumulate in the cells. The grain of wood is due to the orientation of the structural elements. Interesting patterns arise because of irregularities of the grain or the orientation of the grain in relation to the sawed or sliced surface. Knots are the basal ends of branches that become embedded in the thickening trunk.

The most common trees used in the Northeast for timber framing are pine, spruce, hemlock, cherry, and oak. Lumber and lath for wood-lath systems are typically sawn from softwoods such as pine, spruce, and hemlock. When we harvest saplings and branches for wattle-and-daub construction, we use an assortment of deciduous and coniferous species. Woodchips and sawdust for woodchip clay can come from any waste wood that is chipped up in a chipper, either in an industrial setting or on a small scale by renting a chipper.

CELLULOSE

While cellulose insulation is shredded newspaper and other waste paper that is treated with borax, a fire retardant, and blown into cavities using a machine blower, cellulose itself is actually a general name for the type of fibrous molecule that makes up the cell walls of plants. Cellulose insulation is derived from trees that have been made into wood pulp, made into paper, and then repurposed and shredded for use as insulation. Using cellulose insulation incorporates recycling or repurposing of a waste product into an extremely effective building material. For more on cellulose, see chapters 15 and 18; as natural builders we utilize cellulose insulation as a complementary insulation along with natural wall systems made from straw bale, light straw-clay, and others.

STRAW

What we are used to calling "wood" for building purposes comes from trees. However, all of the xylem of vascular plants—including the veins of leaves and the stringlike vascular bundles of palms and bamboo—is actually wood. This is relevant to our understanding of the strength of straw as a building material. As noted earlier in this chapter, the physiological function of wood in a plant is to provide strength for the structure as it climbs higher, which is also true for straw. Straw is the woody stem of cereal grain plants, just as the trunk of a tree acts as the stem of the tree plant.

Straw is the woody, cellulosic stalk of the cereal grain. PHOTO BY KELLY GRIFFITH.

Straw is harvested after the grain and is used in agriculture or disposed of—if not used for building purposes. PHOTO BY STEVE MCSWEENEY/BIGSTOCK.COM

"Straw" is a general term that includes the stems of many and various grasses, most specifically cereal grains such as wheat, rye, rice, barley, and others. Many of the Poaceae, the true grasses, have hollow stems called culms, which are solid at intervals called nodes. These nodes, or "knees" as they are called colloquially, are the points along the culm at which leaves arise.

Straw is, in this way, distinct from hay—a basic thing to understand. Straw is the woody, hollow stalk of the cereal grain plant, while hay is the top of the plant, where most of the nutrition culminates in the seed. When the thresher moves through the field, it is set to a certain height that maximizes the cuts of the plants for the most grain access at the top of the plant. The seeds are what is harvested as grain, because they contain the most nutrition. Hay is the secondary nutrition source, often used as feed for animals and livestock who digest cellulose readily. The remaining stem of the cereal grain, straw, is harvested as a secondary crop, or sometimes even burned in the fields as a waste by-product. Straw is also used for feed occasionally, less so because of its lower nutritional profile, and is also used as animal bedding. This is also why straw is most often available, freshly cut, in the fall—at harvest time. This timing presents interesting challenges for us in the Northeast when the builder requires the straw in the spring or early summer, but the harvest has not yet happened and the straw that is available is last year's.

Most usefully for us, straw is an excellent building material and can act either as the central element or as a component of a natural building assembly. It is a central element in a straw-bale wall system, for instance, while it can also be present as a component in adobe block or cob.

OTHER GRASSES

Other relevant plants in the Poaceae family include bamboo, cattail, and any grasses that can be used for thatching. Bamboo is hollow and grows straight rather than tapering. It grows primarily in Asia but can also grow in North America. Bamboo is widely considered to be a good building material owing to its rapid growth—up to 39 inches in a day, depending on soil conditions—and hence regenerative ability. It can be used for framing, flooring, trim—any use that wood is generally put to. It should be noted that most commercially available bamboo flooring contains formaldehyde binders, and when combined with the distance the material travels to the northeastern United States, it often does not live up to the "green" stamp it is generally considered to have.

Cattail is probably one of the most ancient Poaceae that humans have interacted with. It is these bulrushes that have been identified as associated with grinding stones in Europe dating back 30,000 years. In our region, cattail "fluff" can be harvested as a fiber addition to plaster, and the cattail itself can be used as thatch.

Thatch is one of the oldest known building materials. Thick layers of long-stem grass can be a fairly good insulator and weather protector, and it is easily harvested. Similarly to bamboo, it regenerates rapidly. It can degrade easily, but it can be replaced easily as well. In Europe and the colonial United States, thatch roofs were once prevalent, but the material fell out of favor as industrialization and improved transport increased the availability of other materials. Today the method is enjoying a resurgence of interest. There is a growing interest in "biomass" roofs made from material derived from plant grasses (thatch) or plant-based "tiles" (such as tree bark). Many different types of grasses, cereal grains, and reeds can be used—the longer the better.

OTHER FIBERS

Hemp is a soft, flexible fiber that is derived from the hemp plant. It should claim the spot as the first fiber listed here; however, given that we are writing this from the United States it is unfortunately not available for our work. Our colleagues on other continents, and even in Canada, have access to hemp fiber and hemp board and hemp-crete—many uses for hemp that are relevant for natural building that we do not have access to (Weismann and Bryce 2008).

Sisal is a fiber from the agave plant, which is most often seen in woven rugs. Jute is a thin, long, shiny fiber of lignin and cellulose derived from the jute plant. Jute is also called burlap or hessian; we use burlap as a supplemental fibrous sheet at times in plastering to bridge over dissimilar materials. Coir is coconut fiber. Colleagues of ours in Natural Builders NorthEast (NBNE) use coir fiber as a fibrous mat to bridge over framing or other transitions when plastering, and they swear by it for good ecological and functional performance; we have had success in using it as a replacement for metal or plastic laths.

STARCH PASTE

Wheat paste and rice paste are used in our region as additives to plaster and paint, and occasionally as topcoats. Starch pastes act as additive binders or glues to a plaster or paint mix, and when applied as a topcoat they lend anti-dusting action and increased durability. Starch pastes are derived from flours produced by grinding grain seeds such as rice or wheat. Historically, these pastes were used in ancient civilizations in Egypt, China, India, Persia, and Rome for paper making, art, cosmetics, and food additives. A recipe for starch paste can be found in chapter 20.

Starch is the result of plants photosynthesizing glucose from light energy. Plants store this glucose mainly in the form of starch granules, which accumulate in the twigs of trees near the buds or the top of grasses in the seed heads. Glucose, when in the form of starch, is not water soluble and hence can be stored for longer periods. By storing energy as starch in their seeds, fruits, and tubers, plants prepare for the next growing season.

After grain is harvested, dried, and ground into flour and we mix the flour with water and heat it, the starch molecules become soluble. The more soluble, water-holding glucose molecules then form networks that increase the mixture's viscosity. This is called starch gelatinization. As we cook the starch further,

Wheat paste adds glue-like characteristics to natural paints and plasters. PHOTO BY KELLY GRIFFITH.

it thickens and becomes a paste that can be thinned by adding cool water and then put to a range of uses. It serves as a binder for pigment, as a fine aggregate in paint, and as an additive to plaster and can be painted on top as a wash to mitigate the disaggregation of loose particles—also called "dusting." For more reading on starch paste's uses, see chapter 20.

PLANT OILS AND RESINS

Plants also give us oils and resins that are useful to the building process. Turpentine is derived from a process in the distillation of pine tree resin. It is an organic solvent that is used to thin beeswax or carnauba wax, as a wood finish, or as a thinner for oil-based paints. Synthetic versions of turpentine are widely available, and naturally extracted turpentine is more expensive and more rare.

Linseed oil is the yield from pressing flax seeds. Linseed oil can be found raw or "boiled." Raw linseed oil often will take a very long time to dry, from weeks to months, and should be used with that awareness. "Boiled" linseed oil most often is not boiled (which does thicken the bonds and reduce drying time) but instead has had solvents and drying agents added to it to decrease drying time. Turpentine can be added as a solvent to help drying, but most of the solvents on your average hardware store shelf are petroleum-based, and the drying agents are metals such as arsenic, cadmium, and nickel. These additives tend to reduce the "green-ness" of boiled linseed oil quite significantly.

Linseed oil makes a fantastic sealant for wood and for earth floors when combined with beeswax or carnauba wax as a finish. It can also be added to limewashes and lime paints, in an amount that allows for partial saponification (the lime and linseed oil make soap) of the lime, while the rest emulsifies. Soap is basically the salt from a fatty acid; it results when fatty acids are broken down by a base in an emulsion of two liquids that do not dissolve easily in each other, such as oil and water. Linseed oil can also be added as a topcoat to clay paints and clay plasters, which adds water resistance, deepens the color, reduces vapor permeability, and reduces dusting. Linseed oil is also the main binder in old-fashioned linoleum, and it is being rediscovered and repackaged today as a green building material in the product Marmoleum.

Tung oil is the most plastic-like of the natural oils. It resists water when applied over wood and is the best of the plant-based oils. With multiple coats it dries with a slightly yellowish tint. To our knowledge, not much experimentation has been done to combine tung oil with earth and lime. It could be interesting to explore how its characteristics compare with those of linseed oil in these applications.

Animals and Animal Products

Animals and animal products also play a role in natural building, both in current practice and historically—all over the world in different cultures and regions.

MANURE

One of the most common animal products we use in natural building in the Northeast is manure, which can be utilized in many ways. It is sometimes found as an additive in earth blocks, cob, plasters, and floor mixes. In the colonial United States, and in Britain where lime was common, manure was used in combination with lime to mitigate cracking and to strengthen the mix. Bill and Athena Steen, who run the Canelo Project in the southwestern United States, have also worked to popularize manure as an additive to earth (clay) plasters.

We have had tremendous success adding manure to earth plasters and lime plasters here in the Northeast. The two most common types we use are cow and horse manure. Horse manure is useful because the fiber the horse has eaten has been processed, softened, partially digested, and chopped into small pieces for easy addition to plaster. We tend to use it for interior applications in a final lime-sand coat as a fiber addition to mitigate cracking. Cow manure, on the other hand, benefits from the partially processed cellulose that the cow has eaten, as well as additives from the unique digestive system of the ruminant: the rumen, or the so-called fourth stomach.

Ruminants

The word "ruminant" comes from the Latin *ruminare*, which means "to chew over again." All herbivorous animals need to derive energy from vegetable food, which is mostly cellulose. This is challenging, so they have developed several strategies. The first is to ingest it in large quantities because of its low caloric value—hence their grazing behavior. Ruminants, such as cows, goats, and deer, have developed a digestive system with four well-defined compartments, as well as the accompanying behavior of rumination wherein the animal partially digests and then regurgitates and remasticates the vegetable food matter.

The rumen plays an important role in digestion. The rumen maintains optimal conditions for the bacterial and protozoan life that ferments the cellulose, an important first step to breaking down the cellulose-lignin connections so cellulose can be better digested. These bacteria synthesize protein and essential amino acids from nonprotein nitrogen in the rumen, and they also synthesize B vitamins for the animal (Lewis 1961, 207). The amount of cellulose that can be digested by the ruminant varies from 30% to 80%. One reason this varies so much is the amount of lignin in the straw or hay. Like wood, straw and hay are composed of cellulose fibers and lignin, which is between the fibers. Cellulose that contains more lignin is more resistant to rot and microorganic breakdown, and similarly is resistant to digestion in the ruminant animal. This means that natural builders intent on using manure from the ruminant for its superior quality have access to partially digested longer straw or other grasses—the still-intact pieces being those with more lignin (and therefore resistance to organic breakdown)—for the fiber that is used in plasters, cob, adobe, and other applications.

Cow manure also contains enzymes that are hardening to earthen materials. Cow manure can be used to create a protective coating over lime as it is curing, allowing for carbon dioxide to enter for carbonation while retaining moisture, a necessary state for proper strong lime curing (Holmes and Wingate 1997/2002) (see chapter 17).

Urine, usually acquired by its pairing with manure and/or bedding, is often an unintended additive to adobe, plaster, or earth floor mixes, and to our knowledge, no positive or negative effects are known.

WOOL

The main source of wool is sheep. Wool is quite a good insulator, as anyone who wears a wool sweater in a cold climate will tell you. Similar to the concept of the straw bale, which insulates well owing to the myriad air pockets throughout, wool fibers are crimped and crinkled in such a way that they create air pockets when in a mass. Wool is used as an alternative insulation; producers such as the Good Shepherd Wool Company claim it is R-3.5 to R-3.8 per inch, although independent testing is difficult to find. Known benefits of wool insulation include its ability to retain its insulative abilities even when

Screening cow manure to make manure tea for fine finish plaster.
PHOTO BY ACE MCARLETON.

exposed to water, in marked contrast to fiberglass and cellulose insulation, which do not. It can be purchased in batt form or rope form, which is used to chink between logs in a log structure. It can be quite expensive, however. For small-scale projects that have sheep nearby, acquiring raw wool and washing and carding it before installing it can be a time-consuming yet cost-effective approach.

HAIR

Hair is an animal-derived fiber additive for plasters and earthen or lime floors. Ox, cow, and goat hair were traditionally entrained into lime mortars in the United Kingdom (Holmes and Wingate 1997/2002). Horse hair is the oldest fiber additive to plaster known in the northeastern United States region and can be found by many unsuspecting homeowners who demolish walls in old houses. In fact any of these types of hairs are perfect as plaster additives because they are coarse and not too oily. Human hair, in contrast, is too smooth and oily to be useful in this application.

BEESWAX

Worker bees (always female bees) secrete beeswax from glands on their heads. The production of beeswax is energy intensive: for each quantity of beeswax produced, the worker bees have to consume eight times the amount of honey. Beeswax is the substance that forms the honeycomb (the structure of the hive) where the bees live and their young are born. Historically beeswax was a hugely important material to human beings as it was used for molds, candles, and waterproofing, among many other uses. In natural building today we use beeswax as one type of water-resistant topcoat for earth floors, plasters, and wood.

SHELLAC

Shellac is a resin secreted by female lac insects. It is gathered and sold in dry flakes, which are dissolved in denatured alcohol to make shellac. Shellac largely replaced oil and wax finishes in the nineteenth century, and since then it has itself been replaced by synthetic lacquers and polyurethanes. Shellac is a great option for natural builders and woodworkers who want a nontoxic, nonpetroleum finish that provides superior clarity and gloss with hardness and protection of the base material.

EGGS AND MILK

The eggs of domestic fowl are rich in protein and can be used as a binder to add many benefits to natural finishes. Fine paints can be made by emulsifying oil and egg with pigment (see chapter 20 for more information).

We extract curds from curdled skim cow's milk and blend them into a smooth paste for a binding additive to paints and plasters. Casein is also available for purchase in dried, powdered form. When casein is combined with lime, such as in classic milk paint, the casein, or "milk," reacts with the lime to form calcium caseinate, which is a glue that adds strength, resilience, and water resistance yet reduces the vapor permeability of the wash or mortar. Casein can also be added to a clay plaster or paint and in that case serves a similar purpose: it combats dusting and adds additional binding to the mix, while also somewhat lessening the material's vapor permeability.

BLOOD

Animal blood, particularly cow blood, is a traditional additive to earth floors and adobe blocks. It is the proteins in the blood, similarly to casein, that add hardness to these materials. There is also a tradition in Europe of lime-cement floors that include blood as a hardener: "there are records of the addition of egg white or bullocks' blood to improve the quality of finish" (Holmes and Wingate 1997/2002, 176).

ANIMAL BODY REMAINS: "OFFAL"

In the United Kingdom centuries ago, remains from animal slaughtering, called "offal," were added to lime pits. This contributed proteins, lipids, and other organic matter to the lime, which made the lime set harder. "Proteins introduced into a fat lime putty can assist the carbonation. The reason is not clear, but decomposition of the proteins may produce carbon dioxide throughout the depth of the joints to assist a thorough carbonation. They may also assist in air entrainment. . . . Many protein-rich materials have been used in the past including the use of the lime pits for disposing animal carcasses" (Holmes and Wingate 1997/2002, 662).

Tallow, a clarified animal fat, can be added to limewash to make lime more water resistant and durable, a process that happens because the tallow reacts with the alkaline lime to form a soap, and part of it emulsifies. Tallow does reduce vapor permeability, however, and so is therefore used in the very ultimate limewash or lime coat, and only after the previous coats have been able to carbonate, as this final coat will inhibit carbonation.

Ruminants such as sheep and cows give us milk, manure, and wool for use in natural buildings. PHOTO BY STEVE LOVEGROVE/BIGSTOCK.COM.

Lard is a softer clarified animal fat that can be used in a similar way, as is size (pronounced as it is spelled). Size is a fairly pure form of gelatinous animal glue or comparable vegetable glue that can reduce dusting in a limewash, and also reduces vapor permeability.

A diverse array of plant- and animal-derived materials are used or involved in natural building technologies. This list is not exhaustive but catalogs the most relevant ones used in the northeastern United States and illustrates the abundant contribution of biologic and organic life to our lexicon of building methodologies. For a look at how these resources play a role more broadly in natural building systems through a site-based and permacultural lens, see chapters 2 and 3.

PART 2

Building Science and Performance

CHAPTER 6

Structure and Natural Building

Natural buildings, like any buildings, require a structure that carries and transfers loads including the structure's own weight over time, as well as snow, wind, and other environmental forces—and the weight of occupants and their belongings.

A grounded understanding of the basic physics of structure is necessary in order to be a successful designer or builder. In this chapter we will interweave some basic fundamentals of structure with an evaluation of different natural building systems; for a full primer on structural engineering we will direct you to other texts, as more than an overview is beyond the scope of this book. We will look at different structural systems commonly used in natural building, as well as the basic principles that support them, so as to provide a basis for making informed decisions when designing a structure.

Before we begin to look at the structural performance of different systems, however, it is important to contextualize the impact of structure on the building as a whole. In conventional structures, the structural elements of the frame and roof are largely hidden behind sheetrock, paneling, and siding. There are a few observations we can make from looking at this approach: For one, there is a conscious aesthetic choice to have a structure's role be subtle and/or hidden in most post–World War II dwellings in the United States. For another, the impact of the structural form on the building is downplayed, however fundamental the structure's function actually is. Third, the decision is usually made to align planes of structure, insulation, and services (plumbing, electrical) directly together in "conventional" buildings, which on the one hand creates an efficiency of space, but on the other compromises the ease of maintenance of buried wires and pipes, and even of function—as in the case of insulation when it is repeatedly interrupted and deformed by framing and wiring.

Let's now contrast this approach with a very different system: a straw-bale-wrapped timber-frame building. In this case, the frame is featured prominently in the architectural theme of the building, displaying the craftsmanship of the joinery and the organic form of the massive timbers. The selection of an overtly geometrical form that necessitates significant framing material (consider 12-foot-beam spans of a timber frame compared to 30-foot-beam spans of a steel frame) imposes the form of the structure on the spatial program of the building, dictating the floor plan. By pulling the frame to the visible interior of the building, it is not only easier to inspect for quality, but it is removed from the highly vulnerable transition plane between interior and exterior climates where moisture damage from both condensation and rainfall is most likely to occur. This separation also removes conflict of space between the insulation and the framing, which serves to both reduce conductive losses and to simplify use of a fixed-form insulation material (more on this in chapter 7). Additionally, it removes the need for formwork to support the installation of a flexible-form insulation material, such as cellulose, which requires a cavity (which is frequently provided by the structural framing).

Services have a different relationship with the frame as well in a straw-bale-wrapped timber-frame building. While utility runs may still be situated within a wall cavity, the frame is less likely to be affected by the removal of material in the creation of space for them. In this situation, it may also prove wholly impractical to run utilities through massive beams, and the plan may require running utilities within interior walls of the structure.

It should be clear from this quick comparison that there is no obvious correct approach, and that each approach has benefits and drawbacks under different considerations. Many more relationships between the structure and other elements of the building proper exist, and they will vary from design to design. We will bear this in mind as we consider these different natural building types in regard to structure. What is important is to recognize that structure does not exist in a vacuum; it influences and is influenced by all the other systems in the building as part of a whole, just as our human skeletal system influences and is influenced by our musculature and nervous system.

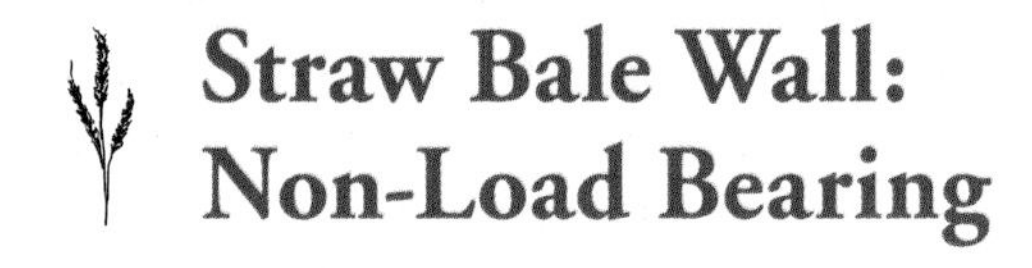

Straw Bale Wall: Non-Load Bearing

When we talk of a non-load-bearing system, we are disassociating any structural role from the bale wall system, except perhaps its ability to hold itself up. The structural requirements of the building must be taken up elsewhere, by a frame.

COMMON MISCONCEPTIONS OF NON-LOAD-BEARING STRAW BALE CONSTRUCTION

One of the most common mistakes in non-load-bearing construction is to attempt to—often quasi-successfully or even unsuccessfully, and usually unwittingly—build a hybrid structure, with loads shared between the structural frame and the straw bale wall. Here is a classic situation: A post-and-beam frame is erected, with temporary braces supporting the posts as the beams are built. Perhaps knee braces are fashioned, maybe let in from behind the frame, or maybe just toe-nailed to the posts and beams, serving slightly more than an aesthetic role. The roof is put on to protect the bales, but the frame is now subject to uplift and lateral wind loads so the temporary bracing is left on until the bale walls are completed, at which point they can finally be removed. What has happened is that the bale walls are being used in a structural capacity to provide shear strength to the frame and resist wind loading. This is not to say that the walls are not inherently capable of performing this role. However, they must be designed accordingly, by virtue of both their own construction (not hard to achieve with appropriate construction detailing) and, more importantly, their connection to the frame, which—in the case of a timber frame—will be limited at best. It is relevant to note that the shear strength comes not from the bales, but from the thick layers of plaster and the width between them (more on the structure of a bale wall is discussed later in this chapter).

In the example of a stud wall, it is a simple mistake to frame the stud walls and install the bales either within the stud plane, plastering the bales over the studs, or adjacent to the frame, hanging siding such as clapboards over the studs. What's wrong with this picture? In conventional practice, sheathing—which provides all requisite strength to resist lateral force—is nailed thoroughly to the studs and plates. If no bracing is added to the studs (such as let-in wood bracing or metal T-bracing) or sheathing applied, then again the bale walls are forced to do the job of providing shear strength—which they are capable of providing in theory but may not be doing in design and practice.

Another consideration of non-load-bearing straw bale construction is that a framing system must be selected that will have no interference with the building's enclosure system. If the frame does bear additional enclosure responsibility—in whole or in part—it must interface with the straw bale wall in an effective manner. Looking at the aforementioned examples, the timber frame has no enclosure system inherent to its form and can readily accommodate a bale wall enclosure system. The stud wall, on the other hand, may handle part of its enclosure (e.g., rain deflection provided by exterior siding) but leave other parts of the system—insulation, air barrier, interior finishing—to the bale wall.

Planes are especially relevant when there is dissociation between the frame and the enclosure system. We find a major structural anomaly in a timber frame wrap that is rarely seen in the world of conventional stick framing, namely, that the frame is set in from the "frame line," or outside plane of the building (so named because that is where the frame is *usually* located) by as much as 20 inches. This has significant effects in the two primary structural transitions in the building: between the foundation and wall framing, and between the wall and roof framing. In the former, we find that a conventional stem wall or thickened edge of a structural slab-on-grade may not have any relationship whatsoever with the structural load paths carried down through the timber frame. At a minimum, additional or extended footings will need to be poured to catch that inset load plane, and the design may get even more complicated from there. In the latter, we find that overhangs provided by the roof structure to protect plastered walls that are already significantly larger than average—24 inches is a convention in the Northeast—are compounded by the 20-inch inset of the timber frame, creating overhangs approaching 4 feet in length. While significant, this may be easily achieved on the eave ends by simply extending the rafter ends.

The gable ends, however, prove to be another story. Supporting not only one but two sets of rafters by framing tie-backs, sub-fascias, and sheathing can be unrealistic, especially on a roof expected to hold snow. The common approach of a timber frame is to run the rafter plates, which catch the bottoms of the rafters, all the way out to support the exterior rafters. This creates significant perforations through the insulated envelope of the structure, however. Not only is wood an inferior insulator, but air sealing around a timber—especially one featuring the checks frequently found in an end-grain exposed beam—is practically impossible. Solutions to this challenge vary widely but must be considered early on in the design process. For example, a lattice-style roof (described in detail in chapter 18) provides horizontal purlins of adequate length and structure to support such an overhang. Stud-framing the gable end is another option, either as Larsen-style wall trusses (see chapter 13 for more information) or conventionally supported by framing as described in the next section.

A final structural consideration in the non-load-bearing straw bale wall system is in the frame itself. Frequently, we use ungraded site-milled or locally milled framing material, which means that all of the quality control is now the builder's sole responsibility. By referencing span tables, we find multiple different designations of quality in both visual and mechanical evaluation categories, each designation dramatically affecting the structural capacity of the member. For example, a piece of lumber with a few spike knots (long knots formed by cutting longitudinally through the knot in milling) along one edge is suitable for use as blocking material. The spike knots will also not be much of an issue if the piece of lumber is used as a stud. It may be passable if used as a beam component, with the knots facing up so as to be in compression under load. It may fail completely if used as a beam component with the knots facing downward owing to the knots significantly weakening the member's tensile stress. Being able to evaluate both the potential weakness of the spike knots and their liability in different configurations throughout the frame directly highlights the importance of this chapter.

Straw Bale Wall: Load Bearing

As we learned in chapter 1, straw bales were first used in the United States by the westward-heading migrants during the turn of the last century. Lacking any other abundant local materials with which to build shelter, such as wood or sod, they were successful in baling prairie grasses, stacking them into walls, and setting a roof on top—so successful in fact that some of these original buildings are still in use to this day. In honor of these early pioneers, load-bearing walls are frequently referred to as "Nebraska-style" construction.

The "load" in question is, of course, made primarily of the roof (a "dead" load, which refers to the loads of the structure itself) and snow load (a "live" load, which refers to dynamic loads imposed upon the structure), as well as any second-story floor framing (dead) and furnishings and people (live). There are other significant loads that need to be resisted, such as lateral loads from wind and seismic forces. When engineering a structure built of wood, steel, or concrete, it is often the loads that are considered to provide the greatest variability when determining an accurate design, whereas the materials have been heavily tested and analyzed and are well understood. This is not the case with straw bales, however. As one of the least standardized materials used in construction today, the variables governing their structural performance are maddening (concerning clear structural analysis, anyway) and include density, size, species, purity/contamination, banding material, orientation within the wall assembly, and moisture content, to name just a few. How, then, are we to make sound decisions about structural engineering using this material? As we detail below, a combination of existing testing data, empirical evidence, and good common sense allows us to develop safe design and construction practices.

It is noteworthy that straw's biologic and cellulosic cousin, wood, is plagued with a similarly dizzying host of variables that affect its structural performance, requiring extensive testing and evaluation to allow for predictable engineering values. As Joe Lstiburek of Building Science Corporation states in his article "Wood Is Good . . . But Strange": "If someone invented wood today it would never be approved as a building material. It burns, it rots, it has different strength properties depending on its orientation, no two pieces are alike, and most cruelly of all, it expands and contracts based on relative humidity. Can it get worse? But, of course. Wood expands differently based on orientation . . . weirder still when wood shrinks and expands, it shrinks and expands differently along the grain than perpendicular to the grain. It shrinks and expands much more at a right angle to the grain, than along the grain. Studs don't get shorter or longer, but they do get thinner or thicker" (2009).

Both wood and straw are unpredictable building materials. PHOTO BY KELLY GRIFFITH.

Despite all these shortcomings, we have still managed not only to develop successful construction practices based around this fickle material, but to create a host of detailed engineering data upon which we can base those practices as a result of extensive testing, evaluation, and field use. If wood can enjoy the success that it has for common structural use by our production-based housing industry, there is no reason to believe that the variables concerning straw's performance preclude it from safe and efficient use.

When considering the structural performance of a straw bale wall, it is important to consider the wall as a "sandwich composite wall system," which is essentially a natural, insulated, stressed-skin panel. This means that the structural capabilities of the wall are borne not by the straw alone—nor by either or even both of the individual plaster skins—but by the entire wall system of two plaster skins firmly connected to a straw bale core. This system is effective for a few key reasons:

- *Thickness of wall.* In evaluating column structures (what is a wall if not an elongated column?), we find that the potential for buckling is reduced with an increased ratio of width to height. Straw bales provide this thickness in spades.

- *Stiff plaster skins.* Compressive loads in the system are largely borne by the plaster skins. The bales are far more ductile than the stiff plaster, and if placed under any load themselves they will deflect under the weight of that load until the plaster picks up the load.

- *Solid connection of plaster to bales.* In order for the skins to effectively work as a system, they must be connected to each other, via adhesion to the straw. Therefore, ensuring quality substrate preparation and plaster application is not just a matter of aesthetics or moisture protection, it is a structural consideration as well.

- *Ductile straw as structural backup.* In the event of failure of the plaster skins, the straw core is able to absorb moderate compressive and shear loading, as has been proven in structural testing of unplastered straw bale walls. The ductility of the straw core of our natural, stressed-skin panel is a great asset in mitigating dynamic forces, such as seismic activity. The straw's low modulus of elasticity and resultant quality of elastic deflection, limited by the plaster application and any compression detailing, allow it to absorb significant amounts of shock without failure—in contrast to the relatively brittle plaster skins with a higher modulus of elasticity, which are able to support enhanced compressive loading. This hypothesis has been proven in rigorous structural testing performed by the PAKSBAB organization at the University of Nevada at Reno engineering laboratory, in which a structural bale wall successfully survived a "shake table test" designed to simulate an earthquake—testing designed in response to the magnitude 7.5 quake that struck Pakistan in 2005 (see sidebar, "PAKSBAB Seismic Testing on Straw Bale Wall Building"). As a system, this built-in resiliency is a strong asset in seismic design.

In order for this system to work, however, proper detailing must be executed. Like most things in life, the devil is in the details. These details are explored further in chapters 14 and 17. Let's look at a few of the most relevant topics, however, pertaining to structural performance.

BALE ORIENTATION AND QUALITY

The density of individual bales as well as, to a lesser degree, how bales are laid into the building—on-edge or on-flat—have significant impact on the structural capacity of a wall. The density of the bales will certainly affect the degree of deformation under initial loading and creep (explored later in this chapter) and the overall quality of the wall system. The orientation of the bales will affect the aspect ratio of the wall and its ability to resist buckling and out-of-plane loads, of greater concern in exceptionally tall or large walls. A bigger issue with edge-laid bales is the adhesion of plaster, which is often more difficult in this orientation. All that said, structural testing has borne out the viability of edge-laid bales in load-bearing wall applications.

BALE PRE-COMPRESSION

Owing to the irregular shape and texture of the bales and subsequent void spaces that are created between courses of bales, as well as variables in the density of the individual bales (as discussed earlier) and their modulus of elasticity, straw bale walls will

PAKSBAB Seismic Testing on Straw Bale Wall Building

In response to a devastating earthquake that struck northern Pakistan in October 2005, killing about 86,000 people, destroying more than 780,000 buildings, and leaving an estimated 3.5 million people homeless, Pakistan Straw Bale and Appropriate Building (PAKSBAB) has been working to develop simple, unique, earthquake-resistant straw-bale building methods that are affordable, are energy efficient, and utilize local labor and indigenous renewable materials.

In May 2009, PAKSBAB conducted the first "shake table tests" of a straw bale building at the Network for Earthquake Engineering Simulation at the University of Nevada at Reno, sponsored by the Earthquake Engineering Research Institute, in which a 14 × 14 × 10-foot load-bearing clay-plastered straw bale building was put on a testing device that subjected the building to simulated ground motion accelerations from the Canoga Park recorded data of the 1994 Northridge, California, earthquake.

The building was built in load-bearing style of site-fabricated straw bales that were narrower than conventional straw bales and placed atop a soil-cement-encased gravel bag foundation. The bales were exterior-pinned with bamboo for stability during construction. Fishing netting was laid below the gravel bags, stretched over the bale walls, and nailed to a wooden top plate. The walls were clay plastered (clay soil, sand, and chopped straw) and lime washed; the roof was built of wooden I-beams insulated with light straw-clay and covered with corrugated metal roofing. Gravel bags were placed over the walls to simulate a light snow load.

The building survived the test, heavily damaged yet intact at twice the acceleration of the Canoga Park record, proving the viability of this construction technology in seismic zones. The authors of the study concluded that the performance of the composite sandwich wall system acted in this way: Compressive loads were carried to the foundation by the stiff plaster skins, which were bonded to the straw core to resist buckling; cracking and spalling of the plaster helped dissipate energy. The fishing netting provided ductility and tensile strength, and also shear strength and overturn resistance at the foundation.

As of September 2011, PAKSBAB has built twenty-five similar structures in Pakistan, using abundant, renewable, local materials that use significantly less energy to manufacture than conventional building materials of steel and concrete, while providing exceptional energy and structural performance. These buildings are "skill-accessible" to the communities in need of the infrastructure, at approximately half the construction cost of seismically safe conventional building alternatives in the region. Learn more about the testing and how to support PAKSBAB's work at www.paksbab.org.

PAKSBAB shake test. PHOTO COURTESY OF PAKSBAB.

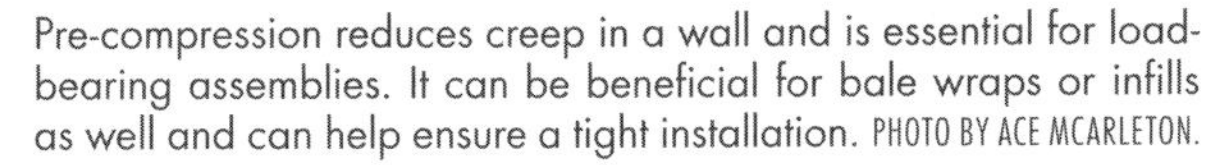

Pre-compression reduces creep in a wall and is essential for load-bearing assemblies. It can be beneficial for bale wraps or infills as well and can help ensure a tight installation. PHOTO BY ACE MCARLETON.

The integrity of the plaster-to-bale connection is critical for structural integrity. PHOTO BY JACOB DEVA RACUSIN.

compress under load. There are different phases of this compression: initial loading, which refers to the period immediately following substantial structural loading; application of plaster skins in non-load-bearing construction; and various points of "creep," which refers to the medium- to long-term distortion of the walls or bales themselves within the walls. Stresses that are a result of settling, during either initial loading or creeping, can result in aesthetic inconveniences, thermal performance losses, subsequent moisture intrusion/accumulation and damage, and, in worst-case scenarios, structural compromises—particularly as a result of deformation of the plaster skins, either in the field of the wall or in boundary conditions (described later in this chapter).

Appropriate pre-compression techniques, as discussed in chapter 14, are very important to help minimize deflection under initial loading to ensure even and uniform load support in structural wall systems and to minimize potential compromise to the plaster-to-bale connection.

PLASTER-TO-BALE CONNECTION

We hope that by this point it has been made abundantly clear that in straw bale construction, as a type of sandwich-composite wall system, the connection between the plaster and the straw is of utmost importance in creating structural integrity, particularly in support of vertical loads.

The function of the combined compressive strengths of the plaster skins and the ductility of the straw core for reducing the potential of the plaster buckling is a product of the connectivity of the plaster to the straw. This is covered in detail in chapter 17. It should be noted that for load-bearing considerations, the addition of mesh reinforcement is generally considered a necessity for structural performance, and this mesh can aid in the plaster-to-bale connection.

PLASTER COMPOSITION

As can be expected, different plasters have different strength properties. Cement-based plasters are superior in compressive strength and are also the most brittle. Lime-based plasters are moderate in compressive strength and are less brittle. Earthen plasters are the weakest and most ductile of the three. Understanding the properties of the plasters is very important in evaluating the structural potential of the wall system, as it is these plaster skins that will bear the primary structural loads, both vertical and lateral, in the wall. There is a good body of testing data available to provide quantifiable values for these different plaster types. Testing has also shown that for relatively simple one- and two-story buildings, appropriately detailed wall systems featuring any of these plaster types are structurally viable under both average vertical and lateral loads and more extreme seismic forces (see sidebar, "PAKSBAB Seismic Testing on Straw Bale Wall Building"). In fact, in testing for in-plane shear loads, a lime-cement plastered bale wall with steel mesh reinforcement was shown to be comparable to a well-sheathed stud wall, while an earth-plastered bale wall with plastic mesh reinforcement was shown to be comparable to a lightly sheathed stud wall. In both cases, the bale walls were more readily able to carry vertical loads even under lateral-load-induced distortion, owing to the aforementioned qualities of ductility, skin-to-core connectivity, and aspect ratio.

PLASTER THICKNESS

The thickness of the plaster skins is a critical component of the wall's structural viability, perhaps even more important than the strength of the plaster. According to Stephen Vardy and Colin MacDougall, Department of Civil Engineering, Queen's University, Kingston, Ontario, Canada: "The plaster strength and thickness has a profound effect on the strength of a plastered bale. The plaster thickness was seen to have a greater effect on the plastered bale strength than the cube strength of the plaster. It was found that doubling the plaster thickness increased the average plastered bale strength 65%, while doubling the plaster strength increased the average plastered bale strength 25%" (2006, 34–35).

STRUCTURAL SYSTEM CONNECTIONS

One of the most important, and most frequently overlooked, elements of structural design is the integrity of the boundary conditions, or areas of connection between different structural systems, such as foundation-to-wall and wall-to-roof. As noted throughout this book, much of the action occurs at the edges, and the realm of structural performance is no exception. Considering the quality of the load transfer and resistance in the boundary conditions is as important as the quality of the individual assemblies themselves. The connection to structural diaphragms such as roofs and floor systems is critical for shear walls to effectively absorb lateral loads and transfer them vertically down to the foundation. This manifests in structures in the straw bale wall-to-roof and wall-to-foundation connection detailing and is generally achieved through the solid connection of plaster reinforcement mesh to bottom and top plates, through tensioning mesh above and below the plates, and fastening with nails or staples. Two basic considerations when designing these boundary conditions are the loading over plaster skins and the plaster skin/mesh reinforcement placement relative to bottom/top plates. In the first case the top plate must be made to bear over the plaster skins, which are designed to support the primary compressive load, and not solely over the straw bale core. In the second, the physical connections are very important, as improperly reinforced plaster edges can result in buckling or crushing, whereas improperly supported skins at the base can result in slippage of the plaster off the foundation, resulting in a loss of load transfer.

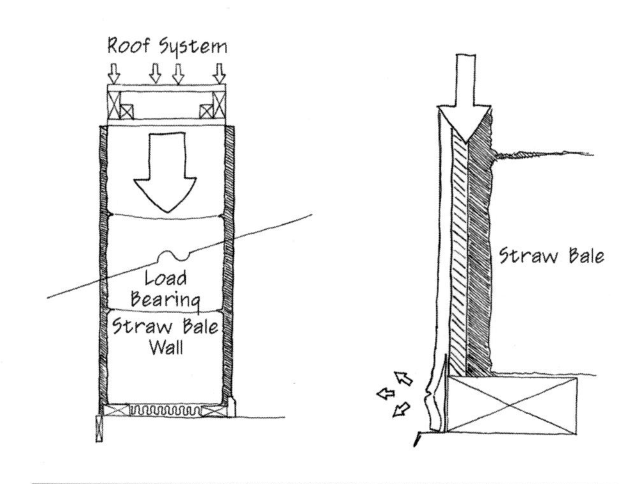

In a load-bearing straw bale wall, the plaster skins receive a compressive load, and therefore good detailing is critical to avoid problems such as those pictured here. ILLUSTRATION BY BEN GRAHAM.

Mass Walls: Load Bearing

Mass walls all have some similar structural properties, be they built of cob, adobe, rammed earth, brick, or stone. The most notable is their tremendous compressive strength; the vertical load-carrying capacity of the materials is very high because of their density. This same density provides a notable distinction from other wall forms—their increased dead load, which must be calculated into any design, particularly of the foundation or any bearing floor systems. Barring any unique peculiarities of the design, these walls are comparatively simpler than structural straw bale walls. Rather than operating as a sandwich composite wall system, mass walls are simply shear walls, providing lateral resistance. Although stone, brick, and adobe are built as an assembly of units, they act as monolithic wall systems that are also able to resist shear loading by virtue of the strength of their joints. The biggest considerations for the structural viability of mass walls will seem familiar by this point, and we will work to highlight specific structural concerns for these wall systems.

One of the greatest benefits of a mass wall is its compressive and shear strength. PHOTO BY JACOB DEVA RACUSIN.

MATERIAL QUALITY AND MIX

The quality of the base ingredients and their proportions in the mix are fundamental issues governing structural viability. The parent materials of the binder (clay) and aggregate (sand, gravel) will have a lot to do with the quality of the finished product. As we learned in chapter 4, there are many different types of both clay and stone, and there is a wide range of performance characteristics for each of these materials: a ball clay binder and granite sand/stone mix will behave very differently than a kaolin clay binder and shale sand/stone mix. For stone and sand, sharp, well-graded materials will also have an impact on the performance.

The mix proportion is perhaps the most important factor in the performance of a wall. As is explained in the context of plasters in chapter 17, and as can be seen in testing results for earthen-wall materials, having the correct balance of binder, aggregate, and fiber (or other reinforcement material) is fundamental to avoid both superficial failures of cracking and dusting as well as larger structural failures.

The strength of the wall relies, in large part, upon the quality of the mix—in this case, cob. PHOTO BY JAN TYLER ALLEN.

In many cases, particularly in the case of rammed earth or adobe, builders will incorporate a percentage of Portland cement into the mix to create "soil cement" or "stabilized soil" for use in a structural capacity, particularly for larger projects or in wetter environments (prolonged presence of moisture weakens clay-based wall systems). This incorporation of cement into the matrix increases the vertical and lateral load-resistance capacity, as well as the ability of the wall system to retain its form and stick to itself (cohesion) under more dynamic and erratic loads, such as seismic forces (discussed later in this chapter).

WALL CONSTRUCTION

How the wall is built will of course affect its structural performance, the greatest liability tending to come in the form of disruptions to the monolithic nature of the wall. In cob construction, this manifests as cobs (lumps of the material akin to unfired, loosely formed bricks) that are poorly knitted into the wall, or when significant drying between courses leads to poor cohesion between lifts. With rammed earth, a lack of sufficient tamping or excessive drying between lifts can similarly create cold joints within the wall. In stone, brick, or adobe walls, the concern in this regard is often with the quality of the mortar joints and the orientation of unit placement (for example, maintaining a "running bond" to avoid continuous vertical joints). In all cases, of course, building plumb and level (or with the correct amount of taper)—and enacting good quality-control measures throughout the construction process—will allow the walls to realize their full structural potential.

Stone walls, particularly dry-stacked ones, bring about additional concerns of stability between the units, given that they are frequently of irregular form. Understanding appropriate construction practices for sound wall building—particularly where there are soil or hydraulic forces on retaining walls—is critical for structural performance.

Dry-stacked stone walls are beautiful and structural and can be made of abundant local materials, but they require a good deal of skill to create. PHOTO BY JESS AHLEMEIER.

WALL FORM

As with any wall, aspect ratio (thickness-to-height) is a key consideration for resisting buckling of the wall. Location, size, and support system of door and window openings are other big concerns. In *Buildings of Earth and Straw*, Bruce King offers a few rules of thumb based on empirical observation of legacy earth buildings. For one, the total length of fenestrations should be less than a third of the overall wall length. For another, piers (vertically oriented areas in the wall between, say, a corner and a window, or between two windows) should be kept to a 4:1 height-to-width ratio. Fenestration openings must be sufficiently supported as well. If lintels are to be used, they must be structurally adequate to support the dead and live loads above and should extend at least 24 inches on either side of the opening. In some regions (e.g., seismically active) or with certain designs, the material choice is critical, as wood may not suit the engineering requirements or be allowed by code.

SEISMIC DESIGN

As with structural straw bale, the design detailing for structures in seismic-prone locations is important. This need for attention is accentuated with mass wall

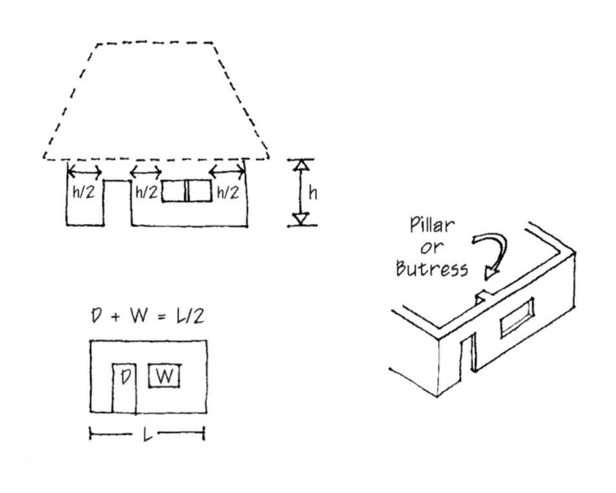

Basic structural rule-of-thumb proportions to consider when designing a mass wall. ILLUSTRATION BY BEN GRAHAM.

structures because seismic forces have a greater impact on heavier buildings. Although it is beyond the scope of this book to delve into the specifics, the general guidelines for seismic design follow those mentioned earlier regarding structural straw bale construction. Appropriate foundation detailing is necessary to fully support the monolithic, enhanced dead-load wall, with positive foundation-to-wall connections. The walls themselves may require embedded steel reinforcement (not unlike plaster reinforcement necessary in load-bearing straw bale construction), and the use of Portland cement stabilization is often necessary or mandated.

Know What You Don't Know

If you find yourself at the end of this chapter feeling flummoxed or overwhelmed, take heart—your feelings are quite justified, as structural design is a complex topic, and the stakes are indeed high. Not having mastery over this subject material need not preclude you from taking an active role in the successful design and execution of structural components of your building. You simply now must look more closely at the issues in question and either learn more yourself about how to provide answers or work with someone who can. At a minimum, this chapter should help you understand what questions to ask. Rather than heading into your next project overconfident that the structure will be fine or terrified that it might not, you are now positioned to develop grounded and well-informed awareness of the creation of a stout building that is able to stand the forces of nature and the tests of time.

CHAPTER 7

HEAT AND NATURAL BUILDING

Humankind has proved to be ingenious in learning how to transfer solar energy into heat. This is true beyond solar hot water panels dumping heat into a boiler, or photovoltaic panels powering an electric space heater. Wood, coal, petroleum fuels—all are products of dead plants and animals who were fueled by energy from the sun, either directly through photosynthesis or indirectly through the food they ate. Although a step further removed, even the formation of uranium for nuclear power and the patterns of wind flow spinning turbines can trace the sun's energy as sources of their existence. Hence, one might think at first glance that our energy future is quite secure, given the heady mix of this still-abundant energy source and our intrepid technological innovation.

Unfortunately, the picture is not so simple, nor so rosy. There are significant impacts from both the harnessing and use of energy. The overconsumption of energy resources—and the ecological consequences of oil drilling, laying of pipelines, mountaintop-removal coal extraction, radioactive nuclear waste, deforestation, social and natural ecological displacement for industrial hydroelectric facilities, silicon and precious metal extraction and processing for photovoltaic panels, and even ridgeline ecosystem impact of wind farms—all present clear and pressing, and often unanswered, questions. The "use" end of these resources is no better: with increased energy consumption comes increased carbon loading into the environment, which exacerbates the anthropogenic contribution to climate change and the unprecedented and unfathomable consequences of unchecked global warming.

It is with this awareness of the tremendous impact that energy production and use have on many levels in our social and natural ecological support systems that we look at the role heat plays in our structures. It is neither realistic nor appropriate to suggest that all inhabitants of cold places simply relocate to warmer climes to address the situation. Instead, the goal of this exploration is to identify how heat, regardless of source, moves in a natural building and how to minimize its loss or gain to maintain a comfortable environment for the occupants of the building, in a way that promotes the long-term durability of the structure.

The two primary avenues of heat loss or gain are conduction through the building envelope and air infiltration. Accordingly, as we look at issues of insulation and air leakage, we will also look at human comfort concerns (draft, warm mass, hot bedrooms) as well as associated fire and moisture considerations, both of which will be explored in greater detail in subsequent chapters (heat and moisture have a very close relationship, as we will soon discover). More detailed discussion of specific heating systems can be found in chapter 21.

Thermal Performance Strategies for Natural Buildings

The rules of thermodynamics apply to all buildings, not just natural buildings. That said, there are particularities in how natural buildings work with

Primary Forms of Heat Loss

Heat moves in three ways: through *radiation*, or transfer through air or space such as the sun shining onto a garden; through *conduction*, or transfer through a material such as through a metal spoon in a hot mug of coffee; and through *convection*, or transfer through a fluid such as gas or liquid (i.e., air, water). The primary forms of heat loss we experience in our buildings are through conduction and convection although heat gains and losses through radiation, such as through windows, are also significant.

Conductive heat losses are controlled by insulation, which keeps heat from leaking through the walls and roof of the building by introducing a material that resists heat transfer. The value of this resistance to heat transfer is represented as an R-value. Interruptions in the insulation plane facilitate conductive heat loss. If you add in all the framing in a standard 2 × 6 stud wall framed 16" on center, the R-19 rating of the fiberglass batts placed therein plummet to an R-13 when evaluating the whole wall—and that's just because of the losses of the studs, assuming perfect installation of the insulation, no windows or doors, and no air leakage. These conductive bypasses are called *thermal bridges*.

Convective heat losses are more difficult to control, as it is difficult to build an airtight building. As the air in the building gets warmer, it expands and becomes buoyant (less dense). Being more buoyant than surrounding cooler air, it rises. We see two things here: stratification within the building between upper- and lower-level air temperatures, and motion driving this stratification. This action fuels convective cycles, which are the pattern of warm air rising and exiting near the top of the building, pulling cooler air into the building, where it warms up, further exacerbating the stratification and fueling the cycle of motion. The strength of this pattern is determined by a number of factors, including the shape of the building, the leakiness of the envelope, and the difference in temperature between the inside and outside of the building. Note that this cycle runs in the opposite direction in an air-conditioned building in a hot climate. There are other such convective cycles that can exist in buildings, such as within walls. This motion creates pressure dynamics in the building, with increased positive pressure as one moves higher in the building and increased negative pressure lower down. This is refered to as the "stack effect" or "chimney effect" (for the driving action behind the draft of a chimney). The stack effect can have negative effects on thermal and moisture performance but can be helpful for inducing ventilation.

There's a saying that it is better to have a tight house that's poorly insulated than a well-insulated house that's leaky. Whereas conductive losses are proportional to the amount of insulation that happens to be missing, convective losses can be much greater relative to the size of the hole due to the pressure dynamics at play. They can also be more uncomfortable, felt on the body as drafts. As we will learn in chapter 8, a leading cause of moisture damage in buildings occurs when water vapor carried into walls through convection accumulates or condenses, concentrating in liquid form in the wall cavity; this happens in both cold and warm/humid climates. Please see appendix A and the references sections at the end of the book for more resources on thermodynamics in buildings.

heat by virtue of their materials, which are distinct from how more conventional green buildings might achieve similar performance goals. Based on an understanding of how heat moves through structures, we can develop a thermal-performance strategy for the natural building.

AIR BARRIERS AND INSULATION

The most fundamental component of a thermal performance strategy is the integrity of the thermal envelope of the building. Looking at a cross section of a building, this "envelope" will ideally incorporate two elements: a thicker plane of insulation (that addresses conductive losses) and a thinner plane of an air barrier (that addresses convective losses). The goal in the development of the thermal envelope is continuity; one should be able to place a pencil on the air barrier plane or insulation plane and trace entirely around the structure without having to lift the tip from the plane. What this exercise of "tracing the line" accomplishes is identification and subsequent control of the transitions, a point we stressed in chapter 6 in the discussion of transitional weaknesses between structural elements. Similarly, the edges and transitions between elements of the building are the greatest vulnerabilities in the thermal envelope. It is easy for the insulation or air barrier to be well executed in the middle of the wall; it is much harder for it to successfully wrap around a tie beam and integrate with the roof assembly in a seamless manner.

What can also be seen by tracing the insulation and air barrier planes in a draft section, especially when evaluated as well in plan view, is that the complexity of the building's form will substantially increase not only the building's surface area-to-volume ratio—increasing its exposure to the cold (or heat, depending on the climate)—but also the potential weaknesses in transitions, such as wall corners. A study conducted by Oak Ridge National Laboratory (Kosney 2004, 3–5) compared, among many features, performance differences between a U-shaped house and a rectangular house for in-cavity (the insulation itself), clear-wall (insulation plus

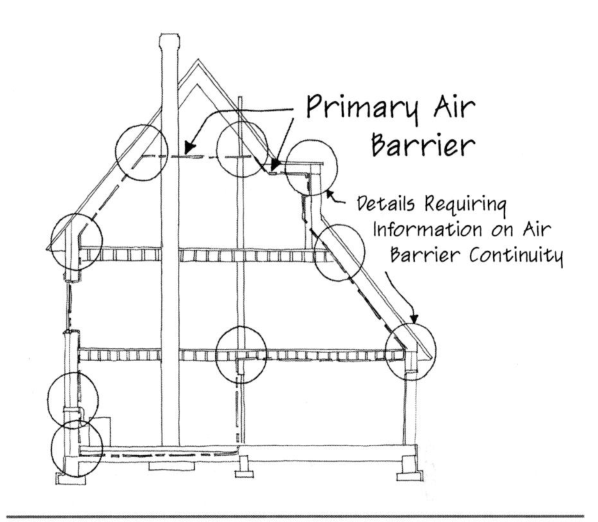

A thorough design process is often needed to ensure a fully contiguous air barrier, especially in boundary conditions. ILLUSTRATION BY BEN GRAHAM.

studs), and whole-wall (insulation plus studs, plates, windows and doors) R-value performance. While in-cavity and clear-wall values were the same for each structure, the increase in form complexity (which includes more doors and windows) decreased whole-wall insulative performance by 15% in the U-shaped house and increased wall-heat transfer rates by 60% owing to the 35% increase in surface area.

Natural buildings rely on vapor-permeable wall systems for their performance. We will delve more deeply into this concept in chapter 8. Suffice it to say at this juncture that the wall, and often roof, assemblies are built to allow drying toward both the interior and exterior of the building. Vapor barriers are generally not used in constructing a natural building; rather, materials that allow vapor

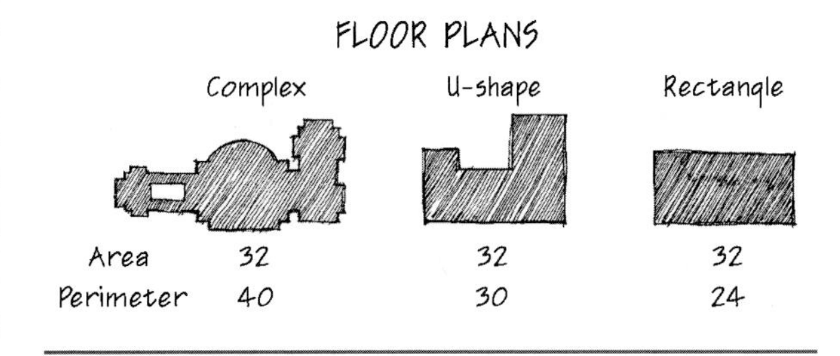

The shape and form of a building, and its relative complexity, greatly affect the cost and feasibility of envelope detailing for energy efficiency. ILLUSTRATION BY BEN GRAHAM.

to pass through wall membranes and liquid to safely diffuse out and evaporate are favored. This vapor-permeable air barrier most frequently takes the form of base plaster. Regardless of the final finish (such as siding or wood paneling), a base coat of plaster is necessary for many reasons, including functioning as the air barrier (see chapter 17). Depending on the climate, there are debates within the building-science community as to whether an air barrier should be installed to the interior or exterior of the building envelope. There are potential liabilities for either strategy. An interior air barrier has to deal with more transitions and perforations and does not address "windwashing," or wind-powered convective loops in the wall, whereas an exterior air barrier must handle difficult roof transitions and does nothing to address interior warm moist air in a heating climate. Our approach, as is commonly advocated by many in the building-science community, is to integrate air barriers on both sides of the wall, thus maximizing thermal performance while enhancing moisture mitigation.

In a natural building, the interior and exterior plaster skins are the walls' primary air barriers. PHOTO BY JOSE GALARZA.

Insulation Options: Straw Bales

Natural building's selection of insulation materials is somewhat limited. Straw bales are a favored material for insulation in cold climates, for a variety of reasons. For one, they offer convenience and good insulation performance. Testing conducted at Oak Ridge National Laboratory in 1998 (King 2006, 187) has been widely accepted as the definitive measure of a plastered bale wall assembly's thermal performance. The results from this study determine an *average* of R-1.45/inch, or net of R-27.5, for a two-string bale on the flat or R-33 for a three-string bale on the flat. Because the orientation of the straw within the wall affects heat flow patterns, the net value is accepted in either orientation (on flat or on edge). These numbers are certainly helpful; however, further testing representative of contemporary construction practices and addressing logistics such as the complete drying of all built-in moisture is needed to be able to arrive at a conclusive number. While, in sum, it does create a very well-performing wall system, it is important to state clearly that straw-bale wall systems are not in and of themselves "super-insulated," and claims of R-50 wall systems have been proven to be inaccurate through more sophisticated testing regimens. We will discuss later in this chapter whether or not R-30 is "enough" for our needs in creating high-performance buildings.

Another benefit of straw bale as an insulator, when compared to many other natural building forms, is its dry installation. Ideally, the bales are very dry before being plastered, and there is a low amount of deeply entrained moisture (maintaining this low moisture level throughout construction, as we will discuss shortly, is a different matter). The modular form of bales is beneficial as well, in that wall systems can be built without the requirement of additional framing to contain the insulation. This modular form cuts both ways, however. As a fixed form of insulation, it can make difficult installation against other fixed forms, such as framing or a roof panel.

Insulation Options: Straw-Clay and Woodchip-Clay

Straw-clay and woodchip-clay constitute another category of insulation materials, essentially cellulose-clay systems. While there are construction details that highlight distinctions between the two (see chapter 15), their essential structure and performance are very similar. The thermal performance of straw-clay wall assemblies, as tested by the Forest Products Lab in 2004, ranged from R-0.9/inch to R-1.8/inch, inversely proportional to density; an early proposal for inclusion in the International Green Construction Code stated a value of R-1.6/inch for walls built to code standards. To date, we have not been able to find credible testing for woodchip-clay wall assemblies, although values of R-1 to R-2 per inch have been suggested. This would seem reasonable, considering softwood's value of R-1.25/inch. This highlights the difficulty in predicting performance values for natural materials—an inherent weakness of nonindustrialized products.

Another benefit of these two materials is their flexible-form nature as insulation. Forming around framing material is no problem, nor is filling in little spots that would be simply impossible with a straw bale; in fact, we use straw-clay to fill in the gaps between straw bales! This serves to eliminate insulation voids that lead not only to conductive losses but potentially to convective losses through *interstitial convective cycling*—convective air currents moving within the wall, which can dramatically reduce the thermal performance of a wall system. Depending on the size and location of the gaps and the temperature differential, convective cycling can increase heat loss by 30% to 50%. Batt insulation such as fiberglass or cotton is particularly prone to creating "chimney" cavities in a wall; CSG, a former Energy Star provider in Massachusetts, has found that 5% void space in fiberglass batt installation is typical, which reduces performance from R-19 to R-11. Full-cavity-fill insulations, of which all the options herein are considered, avoid this problem. As with bales, of course, the form feature of these insulation materials has drawbacks, as well. Framing and containment is required for the use of straw-clay and woodchip-clay, resulting in increased wood usage.

The built-in moisture of straw-clay walls limits the potential thickness of their application, particularly in cold, wet climates. The grass growing out of the straw here indicates that moisture still remains. PHOTO BY BENJAMIN RAY GRIFFIN, GREENER LIVING VERMONT.

The greatest drawback, however, is the moisture level of the material as it is installed. The amount of built-in moisture limits common practice (and aforementioned proposed code) to 12 inches to prevent interior rot conditions from occurring before sufficient drying can be achieved—and even this can be a challenge in a cool and rainy summer (let alone a fall or winter installation, or installation against a sheathed wall, such as in a renovation). This inherently limits the effective insulation value of the wall assembly to R-19.2.

Insulation Options: Cellulose

Cellulose is also commonly used in natural buildings, particularly where a lightweight, flexible-form, dry insulation material is required—roofs, for example. Cellulose insulation features all the advantages of moisture performance, fire resistance, insect repellency, and benign environmental footprint that a builder could desire. Accordingly, we frequently use cellulose in conjunction with other natural wall insulations, in both roofs and walls (for example, cellulose-insulated gable end walls over straw-bale-insulated first-story walls).

The greatest liability from a construction practice is the difficulty of installation in closed cavities, where it can be hard to verify complete installation, and in vertical cavities, in which settling can occur in the event of insufficient density during installation. Professional installation with professional-grade equipment is required to address both of these considerations.

These installation realities are the greatest drawback to cellulose, from a natural building perspective. Purchasing plastic-wrapped packages from the building-supply store and feeding them into the hopper of a large and expensive piece of equipment—to be installed via hose by a well-trained operator—creates a series of points of separation between the builder and the material with which he or she is building. At the philosophical and spiritual heart of natural building lies the importance of developing relationships with place, resource, and process; the industrialized nature of cellulose, while advantageous on many levels, does not serve this goal. Accordingly, we choose to use cellulose as part of an overall strategy for thermal performance in a natural building, but we endeavor to utilize other natural insulation technologies whenever appropriate.

Field Testing of Natural Buildings

Understanding how the properties of different insulation materials play out in real-world scenarios was a chief goal of ours in preparation for this book. While testing has been conducted in a variety of laboratory and individual case studies regarding the properties of different materials and assemblies supportive of our performance criteria for use in natural buildings, we were lacking a more comprehensive testing regimen conducted on a series of buildings that featured variations on similar detailing and thermal performance strategies with verifiable levels of quality in, and documentation of, construction practices—all located within a relatively small geographical radius. In collaboration with colleague Ben Graham of Natural Design/Build of Plainfield, Vermont, and energy analyst Brad Cook of Building Performance Services of Waitsfield, Vermont, we undertook a research project in which we evaluated seven structures in Vermont and New York using infrared (IR) thermography, blower-door testing, and moisture testing (surface scan, pin scan, and probe invasive)—techniques and technologies commonly found in the building-science community.

Blown cellulose insulation is perfect for many situations in natural buildings. PHOTO COURTESY OF NEW FRAMEWORKS NATURAL BUILDING.

Thermal imaging cameras are one of a variety of technologies that can be used to evaluate and understand heat loss in a building. PHOTO BY ACE MCARLETON, COURTESY OF BRAD COOK.

We developed a testing regimen that was applied to all seven structures. The first step was to scan the exterior and interior walls with the IR camera, to develop a baseline of thermal performance. The IR camera "sees" heat radiating from a surface and accordingly shows differences in temperature, highlighting hot and cold patterns indicative of heat loss, air infiltration, or moisture. Next, we set up the blower-door apparatus, which blows air out of the structure until it is depressurized to a set point relative to outside conditions—in this case, –50 pascals (Pa), which is the standard for residential-building performance testing. Computer software was used to control the fan and to log the rate of airflow required to achieve that pressure (measured in cubic feet per minute, or CFM). Subsequent readings were then taken at 5 Pa increments down to –15 Pa to provide a more accurate and consistent result, and a final CFM number was given as a measurement of the leakiness of the building at a pressure of –50 Pa—the higher the CFM, the leakier the building.

On DVD, Chapter 7, see BLOWER-DOOR TEST AND BUILDING EVALUATION

The blower-door test will depressurize or pressurize a building, allowing for clear analysis of air leakage in the envelope. PHOTO BY ACE MCARLETON.

This number was calculated against two metrics: the volume of the structure and the surface area of the structure. Weighed against the volume, a metric of the number of full air exchanges of the house each hour at that pressure was given as air changes per hour at 50 Pa, or ACH50. Because smaller structures have a higher surface-area-to-volume ratio and consequently a larger house will show a lower ACH50, many prefer the metric of CFM/square foot of shell (including floor) of the building, which is arguably a more accurate measurement of the performance of the building's envelope than a metric based on its

3/16/2011 6:07:31 PM
58 deg. IN / 36 deg. OUT / cloudy

3/16/2011 8:03:00 PM
58 deg. IN / 36 deg. OUT / cloudy

The right-hand picture shows where cold outside air is entering once the blower-door test is running. PHOTO BY BRAD COOK.

Energy Modeling for Building Design

As our understanding of how buildings perform increases, so too does our capacity for predicting in the design phase how buildings will perform once they are built. The tools available to designers today are powerful and sophisticated; subjecting a design to their analysis can yield dividends in the lifetime performance of the structure, as well as in compliance with strict efficiency standards of certain rating systems.

This analysis could be as simple as running a heat-loss calculation, in which a series of numeric values are entered into a spreadsheet that catalogs climatic conditions, size and shape of the building, foundation type, window and door sizes and performance ratings, framing type, and performance ratings for wall and roof assemblies. Basic heat-loss calculators can be found for free on the Internet, whereas more thorough and comprehensive calculators may be cost-prohibitive for incidental use.

On the other end of the spectrum, sophisticated 3-D whole-building energy simulation programs have been developed that have the potential to predict with great accuracy the performance of a building. Their greatest weakness, however—and this is the case for any modeling or prediction system—is the accuracy of the input values. To increase the accuracy and ease of access to relevant material and assembly performance data, Oak Ridge National Laboratory has introduced the Interactive Internet-Based Envelope Material Database for Whole-Building Energy Simulation Programs, which, in the words of ORNL scientist Jan Kosny, "links experimental data on thermal characteristics of building envelopes with advanced analytical methods available for thermal and energy analysis" (Kosny 2004, 13). It won't be long before powerful 3-D energy simulation tools, able to interface with extensive databases of performance characteristics, will be in reach of any interested designer, builder, or owner.

volume. The problem with both of these numbers is that they only give an overall performance metric of the structure; to understand where the air is leaking and why, we must look deeper.

To do this, we used the IR camera again to scan the building while depressurized to –50 Pa; at this point, the equivalent of a 20 mph wind was blowing against the building on all surfaces, overwhelming any atmospheric conditions and exposing all substantial convective losses. Now, the picture became clear as to where heat loss was occurring in the building, and a thorough analysis could be conducted on each case study, and compared between cases to expose the patterns of heat loss with our building systems.

A synopsis of the full report aggregating all the data and providing analysis and conclusions based on the results is in appendix D, "Results Summary of Energy Performance of Northeast Straw-Bale Buildings Research." With regard to the subject of heat loss, however, here are some highlights. Note that while this testing was performed on straw-bale-insulated structures, many of the conclusions can be applied to buildings featuring other insulation types. In fact, it is interesting to note that in every case, the largest incidents of air leakage occured in non-straw-bale components of the structure, such as in roof assemblies.

Transitions are a weakness. In every project, transitions and edges were the points of weakness in thermal performance. Specifically, we noticed heat loss between the plaster and window/door rough openings (ROs) and between the ROs and the window units themselves; between the roof and top-of-wall assemblies, both at eave walls and gable end walls; to a lesser extent, between the foundation (either slab-on-grade or basement) and bottom of walls, particularly near posts or where plaster was damaged; where the plaster stopped around the timber frame, particularly where knee braces met posts/beams; and where posts and beams connected.

Air fins work, but quality control can be an issue. The air fins, as discussed in chapter 14, proved on whole to be very effective at controlling heat loss and preserving the integrity of the air barrier in transition from plaster to timbers or other wood elements. That said, in all buildings occasional points of loss were identified, in some cases more than others, indicating potential difficulties of application and the importance of a keen attention to detail. While for the most part air infiltration was minimal—one such structure had a rating of 2.5 ACH50, or 0.15 CFM50/ext. sq. ft. (a very tight house)—we must remember that a relatively small hole can let through a proportionally large amount of hot (or cold) air. Of particular note for quality control is the need to seal between pieces of air fin and ensure complete sealing of the air fin to the framing.

Timber frames are tricky, especially if old and rough. Exposed timber frame structures that interrupt the plaster plane are much harder to seal as a result of the dramatic increase in transitions. The structure in which an old, hand-hewn frame was used proved to be especially difficult to get airtight between the air fins, the stand-aways, and the frame itself. One strategy we have begun to use is to frame the knee braces to the interior of the frame, thus reducing the amount of transition in the air barrier. If the situation allows, a more effective strategy would be to remove the timber frame entirely from the plaster plane, allowing for an uninterrupted plane of plaster (save for windows, doors, outlets, and the top and bottom of wall-boundary conditions), most easily accomplished by setting the frame (or, easier still, a stud frame) inboard of the bale plane.

Walls are adept at dealing with moisture accumulation. When moisture probe tests were conducted on walls that showed signs of visible moisture and, when surface- or pin-scanned, indicated high levels of moisture on the exterior plaster as a result of condensation, the straw just behind the plaster showed relatively low levels of moisture (more about this in the next chapter).

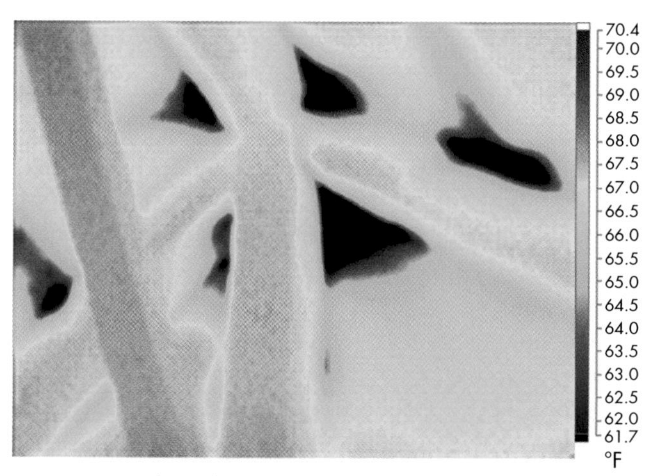

3/10/2011 5:48:40 PM
60 deg. IN / 32 deg. OUT; rain/sleet

Visible Light Image

Sealing between plaster and timber frames is very important to avoid heat loss, and there is a lot of edge to seal in these buildings. PHOTO BY BRAD COOK.

Settling can be an issue. A significant consideration of the performance of insulation in a building is its performance over time. Different insulation materials are subject to deterioration or deformation in different ways. One of straw's greatest negative aspects, especially for non-compressed, non-load-bearing construction, is the potential for settling, or "creep," of the walls, which results in a gap at the top of the wall. In general, we did not notice any significant patterns of creep. That said, in some areas of some buildings, patterns of heat loss along roof lines were noticed, which may be attributable to creep (or could simply be attributable to inconsistencies in the top-of-wall air barrier). This is a phenomenon that has been witnessed by other bale builders and has been especially attributed to walls where the bales were loose, or loosely stacked, or a thick plaster was applied quickly. We try whenever possible to use blown-in cellulose for roof insulation to maintain a direct line of contact between the roof insulation plane and the top of the bale wall, and to apply the insulation after the plaster has dried and the bale walls are stable, to ensure its continuity. That said, when the bales are built to meet a fixed plane of framing or a panel roof, this settling could become an issue.

Common Heat-Loss Conditions in Natural Buildings

In addition to those noted in the previous section, both our building performance testing and extensive empirical experience have revealed other places in natural buildings where we have seen compromises with the thermal envelope. These include:

Timber frame rafter plate or rafter tail protrusions. Two classic examples of substantial thermal bridging come in the roof-to-wall plane, which is already a tricky area to get right. The first is with rafter plates, the big beams that hold the rafters above the wall. It is common practice in timber framing to run these out through the building to catch the last "fly" rafters at the far edge of the overhangs. Unfortunately, this creates a few problems: For one, the beam itself is a thermal bridge penetrating the insulation plane of the wall/roof transition. For another, it makes a lot more edge around which to air-seal when joining the plaster to the plate. Perhaps

Wall-to-roof insulation and air-barrier continuity is very important to detail correctly, for the penalties for inadequate installation are great. PHOTO BY BRAD COOK.

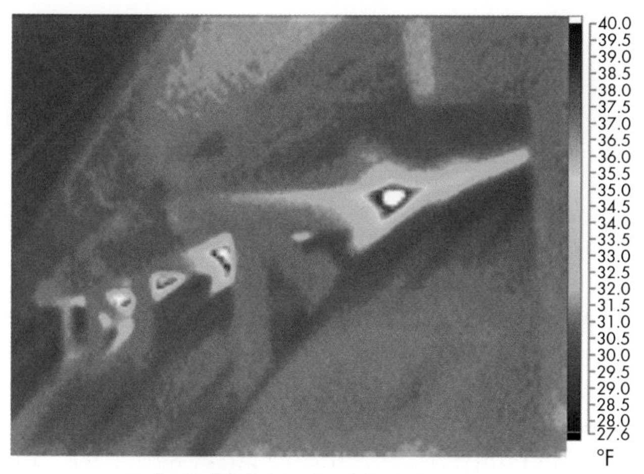

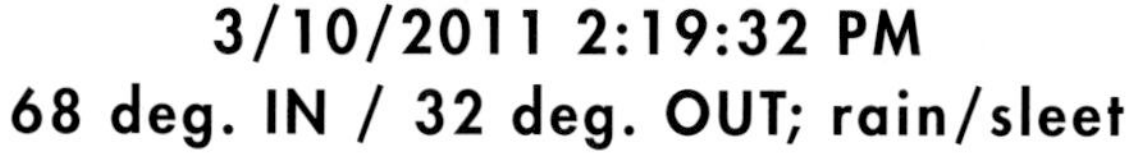

Visible Light Image

Structural framing penetrating the wall plane, such as these timbered buttresses to support the roof eaves, compromises the integrity of the thermal envelope. PHOTO BY BRAD COOK.

worst of all, the exposed end grain of the plate will, without consistent maintenance in perpetuity, absorb moisture, likely causing a crack in the beam, and allow a direct passageway for air exfiltration that is nearly impossible to seal. It is an easy way to catch the fly rafters, but not worth the compromise to the thermal envelope. Find another strategy to support the gable-end overhangs!

Running rafters all the way out to expose their tails can be tricky, too. Although there are ways of trimming out between the exposed tails and limiting their impact on the envelope, all too often this detail is overlooked until too late, and difficult insulation installations and plaster edges need to be reconciled with a series of protruding framing members—again, at a very vulnerable part of the wall. Exposed rafter tails are lovely—just get the detailing right for a tight transition.

Standard stud framing. Any standard 16-inch on-center (OC) stud-wall construction that allows the framing to act as a thermal bridge is problematic. While it is challenging to remodel existing structures to improve this situation, there are numerous solutions for avoiding this thermal transgression in new construction. For one, so-called advanced

Decoupling the framing from running through the entire thickness of the wall reduces conductive heat losses, called thermal bridging. PHOTO BY ROBERT RIVERSONG.

framing technique, also known as value-engineered framing or optimum value engineering, substantially reduces the amount of framing, and therefore the thermal bridging, in the wall. Double-stud construction—either inline or staggered-stud—goes a step further, isolating the framing members from bridging through the wall, and easily allowing for a deeper wall cavity with small-size framing material for greater insulation thicknesses. Both of these techniques are discussed in chapter 13.

Roof assemblies. In our testing of natural buildings, we consistently saw significant air leakage in various parts of the roof assembly. As mentioned, we often found the roof-to-wall transitions (around rafter or top plates, or gable-wall-to-roof connections) to be weak spots in the assembly. These can be remedied in a number of ways, including continuing the flexible air barriers around the framing and sealing to ceiling and wall, or detailing sealing joints from the ceiling and wall barriers to the beam or plate itself. Tongue-and-groove or other decking/framing protrusions were other sources of air infiltration we found; the trick here is to interrupt any jointed wood with insulation, keeping it from running all the way through the wall—this goes for floors, too! Joints in SIP panels were another leaky spot; it has been interesting to hear anecdotally from builders about how difficult it can be to get a good seal between the panels, and even in one well-sealed roof we were able to see, with IR thermography, cold patterns of thermal bridging of wooden I-beams used to connect the panels together. Meticulous attention to detail is the best course of action here. Cathedral ceilings can be a weak spot in a thermal envelope because thermal bridging from the rafters is often unmitigated, and the dimensions of the lumber can limit the insulation thicknesses. The easiest solution is to place a layer of rigid board insulation across the inside face of the rafters before finishing the ceiling,

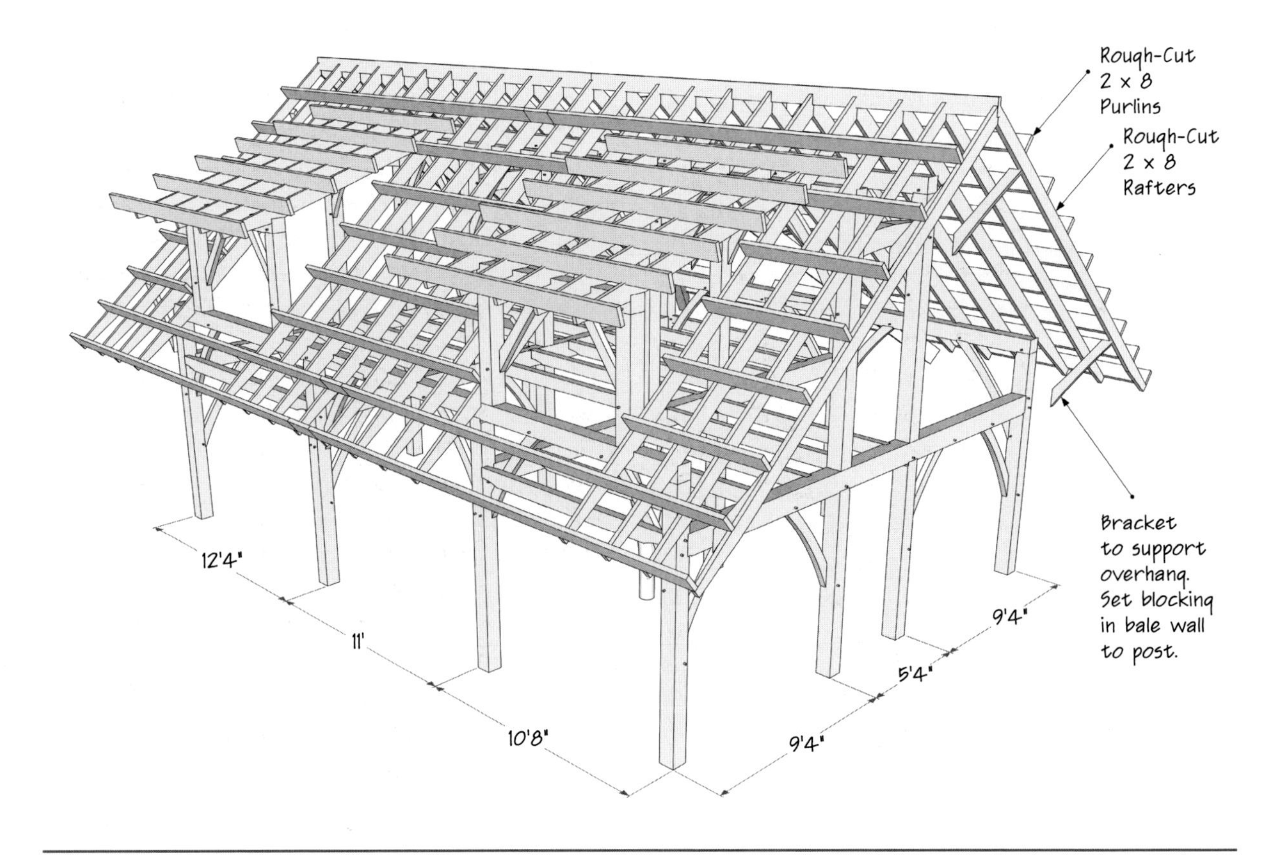

High-performance roofs can involve customized framing plans, such as this lattice-roof design. DESIGN AND ILLUSTRATION BY JOSH JACKSON.

which provides the presence of roof venting and allows the assembly to dry to the outside (see chapter 18 for more on vented roofs). Cathedral ceilings also increase the positive air pressure because of the increased height (which exacerbates stratification in the interior building climate), and they concentrate rising warm, moist air in one area—making air sealing at the peak especially important.

Finally, overhead lights, fans, vent pipes, chimneys, and other perforations through the ceiling let out a startling amount of warm air when not sealed. Again, positive pressure is greatest at the highest point inside the building. Attend to sealing details using gaskets, flanges, sealed boxes to receive appliance installations, or whatever is necessary to control air migration through those perforations, and avoid recessed can lights if at all possible, even ones rated "airtight."

Foundations and foundation-to-wall connections. Similar to the aforementioned roof issues, we have also witnessed, both in our testing and anecdotally, numerous weaknesses in foundation details. Faulty band-joist insulation, regardless of the type, is a regular culprit, as is an abject lack of sealing between foundation stem walls, sill plates, floor-box framing components, subfloors, and bottoms of walls. There are so many transitions here, and each one needs to be found by the builder before it is found by the air current. Finally, under-slab and slab-edge insulation, although becoming more of a standard detail, is still frequently left out of the equation—remember that the thermal envelope surrounds the entire building!

Beyond the R-Value

We truly hope by this point that you can appreciate that there is a lot more to assessing the thermal performance of a building than the factory-labeled R-value of the insulation. When discussing conduction, we learned about how fickle that R-value can be in a real-world scenario (with variables of convection, moisture, and thermal bridging). When discussing convection, we learned that convective losses can be even greater than conductive losses, and that certain types of insulation will invite greater amounts of convection than others. Before we finish our discussion of heat, we should look at the rest of the major factors that govern how our houses keep us at a comfortable temperature.

In assessing a thermal-performance strategy, a valid question to ask is "how much is enough?" Performance does not exist in a vacuum; there are

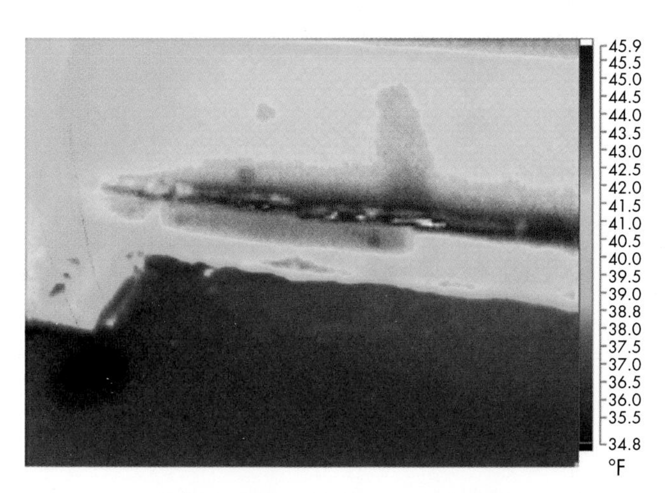

Heat loss is a particular concern with a radiant slab. PHOTO BY BRAD COOK.

Heat and Moisture

While we have mentioned throughout this chapter some of the relationships between heat and moisture, we have exhibited tremendous restraint, saving much of the commentary for its sister chapter, chapter 8. That said, in case you were thinking of skipping that chapter, we feel compelled to highlight a few points to encourage you to read on through the next chapter to fully grasp the relationship between the two.

- *Moisture moves by, among other things, a temperature gradient.* We know that moisture will go from where it is warmer to where it is colder. Accordingly, we can predict where moisture is likely to move in our walls, from season to season and throughout the diurnal (night-to-day) cycle. Being able to predict moisture's movements allows us to stop them—or encourage them.

- *Moisture exists in three phases.* Water is the only substance that exists in solid, liquid, and gas forms in normal atmospheric conditions. Gas moves primarily with air, which means that wherever the hot air goes, so too goes whatever moisture it contains. You may not care about throwing an extra cord of wood in the stove to rationalize skimping on detailing your air barrier, but you better care about all the water vapor flooding into the walls, and condensing once it hits the dew point.

- *Convection-induced pressure gradients exacerbate vapor intrusion.* We learned that warm air rising in a building creates positive pressure at the top of the building (in cold climates), where unfortunately the detailing for airtightness can be the most difficult. This combination of tricky detailing and pressurized warm, vapor-laden air creates a situation in which a lot of air—and thereby a lot of vapor—can enter a roof or wall, with the potential to condense and create a significant moisture problem. Thermal performance and moisture performance are directly related.

- *Ventilation affects heat and moisture.* You've done such a good job sealing up your house, you have now built in indoor air quality and moisture problems! An adequate ventilation system is requisite in any tight construction, be it active or passive (see chapter 21 for more information about ventilation). Not only will you be healthier, but a ventilation system that works well will give the moisture vapor created by washing, cooking, and breathing an easier route to exit than through the cracks in your ceiling.

- *Increased thermal performance reduces drying potential.* As our buildings get tighter and better insulated, the temperatures within the wall get cooler as the heat is no longer allowed to migrate through the wall to the outside. This lack of heat reduces the drying potential of the wall, making it more difficult to recover from periods of condensation. Care must be taken when increasing insulation to ensure condensation is kept to a minimum and climate-appropriate drying pathways are provided.

real implications for design, structure, moisture management, construction phasing, and quality control and verification—and certainly cost—that need to be considered when trying to ratchet up the R-value. True to the values-based approach of natural building, the answer depends not only on whom you ask but on the values particular to each project. In general, however, we tend to fall in the middle of the spectrum, which ranges from the "good enough" approach held by many well-intentioned conventional builders to the extreme performance characteristics of Passive House standards (see chapter 10 for more on Passive House). We believe performance goals should be reasonably attainable in the majority of projects in terms of ease of execution, simplicity of detailing and design, and affordability.

One such method has been presented by Building Science Corporation and is incorporated into prototypes for their Building America program in the "5-10-20-40-60" approach for minimum-assembly R-values: R-5 windows, R-10 below the slab, R-20 for below-grade basement walls, R-40 for exposed enclosure walls, and R-60 for the roof. When using straw bales for insulation, which provide an approximate enclosure-wall value of R-30, or straw-clay/woodchip-clay walls, which are closer to R-20, our options are to allow for a reduced R-value in this component of the building or to innovate different approaches toward increasing performance, such as incorporating other insulation forms (one intriguing strategy taken by Mark Hoberecht of HarvestBuild is to couple straw bale and straw-clay wall systems with dense-pack cellulose walls).

When trying to answer the "how much insulation is enough?" question, it is important to understand that there is a diminishing point of return with increased quantities of insulation—increasing an R-value from 9 to 19 will halve the amount of conductive heat flow through a wall, whereas the same increase from R-33 to R-43 results in increased heat flow resistance of less than one-third. Effects such as increased insulation cost, extra wall framing, footprint increases, usable square-footage reductions, or the use of ecologically harmful materials to achieve higher R-values must be balanced within a larger equation that also includes

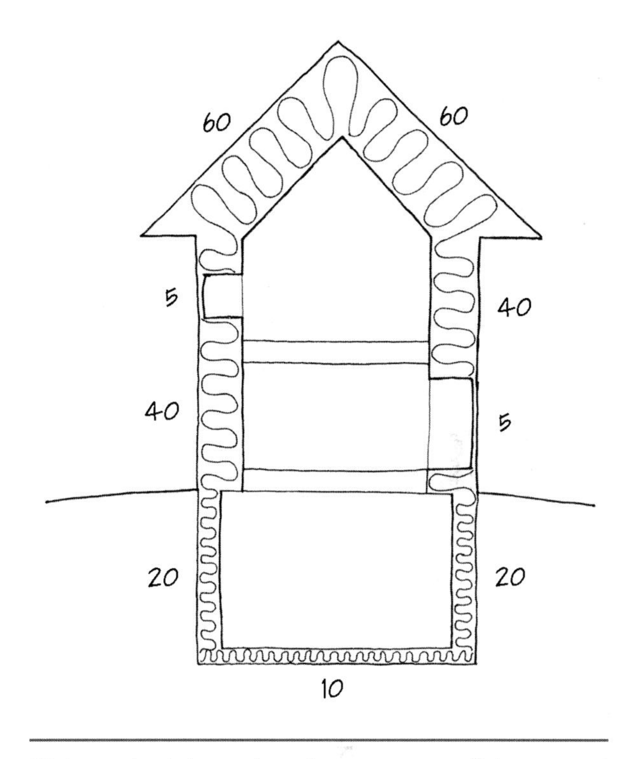

Minimum insulation values for an energy-efficient natural building. ILLUSTRATION BY BEN GRAHAM.

factors of affordability, aesthetics, ease of construction, and material preference. Similarly, we believe an ACH50 rating of 1–3 strikes a reasonable balance of performing very well while being achievable using favored materials and techniques by any attentive and skilled workforce, without dramatically increasing project costs. All that being said, this should in no way discourage the intrepid designer or builder from endeavoring to achieve higher levels of building energy performance. In fact, as the threat of global warming grows with each passing year, we will find improvements in our building efficiency to be an imperative, as our values continue to shift in response to environmental conditions.

The Power of Mass

The concept of thermal capacitance—which is a combination of thickness, conductivity, and thermal capacity, or a material's ability to absorb heat energy

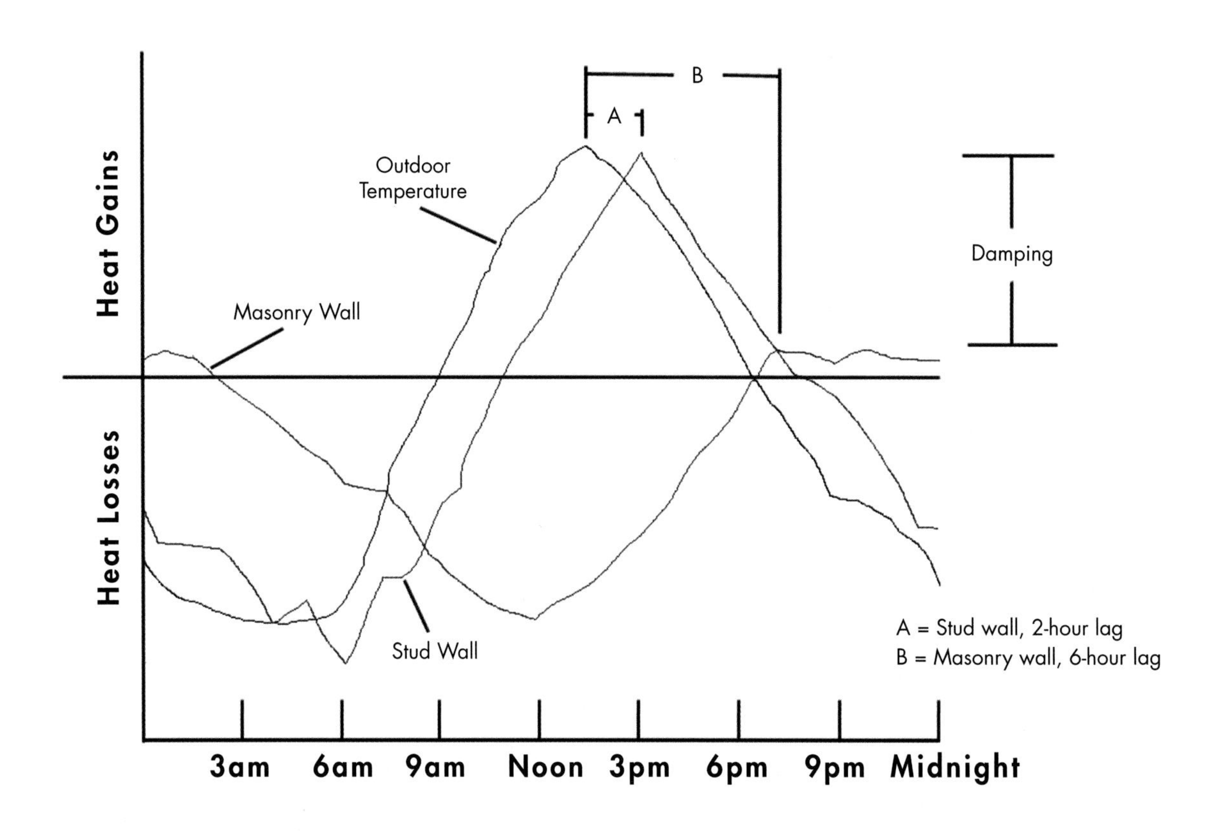

The thermal lag of a mass wall is helpful in managing heat loading in a structure. IMAGE BY JACOB DEVA RACUSIN; SOURCE: BERKELEY SOLAR GROUP AND CHARLES ELEY ASSOCIATES, *ENERGY BASICS BOOK* (CONCRETE MASONRY ASSOCIATION OF CALIFORNIA AND NEVADA AND THE WESTERN STATES CLAY PRODUCTS ASSOCIATION, 1986).

without a resulting change in temperature—is important for understanding how mass works to control heat in our structures. Mass acts as a big thermal battery, or electrical capacitor. When cool, it can be charged with heat, able to pull a whole lot of BTUs out of the surrounding environment without affecting its temperature very much. When the surrounding environment is cool relative to the mass, the mass has a whole lot of heat to release back out over a long period of time. Therefore, mass serves to dampen and delay swings in temperature; mass brings time into the equation.

It is this "time" piece that is the other key variable to understand when thinking about how mass behaves with heat in a building. The time it takes for heat to move through a wall or roof assembly is called the "thermal lag." Using a plastered straw bale wall as an example, this thermal lag has been tested, as referenced in King's *Design of Straw Bale Buildings*, to be approximately twelve hours—quite convenient when considering that this places the heat cycle of the wall directly opposite the diurnal heating cycle!

This same effect is realized within a structure and is a hallmark of passive solar design—sun radiates through a window, charging the floor mass, which retains the heat and keeps the building from overheating during the day (see chapter 10 for more on passive-solar design strategy). At night, the heat source disappears, heat flows out the building (including a reversal of traditional heat transfer out the window), the interior climate cools, and the mass of the floor radiates its heat back into the building. (*Note*: because the heat goes into and leaves the same side of the floor, the optimum thickness is 4–5

Windows—Winners or Losers?

Windows do many things for us in a building—let fresh air in, offer us a view, let light in, allow us to connect to our environment from within the building. However, they are also weak links in the building's thermal envelope. So, how do we design windows to address heating and cooling concerns?

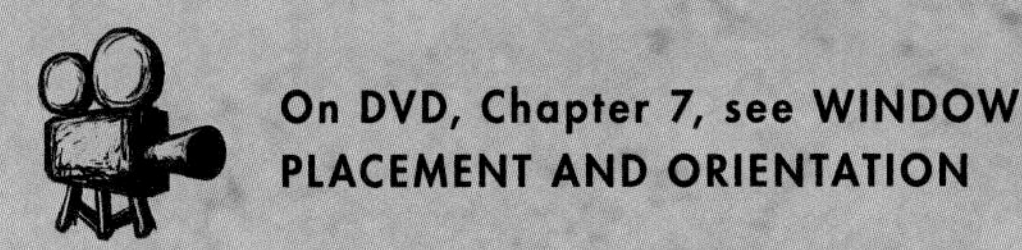

This consideration is part of developing a passive solar design strategy, as described in chapter 10. In this approach, select windows—primarily those on the south face of the building, sized so as to avoid over-glazing—are specified to allow in a lot of sunshine. The metric to use when comparing windows is the solar heat gain coefficient, or SHGC. But as we know, when the sun isn't shining, those windows are big heat losers. The value for heat-loss performance is measured as a U-value, which is the inverse of the R-value (U=1/R). Accordingly, fewer windows are desired on the north walls, and they should have a low U-value; windows on the south often strike a balance between low U-values and high SHGC values. Robert Riversong of Riversong HouseWright notes, "The best cost-benefit ratio for a cold New England climate is found in a double-glazed low-E argon-filled (U-0.30) aluminum-clad wood or fiberglass window with a high (0.50) SHGC."

Cost ends up being a large factor, as well. There is a considerable debate among energy-efficiency practitioners concerning the value of high-performance windows; technologies such as insulated sashes, complex triple-pane glazing configurations, low-emissivity films, noble-gas injections, and more are coming to market every year, with increasingly lower U-values. Unfortunately, they also come with increasingly high price tags, and availability and lead times can be significant barriers. Whether or not the money invested in ultra-high-performance windows can be justified by the performance return is something that will need to be carefully considered in the program of the building. Riversong recommends, "The *minimum* efficient window would be double-glazed low-E argon-filled with nonconductive frames and made with sufficient quality to remain durable and airtight for many years."

inches.) In another heating scenario, a masonry heater is charged with a small wood fire; due to the efficient combustion design of the heater, the majority of the BTU output of the fire is retained in the mass of the heater and is radiated slowly and steadily into the building over a prolonged period of time.

The "mass effect," as explained above, can improve the thermal performance of a wall system beyond the rated R-value by a factor of 1.5 to almost 3, when incorporated into a wall assembly in such a way that it can have a measurable effect on the interior climate. What this means for us in moderate to cold climates is that the mass must be (1) contained within a well-insulated envelope; (2) exposed to the interior of the building; and (3) thick enough (optimally 4 inches) to store and release significant amounts of heat. The effect of mass on the overall thermal performance of a wall system is referred to as

Masonry heaters are beautiful, efficient, comfortable, and clean burning. PHOTO BY ACE MCARLETON.

the "dynamic benefit for massive systems" (DBMS) (Kosney 1999).

While the value of mass is best realized in regions with a high diurnal temperature swing (i.e., the southwestern United States), its effect is still relevant even in cold climates, and it has been shown in testing conducted by Oak Ridge National Laboratory that the DBMS, and therefore the overall performance of the building, is greater in systems where there is direct contact between the mass of the wall and the interior of the building, within the insulation plane. Testing conducted between four wall systems—an exterior-insulated concrete wall, an ICF wall (insulated concrete form, which is a sandwich of insulation-concrete-insulation, and, hence, sometimes referred to as ICI), a CIC wall (insulation in the middle of a concrete wall), and an interior-insulated concrete wall—conclusively showed thermal performance benefits for the exterior-insulated and CIC walls, where concrete was exposed to the interior, over the other two systems (Oak Ridge National Labs and Polish Academy of Sciences, 2001).

It is relevant to note that we are discussing the use of mass in concert with insulation. Uninsulated mass wall systems work very well in hot climates to control excessive heat gain in buildings if thick enough to reradiate their heat back to the atmosphere when the nighttime temperatures drop before the heat can work its way inside. This benefit is not realized, however, in hot climates that maintain temperatures above average human comfort conditions (68°F–78°F), nor in moderate or cold climates where the mass may well be more inclined to simply continue to pass the heat, albeit more slowly, to the outside rather than keep it contained within the building envelope. While well-designed uninsulated mass-wall structures can be built to perform well in moderate to warm climates, careful consideration of heat loss potential and the use of conservative performance values of the wall material are recommended to avoid the construction of an underperforming building.

CHAPTER 8

Moisture and Natural Building

Water is truly a magical substance that is the foundation of biologic life on our planet. It is also the most destructive element we face in creating durable shelters. This is as it should be: as stated in the second law of thermodynamics, all complex systems in the universe—people, ecological systems, buildings—are driven by entropy toward disorder and breakdown. Water is simply doing its job as a primary vehicle through which entropy can do its work. For us to be able to build responsibly in any climate, we must design for water from the beginning, not leaving it as an afterthought or secondary consideration. And to do that, we must understand what water is, and how water works. One of the most important distinctions in how natural builders approach the challenges water presents to structure durability is that we try to work with water, rather than attempting to uniformly defend against it. We do this by understanding how water interacts with buildings—how it moves, where it shows up, and why. In this chapter we will explore how natural builders are striving to foster a successful relationship between water and durable structure.

Moisture Performance Strategies for Natural Buildings

Just as in the last chapter we looked at creating a thermal-performance strategy as part of a holistic approach to our building's response to unwanted heat loss or gain, so too must we create a holistic moisture-control strategy. To do this effectively, however, we must look more closely at the vectors for moisture exposure in our building, from construction through use. Understanding water's sources and patterns of movement in structures is the foundation of our strategy. Let's look at the points of moisture entry throughout the building life cycle.

SOURCE OF ENTRY—BUILT-IN MOISTURE

The first place moisture starts in a building is with the materials themselves. In conventional construction, these materials are dry when purchased—sheathing, bagged insulation, drywall. In the world of natural building, this is not as often the case. Green lumber, as discussed in chapter 13, is very wet when purchased, especially green timbers—far more so than the kiln-dried lumber used in conventional wood-framed construction. Straw is supposed to be quite dry when sourced from a supplier, and may well be. It needs to stay dry, however, through transportation (which can be dodgy in an open trailer!) and while in storage on-site. Site storage of materials for straw bale construction is indeed unique. In conventional construction, the shell is "dried in," or enclosed and protected from precipitation, long before moisture-sensitive insulation is brought in, wrapped in plastic and ushered from the delivery vehicle to the building. Bales, on the other hand, help create the enclosure and are trucked in en masse without the benefit of protective wrapping. Therefore, considerable precaution must be taken in scheduling and providing logistical support for receiving bales and keeping them dry until they can be installed.

There are many opportunities for moisture to enter during construction, which must be avoided. PHOTO BY JDONABELANDEWEN@FLICKR.

Keeping the bales as dry as possible therefore frequently involves, among other measures, building and drying in (waterproofing) the roof first, before the walls, which is why wet-climate natural builders tend to favor non-load-bearing construction in order to protect the work area. The same comes into play for cob, adobe, and straw-clay. Any wall system that is subject to vulnerability during construction must be appropriately cared for, either by pre-installing the permanent or temporary roof or by meticulous top-of-wall tarping or other protection mechanism.

The use of clay, which is invariably wet-processed to facilitate incorporation into the mix—be it plaster, cob, straw-clay, or woodchip-clay—is another major contributing source of built-in moisture. While clay controls moisture exceedingly well as a component of wall systems once it has dried (more on this later), its use in straw-clay and woodchip-clay makes for such a wet material that the application thickness of such mixes is limited by their built-in moisture content to 12 inches. The plastering of an otherwise dry wall system introduces a large spike of moisture that can present challenges if applied heading into a cool and rainy fall when climatic drying potential is reduced. This problem is exacerbated if the construction process is still at an early stage and the plaster does not receive the benefit of mechanical drying or ventilation. Cob construction also creates a tremendous amount of built-in moisture, which again can be a limiting factor if installation does not move quickly enough and a short drying season prompts potential drying issues. This is one reason pre-dried adobes are attractive as a mass form.

Curing concrete is another large source of built-in moisture, as is curing lime. In fact, lime needs to be misted repeatedly to be kept damp throughout its twenty-eight-day curing process. This can be difficult to achieve but does not present moisture problems on the exterior, as the majority of the water is likely to evaporate into the atmosphere rather than travel further into the wall and concentrate enough to cause damage. On the interior, however, spraying many gallons of water directly onto the walls in an enclosed, unvented, unheated or minimally heated structure in a cool, damp fall for weeks on end can be a real problem, and moisture loading and drying potential both *must* be considered in the construction scheduling.

SOURCE OF ENTRY—DETAIL AND DESIGN

The next major source category for moisture comes once the building is built—how does it respond to moisture vectors within and outside of the structure? As this is the largest source of entry, we'll break it apart by water phase.

Liquid—Precipitation

The most obvious source of moisture intrusion is from precipitation—rain, snow, and the like. Snow tends to be an issue only over horizontal surfaces, where it remains on the surface to be released into the wall upon thawing. Snowbanks piled up against a wall do not tend to cause as much damage as rain, as once the layer of snow in contact with the wall is absorbed, the rest of the bank is now separated from the building. That said, repeated wetting and reduced drying from snow piling can be an issue.

Rain, however, is a mighty force that on the currents of wind can blow sideways, backward, and

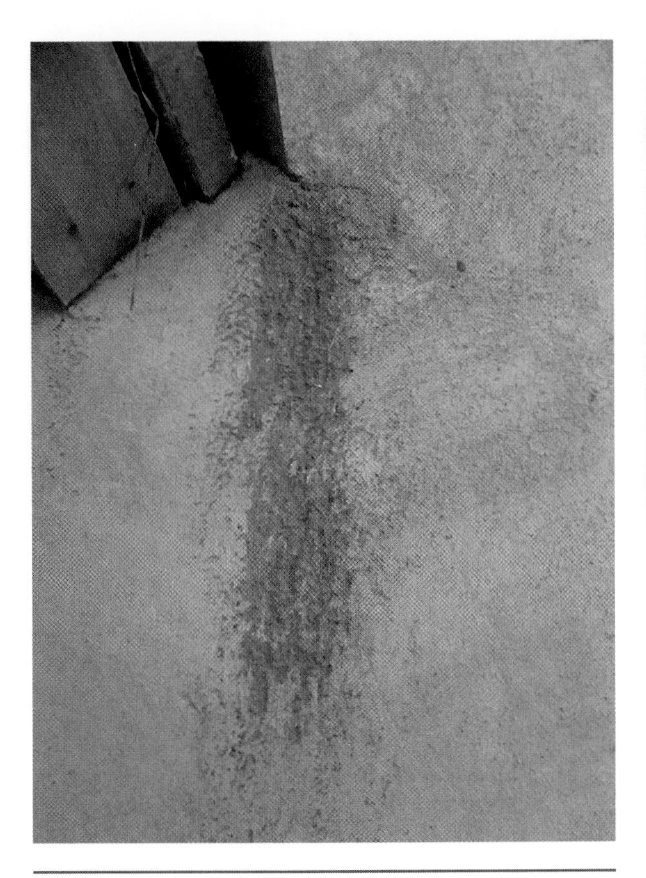

Flashing must be installed correctly to avoid erosion or other water damage, such as the plaster erosion caused by poor sidewall flashing pictured here. PHOTO BY JULIE KROUSE.

Consider splashback effect on walls above a shed roof. PHOTO BY ACE MCARLETON.

upside down. Wind-driven positive pressure applied to the face of a wall can create a negative pressure environment on the leeward side—which can induce a moisture drive in that direction, in addition to the kinetic energy of the wind. Flashing and material connection details must include considerations of water moving horizontally, or even upward, in addition to moving with gravity. These details must also address the powerful surface tension of water, which will allow it to run along the underside of a horizontal surface and potentially into a gap, without dropping. Flashing details are presented in appendix B of this book.

In addition to the obvious risk of the moisture working its way into the building, the erosive action of water, especially on plasters, must be considered, in that problems that may not occur initially can develop over time without proper maintenance. It should be assumed that, despite any measures taken to the contrary, some water will still find its way into the assembly, eventually.

Splashback is another strong erosive effect from water, in which water dropping off an eave onto the ground bounces back up against the sidewall. This can not only lead to repeated wetting of an area, inhibiting drying and increasing moisture drive into the wall, but can rapidly erode the plaster. Its effects are strong: we've seen splashback 30 inches high up a wall from a drip line 24 inches from its face! This may be easy to control at the primary eave line through gutters, water-dispersing ground cover, or bottom-of-wall protection, but what is frequently neglected is splashback off projecting roof planes onto second-story walls. This is very important to consider in the design phase as maintaining plaster in these locations can be difficult if access onto the roof is not practical. Raising the base of a moisture-sensitive wall up a good 18 to 24 inches above a moisture-durable wall (such as concrete) is standard practice to address this issue, discussed in chapter 14, as part of the "good shoes and good hat" approach.

Liquid—Ground Source

Surface water pooling against a poorly treated concrete foundation wall, or groundwater soaking the wall from below, taps into a very large moisture source. Through capillary action, the porous concrete is able to wick moisture up a very high distance—up to 10 kilometers, or 6 miles, if given a tall enough wall!—and therefore any ground-linked concrete should be considered a moisture vector, and any moisture-vulnerable material it contacts is at risk. The solution is to create a capillary break either around the concrete itself, to isolate it from the liquid source, or between the wall and sensitive parts of the building (or, to be redundant, both). Capillary breaks include materials whose pore sizes are either too small, like most plastics or paint-on rubber membranes, or too large, like gravel, to allow capillary suction to occur.

Liquid—Plumbing/Roof Leaks, Flooding

The problem with any kind of leak—especially a roof leak—is that where you see the water is rarely where the water is getting in, because water is highly mobile. So, no matter how well you have detailed your walls for every other source of entry, all it takes is one clumsy move by the roofer to subvert all of that hard work. And depending on what path the water takes, you may find that visible signs of infiltration don't appear until after the core of the wall is saturated. Accordingly, at the tops of walls, detailings that range from the use of solid-surface materials such as plywood to the installation of air barriers to encasement in base-coat plaster (our preference, when possible) will all serve as good backup safety measures. One of the benefits of plastered walls is that because plaster is a hygroscopic material, it shows discoloration when it is wet, which can be a clear indicator of a problem in need of attention.

Plumbing leaks can be catastrophic—an entire well can empty into your wall at 40 PSI if all goes wrong—and should be very carefully considered when designing the plumbing layout of a building. Do not put water leads or drains in an exterior wall!

Toe-ups are essential parts of a moisture control strategy. PHOTO BY JOSÉ GALARZA.

Vent stacks, too, may handle a whole lot of moisture from condensation, making them a potential liability as a moisture source if placed within an exterior wall. Even if your plumbing is kept well away from your exterior walls, all it takes is an overflowing toilet, busted pipe, or spilled bucket of mop water to flood a floor. If moisture-sensitive insulation extends to the floor, this could be a problem. Even plaster that runs down to a puddle will wick moisture up into the wall. We use toe-ups (described in chapter 14) and baseboard trim as "flood insurance."

Vapor—Diffusion

Vapor can move into or out of vapor-permeable building systems by diffusing through the wall membrane itself, depending on the direction of the vapor drive. The strategy of using a polyethylene vapor barrier not only attempts to halt air infiltration through cracks and gaps but also strives to reduce vapor diffusion through drywall or other vapor-permeable finishes. The problem is that vapor drives go in both directions in cold and moderate climates, and a vapor barrier serves only to impede drying via diffusion—it does not stop diffusion as an entry source. Placing strong vapor retarders or barriers on both sides will only ensure that the walls cannot dry out at all. Remember that a very small hole will allow a large volume of vapor-laden air into the wall system, where it will then be trapped.

We have identified that diffusion-borne moisture is in and of itself insufficient to cause moisture damage in the walls and ceilings of well-detailed buildings; diffusion's potential as a drying mechanism outweighs its liability as a moisture vector. Therefore, we support the use of airtight, vapor-permeable wall and roof systems and shun the use of class I or II vapor retarders (those performing as, or close to, vapor barriers) in any locations except in foundation systems. (Note that we prefer to use the more accurate term "vapor permeable" to the more confusing "breathable" when describing these assemblies, as "breathable" to many implies air movement, which we certainly wish to avoid.) Looking at diffusion and absorption as different sources of moisture entry and/or exit highlights the efficacy of Tyvek and other similar house wraps whose pore spaces are too small for absorption through capillary suction of liquid water, yet large enough to allow diffusion of individual water vapor molecules, enabling drying.

Vapor—Air Leakage (Air Barriers)

In chapter 7 we discussed the importance of recognizing the pressures that buildings face in keeping moisture vapor out of the walls, as, after rain, this is generally the second largest moisture vector in a building. To summarize:

- Temperature stratification in buildings in cold climates creates a positive pressure environment in the upper half of the building, and negative pressure in the lower half (the reverse in warm climates with air-conditioning, also creating a difference in temperature between interior and exterior climates).

- Warm air, charged with a lot of water vapor from cooking, respiration, and so on, is forced into the wall. In a 70°F room with 40% relative humidity (RH, the measurement of the amount of humidity in a given volume of air at a given temperature relative to its carrying capacity), a 1-inch-square hole in a 4 × 8 sheet of drywall will allow 30 quarts—that is 7.5 gallons—of water to pass through in an average cold-climate heating season. That is over 100 times the rate of diffusion through the entire 4 × 8 sheet (Lstiburek 2004, 18)!

- The air cools as it reaches the exterior plane of the wall (in cold climates), and as it cools, the RH increases until it reaches 100% (saturation of the air's carrying capacity, the

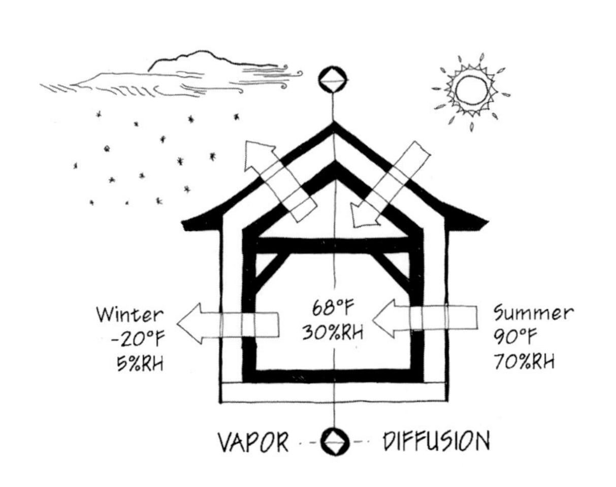

In winter in a moderate to cold climate, vapor drive is primarily to the outside, whereas in the summer the drive is primarily to the inside. ILLUSTRATION BY BEN GRAHAM.

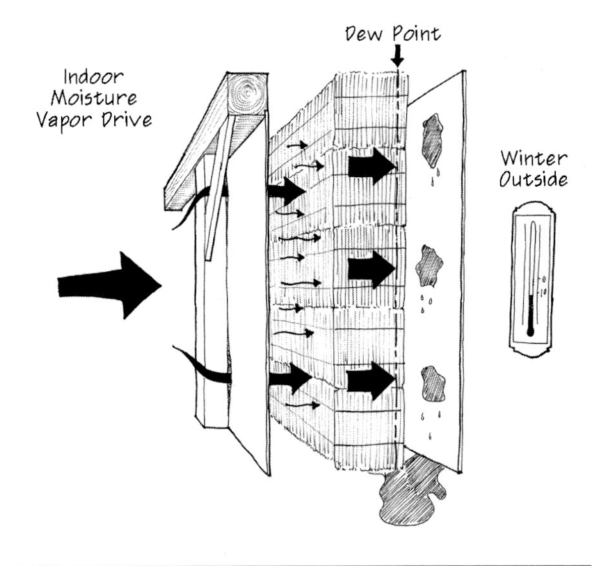

Vapor carried into the wall system through convective loss condenses when it reaches the dew point, causing water damage in the wall. ILLUSTRATION BY BEN GRAHAM.

temperature called the dew point), when the vapor condenses into water. The higher the RH of the infiltrating air and the colder the temperature in the wall, the closer to the middle of the wall the dew point will be reached, and the closer to the middle of the wall the harder it is for the moisture to dry out in the event that condensation forms in that location. That said, vapor requires a cold impermeable surface upon which to condense, so absent of metal or framing in the wall the condensation will form predominantly behind the exterior plaster skin.

The take-home message here is that moisture moves very effectively in vapor form through air migration. We must take the design and execution of air barriers very seriously not only for heating performance but for moisture management. We must also consider the difference between an air barrier, which stops airflow, and a vapor barrier, which can ultimately reduce drying through diffusion. In natural building systems that rely on plaster as an air barrier, we must look critically at the transitions between windows, doors, timbers, floors, and ceilings to ensure continuity of the air barrier throughout the whole building (see chapter 17 for detailing information).

Vapor—Ground Source (Foundation Detailing)

Vapor migration through concrete foundations, or evaporation from absorbed liquid, can be a huge source of moisture in a building. In the Northeast, with its high water tables and abundant groundwater, the amount of vapor rising off the ground at any given point can be comparable to that rising off an open body of water. Just pack your tent up one sunny morning and notice the only wet patch of grass is where you just slept. The moisture source is ground vapor. Now, dig a big hole in the ground, line it with concrete, and stick an airtight building on top—it's like building a house on top of a pond! To avoid trapping large amounts of vapor—as well as soil gases such as radon—in a building, a vapor barrier must be installed below the slab—and ideally over or under the footing.

SOURCE OF ENTRY—USE AND PERFORMANCE

In addition to the building's basic design and detailing characteristics, how its mechanical and human-operated control systems perform and the occupants' use patterns will also have a significant effect on moisture loading.

The ventilation system, as discussed more thoroughly in chapter 21, is a fundamental part of any building that features an air barrier. Although not really a "source of entry," it is important enough to merit inclusion in the conversation. Disregarding for the moment (and please, only for the moment!) the concerns of occupant health, this ventilation system is a key part of a moisture control strategy; it is the other half of the "keep moisture out of the walls" approach, which is to "give the moisture a safe place to go." Both mechanical and passive systems, if designed correctly, have advantages and disadvantages, but they need to be, first, in place, and second, in use, for the system to work. Planning on using operable windows for ventilation? You best remember to operate them on those –10°F days in the middle of winter, lest you run the risk of moisture accumulation in your building, and ultimately in the parts of the envelope you don't want it going. Is the power out? So is your mechanical ventilator.

How the building is used can dramatically vary the amount of moisture sourcing in the building. Is your building housing a brewery? A pasta factory? A line-dried laundry-drying facility? These increased, point-source moisture-loading scenarios must be factored into the design, to ensure they are not overwhelming an inadequately specified system. While the examples given are tongue-in-cheek, identifying point-source areas such as kitchens and baths and designing accordingly is relevant.

A much more common, point-specific use load can come from frequent mopping, wiping, splashing, or cleaning. High-use areas subject to repeated or prolonged wetting should be protected by surfaces with reduced absorption capacity that will give the moisture a safe place to drain; be careful how your plastered exterior wall interfaces with that walk-in shower!

Moisture-Control Strategy

Taking a step back from the source-by-source approach to moisture issues and solutions reviewed above, what we can generally understand is that moisture sourcing is quite varied, moisture movement is very complicated, and moisture issues affect and are affected by multiple other systems throughout the entire structure. Accordingly, to deal with moisture in a comprehensive and meaningful way, a holistic moisture control strategy must be developed for the building. In *Design of Straw Bale Buildings*, John Straube outlines a well-researched and highly practical strategy informed by the concept of the moisture balance (King 2006, 145–59). This model illustrates that there are three elements at play in dealing with moisture pressure—wetting (and more specifically, the avoidance of wetting), drying, and safe storage of moisture within the building assembly—and that these elements all must be balanced for the moisture-control strategy to work. As long as the rate or timing of wetting does not exceed the rate of drying by more than the safe storage capacity of the building, then short-term accumulation will not become a long-term problem.

In historic buildings, what was not achieved through wetting-control mechanisms was achieved through strong drying potential (aided by significant heat loss) and adequate material storage. Whereas modern assemblies tend to put most of their faith in wetting control—utilizing many hydrophobic materials incapable of storage, which limit drying potential—natural buildings are ideally constructed using the more "balanced" approach evidenced in historic houses (many of which are still with us) while the associated heat-loss issues are simultaneously addressed. With this balance of wetting control, drying, and storage in mind we can create a clear moisture-control strategy, outlined in order of importance: (1) design; (2) deflection; (3) drainage; (4) deposit (storage); and (5) drying.

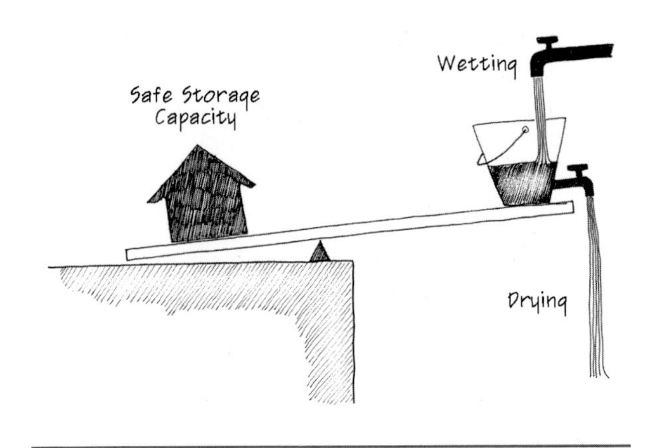

A successful moisture-control strategy relies upon a balanced approach to moisture management. ILLUSTRATION BY BEN GRAHAM.

DESIGN

Straube lays it out very clearly: "The best moisture control strategies always involve designing problems OUT—not solving them after they have been needlessly designed into the enclosure" (King 2006, 144). We cannot emphasize this point enough. The easiest, least expensive, and most successful approaches to moisture management come by thinking about moisture issues early on in the design process, and allowing the design to respond to those issues. They will not go away, and they will only get harder to deal with the longer we wait to address them, so we want to affirm once more how important it is for moisture management to start *in the design phase!* The specific techniques are provided in the sections that follow.

DEFLECTION

The easiest water to deal with is water that doesn't get into the building in the first place. Deflection of water can be addressed in a number of ways depending on the context. To begin with, siting is critical: siting the structure in a low-exposure area will dramatically limit the amount of wind-driven rain, which is much more liable to infiltrate the building than downward-falling rain.

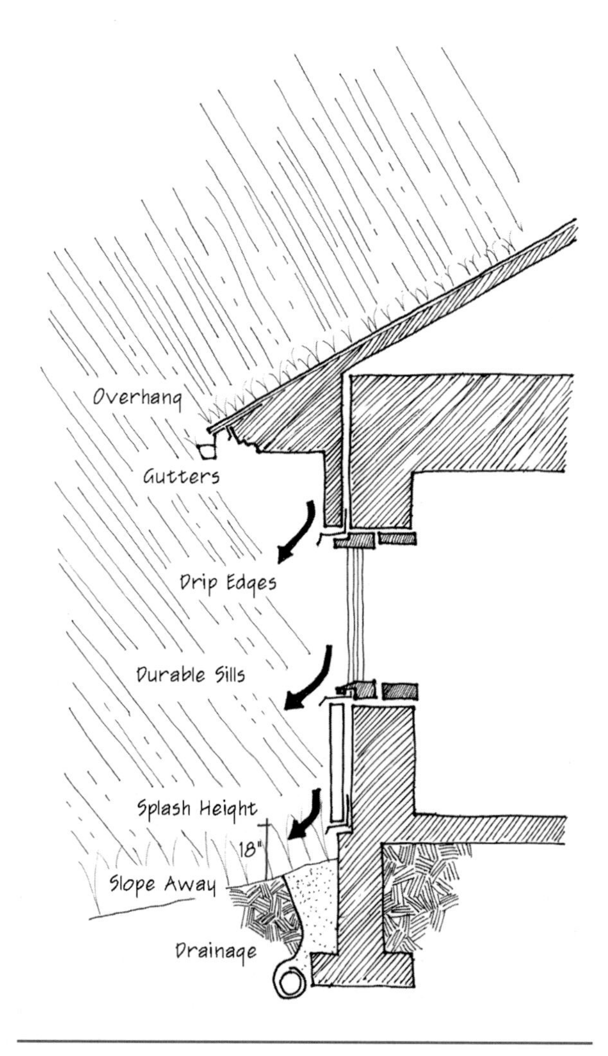

The shape and detailing of the structure will greatly affect the exposure of its wall surface to moisture. ILLUSTRATION BY BEN GRAHAM.

The shape and profile of the structure itself will also significantly affect the amount of rain it is exposed to: the simpler the form, the fewer transitions and potential weaknesses and less surface area through which rain can penetrate. If there is a dominant wind vector from a specific direction, a tall gable-end wall provides much more opportunity for penetration than a large saltbox roof plane that faces the weather. Taller buildings receive significantly more moisture than shorter buildings, and larger overhangs also make a significant difference in rain exposure.

Zeroing further in on the building, wall protection and flashing issues are the next place to address rain penetration. For a natural building, this starts with the "good shoes, good coat, good hat" approach (as described in chapter 14): raise sensitive parts of the wall up off the ground on a high foundation (good shoes) and protect them well with generous roof overhangs (good hat), which will further minimize their exposure to moisture.

To continue the metaphor, a "good coat" is also a necessity. We favor the use of lime-based renders (exterior plasters) when a finish plaster is desired because of the combination of durability and vapor permeability. Lime plasters, however, are very porous and absorptive finishes; filling the surface pore spaces with multiple coats of limewash or using a silane/siloxane or silicate-based finish (as described in chapter 20) have both been tested and proven to reduce moisture sorption without limiting permeability.

A further layer of protection can be achieved by installing wood siding over the base coat of plaster. In many situations, it is easier and often less expensive to install siding than it is to finish plaster (particularly in terms of labor and flexibility of weather), and while the durability of wood versus a quality plaster is debatable, repairing individual pieces of siding is easier than patching plaster in a lasting fashion and blending to a matched finish. The greatest strength of siding systems, however, occurs when they work as rain screens, with a drainage plane between the siding and plaster (discussed later in this chapter).

Regardless of the final finish, utilizing appropriate flashing details at all penetrations—windows, doors, mounted fixtures, lights/electrical boxes, vents—is mandatory. The point of flashing is to have physical barrier materials that permanently shed water away from the interior of the building, often relying on lapped installation to keep moisture from finding its way in through cracks via surface tension or pressure. Caulk is not the same as flashing, and it is a poor substitute for adequately designed and installed flashing systems. Many basic flashing details are presented in chapters 14 and 17 and in the designs found in appendix B.

The surface texture of the wall also affects moisture issues with regard to durability, particularly with plaster finishes. In the United Kingdom, traditional lime-plaster finishes over earthen cob walls were of a style called "splatterdash" in which very coarse aggregate (⅜ inch maximum size) was "harled" or hurled against the wall with a curved trowel, lightly passed over with a darby (a long straight edge) to cut off the high spots, and limewashed. This highly textured surface helped to break up the surface tension of water rivulets running down the wall, dispersing the water over a wider surface area and reducing erosion of the plaster. Furthermore, these older buildings featured sloped exterior windowsills that projected beyond the plane of the wall, in contrast to today's modern windows that frequently feature little or no sill.

Deflection mechanisms specific to vapor rely primarily on air barriers, on both the interior and exterior when possible, although issues of climate and practicality may dictate placement of a single air barrier. Air barriers were discussed at length in chapter 7. Ground-source moisture deflection is achieved through both vapor barriers and capillary breaks, as discussed in regard to identifying the sources of entry earlier in this chapter.

DRAINAGE

In dealing with liquid moisture, the most effective initial strategy is drainage. Good drainage begins with the site, ensuring that surface moisture is drained away from the building. Appropriate drainage must also be integrated into the foundation design—some foundation systems, such as the rubble trench, rely upon groundwater drainage to avoid frost heaving (discussed in chapter 12). Incorporating gutters and downspouts or rain chains—possibly for rainwater catchment—is another drainage strategy that will help safely direct moisture away from the building.

In general, there is no faster way to deal with bulk loading of liquid moisture into the building than to drain it from the assembly. To do so, however, may not always be practical; to use drainage as part of a strategy, we must contextualize this fact with an understanding of what the wetting mechanism is, and where the wetting is occurring. Rain penetrating a plastered surface, for example, will be absorbed readily by the hygroscopic plaster (presuming a clay- or lime-based coat); by the time there is sufficient moisture concentration to trigger drainage potential, moisture damage will be inevitable.

Infrared thermography can indicate the presence of moisture by showing the temperature differences between the dry and moist areas of the wall. PHOTO BY BRAD COOK.

If, however, a siding system has been placed over the walls as a deflection mechanism, then providing an opportunity for rain that penetrates the siding to safely drain without being absorbed directly into the wall assembly behind is a terrific option, one that is regularly employed by builders of all persuasions, natural and otherwise. If this drainage cavity can be fully vented to atmospheric conditions at the bottom and top of the wall, then a "vented rain screen" has been created—one of the most durable and effective moisture-control techniques available. In addition to providing a drainage plane for the assembly, a convection loop can occur in the air space that will encourage diffusion and evaporation from the surface of the wall material behind (in this case, plaster), further enhancing the drying potential. See the sidebar "Rain Screen Detailing" for further considerations on a well-executed vented rain screen.

Rain screen siding is one of the most protective wall systems available for the natural building. PHOTO BY JACOB DEVA RACUSIN.

Rain Screen Detailing

Rain screens are great—they can be aesthetically pleasing, low-maintenance, easy to install and repair, highly durable, and exceptionally effective as protection and drying enhancement mechanisms for a base-plastered wall assembly. That said, the details must be right, and the devil always lies in the details. Here are a few notes for successful rain-screen design:

- If installing horizontal siding over a cellulose-insulated sheathed stud wall, simply strap out the sheathing 16–24 inches OC with 1 × 4 or comparable vertical strapping.

- If siding over a stud wall insulated with woodchip-clay or straw-clay, there are a couple of different approaches. Either set your exterior laths on nailers in between stud bays, set back from the face of the studs so that the base coat is held back from the face of the studs ½–¾ inch (which may compromise the integrity of the air sealing on that side of the wall); or lath over the stud wall, and then install pre-air-finned 1- to 1½-inch nailers over the face of the studs. Plastering between the nailers is a reasonable attempt at air sealing while maintaining a comfortable gap between the plaster and the face of the nailers. In either case, be sure to use a siding product that can accommodate the span of the studs, or else solid sheathing may be required before the siding is installed.

- For straw bale walls, if there is an exterior stud wall, follow the recommendations laid out above for woodchip-clay or straw-clay walls. For interior timber frames, the wall will essentially have to be stud-framed to the exterior to receive the siding. This wall need serve no structural role other than resisting out-of-plane shear loads and being self-supporting. However, given that the nailers can't be hung on the plastered straw bales, close to full-dimension 2 × 4s will be needed for stability when spanning from floor to floor. On multistory buildings, it is important to tie back into the internal frame at each story. If these studs are to serve as the air space—bales laid to the interior face—and the windows are hung to the exterior, note that a substantial break in the thermal envelope will be created, and it will need to be rectified.

- The gap behind the siding wants to be at least ½ inch, practically speaking (which can be difficult, if the siding is going over an irregular plastered surface), to allow for sufficient drainage and airflow. If installing the nailers for hanging the siding after the bales are installed and plastered, be sure to mind your planes while laying up the bales, lest you find yourself with a big conflict! Pre-installing siding nailers, while cumbersome for baling and plastering and a potential compromise of the exterior air barrier, will ensure a good gap and a flatter wall.

- The drainage plane must be kept clear—no horizontal nailers supporting your vertical board-and-batten siding, for example—and should be able to drip clear to the ground, or be flashed safely out of the assembly.

- The bottom and top of the wall should be well screened to avoid pest entry behind the siding. We like to pre-install window screen or metal mesh hanging long at the bottom of the wall, then wrap it up around the bottom of the nailers once they are installed and secure to the face of the nailers temporarily until the first piece of siding is installed. At the top of the wall, although it is tempting to vent into the roof vent, this should be avoided so that a firestop can be created between the walls and the roof, and to keep sun-heated air from melting a snow pack on the roof and causing icicles or ice damming. Various off-the-shelf and site-built wall-venting solutions are available, which generally

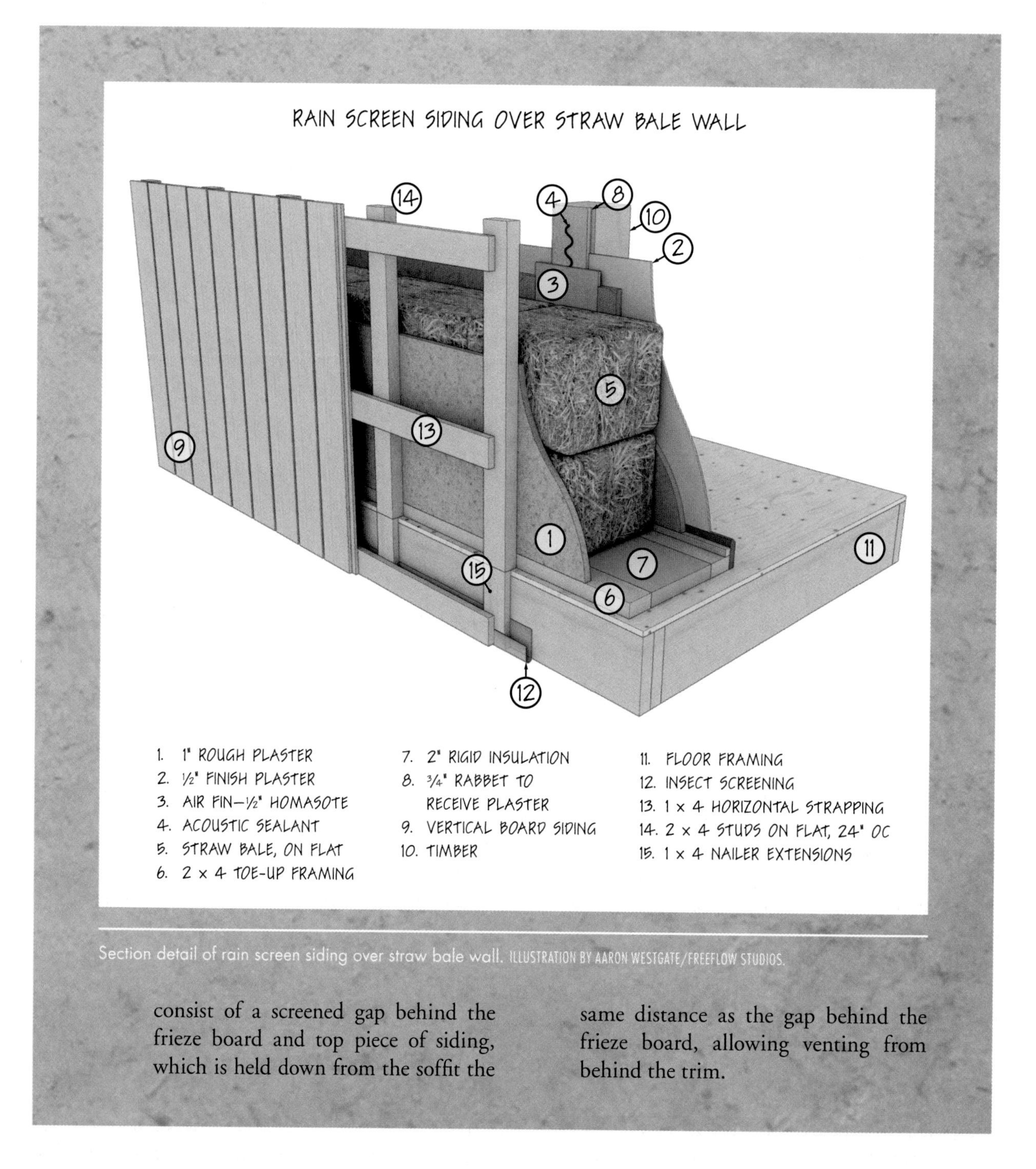

Section detail of rain screen siding over straw bale wall. ILLUSTRATION BY AARON WESTGATE/FREEFLOW STUDIOS.

consist of a screened gap behind the frieze board and top piece of siding, which is held down from the soffit the same distance as the gap behind the frieze board, allowing venting from behind the trim.

For stuccoed walls, a similar multilayer "drain screen" can be used, which builds a drainage plane in behind the outer layer of stucco before it hits the substrate. This is achieved by installing a double layer of house wrap and plaster lath between the base and finish layers of plaster.

Windows are great places to create drainage potential. Building solid window pans below the units and allowing them to drain fully and safely out of the wall is the most secure way to address one of the greatest weaknesses in the envelope.

Some builders suggest the use of a gravel-filled bottom-of-wall toe-up in straw bale buildings, often complete with exterior weep holes, to allow for drainage. As we mentioned earlier, by the time there is sufficient liquid concentration as to run down and pool at the bottom of a wall, moisture damage has already occurred. This fact alone need not preclude use of this detail; what does, however, is the tremendous conductive loss built in at this place in the wall, which will in turn lower the temperature of the wall, pushing the dew point further toward the interior, and increasing the likelihood of condensation, thereby creating the very problem we are seeking to address. If you are going to include bottom-of-wall drainage, the exterior of the toe-up should be lined with insulation, although this may interrupt weep hole drainage to the exterior. Perlite and rock wool are options for insulating drainage materials.

DEPOSIT (STORAGE)

Though "deposition" is commonly included as a component of "drainage" in moisture-control strategies, we feel this element of the strategy is one that separates the natural building approach to moisture mitigation from conventional construction in two important ways. For one, we design our buildings intentionally from the outset to be able to hold significant amounts of moisture for short periods of time, a practice that is based on our understanding of the wetting potential of our assemblies (as discussed earlier in this chapter). For another, in taking this approach we acknowledge and embrace the reality that building a moisture-proof wall or roof assembly is an impossibility, at least in our cold and wet part of the world—and from the best of our understanding in most other places as well save for hot, dry climates. It is really this latter philosophical characteristic that, in part, helps define natural building: a design approach that guides the creation of truly climatically responsive and adaptive structures.

The primary materials likely to be found in natural wall enclosures are either cellulosic—wood and straw—or of a few select geologic materials—lime and clay in particular, and cement stucco in some cases (outdoor stone walls are not germane to this discussion). The benefit of all these materials (stucco and most stone notwithstanding) is the combination of their hygroscopicity and their resiliency. While it might seem counterintuitive to want our building materials to attract moisture, if we consider that moisture will be there regardless of the materials' nature, it serves us well to utilize materials that can safely adsorb and absorb moisture and hold it for a period of time without immediate damage. Also, just as thermal mass moderates daily indoor temperature swings (as discussed in chapter 7), hygroscopic materials help moderate changes in indoor humidity, maintaining a more uniform and comfortable environment.

The moisture content (MC) threshold before mold can begin to develop (20%, or 80% RH) is similar for straw and wood—which, including cellulose insulation, are the primary materials found in heated natural buildings in cold climates. Rot and decay threshold conditions are higher still (28% to 30% MC or 95% RH for wood, slightly less for straw owing to increased surface area). We have found through moisture testing (conducted ourselves and by other builders and researchers) that this threshold falls comfortably higher than the range of seasonal fluctuation and incidental wetting, *providing that the other components of the moisture control strategy are working effectively*. In other words, if your design is sound and your detailing is tight, the amount of moisture that might be expected to find its way into a wall from one source or another over the course of a season can readily be handled by wood and straw without triggering mold growth and decomposition. An additional advantage of straw over wood or cellulose is that the large pore size of the tubular stalks reduces capillarity (wicking potential) while still providing good thermal resistance. This means that even if localized liquid wetting occurs on the exterior face of the bales, they will be less likely than other cellulosic materials to wick that liquid into the wall cavity. Empirically, we have verified this with moisture testing, which repeatedly shows that elevated MC percentages drop off rapidly moving from the exterior into the center of the wall cavity.

We also look to the plaster coat—which is already protecting the wood and straw encased within by deflecting both liquid and vapor moisture—to further serve as an enhanced storage mechanism. We can do no better in this regard than to use clay for the job. Clay plaster has four times the water absorption rate of cement stucco, and while much of this can be attributed to pore size, a fundamental component is the incredible relationship between clay and water molecules. As noted in chapter 4, clay's cationic (charged) nature makes it highly hydrophilic due to the fact that it bonds electromagnetically with the polar water molecules quite readily. Coupling this with clay's high surface area (its molecular form is platelike), we realize that as designers, we are using one of clay's greatest strengths—its ability to attract and hold large amounts of water without deformation—to support one of straw and wood's greatest weaknesses: susceptibility to rot in prolonged presence of elevated moisture levels. It is primarily for this reason that we strongly advocate the use of clay-based plaster base coats in wall systems (see chapter 17).

Walls built of earth consisting of clay soils have an increased capacity for moisture storage, with a greatly reduced threat of decomposition—although deformation and structural performance will be compromised. A big caveat for any assembly, however, is that safe storage of moisture must be directly linked to a drying mechanism that allows the moisture loading to be reduced within a reasonable amount of time; this is the "safe" part of safe storage.

DRYING

We now examine the final stage of the moisture-control strategy. Appropriate design detailing has been incorporated, but some moisture has penetrated the deflection mechanisms of the assembly; the moisture has been adsorbed and absorbed by hydrophilic assembly materials; now, we must get that water to leave. There are different approaches to drying out or otherwise getting rid of water, and many considerations, based on the form the moisture is in and where in the structure it is located.

The action of drying is to encourage liquid moisture to evaporate and leave the assembly as water vapor. One of the greatest liabilities of a plastered wall assembly that connects directly to moisture-sensitive interior components is that it breaks a cardinal rule of "reservoir cladding" systems (those that absorb water, like brick or plaster), which is to provide an air gap behind the cladding, as in the rain screen, to reduce moisture transfer into the wall. Here is where we rely on the storage capabilities of the base coat, being more hydrophilic than the materials it protects, as well as the finishing of the plaster to reduce absorption as much as possible without inhibiting the vapor permeability necessary to ultimately allow the evaporation of the stored liquid. To this point, John Straube's published test results indicate that finishes such as limewash and siloxanes will reduce absorption while maintaining adequate permeability, sodium silicate and wheat paste finishes will have no measurable effect on moisture properties, and oil paint and heavy applications of linseed oil will reduce both permeability and absorption. (See Straube's 1999 report "Moisture Properties of Plaster and Stucco for Strawbale Buildings" for detailed analysis of his study.)

Moisture drives in walls are very erratic, can change over the course of a day, and can even happen in two directions simultaneously. This highlights a couple of things: first, we need to ensure that we can allow adequate vapor migration through both sides of any wall or roof assemblies; and second, we need to consider where primary and secondary drying potentials exist and encourage drying in that direction. For example, in cold climates (and this is more relevant the more extreme the cold), it is best to ensure that the exterior wall surfaces are permeable enough to allow for primary drying, while interior surfaces, though still permeable, can be less so in order to reduce diffusion from the primary vapor drive coming from the interior of the building and encouraging drying to the exterior. In simpler terms, the exterior should be more permeable than the interior, but the interior should still be permeable enough to allow diffusive drying. This is as simple as adding additional coats of plaster or paint to the interior. It must be made clear here that using an interior vapor barrier in cold climates serves to inhibit drying potential rather than aid in reducing moisture issues in wall assemblies—especially during the summer when the primary vapor drive is toward the interior of the building. This has been shown using sophisticated moisture modeling software developed by Oak Ridge National Laboratory to contrast cellulose-insulated walls with and without installed interior 6 mm polyethylene vapor barriers (National Fiber n.d.).

We now have identified the primary reason that we do not advocate the use of stucco as a plaster in cold climates, though it was copied in early northeastern straw bale structures from precedents set in the desert southwest, and in fact we identify its use as a significant factor in many of the wall system failures we have seen in our region. There is also plenty of documentation in the United States and in Europe of the rapid failure from moisture damage of decades- or centuries-old earthen walls upon recoating them with cement stucco. In contrasting the material properties, the reasons become obvious. We have already noted in the previous discussion of storage that stucco has a very limited moisture sorption capacity, which means that it will saturate rapidly and pass any additional moisture into the substrate materials, which are more hydrophilic than the stucco. An assessment of permeability values shows that cement stucco is only 1.16 perms, placing it just barely above the categorization of a vapor barrier, which is less than 1 perm. By contrast, a lime-sand plaster is approximately 9–12 perms, whereas an earthen plaster is an incredible 16–20 perms. It is interesting to note that using a 1:1 cement-lime mix increases permeability by nearly sevenfold, to 7.12 perms. This is not to suggest that the use of cement stucco commits a building to a lifetime of moisture failure; rather, it is a cautionary statement that asserts that when using a cladding material that will inhibit drying and otherwise not support moisture management, moisture damage is more likely to occur.

Additionally, concerns have been raised about the permeability of hydraulic limes (a recommended manufacturer, St. Astier, sells three products in increasingly greater strengths of hydraulicy: NHL2, NHL3.5, and NHL5). We commissioned a report by Robert Riversong of Riversong HouseWright to compare testing data provided by multiple sources on a wide variety of non-hydraulic and hydraulic lime (the latter provided by the manufacturer), cement, and earth-based plaster materials. It was concluded that, when equalized for units of measure, the NHL limes all performed comparably in terms of permeability with the non-hydraulic limes (a range of 10–12.9 perms), while the NHL5 proved to have a third of the absorptivity of the NHL2; this would point toward the applicability of a hydraulic lime as a suitable finish for vapor-permeable wall systems, one which we have successfully utilized ourselves.

Smarter Buildings

The most common criticism we hear of natural wall systems, and straw bale walls in particular, is that they are prone to rot and moisture problems. This criticism is understandable. After all, a straw bale does conjure images of biologic activity, and there is precedence of failed structures. However, a closer look at any of these failed structures—and we have looked at quite a lot of them—clearly indicates patterns of poor design, poor detailing, poor execution, and in general a poor understanding of how moisture works in a building.

This is not surprising. How many of these buildings were designed and executed by professional straw bale builders with a nuanced understanding of hydrodynamics? How many case studies and regionally specific data sets were their designers and builders able to reference at the time? How widespread was the information on heat and moisture properties of insulated buildings fifteen to twenty years ago? Furthermore, how much better have their conventional counterparts faired? How many incidents of mold and rot in fiberglass-insulated 2 × 6 OSB-sheathed walls exist in the region, how many have passed inspections and been insured, and how oft-repeated are these obviously flawed designs?

Natural building in this region—and in fact in all regions—was initiated largely by owner-builders, amateur enthusiasts, and a handful of innovative and thick-skinned professionals, many of whom did not go on to build a second structure, let alone a third or fourth. A generation later—that is all it has been—we have climbed very quickly up a very steep learning curve—on the backs of those who took risks on those first structures—to learn from some of the early failures and successes. Many lessons have been learned as well from the many failings and achievements of conventional buildings around us to understand how to design a building that not only adequately deals with moisture, but excels at dealing with moisture, intentionally by design. This has been achieved with the support of many in the building-science community who have turned their resources and critical eyes toward natural building materials and techniques, giving us not only valuable information on how to improve these systems, but confirmation that they are well worth improving upon.

This confirmation comes not only in the form of theory, modeling, and analysis of assemblies based upon the moisture-performance properties of these materials, but as a result of field testing and empirical observation in many different structures in many different regions. The Canada Mortgage and Housing Corporation released a study in which field testing of twenty-two straw bale houses across Canada (British Columbia, Alberta, and Nova Scotia) were evaluated for moisture performance. The results of the testing showed that the moisture content of these wall systems—all of which were described as being stuccoed—fluctuated seasonally within a safe range, unless there were identifiable design or application flaws that caused increased wetting in particular areas. One exception, on the rainy and very humid west coast, showed elevated moisture levels with only high climatic relative humidity levels as an identified source, potentially, yet inconclusively, indicating a liability of stucco finishes on natural wall systems in high-humidity regions. (Canada Morgage and Housing Corporation 2007) This testing corroborates the results of numerous other tests from around the United States, from California to New York.

Our own testing conducted on over a dozen structures over the last ten years (including those featured in the research project referred to in chapter 7 and in appendix D), using a Delmhorst hay probe moisture meter, has confirmed these and other published results as well as those reported by builder colleagues who have been testing and monitoring their own projects throughout the Northeast. Moisture contents tend to range from 5% to 14% seasonally (not adjusted for temperature factor and hay-to-straw conversion, which are purported to raise and lower the results a few percentage points, respectively) and depending on the location of the wall, the exposure to rain, and the type of finish, with higher readings associated with identifiable wetting sources.

On DVD, Chapter 8, see PROBE MOISTURE TESTING

In our 2011 research project, we identified areas of saturated exterior plaster in a few of the projects. With one exception—a large crack high on a gable wall identified as an area of significant air leakage, where readings behind the plaster read 19%—the results invariably showed moisture contents of 12% to 15% for buildings over a year old, even when surface and pin scans and visual inspection of the plaster registered near-saturation levels. In one structure that had received plaster on the interior just a few months prior and was heavily loaded with construction moisture and moisture from being inhabited, high moisture readings in the 20s (and one in the low 30s) were found directly behind the plaster. Readings in the center and interior edge of the bale dropped down to the upper and lower teens, respectively, indicating that the vapor drive was moving to the exterior, and the moisture was not dispersing evenly throughout the entire thickness of the wall but was concentrating toward the exterior.

There is still a pressing need for greater research and published data on moisture patterns in natural buildings of all types, particularly those using earth- and lime-based plaster finishes, and especially longer-term surveys that identify moisture migration patterns over the span of seasons. That being said, there is a large enough body of data, supported by scientific theory, that points conclusively toward the efficacy of natural building systems, as we identify in this book, to readily handle moisture pressures, even those of cold, wet climates. We are turning the greatest weakness of the technology into one of its greatest strengths, utilizing materials that have been used for thousands of years. The greatest liability to the system is that it requires critical thinking, careful analysis, and nuanced detailing—three ingredients often lacking in most building efforts. If building a more resilient building is harder—and it is that—then it certainly is worth the trouble.

CHAPTER 9

Fire, Insects, and Acoustics

There are other considerations worth examining in the scientific evaluation of buildings, and that merit at least a cursory understanding of how they relate to the natural building systems we discuss in part 3. To broaden the context we are developing to evaluate these different systems, we will conclude this section by looking at issues of fire, insects, and acoustics.

Fire and Natural Buildings

Given that earthen and stone building assemblies are highly resistant to fire damage, and that there is abundant existing published information on wood performance in fire conditions, we will focus here on straw. In addition to "Don't they rot?" the most common question people ask about straw-insulated wall systems is, "Don't they burn?" Some will even follow up with an anecdote concerning an uncle's barn that spontaneously combusted as a result of loosely piled damp hay undergoing exothermic decomposition. The quick answer is that, quite to the contrary, plastered straw-bale and straw-clay walls are superior to standard fiberglass-insulated stud walls with regard to fire resistance. There are three ingredients needed for fire to ignite and spread: spark, fuel, and oxygen. While loose straw on a jobsite is incredibly flammable, straw compacted into bale form, or covered in clay slip and compressed into wall forms, contains dramatically less oxygen to the extent that once the exterior layer of straw has charred, the wall smolders in a low-oxygen state. In fact, fire testing on unplastered bales has confirmed their performance to be comparable to that of heavy timbers in their high fire-resistance capabilities.

Formal ASTM International (formerly known as the American Society of Testing and Materials) fire-rating testing was commissioned in 2006 by the Ecological Building Network on two walls: a two-string flat-laid straw bale wall plastered with two ½-inch earth plaster coats, and a two-string edge-laid straw bale wall plastered with two ½-inch lime-cement stucco coats. The testing consisted of two parts: In the first part, the 10 × 10-foot sample walls were subjected to a 1,700°F blast furnace for a prescribed period of time, and thermocouples on the other side of the wall measured temperature increases across the assembly. In the second part, a fire hose was then turned onto the burned wall for a short time. To pass the fire test, the average temperature increase must be less than 250°F, and any individual sensor temperature increase must be less than 325°F. To pass the hose test, the stream of water must not pass through the wall.

The earth-plastered wall successfully passed a 1-hour fire resistance test (1 hour in the burn chamber, 1-minute hose test). A few interesting observations were noted in the results. Cracks formed in the earthen plaster, allowing for increased oxygen and charring of the straw. Charring in the bales themselves was only 3–4 inches in depth, whereas charring in loose straw chinking between bales (no clay slip) was much deeper. These observations point to the importance of performing regular crack repair maintenance on walls as a fire protection mechanism, as well as to the need to attend to good quality control in chinking voids in walls with straw-clay. An additional interesting note was the partial vitrification of the earthen plaster, which increased its resistance to water erosion.

The lime-cement plastered wall successfully passed a 2-hour fire resistance test (2 hours in the burn chamber, 2.5-minute hose test). Cracking and charring patterns were similar to those of the first test. A concern in testing the edge-laid straw bale wall was the potential for the plastic binding strings on the bales to melt, causing the wall to deform and creating a conflagration of fire with increased exposure of the straw to oxygen. While the strings did melt, the wall held securely together, indicating the importance of good compression detailing during construction to ensure a stable wall under fire conditions (King 2006, 179–82).

While these walls have successfully passed code-recognized testing, there are plenty of case studies of fire damage or catastrophe in straw-built walls. In King's *Design of Straw Bale Buildings*, contributing author Bob Theis tallies fourteen different reports of fires during and after construction. The largest cause was construction activity, such as soldering or welding, while candles, improperly detailed combustion units (heaters, fireplaces), and electrical fires were other accidental causes identified. Of the six buildings that were plastered at the time of the fire, only one (in which the fire began in the roof framing) was a total loss. Of the eight buildings that were unplastered at the time of the fire, six were a total loss. The clear take-home message is that both fire danger and the potential of catastrophic damage is great during construction conditions, and appropriate fire-safety measures such as building protection during fire-prone construction activities and easy access to fire extinguishers is critical (King 2006, 176–77).

A 2011 fire at the home of a former New Frameworks Natural Building consulting client in New Hampshire made clear the real-world issues of fighting a fire in a straw bale house. The fire was purported to have been started by an electrical space heater on the second story, then burned through the pine ceiling boards and ignited the foam in the panel roof, which in turn fell into and ignited portions of the exposed timber frame. The straw bale walls themselves were fully intact while the fire was being extinguished. However, owing to concern of lingering "hot spots" within the wall (it is unknown whether this concern was verified by inspection), the straw walls were torn down with assistance of an excavator. In some firefighting communities there exists a distrust of natural wall systems as a result of a lack of familiarity with their construction and performance. In the words of one volunteer firefighter involved with the fire: "I think making the local fire departments aware of where [natural buildings] are in town, and what the benefits are, will be a big plus. It was a big shock to our firefighters when they used a chainsaw to cut a wall and found straw within . . . firefighting is a dangerous job and the folks that do it do not like any more surprises than they normally get."

Cellulose insulation performs effectively in the presence of fire as well. Similar to a straw bale, cellulose chars to a depth of approximately ⅛ inch, after which point very little further damage occurs. In fact, we have had the experience of melting a penny with a pencil torch while holding it on a small nest of cellulose insulation!

Borate—a relatively nontoxic mineral used industrially for fire, pest and insect, and microbiological

Cellulose insulation provides superior fire resistance, as evidenced by the melting of a penny with a blowtorch while held on a nest of cellulose. PHOTO BY KELLY GRIFFITH.

resistance—used in the formulation of cellulose is what gives the material its fire resistance, as opposed to toxic halogenated fire retardants, as discussed in chapter 2. Cellulose has little to no rating on the Smoke Developed Index (SDI), an ASTM-rated test that values the amount of smoke produced in the presence of fire, which is the most significant hazard to occupants and public safety officials in the event of a fire. Compare this to fiberglass, with an SDI of 50, or spray foams, with an SDI of 300–450; foams are also very flammable once code-mandated fire barriers have been compromised, as was clearly indicated in the ignition of the foam roof panels in the New Hampshire case study discussed earlier.

Insects and Natural Buildings

It is not surprising that natural materials are occasionally host to insects in buildings, just as they are in their source environment. Fortunately, good construction practices and basic maintenance will make your building an unsuitable habitat for insects.

In earthen construction, a variety of different soil-borne bugs, from certain termite species to mud daubers and other wasps to mason bees, have been known to take up residence in cob or adobe walls. The simplest and most effective protection for these walls is a durable lime-based plaster. The alkalinity of lime is repellent to both micro- and macro-biological activity, and a finish of multiple coats of limewash will reduce moisture absorption into the wall surface, further discouraging insect habitation. Furthermore, separation from the ground is a key design feature that not only cut off access to the wall from crawling insects, but also increase longevity of the walls by reducing moisture exposure.

In straw construction, one of the primary vectors is pre- or during-construction contamination of the straw itself. Sourcing straw that is, first, clean, without weeds and abundant quantities of seed heads, and second, dry, will go a very long way toward reducing or eliminating insect contamination. The insects that have most commonly been found in straw construction projects are beetles, book lice, and mites—very tiny creatures that feed on mold and other fungi. Because of their small size (generally less than ⅛ inch), they are able to crawl into and out of very small cracks in wood, windows and doors, and plaster transitions. The most important thing to realize about insect presence in straw construction is their association with moisture: because their food source is borne of straw decomposition in the presence of moisture, they can be seen as bio-indicators of moisture in the wall that could be causing problems much worse than a windowsill full of little bugs.

We have seen a number of times, and have heard corroborated in colleagues' experiences, an outbreak of tiny insects shortly following plastering, lasting upwards of six to twelve months, frequently seen in "truth windows" (glass mounted onto the interior of natural walls exposing the insulation material within, proving the "truth" of the wall's construction) or flying and crawling around windows (particularly on the south side of the building). Invariably, within a year's time, as the construction-related moisture dries from the walls, the insects vanish. A sustained presence of insects would indicate a sustained presence of moisture and trigger moisture testing of the wall to determine the source of their food.

In wood construction, termites and carpenter ants are frequently the greatest threat. Considering that ground contact is the primary vector for the most common type of termite exposure, simple flashing techniques at foundation/wall transitions and keeping siding, framing, and other wood components separated from grade are effective control mechanisms. Cellulose insulation—a wood product—is undesirable for insect habitat owing to the borate additives, which are pest repellent and mold suppressant. Carpenter ants are different than termites in that they live in moist wood but do not consume the wood as a food source. Control measures are similar as for termites, including the mitigation of moisture issues in wood components to discourage habitat.

Acoustics and Natural Buildings

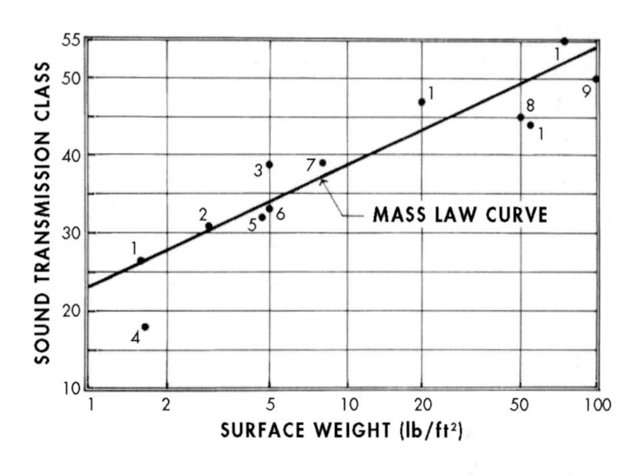

The mass law correlates the mass of a wall and its relative resistance to acoustic transmission. FROM M. DAVID EGAN, *ARCHITECTURAL ACOUSTICS* (FORT LAUDERDALE, FL: J. ROSS PUBLISHING, 2007).

We are fortunate as builders of natural wall systems to be working with assemblies that are inherently highly resistant to sound transmission. In the case of earthen walls, the high levels of mass, combined with the porosity of an earthen surface, work to reduce sound transmission through the wall. Plastered straw bale and straw-clay walls not only utilize the mass of the plaster and porosity of the surface material to absorb sound, but also benefit from the ductility of the straw core.

The acoustical mass law states that in solid panels, the magnitude of sound transmission loss *within certain frequencies* is controlled entirely by mass per unit area of the wall. Furthermore, the law states that transmission loss increases 6 db for each doubling of the frequency (meaning that lower frequencies transmit better than higher frequencies), or by doubling of the wall mass per unit area, up to a plateau frequency. Accordingly, the conventional approach toward sound isolation is to incorporate mass into the wall. Earthen walls, such as cob, adobe, rammed earth, and earth bag, have exceptionally high amounts of mass per area of wall, which make them highly effective acoustically. One test result showed that a 12-inch rammed earth wall had a Sound Transmission Class (STC) rating of 57, which is very effective (in contrast, a typical interior wall of ½-inch drywall on wood studs has an STC rating of 33; considering the logarithmic nature of the scale, this equates to a sixteen-fold increase in performance) (Dobson 2000, 7). It is relevant to note that the mass law applies only to a range of frequencies within the audible range—in very low and high frequencies, stiffness of the material will play a more influential role.

Straw bale walls offer an alternative approach to sound isolation: a dampened cavity of straw, sandwiched between two membranes of sufficient mass, which are moderately stiff if built of earth (relative to cement-bound mass). Testing results on two-string earth-plastered straw bale walls have shown to be rated approximately STC 55 (King 2006, 197–98). It is notable that the walls reduced more of the low frequencies than comparable mass walls; as indicated by the conditions of the mass law, the combination of mass of the plaster and the ductility (reduced stiffness) of the straw core reduces sound transmission across a wider range of frequencies than walls relying upon mass alone. The use of earth plasters will improve performance relative to cement or lime plasters due to their lower modulus of elasticity (measurement of stiffness). Additionally, applying plasters of differing thickness to either side of the wall (i.e., 1 inch on the interior, 1½ inches on the exterior) will reduce the potential of coincident reverberation (bouncing sound) between the plaster skins, which can increase the amplification of particular frequencies. Accordingly, earthen plaster finishes will help dampen a room that is too "live," as can be experienced in wallboard-finished rooms, particularly those of certain aspect ratios without carpeting or much furnishing to help absorb some of the sound waves.

We have favored the use of plastered straw-clay and woodchip-clay walls as natural interior partition walls to provide superior acoustic isolation within a structure, such as surrounding bedrooms and ground-floor bathrooms. Empirically, our clients testify to their outstanding performance. Theoretically, this

Straw-clay and woodchip-clay walls are terrific for acoustical control. PHOTO BY ACE MCARLETON.

can be attributed to the same properties of a mass-encased dampening core that a straw bale wall offers.

Regardless of the wall material and its position as an interior or exterior wall, achieving good acoustical performance relies upon proper detailing of the transitions, including windows and doors. Just as with moisture, thermal, and structural concerns, it is the boundary conditions of how the assembly is connected to the other elements of the structure that matter as much or more than the assembly itself.

PART 3

Natural Building Practices

CHAPTER 10

DESIGN

Were we to define natural building in one word, that word would be "relationships." The practice of natural building not only honors but encourages thoughtful and multifaceted connections in the process of creation. Nearly all other design and construction modalities—including conventional green construction—are overtly product-oriented. The process of working with materials, working with people, and engaging in craft serves the efficient creation of the end product of a building. Even buildings intentionally designed to evoke certain experiences and support the well-being of the inhabitants are still focused on the completed building. This is not a bad thing; efficiency is good, and if you are less interested in how the building is built than what the building is after it is built, then this approach is perfectly valid. Natural building is more than a product, however. This is not to say that there is no regard for the efficiency of the process, or that the product is secondary to the process, but rather that the process has a recognized and stated value in and of itself, and as such the project is designed not only around the finished product, but around the process through which the end result is realized.

Natural design—for our purposes, defined as the design of a natural building—recognizes that there is value in labor-intensive practices that feature on-site manufacturing of building products from raw natural materials. There is value in the sourcing of these materials from specific individuals and companies that are responsible members of the local community. There is value in involving the clients not only in the design process, but in the construction process as well—even if it is only to help source materials and help clean up the site. There is value in bringing members of the clients' community into the process as well, whether as enthusiastic spectators or as skilled workers. There is value in the handcrafted thing, in the imperfections left by chisel and trowel, in the marks of a human's hand in the artful construction of the building. What each of these values represents is a relationship—between the designer, the builder, the client, the community, the local economy and industry, the site, the region, the forest, the field, the earth.

Within the design process, we see further levels of relationships created between the many elements of the program. Good design is the harmonious solution to the problem of how to consider as many influences on the project as possible. We see each of these influences—site, structure, self—as being in relationship with the others. In any healthy relationship, we will find that healthy conflict—that defines the edges of the forces in opposition—is the substance of a strong relationship. Recognizing the inherent conflict, for example, between the cost of building a handcrafted timber frame on a shoestring budget and the value the frame contributes to the project is important to solving the problem of whether to use a timber frame as the structure of the building—when the conflict prompts conversation and introspection that help define the edges and prioritize values. The frame, the budget, the builder, the owner, the designer, the forester, the forest—all dance between conflict and resolution, all in relationship. How we go about the task of managing and organizing these many connections is akin to the process of managing and organizing the many relationships within the design program. By first developing a framework in which we recognize the role these many relationships play in the project, we can then set priorities, encourage harmonious relationships, and discourage unnecessary conflicts, allowing the process to move toward the end goal of occupancy.

Holistic Design System

We begin by developing the framework in which the elements of the program and the human and physical resources involved in the project can be organized. There are a number of different examples of design processes that foster the goal of encouraging and managing relationships: holistic resource management and whole-systems design are two such practices that have been developed for agricultural and systems design, respectively. We will use the somewhat generic term "holistic design system" to describe our similar approach. The words "holistic" and "system" are important here. Holistic refers to the comprehensive nature of our approach, as described earlier in our attempt to cast as broad a net as possible to consider all the potential influences on a project. The system is where we are headed now in our conversation: looking at the relationships between these influences as they work together to create something far greater than the sum of the parts.

Terrific examples of a holistic design system can be found in any natural ecosystem, and we consistently turn to natural systems for inspiration as a model for our processes. The economic success of an ecological system (such as a forest) can be valued by its productivity for market-valued products, just as the economic success of a building can be valued by its cost per square foot—a question we are asked more often than perhaps any (and which we will address in the next chapter). The success of a given ecological system as life-sustaining habitat that is integrated within a larger biotic scheme can also be valued by its diversity of relationships (biodiversity), which is a major factor determining its biotic efficiency (the amount of life that can be sustained within the system) as well as the resiliency of the ecosystem—its ability to absorb stress without collapse. We assert that the success of a building should be valued by the same metric.

We are encouraging relationships—akin to biodiversity, if you will—in our building processes in order to, first, have as positive an effect as possible on the site and in the region and, second, increase the potential for both the process and the product to respond gracefully to stress. Examples of the former include supporting the local economy, providing access to safe shelter for housing-insecure individuals, empowering owners to have active roles in creation of their own habitat, and investing in industries that engage in sustainable land-management practices in the region. Examples of the latter include sharing responsibility and increasing coordination and cooperation within the design process between all major players, fostering realistic and responsive budget development and management, and utilizing materials and assembly systems that are durable, are responsive to microclimate conditions of the site, are easy and inexpensive to maintain and adapt over the life of the building, and include redundancies of protection in critical performance areas such as structure and moisture management.

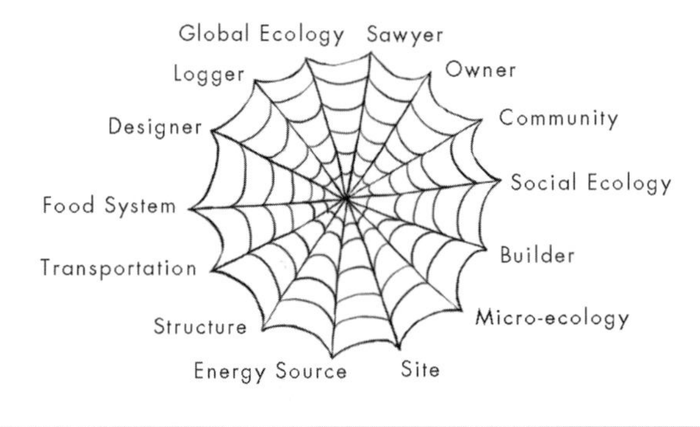

Natural building seeks to cultivate relationships between the many different entities involved in the creation of a building. ILLUSTRATION BY JACOB DEVA RACUSIN.

Values-Based Design

Returning to our point about conflict within relationships instigating the design process, we find there is a need to set a schedule of priorities to organize the many competing agendas in a building project. It is not enough to invite complexity into the project by fostering all these different relationships; we must also take responsibility for creating structure and order around these relationships to ensure that the conflict is controlled, minimized, and positive. Clearly identifying the priorities of the project—such as budget, environmental sensitivity, performance requirements, and aesthetics—is the heart of the program, and the program will be instrumental throughout the project in negotiating these relationships.

The priorities established in the program should reflect the values of the designer, builder, and client, and we must encourage the creation of a program that clearly defines and communicates this set of values through the elements presented therein. This process of identifying and strongly articulating values may be seen as selfish or judgmental by some clients. It is important, however, for the clients to feel safe and encouraged to be clear about defining their values, for it is only then that their values can be served by the design, yielding a product that best supports their needs and inclinations.

This is an important aspect of any design methodology, and we feel it is particularly important in natural design to embrace a values-based design process. Success is defined not only by metrics of performance or a points-based rating system, although those may help in supporting a successful design. Success is ultimately defined by whether a building succeeds at being an expression of values that have been sculpted, refined, and defined by a comprehensive process that promotes a diversity of relationships during the construction phase and throughout the useful life of the building.

The Golden Triangle

Prioritizing elements within a program is a difficult task, especially with a good program that is fully developed with many items to consider. There are many different approaches to assigning a priority schedule to a program, including standard important strategies such as identifying the fixed elements (such as site considerations) and working from larger elements down toward smaller ones. Neither of those are comprehensive prioritization tools, however. One concept we have found helpful is that of the "golden triangle of building."

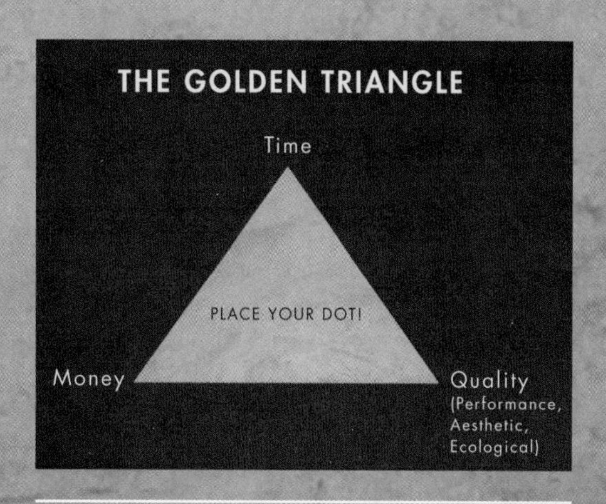

You can choose to prioritize time, money, or quality, but not all three at once. ILLUSTRATION BY JACOB DEVA RACUSIN.

The graphic shows a triangle with each of the three points holding a different value: money, time, and quality. Quality can be defined as quality of aesthetics, performance, durability, environmental sensitivity—any quality you choose to apply. When trying to value any given decision, apply the golden triangle to your decision by placing your finger on the triangle near the values that have guided your decision making.

Using the approach for the whole project as an example, choosing a midpoint along the left side will indicate the priority that the house be built quickly and cheaply, and that quality is much less of a held value. If you put your finger at the midpoint on the right side you value a house that is built quickly and of high quality, but may cost a lot to achieve those values. Choosing a point along the bottom side indicates that building a house inexpensively and of high quality is of higher priority, and that it may require a large investment of time—most likely the client's or some other unpaid labor—to realize those values.

In reality, the finger often gets placed somewhere in the middle of the triangle. The golden triangle can be used for prioritization in making decisions large and small, as well as in analyzing causation of decisions. For example, prioritizing quality of the shell performance at the expense of cost will allow consideration of finishes that will require a devaluation of either time or quality. Similarly, an investment of time in an owner-built handcrafted timber frame may require a cost investment in mechanical subcontracting to keep to a construction schedule. While it is by no means a foolproof evaluation technique, it can be a helpful tool when faced with making thousands of decisions, often under great time and budget pressures.

Integrative Design: Best of All Worlds

Integrated design is a collaborative design process in which all the major participants of the project—the designer, the builder, the owner, the engineer—are involved during the design process with the goals of making the construction process more efficient and enjoyable and improving the accuracy and efficiency of the budget. We take liberties with the phrase "integrat*ive* design"—which is becoming commonly used as a more accurate replacement for "integrat*ed* design"—and expand that definition to include a more philosophical integration of any of the "major participants" involved in creating a quality building. In the world of health care, "integrative medicine" is a practice in which Western allopathic medicine is supported by other modalities, including Ayurveda, herbalism, traditional Chinese medicine, homeopathy, and bodywork, to name a few. The result is a holistic approach to care that addresses the health of body, mind, and spirit through the harmonious integration of a variety of different healing modalities that maximize the benefits of each system while minimizing each system's drawbacks. We subscribe to a similar "integrative design" approach, in which we include elements of many different building modalities in service to a holistic design system, increasing the diversity of the design process. Following are three examples of harmonious integration between seemingly unrelated, or even conflicting, elements.

NATURAL MATERIALS AND BUILDING SCIENCE

The world of natural building, as narrowly defined as "involving the use of natural, unrefined, nonindustrial materials," is widely criticized for poor performance and a lack of awareness or response to basic principles of building science, particularly involving heat and moisture dynamics. This criticism is, on the whole, well deserved, as there are far too many examples of failed or underperforming projects whose problems are rooted in a lack of attentiveness to these basic principles of physics in the design process (condensation and air leakage are two widespread and classic examples, especially in cold climates). A lack of professional design support, a dearth of research and information concerning the performance characteristics of natural materials and assemblies, and a proportionally high number of owner-designer-builders have all contributed to these failures.

The application of building-science principles that have led to high-performance, superinsulated structures that promote rampant use of ecologically devastating foam and other industrial materials without any consideration of life-cycle impacts and ecological payback for the use of these materials is equally limited in its approach toward reaching a truly sustainable built environment. Members of the mainstream building-science community do not generally consider the relevance of more ecologically sensitive materials because they fall outside the spectrum of conventional architectural practice, and therefore, until recently, they have not been able to offer much guidance to natural building practitioners on improving best common practices.

We have seen a growing relationship and information exchange between both communities in the last two decades that is proving to be mutually beneficial. Natural builders are finally able to access relevant data and theory pursuant to their chosen materials and assemblies, resulting in buildings of increasingly higher quality, often outpacing performance characteristics of moderately energy-efficient green buildings. Our ability to work with Brad Cook, an established building analyst in Waitsfield, Vermont, to assist in conducting heat loss and moisture performance testing on a series of natural buildings in 2011 (see appendix D for more information) exemplifies the successful integration between the natural-building and building-science communities. We gained valuable information concerning the performance of these buildings and the efficacy of our detailing, and Cook was able to witness firsthand the quality of both heat and moisture performance in these structures, providing him with valuable solutions to building-science problems. As the building-science community sees from both laboratory and empirical testing the high-performance characteristics of good natural design, its embrace of these technologies as viable solutions helps solve the ecological impact problem that is currently incorporated into the design of the majority of today's high-performance buildings.

SOCIO-ECOLOGICAL SENSITIVITY AND MODERN TECHNOLOGY

In assessing the integration of natural building technologies and high-performance structures, we can discern that there is no cold-climate natural building that does not include strategic and appropriate uses of industrial materials to obtain specific performance goals (the same may be true of those in warm climates). We use many tubes of acoustical sealant as an integral component of our air-barrier systems when managing the transition between plaster and wood. Spray foam is still used to insulate cracks and crevices, such as around window casings. Plywood, bituminous water barriers, steel roofing, concrete—they all find their way into our buildings, as their inclusion is easily paid off in durability, thermal performance, or other metrics of quality that the palette of natural materials simply cannot provide.

Building performance and durability are only a few categories that illustrate the need for incorporating industrial materials into the natural building approach. Using the products of an industrial system to their other greatest advantages—saving time and labor, which often reduces budget—is another valid reason. Roof trusses and concrete foundations are great examples of this: given limited time and labor resources, the investment in an easier installation of a foundation or roof system can free up resources to invest in more labor-intensive wall, floor, or finishing options.

Finally, access to materials in many situations is easier when working within the industrial system. Picking up 2 × 6s at the local supply center requires much less time and effort than locating and driving to pick up 8 × 10-inch 24-foot pine timbers at a local sawmill. Rigid foam insulation is still the material of choice below grade, until mineral wool board is more widely distributed or other product offerings are brought to the marketplace.

GREEN BUILDING AND NATURAL BUILDING

All of this builds up to the larger conversation about the rift that has long existed between the natural-building and green-building communities. Both groups of practitioners are faced with and are responding to the same dire ecological and social problems of our day, to one degree or another. Both communities have valuable approaches and technologies to bring to the table in the pursuit of viable solutions. Yet the focus of the conversation has been on identifying each other's supposed weaknesses: natural building is lacking in design and doesn't perform well, and green building ignores social problems and overemphasizes energy efficiency above all other environmental issues.

Our approach of integrative design seeks to bring together the best that these two modalities have to offer—as well as other methods within the world of design. We have been successful in interweaving non-industrial materials with appropriate performance detailing, systems evaluation and diagnostics, and industrial technology and product development—outcomes of green-building practice—in a process-oriented, ecologically considerate, and owner- and community-empowering approach to design and construction. This integration is the hallmark of natural building. Given the scope and immediacy of the problems discussed in chapter 2, it is our belief that we do not have any choice but to focus on the best of all worlds in finding solutions.

Design Modalities

There are many different approaches to design from which the designer of a natural building can draw inspiration and information, a sampling of which are listed below. Each of these are design modalities that, in and of themselves, can firmly guide the parameters of the design or, when used in concert and combination with other strategies, serve as useful and informative tools to help inform the design process, from program through design development.

PASSIVE SOLAR

Elements of passive solar design are nearly ubiquitous in green-building design today, at least in its most basic form of southerly oriented windows (in heating climates), an approach called *sun-tempering*. True passive solar design, however, is much more complicated. The goal of passive solar design is to maximize the heat input from the sun as a heat source for the building, without overheating the building during the cooling season. Accordingly, in addition to simply orienting windows to the south, there are many other factors in the design of the building that must be considered (more discussion on heat and passive solar design can be found in chapter 7). There are many good modeling tools available to assist designers, such as the Energy-10 program developed by the National Renewable Energy Laboratory (NREL), a laboratory of the U.S. Department of Energy. Further resources can be found at the end of the book in appendix A.

BAU-BIOLOGIE

Bau-Biologie, a German design practice also known by its English translation of "building biology," is the study of how buildings relate to both human biology and Earth ecology. According to the Institute of Bau-Biologie and Ecology—the organization representing Bau-Biologie in America—Bau-Biologie is defined as "the impact of the building environment on human health, and the application of this knowledge to the construction, or modification, of homes and workplaces, and the holistic interaction between human life and other life-forms with our environment." Bau-Biologie was founded in post–World War II Germany, when the human health effects of reconstructed houses that were built rapidly using toxic materials that contributed to sick-building-syndrome-type ailments became too numerous and severe to ignore.

The nature of the materials used in a building is a primary factor in the building's impact on human health. In an evaluation of a list of building materials, from the natural to the synthetic and industrial, against a series of sixteen different criteria (including biologic/ecologic health, radioactivity, hygroscopicity, and odor), natural materials of timber, stone, straw, and earth consistently came out on top as "flawless" or "nearly flawless," making them compatible with the principles of Bau-Biologie (HarvestBuild Associates n.d.). This clearly indicates strong overlap between the goals of Bau-Biologie and those of natural building.

BIOPHILIC DESIGN

Biologist E. O. Wilson brought us the concept of biophilia in his book of the same title, in which he explored humankind's affinity toward life and living systems, and the role of this affinity as a defining characteristic in morality and the human experience. The concept of biophilia can be brought to practice in many different realms; the built environment is certainly no exception. Biophilic design centers around the incorporation of the natural environment into the built environment, producing a positive influence on psychological, physiological, and social well-being. The underlying philosophy of biophilic design is quite compatible with natural building in that there is a recognition that humans are part of an ancient ecological system, and that we have evolved to respond favorably to elements of our ecology that have sustained and nurtured our species since its inception. Applied to the built environment, biophilic design recognizes our buildings as our habitat and seeks to promote our well-being by patterning our buildings after the ecological features toward which we are inclined.

The benefits of biophilic design have been tested and verified, and they range widely. They include decreased recovery time from surgery and illness, enhanced cognitive function in children and adults, improved well-being in occupants, improved worker satisfaction and performance, reduced blood pressure and anxiety, and much more (Heerwagen and Hase 2001, 30–36). A second benefit of biophilic design is the occupant's cultivation of a stronger connection to the natural environment, which in turn will create a more ecologically minded consciousness and spur more environmentally friendly action and decision making. The potential for biophilic design to improve the lives of users and occupants of our buildings is tremendous, as is the potential for users and occupants to develop a relationship with their ecology and work to combat ecological problems. The practice of natural building can readily be integrated with biophilic design to achieve these objectives.

REGENERATIVE DESIGN

If ecological design can be loosely defined as design that minimizes ecologically destructive practices and recognizes that the health of our ecosystem holds value, then regenerative design takes the next step in creating designs that enhance the health of our ecosystems and encourage us to repattern our lives in new ways that allow us to live in greater harmony with the world around us. Rather than "do no harm," the approach is to "create positive change."

In order for this approach to design to exist in any meaningful way, it requires changes in the way we conduct business, develop towns and cities, and live in our buildings. While ecological design can often be achieved without significant changes in the lifestyle or consciousness of the occupants or users, regenerative design recognizes the human and social elements in building design that must be addressed to achieve deep sustainability. Such cutting-edge approaches are difficult to employ on a massive scale and are inherently challenging on both a personal and societal level. Addressing root causes of the ecological impact of the built environment can be difficult to both identify and address. Examples include our culture's disassociation from and subsequent lack of compassion for the ecological systems we damage (out of sight, out of mind) and existing community-scale development that is expensive and complicated to alter. However, such difficulties should not deter efforts at implementing these critical methodologies.

Design Rating Systems

There are many dozens of ratings systems in the world of building and design. Some are specific to a given material, such as the Forest Stewardship Council's (FSC) rating for sustainable wood products. Some are performance-based for the building as a whole, determined by testing to verify performance-standard compliance, while others are prescriptive, mandating design features of the building such as minimum insulation requirements. Most incorporate a combination of the latter two approaches. While building-rating systems are voluntary (as opposed to codes, which are not), there can be many benefits to using them, including incentives, marketing and publicity, and improvements in the design and performance of the building. If a building is to be subjected to a ratings evaluation, it is critical to identify this in the design phase and understand the conditions of being certified. Here are just a few of the many different ratings systems available for consideration.

LEED

Standing for "Leadership in Energy and Environmental Design," the LEED program features a series of different certifying tracks for different types of projects (categories include homes and health care) to promote sustainable building and development practice. The standard is built on a point system in which a certain number of points are awarded for implementing different features, such as construction-waste reduction or use of regional materials, in each of several key areas.

Since its launch in 2000, the LEED program has grown tremendously and been received enthusiastically by industry and the marketplace alike. One of LEED's greatest successes has been its power as a branding tool. The LEED program has greatly increased cultural awareness of, and appreciation for, the environmental impacts of our buildings and can be credited with popularizing "green" building in the mainstream, not unlike the success of the USDA's National Organic Program (NOP) in promoting organic food. Yet, as for the NOP, there are plenty of criticisms of the LEED program.

One of the chief complaints is the ability to "game the system" or "play the numbers" by crafting designs that optimize point accumulation, rather than optimizing systems integration, maximum potential efficiency, or any number of other more appropriate goals. Accordingly, LEED buildings can be built that are only marginally better, in some categories, than conventional projects. An example of this includes select LEED-certified buildings that were shown in post-construction commissioning not to exhibit substantial energy reductions compared to conventional counterparts. Another critique centers on the relative value of human health in the rating system, and the balance of the rating system in general. There is a very high priority placed on energy efficiency, which some feel serves to diminish the importance of other elements of a green building. Another criticism relates to the concept of simply "doing less harm," which is what LEED generally accomplishes, as opposed to "creating positive change" (a tenet of regenerative design). It is possible to be in a LEED building without ever knowing that it was built as a LEED building. While to some this is a distinct advantage, to those looking to see rating systems push the design envelope further toward a more sustainable future, this tendency builds in the potential for complacency in thinking that if certain materials and design details are incorporated, then we will have done "good enough" in "going green" and sleep with a clear conscience. This complacency will keep us from achieving more holistic changes to how our built environment affects its inhabitants and surroundings.

Love it or loathe it, the LEED program seems destined to heavily affect the future of green building for the next decade or more, and it will be increasingly relevant for consideration in the design of the natural building.

PASSIVHAUS (OR PASSIVE HOUSE)

Just as they brought us Bau-Biologie (see earlier in this chapter), so too have the Germans brought us the Passivhaus design movement, administrated by the Passivhaus Institute (PHI). While referred to as "Passive House" (the English translation, as used by the now-unaffiliated Passive House Institute US) in North America, the use of the German spelling helps to both distinguish the unique nature of the program and avoid confusion with "passive solar houses." The Passivhaus system challenges designers to create buildings with exceptional energy performance standards. The focus of the Passivhaus system is quite specific to this topic, and the program makes no apologies for this (in recognizing global warming as the major environmental crisis of our day, this is understandable).

The Passivhaus standards are performance-based. The requirements are that the building (1) be airtight to less than 0.6 air changes per hour at 50 pascals pressure (ACH50; see chapter 7 for description of this standard); (2) use no more than 4.75 kBtus of heat per square foot per year; and (3) use no more than 38.1 kBtus of total energy per square foot per year. The "passive" part of Passive House refers to the fact that these buildings are so energy efficient that they do not require the use of a full heating system; all the heat demand in the building can theoretically be provided by the ventilation system alone, perhaps with the additional support of an in-line boost heater. An online design tool helps greatly in crafting a design to achieve these very high standards (see the reference section for this chapter for the website address).

The Passivhaus movement has realized tremendous reductions in building-generated energy consumption and is by far one of the highest standards available for achieving energy performance. There are a few complaints, beyond potential confusion surrounding the name. For one, the use of the ACH50 standard penalizes smaller buildings, whose volume-to-shell area ratio is much lower than that for larger structures. For another, there is no scaling of the standards respective to climate, penalizing cold-climate buildings that must overcome harsher environmental conditions to meet the same standards as those in warmer climates (although there is a compelling counterargument that those in cold climates should not be allowed a greater degree of energy consumption just by virtue of their location). Finally, the reliance on use of industrial materials—in particular, foam—to achieve performance goals falls short of a comprehensive approach to sustainable design, given what we know about the impact these materials have in our environment, today and long into the future.

More and more Passive House buildings are being built in North America each year, a trend we expect to see continue as awareness about climate change and improvements in technology and design grow together. In 2011 the Passivhaus Institute (PHI), the original organization of the Passivhaus standard in Germany, severed its relationship with the Passive House Institute US (PHIUS, the former U.S. chapter of PHI), indicating further developments in the field of Passive House implementation in the United States in the years to come. Following successes in Europe, some American designers and builders are already working to incorporate natural building techniques into Passive House design. Mark Hoberecht of HarvestBuild Associates in Ohio has created two such systems, featuring both straw-bale and straw-clay wall systems in concert with blown cellulose insulation. As we continue to learn about the performance characteristics of these materials and improve our best common practices, we expect to see more such commingling between these two worlds.

LIVING BUILDING CHALLENGE

If a problem with the LEED program is that it does not go far enough to encourage changes in the way we design, build, and use our buildings, then the answer may lie in the Living Building Challenge (LBC). Developed (in 2006) and administrated as a project of the International Living Building Institute by the Cascadia Green Building Council (a charter chapter of the USGBC), the goal of the program is well identified in the name: to challenge individuals and communities to make transformational and fundamental changes in the creation of their buildings. The LBC is similar to LEED in the identification of several key areas of evaluation: site, energy, materials, water, indoor quality, and beauty and inspiration. A fundamental difference, however, is that the categories within each area are not point-valued—they are all mandatory requirements.

Some of the requirements are challenging indeed. For example, buildings must achieve net-zero energy and water usage. In the materials section of the guidelines, there is a "red list" of excluded materials such as PVC and halogenated flame retardants, which may be impossible to avoid; in those cases, documentation of written contact with manufacturers requesting development of products without use of the excluded materials is required for exemption.

This level of challenge has made it difficult for builders/designers to meet the program's performance standards for their projects. In October 2010, the first two projects were certified following analysis of a year's worth of performance data to ensure compliance. Criticisms of the program relate primarily to the difficulty in achieving the standards, by virtue of cost, logistics, or code prohibition (or potentially all three). That said, such criticism does support the intended goal of the program: to challenge what our built environment can be, and indeed must be, in order to realize a future built environment in balance with the natural environment.

CHAPTER 11

Before Construction

The design phase may start in a conceptual and creative place, but administrative and logistical considerations quickly come into play. In this chapter we will discuss some of the different issues that must be addressed as a part of the design process that, while not necessarily the most inspiring, are fundamental in guaranteeing the project's success, and can even be enjoyable—or at least satisfying.

Budgeting and Finances

The practice of natural building is closely associated with a mythology of being a uniquely inexpensive building process, through which one can construct a "dirt-cheap" building that is beautiful, features a high level of insulation, has a minimal environmental impact, and is fun and oh so easy to build. As we learned in chapter 10 in evaluating the golden triangle, one cannot have qualities of affordability, rapid construction, and quality all maximized simultaneously—a balance between the three must be sought. Taking a realistic look at the financial picture of a natural building is essential to dispel this mythology and ensure the development of an accurate budget—whether large or small—that is well matched to the design. While it can be very difficult to discuss money with clients—or, as an owner-builder, to maintain an awareness of your own financial situation—candor, frankness, honesty, and discretion are very important qualities to bring to the process. As designers and builders, our responsibility in this process is to understand and communicate the cost of our building processes.

HOW MUCH DOES A NATURAL BUILDING COST?

The question of cost is one of the most frequently asked questions we hear. It is often in reference to a "straw bale house" or "compared between straw bale and straw-clay." Our answer, which is often—and understandably—received as a cop-out, is, "It depends." This is not only a valid answer, however, but an important, nuanced one for those who are actually interested in receiving an honest answer. The reason that conventional construction is so simple to represent in an accurate cost-per-square-foot number is that the materials and assembly systems are so industrialized, and the requisite labor inputs therefore so standardized, that predicting cost is much easier—the same building, built again and again from the same standard materials, can be expected to cost the same amount (although cost overruns and mis-estimations are unfortunately extremely common in "standard" construction). Natural buildings, on the other hand, are anything but standard, and the wide range of differences from one project to another makes simple broad-stroke cost estimating even more difficult. Let's look at some of the factors affecting pricing of a natural building.

Percentage of Building

Many people think of a natural building as being defined by the wall system (straw bale, cob, adobe). In truth, walls represent only a small portion of the overall cost of a building. Trying to associate total construction costs with a characteristic of the wall construction is very misleading, and unhelpful in understanding how price is associated with design decisions.

Labor Costs

Our invoices for construction services generally break out to be 25% to 40% materials and 60% to 75% labor. Unlike industrial building systems, which carry substantially higher materials costs (reflective of the industrial infrastructure required for their production) and lower labor costs (a chief intention of industrial building), in natural building systems the materials are incredibly inexpensive—often free—yet make for a much more labor-intensive process; after all, the builder is also the manufacturer in a natural building project.

What this means is that seemingly minor changes in design can have very significant repercussions in cost by virtue of the amount of "fiddle time" required to execute the details. An example of this would be a top-of-wall detail. Option A entails running straw bales up to the underside of the rafters, plastering to a soffit, and having the wall cavity/roof transition area be insulated with the blown-in cellulose installed in the roof. Option B entails beveling straw bales to the angle of the roof pitch so as to be able to form a tight connection to an insulated roof panel, notching the bales to fit around the projecting exposed timbered rafters, and installing well-sealed air fins and plastering around the rafters. The difference in cost for straw bale and plaster work between these two options can easily be thousands of dollars.

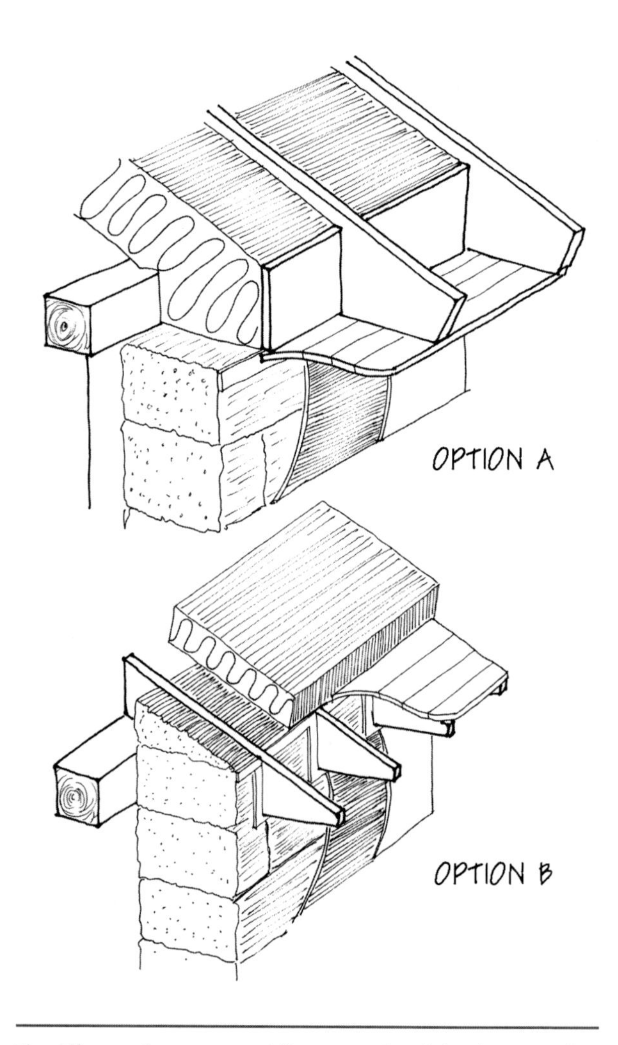

The difference between two different top-of-wall details can result in hundreds, if not thousands, of dollars in cost. ILLUSTRATION BY BEN GRAHAM.

Site Considerations

Another major cost driver is site logistics. This is not unique to the world of natural building, but it is amplified by both the "manufacturing" side of working with natural materials and the unique set of logistical needs that may not otherwise be anticipated by those accustomed to more conventional forms of building. A well-protected building site (needed to keep sensitive materials such as straw dry from storage through modification to installation) takes a good bit of forethought. For plastering to be efficient, the site must be well arranged to keep the movement required for processing, mixing, and transporting to the wall as efficient as possible. The plastering process itself involves having complete access to every surface of the wall, including not only a place to stand, but a setup for waist-high mortar boards directly behind the plasterers. This requires pipe staging or a comparable setup around the perimeter of the building on a solid and even grade—not necessarily an easy thing to accomplish on a large building.

Proper scaffolding setup is one of many details that need to be provided for an efficient and cost-effective plastering process. PHOTO BY ACE MCARLETON.

Failure to provide adequate setup has a direct and notable financial repercussion; we saw a 30% increase in cost for exterior base plastering as a result of notably poor site preparation on a recent project, which could have been avoided.

Nonstandard Construction

Our company may be unique in this regard, but in general the "trade" of natural building is particularly notable for the gray area between "contracting" and "consulting." By this, we refer to the extent of involvement of clients, their friends/crew members, and their community in the execution of the work in addition to or instead of a hired crew. It is rare for a general contractor to work directly alongside a client or to allow a client to reduce the cost of the project by sweeping up, organizing the site, and making runs for materials; it is even rarer to find this option with a subcontractor. By contrast, in nearly every project we do we include the client and/or the client's friends, relatives, or colleagues in our "crew" to some capacity. Estimating for unknown amounts of labor input by people with whom we have never worked is a difficult task, indeed. Here are three ways that we will commonly work with clients, each of which involves a different pricing structure:

- *Consultancy.* We charge an hourly fee as consultants to help train and oversee owner-builders or other contractors and their crews to execute the work. Sometimes this involves a visit or three, sometimes it involves full-time supervision.

- *Consultancy/contracting with support.* It is very common for us to work as contractors with our own crew and have the owner-builder or contractor working with us, prepping ahead, cleaning behind, or even working directly alongside us, which expands our work in supervision and instruction.

- *Full contracting.* There are some clients who prefer that we execute the entire work ourselves, from scratch to finish. This makes estimating labor a bit easier—given all the other variables to reconcile, as mentioned above.

If you are a contractor, you should consider the extent to which you are willing and interested in working with clients who want to be engaged with your professional practice. While we find it very rewarding and enjoyable to work with our clients in these different ways, this is far from a universally held opinion, and developing your own boundaries is important. For clients, consider carefully how you would like to be involved, and if you desire a more hands-on role, make this very clear to your contractor up front. You may have to search around a bit for a contractor who is willing to work with you in this way. As in any collaborative process, clarity of expectations and division of roles and responsibilities is critical.

REAL NUMBERS

It is all well and good to have an understanding of the variables that make pricing difficult. But how do we get from there to a closer picture of how the numbers play out? In our cost tracking, we have found that contracted natural buildings tend to cost anywhere from $150 to $300 per square foot. The smaller number represents a very simple design with efficient (read: inexpensive) details that involves subcontracting of key features, but that is largely an owner-contracted and owner-built project, with no value assigned to owner labor (note that this number can certainly go down as the subcontracting is reduced). The larger number represents more complicated and intricate details such as exposed rafters or interior timber frame or irregular shell design, decorative elements such as niches and tile inlays, precision aesthetics such as plumb walls and highly accurate window and door reveals, and finicky transition detailing.

The level of detail and complexity in a wall, including decorative inlays, niches, and edge transitions, has a significant effect upon cost. PHOTO BY KELLY GRIFFITH.

TABLE 11.1. COST OF CONTRACTED STRAW-BALE-WALL COMPONENTS

Building system	Cost per square foot of wall
Air-fin installation, straw bale installation, and plaster preparation (not including straw bales)	\$3.70–\$8.00
Rough plaster mixing and installation	\$5.50–\$9.00
Finish plaster mixing and installation (not including paint)	\$3.40–\$9.50
Total wall costs (for two-coat system)	\$12.60–\$26.50

Note: The price range reflects broad diversity in design detailing, level of finish, site logistics, and client involvement. Description of plaster preparation can be found in chapter 17. Rough plaster is clay-based, applied approximately ¾ to 1 inch thick. Finish plaster is clay- or lime-based, applied approximately ¼ to ½ inch thick.

Source: New Frameworks Natural Building, LLC

Following similar standards of range, we can look at the costs of our wall system components as case studies for cost analysis; see table 11.1.

We have found other builders' estimates for straw bale construction and plastering in our region to be comparable. As can be seen by the range, the quick answer "it depends"—as vague as it may be—really is the most accurate. Once developed plans and a clear plan for execution are in place, these numbers can be more accurately predicted.

We are often asked to make a whole-house comparison between a natural building (often straw bale) and a conventional one. Making a true apples-to-apples comparison is incredibly difficult as there are so many variables. For example, a straw bale wall can be built to a timber frame, a conventional stud frame, or a hybrid or innovative frame, or it can be structural unto itself. A natural wall system may dramatically affect electrical layout and installation, or it may not affect it at all, depending on minor tweaks of the design. The footprint may increase to accommodate a thicker wall, but extra framing (as for a double-stud wall) may be avoided in the case of straw bale walls. Ultimately, we have found that if cost comparisons between natural building options and conventional options are so important as to need to chase these variables down to conduct an "accurate" comparison, there is a strong likelihood that the other values of natural building are not being weighed heavily enough, and a more conventional and predictable approach is likely to be in that client's best interests (see discussion of value in next section). For a closer look at a budget breakdown for one owner-built straw bale project, please see appendix E.

VALUE

No conversation about budgeting and costing is complete without a discussion about value. One can track and compare staggering amounts of financial data, but without a clear understanding of the value received for any given cost, the information is useless. Here are a few important considerations in assigning value in the context of natural building.

Value in Process, Not Just Product

Natural building is unique from most other forms of building in that there is an inherent value not only in the product—the building, built well, on time, on budget—but in the process as well. It is imperative to understand that, in the world of natural building, there is real worth in the experience of working with natural materials—from their harvest and procurement to their manufacturing on-site into usable building materials to their final installation in the building. It is because of this recognized, process-oriented value that a culture of bale raisings and plaster parties has remained strong (as has the popularity of natural building among the owner-builder community), even as the professional trades of natural building have expanded in the last few decades.

Natural building can be an exercise in community building when the community builds the walls of a home. PHOTO BY BENJAMIN RAY GRIFFIN, GREENER LIVING VERMONT.

Value of Craftspersonship

There is little craftspersonship required or displayed in a painted drywall wall. There may be terrific workpersonship, especially in a challenging remodel environment. But craftspersonship—the magic combination of skill, passion, and artistic sensibility—is difficult to achieve in the world of industrialized building these days. There was once a time, before the industrialization of our built environment, that craftspersonship was a value more commonly held in our culture and reflected in our structures. The commodification of buildings in the 1950s changed this, however. Often reserved for expensive custom woodwork, tiling, or other finish features, craftspersonship today is frequently relegated to the buildings of the wealthier communities of our society. As a hand-built endeavor, the natural building offers opportunities for craftspersonship and the beautiful mark of the human hand in the finished product as part and parcel of the building's form. This is not to say that natural buildings are inherently craftsperson-quality products—remember, skill is a critical ingredient. Rather, natural buildings readily hold the potential for craftspersonship—a value that often carries a hefty price tag in conventional buildings—by virtue of the refinement of raw natural materials into a finished building.

Triple Bottom Line

Emerging from progressive financial circles is an economic standard that has been gaining widespread acceptance as being more comprehensive and just than the conventional profit-based standard: the triple bottom line. We are all aware of the "bottom line," in which profit dictates viability of a program; the triple bottom line considers people and place in addition to profit. "People" refers to the value of social equality, such as fair trade measures. "Place" refers to the value of ecological sensitivity, such as low embodied energy or a small toxic footprint. "Profit" is thereby placed within an appropriate context when assessing the value of a given process or product. Acceptance of this new form of analysis greatly supports what we have known for a long time: the value of natural building extends far beyond its potential as a cheap alternative to conventional constructions and is held as well in its power to change lives and heal the impact we have inflicted upon our world (Savitz 2006).

Assessing Resources

A comprehensive analysis of available resources is very helpful when compiling a budget and developing a design. "Resources" come in many forms—financial, physical, informational, social—and having a clear picture of them may well provide opportunities to reduce the overall cost of your project.

FINANCIAL: CASH AND DEBT

Beginning with a keen understanding of your financial picture is an important place to start. This is not a time to be optimistic. Be conservative when assessing your finances, both cash (or things that can become cash) and debt (your ability to carry a debt load). If carrying a loan is required, there are a variety of different lending agencies beyond the national chain banks. Alternative lending agencies such as credit unions are potential loan sources for less-standard projects such as natural buildings, in that often the loans they provide are not repackaged and sold on the national market, allowing them greater flexibility. Affordable housing agencies and community development organizations can also help secure or provide loans. Beware the terms of the conventional construction loan, if you do choose to go that route—restrictions on the design and utilities of the structure, as well as on the time by which the building must be completed, may prove to be unacceptable for your project.

Using salvaged materials, such as this metal roofing, can be integrated in form and function into a building while reducing the waste stream and saving cost. PHOTO BY NICK MOREHOUSE.

PHYSICAL: TREES, DIRT, ADAPTIVE REUSE

One of the hallmarks of natural building involves the incorporation of abundant regional natural materials into the building. To this end, having a strong understanding of what resources are available to you in your region is critical to be able to successfully incorporate them into the building. This survey should begin early on in the design phase, so as to be able to best reflect what your region has to offer. For example, here in the Northeast, we have an abundant wood resource; this may incline us toward construction of a wooden frame. If you are interested in pulling this wood off your property, you must be sure you have the trees to support this—timber frames require large timbers. Even if you do have the timbers, you need to make sure that their harvesting will not lead to high-grading of the forest (removing all of the larger, higher-value trees), leaving it weaker and more susceptible to damage from storms. If this is not an option, then you will need to either source a mill in the region for your timbers or perhaps switch to a stud frame or structural masonry or straw bale wall to take advantage of the lumber that is available to you from your property. Clay-rich soils, animal manure, stones—there are many opportunities to access physical resources from a site and its surrounding region.

Salvaged materials are a great option, as their use preserves the ethics of natural building when sourcing products that cannot be readily produced by natural materials. Because of the highly customized nature of natural building, it is easier to incorporate salvaged materials without an additional cost, compared to conventional construction models in which any nonstandard material comes with a price for customization. It does take a keen and patient eye to buy quality salvaged supplies. While screaming deals are certainly there to be found, be prepared to pay for quality. Classified ads, reputable salvage yards, and flea markets and garage sales are all potential sources for building materials. Just remember: *caveat emptor*! If performance is mission critical—such as with a window—holding out for high-quality salvaged materials is important; this might be a good place to buy new, if energy efficiency is a high priority. In the end, all the cost savings of a cheap window or door will be quickly lost as a result of increased operational costs (heating or cooling) if quality can not be ensured; we found in our research project in which we conducted thermal performance tests on seven different straw bale buildings in the Northeast that a structure incorporating all salvaged windows suffered from high levels of air leakage in the window units themselves. Taking a risk on cabinets, however, has lower stakes.

INFORMATIONAL: EDUCATION, RESEARCH, PUBLICATIONS

Whether you are an owner-builder or a professional, taking advantage of the breadth of published resources that are easily accessible in this digital age can provide information in the comfort of your own home (via a computer) that will empower your incorporation of a new technique, material choice, or design approach—and don't forget the resources available at your local library. See the references and resources at the back this book for further reading. Additionally, experiential information can be gleaned from a variety of educational institutions. The Yestermorrow Design/Build School of Waitsfield, Vermont, where we are instructors, is a terrific example of a school that offers a wide array of curricular options, including courses on many of the topics covered in this book.

SOCIAL: SWEAT EQUITY

Because natural building is such a labor-intensive practice, there is a tendency to devalue sweat equity—the labor you put into a project to offset cost. This labor has great value, however, and must not be wasted. One way to value sweat equity is to draw an equivalent to earning potential. For example, if you earn a good salary for a job you love, taking lots of time off from this job to do work that could otherwise be done by an inexpensive laborer does not play out in cost-benefit analysis, unless there is another compelling reason to engage in this work. In this situation, the ability to repay a monthly loan is strong, and therefore your debt resource may be higher. On the other hand, if you do not have a steady, well-paying, or fulfilling career, then it may be a better investment to do more of the work yourself, increasing skills, experience, and portfolio work in the process. Debt-carrying capacity is lower, so a pay-as-you-go strategy may make more sense.

It is common in labor-intensive, community-oriented natural-building projects to incorporate lower-skilled volunteers, work traders, or labor exchangers. Indeed a bale raising or plaster party is a terrific opportunity to bring members of the community into the project. That said, you often get what you pay for with labor, and having a whole bunch of volunteers on-site with no training, who have no greater incentive than their own interest to take ownership in the work, can often create as much work in oversight and fixing mistakes as it saves. If you are bringing volunteers into the project, the goals of their inclusion should be clearly stated. Is the goal to save money? If so, they should be selected very carefully, given specific tasks (often menial and repetitive), and organized in such a fashion so as to be efficiently supervised (primarily by reducing their numbers and minimizing turnover). Is the goal to spread the word about natural building and bring the spirit and energy of the community into the project? In this case, keep the production expectations very low and the complexity of the project even lower, and be prepared to manage a lot of people who are there primarily to enjoy themselves and learn what the project is all about. Both of these goals are valid. The important thing is to make sure they are clear, because they are often in conflict.

The decision to be an owner-builder or owner-contractor is a huge one, and there are many, many considerations to evaluate when making this decision. A gray area often exists between being an owner-builder and an owner-contractor, and navigating the distinction between these two roles can also be challenging. In this book, we assume that you have heavily considered this decision and have chosen to go forward as an owner-builder or owner-contractor. If you are not so convinced—or even if you are, and feel confirmation of your decision is justified—refer to the suggested further readings in appendix A for some excellent perspectives to ensure this path is indeed the best one for you.

Codes, Permitting, and Insurance

Building codes are often seen by designers and builders as an obstruction to overcome in making progress on their project. A healthier and more productive approach may be to approach codes respective to their intent, which is to create a set of guidelines for safe and appropriate construction. While it is true that codes can be onerous and frustrating, especially for innovative and developing technologies that have yet to gain full acceptance in the code community, they do keep a lot of people safe and help avoid a lot of building mistakes. Certainly the more restrictive the code, the more likely conflicts and problems are to occur with innovative practices. Jurisdictions without mandated building codes provide ample opportunity for unfettered exploration—as well as for the building of poor-quality structures. Ultimately, most of the potential ease or difficulty lies with the proclivities of the code official, who has the ability to take an ignorant or narrow-minded interpretation of the codes or to invite an open-minded and logical discussion into the application process.

Codes vary widely across North America. While there are national standards in the United States and Canada, they are further refined and enforced by state or province, county, and/or municipality. This jurisdictional specificity is responsive to circumstances such as seismic conditions and climate that vary from region to region (e.g., California's seismic engineering requirements, and frost-depth minimums for certain foundations in New York). Accordingly, a design that is compliant in one state may be well out of compliance in another. Some states, such as Vermont, have no statewide or county building codes, and therefore the only building code jurisdictions that must be respected are those of the few cities in the state (the state's towns do not have building codes). Vermont does, however, have a mandatory statewide energy code that must be considered in the design, construction, or renovation of a building.

Code inclusions for natural building technologies vary widely as well within established codes. In the case of both earthen (such as adobe and rammed earth) and straw bale, there are regions of the United States, such as the Southwest, that have specific code inclusions that mandate compliance with prescribed code regulations. This has the benefit of ensuring successful permitting of a project, providing that there is full compliance with the code. The obvious downside is that restrictive code requirements force the hand of the designer and builder in uncomfortable ways; a clear example is the provision in New Mexico's straw bale code that restricts the use of load-bearing straw bale construction. Many other jurisdictions have guidelines for appropriate construction practices in their codes that may influence conversations with code officials but are not requirements for compliance. A few restrict the use of certain techniques entirely, while most make no mention of them whatsoever.

The majority of jurisdictions do not have a full restriction or a formal required code and instead feature a total absence of code requirements for natural building. This means that conversations with code officials can be expected upon submission of the permitting application. Herein lies the heart of the process of gaining permitting approval for a natural building project: developing a strong relationship with the code official. Behind all the rules and regulations, there is a person with whom you as an applicant will be working to understand how the project can navigate the code requirements of your jurisdiction; the nature of this relationship—adversarial or amicable—will greatly affect the outcome. Here are some tips on how to work with your code official:

- *Know the code.* Take the time to familiarize yourself with the code for your jurisdiction by contacting your local building department. Having a strong understanding of the language and substance of the code will not only help you understand the parameters of your design but enable you to have an intelligent and well-informed conversation with your code official, which is valuable to your success.

- *Begin a conversation.* In chapter 10, we discussed the integrative design process, which centers on the inclusion of all major parties in the design process to allow real-time response to concerns and inputs from all vantage points. Include the code official in this process, and start early so his or her concerns can be understood and actively addressed in the design phase, rather than after the design has been completed. While a code official does have power in the relationship, approaching the interaction by engaging in proactive dialogue in a language he or she can understand is by far the most effective strategy in presenting your case.

- *Create good plans.* Drafting is the visual language of code officials, and having a complete, well-detailed, and well-notated set of drafts will keep ambiguity at bay. The greatest concern of the official is the unknown, so graphically identifying as many of the details surrounding unfamiliar techniques as possible will help the official be more comfortable. Understanding the code and the official's concerns enables you to make inclusions in the drafts that point out your intention to build a code-compliant building and make it easier for the official to say yes.

- *Cite precedent.* Being able to cite existing precedent—ideally within the same jurisdiction, or in regional jurisdictions—can go a long way to help demonstrate viability of the technology. An existing case study can prove the effectiveness of an unfamiliar building modality. If you can find another code official supportive of the natural building technique you are using, request his or her permission to be used as a resource when communicating with your code official.

- *Get stamped.* You may well find yourself in a jurisdiction that requires the stamp of an architect or an engineer. If you are not already planning on working with one of these professionals, then understanding whether or not this requirement applies to you will be especially important. Connecting with a professional who has experience in working with natural building technologies will be very helpful in ensuring that you not only receive the stamp but also receive positive feedback and suggestions for design improvements from an experienced professional.

- *Educate and explain.* Be prepared to answer many questions concerning your project from a skeptical code official. To be able to influence your official toward finding favor with your plans, you must have done your research and be able to provide not only good explanations but strong studies or sources that back up your claims. Refer to the resources for part 2 of this book for information concerning structure, thermal performance, moisture, and fire prevention.

- *Have a positive attitude.* As we mentioned at the beginning of this section, it is very important to approach the conversation with your code official not as an adversary but as a partner. Her or his job is to ensure that buildings get built to the code specifications of that jurisdiction, not to prevent buildings from being built. Your job is to help make that happen by making your project easy to approve. Expect to be met with skepticism and doubt. Rather than acting defensively, use this opportunity to learn as much about your official's concerns as possible, so that you can continue to provide helpful information and refine your design appropriately.

- *Be persistent.* An initial rejection is not the end of your project. Continue to work with your code official to try to find ways to answer his or her concerns to reach a successful outcome. If this is not possible, discussing your case with more senior officials in the building department or within the municipality or state government may help move your application forward. A formal appeal process may ultimately be necessary; at some point, enlisting support from a lawyer or consultant who specializes in building-code appeals may be a worthy investment to help usher your project through an onerous process. Research all the steps of the appeal and how the process works, and be sure to prepare yourself accordingly to defend your case well. A legitimate plan must be approved before you can start building.

Other Permitting Considerations

Code compliance of your building design is only one permitting hurdle to leap. Depending on whether you are in a rural or urban environment and on the regulations of your region, you will need to secure a host of other permitting approvals and tackle a series of other administrative logistics to move toward construction.

- *Local zoning* ordinances are designed less around the nature of the structure and more around development patterns for the region; considerations will range far and wide, including height of the building, location on the site respective to the edges of the property, and type of use (residential, commercial, industrial), among many others.

- *Energy codes* are in place in many states and may or may not be tied to a building code, if one exists. These codes are frequently a balance of prescriptive and performance requirements, and they cover many parts of the building's design, including insulation, air barrier, combustion equipment, ventilation, window and door performance, and more.

- *Septic/wastewater* permits will likely be required, if hookup to a municipal sewer is not available. This requires digging test pits on the property to identify suitable soils for supporting a septic system. In Vermont, statewide mandatory septic regulation is far more restrictive than building codes.

- *Water-supply testing* for a private well or spring may also be required; some jurisdictions may not allow accessing water from springs, requiring a drilled well.

- *A clean property deed/title,* beyond being a smart idea for your assets, may be a requirement for permitting.

- *Establishing a new access* such as a driveway will likely require separate approval from the road commission or other similar authority, frequently as a component of the zoning process. If legal rights-of-way need to be secured to allow access to your site, this adds an additional layer of complexity (and often expense).

- *Environmental sensitivities* in certain areas may affect permitting of your site in the zoning process, whether mandated on a municipal or state level. Examples include high elevations, steep grades, and wetlands.

- *Be patient.* Approval will take time—days, weeks, even months. Code officials are not likely to respond favorably to feeling pressure to comply with your construction schedule, so be prepared to summon extra stores of patience to methodically follow through the permitting process in the time it takes. After all, there is little else you can do but be patient.

There are many different efforts under way to bring natural building techniques into building codes. At the time of this publication, a section on straw bale construction—including plastering—is working its way through the process of inclusion into the next version of the International Building Code (IBC). Adobe construction—referred to as "unfired clay masonry"—is already included in the Uniform Building Code (UBC), a national building code in the United States. Organizations such as the Development Center for Appropriate Technology of Tucson, Arizona, have been working tirelessly to remove barriers to code acceptance for natural and green building techniques in the building-code-development community.

All of this work will ultimately make it easier for projects to receive permits, without the required expense of an architect or engineer's stamp. It also means that there will be a greater chance that further development of these building techniques will be stymied by outdated and poorly crafted code mandates. In their book *More Straw Bale Building: A Complete Guide to Designing and Building with Straw*, Magwood, Mack, and Therrien state, "While codification [of straw bale buildings] could make approvals simpler to obtain, it also poses the risk of freezing the technique before adequate experimentation leads us to sound standard practices." They point out that the current prescriptive nature of most natural building codes—mandating specific construction practices—will someday yield to performance-based codes, allowing greater flexibility in application to achieve the same results (Magwood et al. 2005, 148–49). As more of the double-edged sword of codification's blade is exposed, we as natural builders will do what we've always done best: adapt to the circumstances in which we are working. In the meantime, we will be continuing to advocate for well-worded and flexible code language, and enjoying the freedom currently offered us by a relative lack of code standards.

Homeowners' insurance is another logistic that can be more difficult for the owner of a natural building than for the owner of a conventional one. Ultimately, the same basic considerations for financing and permitting apply to the world of insurance. Insurance agencies are a risk-averse lot, and fear of the unknown is a big issue for insurance agents when considering alternative construction practices. While it is critical to be fully honest and forthright concerning the nature of the building, this may not be the correct audience with whom to tout the innovative and experimental techniques employed in the building. If concerns arise, be prepared to cite precedents in your region; find other insured buildings in your community or region as examples. Providing solid documentation addressing particular concerns is also helpful, and doing so in a respectful and constructive manner, as with code officials, is an important tool to gain acceptance from your insurance agent.

CHAPTER 12

Foundations for Natural Buildings

Given that the foundation of a building is the first element to be constructed, as well as the interface between the structure and the earth that supports it, its importance is difficult to overstate. After all, an oversight in something as seemingly insignificant as the footing drain can cause major, long-term damage—aesthetic and/or structural—and fixing a problem eight feet underground below a house is by no means a simple proposition. It can also be quite terrifying for the owner-builder—after all, you are essentially pouring tens of thousands of dollars into a hole in the ground, and if you are off by a few inches in the design or execution, the ramifications can be severe throughout the rest of the construction process. Therefore it is hardly surprising that when it comes to innovative options for a foundation system (or even proven alternative ones), owner-builders and contractors alike are quite conservative about taking risks. This is understandable, yet unfortunate, as the cold-climate convention of a poured-concrete, stem-wall basement is by no means a one-size-fits-all solution, nor is it often the best option even when a basement is a desirable feature of the structure. In this chapter, we will begin by examining what the roles of the foundation are in a structure; this will provide us the background needed to then look at the different foundation options that are best suited to a natural building.

In order to evaluate the efficacy of a given foundation system, it is important to first identify all of the things that a foundation accomplishes for the building. Listed in the next section are some of the functions a foundation can serve depending on the nature of the structure. Not all of the foundation systems we will look at will serve all of these roles, nor are all of these roles relevant in every building. Therefore, evaluating the foundation component of the structure with a comprehensive understanding of how it fits into the building and the site as a whole is crucial.

Functions of Foundations

A building's foundation is important for:

- *Anchoring the building in a stable, secure fashion to the ground.* Part of a good foundation's job is ensuring the structure sits upon undisturbed or fully compacted soil that is suitable for the loads it is asked to support. An understanding of the structural properties of the soil is necessary for the proper design of the foundation, as a sandy soil will behave differently than a clay-rich soil. A sandy soil may be more prone to settling, whereas a soil that is very rich in clay may be susceptible to expansion and heaving in the presence of elevated groundwater levels. Foundations built on a slope have additional considerations in their relationships to the ground, including lateral forces of earth against bermed walls and stabilization of bank and erosion of slope.

- *Distributing structural loads safely to the ground.* In designing the structure of a building, the load paths of the roof, the building itself, any particularly heavy elements such as a masonry chimney, and any other loads that gravity is tugging upon that could affect the structure should be mapped to the ground, or they will be borne by parts of the building that might not be designed or equipped to carry them. Therefore, a function of the foundation is to receive these loads and spread them safely onto the ground.

- *Establishing an accurate footprint of the structure.* If your foundation is not built to the correct size for the structure, at best you will be looking at a big inconvenience in figuring out exactly where the posts or the frame walls are supposed to be located, and how they are to be supported. A foundation that is not plumb, level, square—or if its dimensions are incorrect—will result in exponential challenges if not caught and remedied straightaway.

- *Helping to tie the structure together at the base.* Not all frames require the use of a "structural foundation" to help tie the building together, but some do. For example, the timbered sill of a traditional timber frame in many cases was both a component of the foundation and a structural element of the frame.

- *Protecting vulnerable parts of the building from ground or surface water.* Water is a major factor in why buildings have problems, and a significant source of water intrusion into structures is from belowground or along the top of the ground. Both of these sources need to be properly mitigated in the design of the foundation, both for the integrity of the foundation itself and to prevent moisture damage in the structure as a whole. Not only is degradation due to moisture an issue, but so is hydrostatic pressure of groundwater pushing against, for example, a basement wall.

- *Providing an interface between the earth and vulnerable parts of the building.* Running walls too close to the ground will invite moisture damage, insect infestation, and other problems. Some foundations, like stem-wall foundations, can be built to rise up above grade; others, like shallow, frost-protected slabs, may not be able to accommodate this need without additional measures, such as moisture-resilient cladding along the bottom of walls sitting on the slab, close to grade.

- *Resisting movement from frost heaving.* This point is of particular importance for cold climates. The pressure exerted by the expansion of water as it freezes is tremendous—9% by volume with a force of 150 tons per square inch (Velonis 1983)—enough strength to simply push a building out of the way. Freezing water can (and usually will) fracture concrete, split timbers open, and knock buildings out of plumb. Choosing a foundation system that will render the structure immune from the forces of frozen groundwater is critical in cold climates, and it must be designed to accommodate the depth of freezing of a given region, known as the "frost line."

- *Creating interior space.* Below-grade space enclosed within a foundation system can be very economical, particularly as you push further north. If you are already pouring frost walls (concrete walls designed to go below the frost line in the soil to resist uplift from frost heaving), simply digging another couple of feet down and excavating the interior space can provide an entire extra level of functional space within the structure. Finishing, insulating, and heating or cooling basement space can be complicated, and there are many options, so this should ideally be designed before the foundation is constructed.

Key Considerations of a Foundation System

There are a number of issues to consider when evaluating the suitability of a foundation for a natural building. Though the standard concrete-walled basement may be appropriate, there are also reasons why it might not be. For starters, we are trying to keep our impact as light as possible while still adequately providing a functional structure. This means that digging out a full cellar hole when a simple trench might otherwise suffice for the building results in a lot of excess excavation, site disruption, diesel consumption, and carbon emission. It also means that a whole lot more cement is being manufactured, which as we discussed in chapter 2, has a relatively high embodied energy and significant toxic footprint and should be minimized in its use. The insulation of a standard concrete stem wall is generally a 2-inch piece of foam glued to the outside walls. However, if thermal performance is a criterion of the foundation system, this standard should be raised to meet the performance expectations of the structure as a whole, taking into account the amount of insulation, the type of insulation, and the integration of the insulation into the overall system (i.e., placement, protection). Regardless of the form of the foundation, when it comes to natural buildings, here are the key considerations.

CONTINUITY OF THE THERMAL ENVELOPE

We discussed in chapter 7 the concept of a "thermal envelope" and its importance as part of a strategy for energy-efficient design. We generally conceive of this "envelope" as wrapping around the walls and roof of buildings, to keep the cold air out. While this is indeed where the majority of the heat loss in the building occurs, it is just as relevant to ensure the envelope wraps entirely around the foundation, as well. As mentioned above, in the standard northeastern basement design, this envelope generally consists of 2 inches of extruded polystyrene (XPS) foam (nominal R-value of 10) affixed to the concrete walls after the forms have been stripped and the walls dampproofed with tar or wrapped with a moisture-resistant barrier (MRB). This is usually where the insulation ends; the footings and most frequently the floor slab are not insulated, nor are the transitions up to the rest of the building, such as the sill plate (more on this later in this chapter). One strategy to make sure the thermal envelope encases the entire structure, including the foundation, is to continue this plane of foam insulation under the floor slab (sub-slab insulation). Providing the insulation board is of suitable compressive strength to handle the load, it can be used to wrap the footings as well, thereby maintaining an unbroken thermal envelope below the foundation. Please note that super-insulation strategies that involve sub-footing insulation may have significant structural implications, especially in seismic areas, and such details must first and foremost be structurally sound. Consult local code provisions or the services of an engineer if in doubt.

If reduction of the use of foam insulation is a project goal, switching to an alternative insulation material may be a good option; it should be noted, however, that the insulation chosen must be suitable for below-grade installation, and if it is to be placed under footings, it must be structurally stable. Mineral wool board is an excellent option. Mineral wool, which is found in a wide variety of rigid board, batt, or pipe forms, is made of basalt rock, mineral slag (postindustrial mineral waste from the steel and copper industries), and binders such as phenolic resins. Mineral wool board offers exceptional fire resistance (perhaps not as important below-grade) and an R-value of approximately 4/inch (slightly below XPS foam board, which is approximately R-5/inch). High-density products are available for sub-slab insulation. Mineral wool board has outstanding moisture-resistance properties: it has a very low water-sorption rate; does not rot, mildew, or deteriorate in the presence of water; frequently acts as a drain board, encouraging water to drop to a drain pipe (when installed vertically); and can be designed to perform as a vapor-resistant barrier. An additional benefit is deterrence of pests, such as mice and termites.

Regardless of what is used for insulation material, maintaining the integrity of the thermal envelope is the most important consideration. To achieve a high-performance energy-efficient foundation system, an unbroken plane of insulation should decouple the foundation—and the interior climate with which it is associated—from the exterior conditions of either freezing air or, at best, approximately 50°F soil temperatures to minimize heat loss from the building. If you are designing for a heated slab, this sub-slab insulation is all the more important to avoid heat loss from the slab into the ground or air along the slab edge. It is one thing to plan to install insulation boards to a wall after it has been poured. It is another layer of forethought to lay down sub-slab insulation before pouring the floor slab—which is generally done after the wall pour in standard basement construction. An even greater degree of planning and design is necessary, however, to insulate entirely below the footings. Not only does this involve the complexity and difficulty of incorporating insulation at the earliest stages of formwork for the foundation, but it adds the logistical pressure of working with subcontractors—a common feature of many foundation phases of construction—to ensure design compliance and quality execution, if you are not installing the insulation yourself. This also speaks to the need for additional design detailing to find the appropriate insulation type and quantity and ensure that it can be installed in an efficient and effective manner that does not compromise structural integrity.

So why bother? Knowing that significant heat loss can occur in uninsulated foundations, that the embodied energy and cost of insulation are easily recovered within its life span of service in the building, and that retrofitting a foundation with insulation is very difficult, designing for contiguous insulation in the foundation is well worth the effort.

Footings are only one common thermal bridge in foundation design. Making the transition from the foundation to the rest of the building—be it via stud walls, a floor box, or toe-up for a natural wall system—is another area where breaking the insulation plane is all too easy to do. In standard construction, a sill plate is commonly bolted to the top of a foundation wall, upon which a floor box is built, which in turn supports the walls of the house. This presents multiple transitions within a very small area, and therefore multiple opportunities to lose insulation continuity. Knowing how these transitions are made is important; see appendix B and the figures in this chapter for examples of foundation-to-wall transitions.

Another consideration regarding insulation is where it is located within the foundation system. Let's go back to our standard basement foundation. Again, the norm is to place 2-inch foam board on the exterior of the building. This is an effective strategy in terms of insulation placement for performance's sake: keeping the body of mass of the concrete walls within the insulation plane of the structure is a good plan. For longevity and durability of the foam, however, issues arise concerning physical impact, UV degradation, and damage from mice and insects, if the foam is not adequately protected. Another strategy is to place the insulation on the interior, or to build an insulated interior wall. This may help protect the insulation from the elements, but not only is the mass pushed outside the conditioned space in the building, moisture issues can arise as a result of condensation forming on the exterior (cold) side of the insulated wall and being trapped between the concrete wall and the insulated wall. Other insulation strategies such as insulated concrete forms and "sandwich walls" will be discussed later in this chapter.

ENHANCED PROTECTION OF WALLS

In addition to attending to the foundation-to-wall (or floor) details for the sake of continuity of insulation, it is also important to design this transition to ensure adequate protection of the moisture-sensitive wall system. Remember from our list of foundation functions above that one of the roles a foundation plays is to provide an adequate interface between the ground and vulnerable parts of the structure; this is where that function comes into play. A few components of this functionality may include raising the building an adequate height off grade, providing a capillary break between the ground and the building, and attending to appropriate alignment of planes and flashing details. In the world of natural building, we generally don't have the luxury (or liability!) of running vinyl siding down to within inches of the ground. Were we to attend to our exterior plastering details in this manner, we would see a much greater maintenance pressure upon the building than if we were to modify the design to terminate the plaster, say, 24 inches off grade—our normal practice. That said, many natural buildings employ wood siding or even cement composite siding installed as part of a rain-screen system protecting a natural insulative wall. These provide other opportunities for making the foundation-to-wall transition such as lapping over the edge of the floor box or even the top of the foundation wall.

If we are using a foundation system such as a slab-on-grade or rubble-trench foundation that does not inherently provide stem walls rising an indefinite distance above grade, then we must design a different protection mechanism into this transition. It could be masonry, such as a block or poured stem wall atop the trench or slab; quite often this detail is addressed by a built-up floor box, knee wall, or other carpentry-based solution. We will discuss this further in the sections of this chapter to follow.

WALL THICKNESS AND LOCATION OF LOAD PATHS

One of the trickiest parts of natural building design can be the different ways of associating or dissociating insulation and the structural, exterior, and interior planes in the building. A classic example of this is the straw-bale-wrapped timber frame structure. Let's say a timber frame is exposed to the interior space. Add 21 inches of wall thickness (18 inches for bales on flat, plus 3 inches for interior and exterior plaster). Now, let's say you want to hang exterior siding to protect the walls, so add another 4 inches for clapboards and custom rough-milled 2 × 3s. Now we are looking at an exterior plane that may well be a solid 2 feet past the frame, which is where the structural plane is located. If you have a foundation that is oriented respective to the "frame line," you may find that your structural bearing points—the footprint of the timber posts—are landing on little more than a nonstructural slab or unsupported floor. The same condition can exist in a double-stud wall. If you have a 16-inch double-stud wall infilled with, for example, woodchip-clay, it is essential to understand which of those two walls, or a combination of the two, are supporting the structural loads of the building—and to then design your foundation walls and footings to carry this load safely all the way to the ground.

A number of strategies can be employed to assure this is accomplished, depending on the foundation type. Fundamentally, all load-bearing points must be fully supported by adequate footings. In a post-and-beam structure, it is quite common for there to be load-bearing interior walls or posts to carry the loads from the first floor down to the slab; below the slab, footing pads are poured to reinforce this load and take the pressure off the slab itself. Often, though, the exterior basement walls are designed to carry the perimeter structural loads. If those loads are out of plane from the exterior—as in the case of the straw-bale-wrapped timber frame—the footings need to be wide enough to carry both the basement wall and the perimeter structural loads. This can make for 36-inch footings or larger—quite different from the standard 16-inch footing below the average poured 8-inch concrete stem wall! Additionally, just as interior posts must carry the first-floor loads to the slab (and sub-slab footings), perimeter posts must now be placed in the basement. Often these are wood or metal posts; sometimes concrete pilasters are formed and poured, either concurrent or subsequent to the basement wall pour.

Another strategy is to place the basement wall below the structural plane—more standard with conventional construction—and find another way to support the insulation plane. This could mean cantilevering the floor deck to support the bales; in

BASEMENT FOUNDATION FOR TIMBER FRAME WITH STRAW BALE WRAP

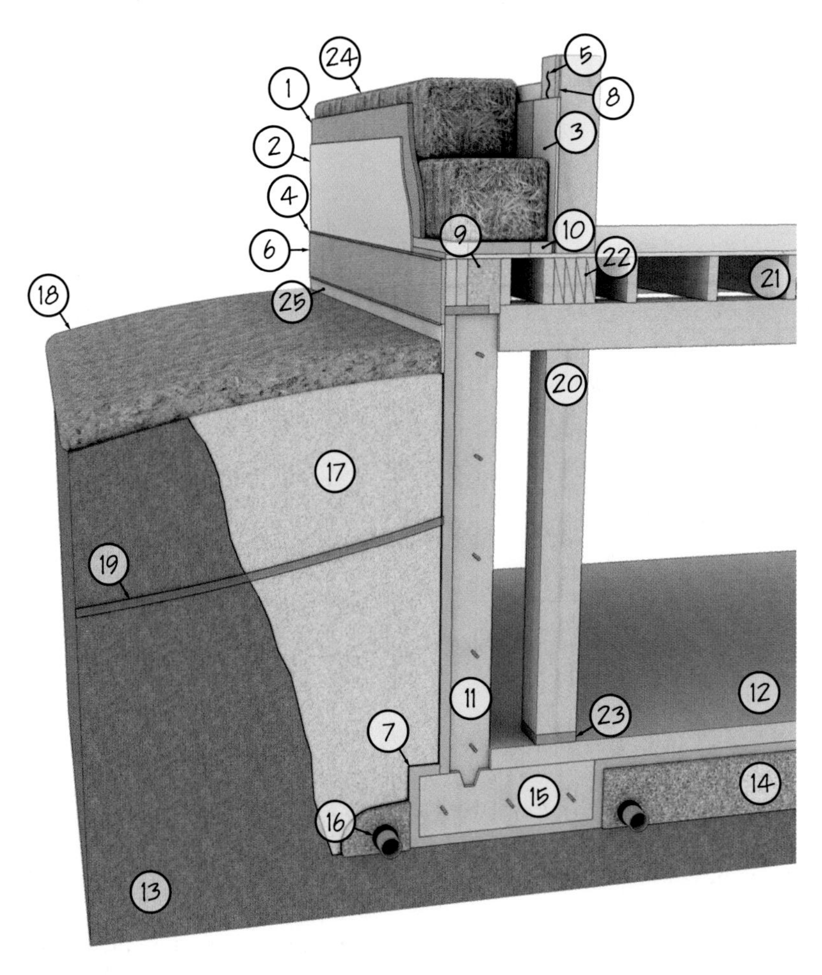

1. 1" ROUGH PLASTER
2. ½" FINISH PLASTER
3. AIR FIN—½" HOMASOTE
4. METAL FLASHING
5. ACOUSTIC SEALANT
6. TRIM
7. 2" RIGID INSULATION
8. ¾" RABBET
9. CELLULOSE INSULATION
10. 2 x 4 ON FLAT, TOE UP
11. CONCRETE STEM WALL
12. 4" CONCRETE SLAB
13. UNDISTURBED/COMPACTED NATIVE SOIL
14. WASHED CRUSHED STONE
15. FOOTING
16. 4" DRAINPIPE (wrapped in filter fabric)
17. POROUS BACKFILL (sand)
18. CLAY CAP
19. FROSTLINE
20. TIMBER
21. FLOOR FRAMING
22. SOLID BLOCKING
23. 1–2" ROT-RESISTANT PAD
24. STRAW BALE, ON FLAT
25. STUCCO

Foundation details may need to change when the structural plane of the building is not in line with the perimeter, such as with a straw-bale-wrapped timber frame. ILLUSTRATION BY AARON WESTGATE/FREEFLOW STUDIOS.

the case of a double-stud wall, it could mean designing a Larsen truss system, in which the outer stud walls are hung off the inner stud walls (see chapter 13 for more information on double-stud walls).

EFFECT ON INTERIOR CONDITION

If you are planning on creating a foundation that incorporates a basement, it is critical to consider what the implications may be for the intended uses of that space. For example, if you want to include a root cellar in the basement, the space will need to be kept at approximately 32°F to 34°F and 90% to 95% humidity, depending on what is being stored, with provisions for ample drainage and walls that can easily be scrubbed and sanitized. Obviously, keeping a cabbage fresh until March is a very different use condition than typical living standards: 68°F and 35% humidity, soft floors, and pretty walls. Does this affect your foundation design? Absolutely! Not only will that root cellar need to be isolated (heat and moisture transfer) from the rest of the basement, but it will need to be isolated from the rest of the house above it. Any structural elements will need to be able to stand up to those conditions (maybe not the best place for pine posts). The sub-slab, wall insulation, and even the slab itself may not be utilized here to encourage cooler and damper conditions. Ventilation, drainage, and utility supply will all need to be considered.

How about the rest of the basement? If it is to be used as living space, then the foundation needs to take on additional roles to allow for the creation of a healthy and comfortable interior environment. As anyone who has been into their grandparent's musty, dank, dungeonlike basement to fetch a jar of canned peaches can attest, this is not a given. The room will need to be warm underfoot, well ventilated, well lit, and attractive, with well-controlled humidity and temperature; all of these needs will have to be met by the performance of the foundation walls, drainage detailing, insulation, slab detailing, and more. Air quality is another concern; toxic materials such as concrete sealers, or even rubber tires used as formwork (see sidebar "Rammed Earth Foundations" in this chapter), are a danger if considered for their role in the foundation without regard to their effect on the interior climate, and then can be exacerbated in a basement where passive ventilation may be more difficult to achieve.

Types of Foundation Systems

With these considerations in mind, we can now evaluate options for foundation systems from a well-informed perspective. Given that many of the systems we employ are either nonstandard, experimental, or used out of the historical context in which they were originally practiced, it is all the more important that we ensure our designs are well vetted to avoid design conflicts. Of note, it is important here to restate the importance of prioritizing structural performance above all else (i.e., omitting sub-footing insulation when it may compromise structural integrity). Let's explore the options.

FROST-WALL FOUNDATIONS

Frost-wall foundations are the most common forms of foundations in the moderate and cold climates of North America. In many of these foundations, the frost walls are dug down to 7 or 8 feet and the interior of the footprint is excavated to create a basement. In addition to the convenience of the basement, a major reason frost-wall foundations are so popular is their efficacy. Commonly believed to be the most foolproof foundation system, in frost-prone regions where liability for failure from frost heaving runs high, reliability is perhaps the most valued feature of a foundation system. Unlike other systems that will be explored later in this chapter, there is no reliance on unseen insulation or drainage systems—footings are poured below frost line, a wall is poured on the footings, and the builder and owner sleep well at night knowing the foundation is fully anchored below frost depth.

The greatest liabilities of frost walls are their cost and impact in both excavation and material

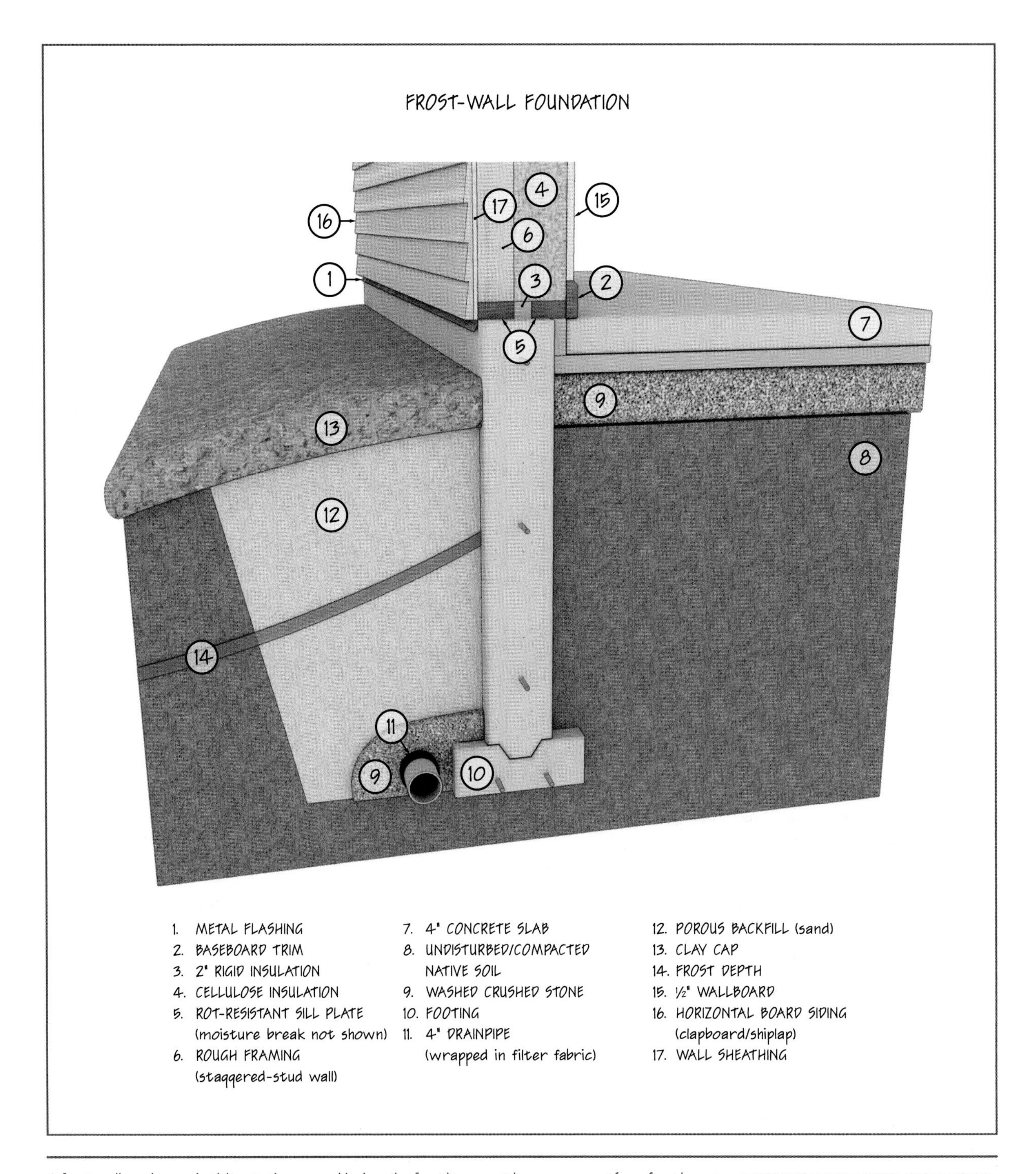

A frost wall anchors a building to the ground below the frost line, avoiding movement from frost heaving. ILLUSTRATION BY AARON WESTGATE/FREEFLOW STUDIOS.

Cement Reduction Strategies

Given the large carbon footprint carried by Portland cement—the binder in concrete—it is worthwhile from an ecological perspective to not only reduce the amount of concrete in our buildings, but to reduce the amount of Portland cement in our concrete. A simple and common approach is to utilize "supplementary cementing materials," additives included in a concrete mix that reduce the amount of Portland cement needed to achieve desired structural requirements. One of these additives is fly ash, the by-product of coal combustion sourced by scrubbing the smokestacks of coal-fired plants. Fly ash can reduce cement content in concrete by 10% to 30%, while making the concrete stronger and easier to install (Headwaters Resources 2005).

Other common additives include blast-furnace slag, a by-product of iron production, and silica fume, which is a by-product of silicon and ferro-silicon metals (EcoSmart Concrete 2004–2008). Companies such as EcoSmart are making information and technical support more readily available to increase access to these strategies to lighten the impact of the concrete we must use.

Magnesium-based cements offer another opportunity to reduce our reliance on Portland cement. The benefits of magnesium cement are many, when compared to Portland cement. Reduced kilning temperatures in its manufacturing result in significantly lower embodied energy and carbon footprint. A faster set time and harder end product improve workability and application. Magnesium cement's ability to bind successfully with cellulosic materials, such as woodchips and straw, provides greater opportunities for integration into natural building systems and allows for incorporation of inexpensive and insulating cellulosic materials to offset overall cement usage and improve insulation performance of the material (see discussion of ICFs later in this chapter). Finally, magnesium cement does not cause the same cellular disruption in the human body that Portland cements are alleged to cause. George Swanson of Swanson Associates, based in Austin, Texas, asserts that studies in Germany have purportedly shown that direct exposure to Portland cement (e.g., standing on a cement slab floor) may cause interference in cellular functioning and that the results of these studies have contributed to regulations being added to German labor law that limit standing exposure to Portland cement floors to one hour maximum per day (we, the authors of this book, have been unable to verify these studies).

consumption. The colder the climate, the further you must dig to reach below frost depth, and that cavity is then filled with Portland cement–based concrete, increasing the carbon and toxic footprint of the building. If the reliability of a frost wall is desired, a number of strategies can be used to reduce the impact of a frost-wall foundation.

If a full basement is not needed in a design, then a simple frost wall without a basement will represent a savings in both excavation impact and concrete because the interior of the footprint need not be dug out. Another way to lighten the ecological load is to include additives (known as supplementary cementing materials) in the concrete that reduce the Portland cement content—the ingredient most responsible for the hefty carbon footprint of concrete (discussed in chapter 2). See the sidebar "Cement Reduction Strategies" for more information.

AAC Block

An alternative material at our disposal to improve the standard frost wall is the autoclaved aerated concrete block, or AAC block. The Swedes have used AAC for almost 100 years, although it has been in use in the United States only for the last couple of decades. AAC is made by combining Portland cement, lime, silica sand, fly ash, and aluminum in a mold. Reactions between the cement and the aluminum cause the formation of tiny hydrogen bubbles, which causes the concrete to expand fivefold beyond original size. After this reaction, the block is cut to size and cured, leaving a block that is lightweight, airtight, nontoxic, nonorganic, lower in cement content, and with an R-value of approximately 1.25/inch (National Association of Home Builders Research Center n.d.). Concerns have been raised about the porosity of AAC block and subsequent vulnerability to moisture; therefore, adequate drainage and moisture-barrier detailing must be employed if using AAC block.

Foam-free ICFs are an opportunity to reduce both petroleum and concrete content of a foundation wall without sacrificing structural or insulative performance. PHOTO BY JAN TYLER ALLEN.

One of the distinguishing features between frost walls and the systems described later in this chapter is how they keep the building from shifting due to frost heaving. Again, since the shifting and heaving are a result of the force of water expanding upon freezing, strategies to avoid frost heaving all entail preventing water from freezing below the foundation. With a frost wall, this is essentially achieved by displacing water below the building with concrete; the most appropriate designs are accompanied by footing drains that remove any groundwater that may be inclined to accumulate nearby.

Insulated Concrete Form (ICF) and Concrete Sandwich Wall (CIC) Foundations

These two strategies are advances upon the standard frost-wall foundation system and are designed for improvements in wall performance—specifically, thermal, acoustical, and durability performance. While these walls are frequently targeted for use in whole-house construction, for systems employing full basements, and especially walk-out basements with one or more walls exposed to atmospheric conditions, their increased performance can readily justify their use. ICF walls and sandwich walls differ by the placement of insulation in the wall system and the method of construction. We will look at these wall systems specifically in the context of use as insulated foundation walls, not as whole-house wall systems.

ICF walls are aptly named: insulated stay-in-place forms, either in block or panel form, are built onto the footings to form the walls of the foundation; the cores are reinforced with rebar and poured with concrete. Whereas the conventional poured concrete stem wall involves the erection of temporary panel wall forms that are removed after the wall cures, ICFs are permanent forms that either are constructed primarily of insulation or incorporate insulation inserts into the formwork, and they are left as an enduring part of the wall. A major benefit is the ability for the owner-builder or his or her contractor to easily build up the wall, as the ICFs

One Builder's Experience with ICFs

Ben Graham of Natural Design/Build in Plainfield, Vermont, has worked with a variety of different ICF systems and shares his experience:

> Working with Rastra [a brand of ICF] was hefty work but made for a simple and flexible system. I chose Rastra for a particular project because I could stand the 15-inch pieces upright and bevel the sides to create a circular curve I needed. The insulation value was also appealing. The bracing needed was minimal, even for a 10-foot pour. On my first try with guidance from the representative, I was able to pour a consistently rising wall with no problems. Windows and doors were cut out after gluing them with adhesive foam and locking the corners with large metal staples. The only downside was all the bits of foam left on-site and the trucking and carbon costs in shipping [the ICFs] from Arizona.
>
> In a house design that used Durisol instead of Rastra, the cost factor took precedence, and I was interested to use the mineral wool insulation with the recycled wood, rather than the foam. While I did not actually work with the Durisol, it was reported as a little more fussy in getting all the blocks level. Because the blocks are smaller and they rely on weight for stability, they can only be poured in 4-foot lifts, making the process a bit more involved. I also know that the plasterers had issues with the reaction to lime plaster that I don't think would have occurred with the Rastra.

are lightweight and easy to move and do not require the use of specialty equipment. This allows for walls to be built quickly while retaining greater control of labor and scheduling, which is frequently lost when subcontracting out wall forming and pouring.

The majority of ICFs use XPS or EPS foam as their insulation of choice; there are many different styles of ICF, and each manufacturer has its own system, which may or may not support the needs of your structure or building process. As natural builders, we are always trying to find ways to reduce our use of petrochemical foam. Therefore, we prefer to source an ICF made with a high quantity of recycled foam to reduce virgin foam manufacturing. Even better would be to source an ICF that doesn't use foam as the insulation. There are a few manufacturers, such as Faswall and Durisol, that manufacture an ICF block that is made primarily of mineralized wood chips, bonded with a small amount of cement, and insulated with mineral wool inserts; these ICFs come as close as one can to upholding the principles of natural building, while taking advantage of the benefits of the insulated concrete form strategy.

Concrete sandwich walls can almost be thought of as the inverse of ICF walls. Whereas an ICF wall features two insulative wall planes with a concrete core (ICI), concrete sandwich walls feature a plane of insulation embedded within two concrete skins (CIC). This approach to insulating offers a few benefits. While the insulation may be comparable to or even greater than that of an ICF wall (there is flexibility in designing the thickness of the wall and the amount of insulation to install), the CIC wall offers the benefit of taking greater advantage of the mass of the concrete wall to offset heat loss. As explained in chapter 7, the effect of mass on the overall thermal performance of a wall system is referred to as the dynamic benefit for massive systems (DBMS), and it has been clearly shown that the DBMS, and therefore the overall performance of the building, is greater in building systems where there is direct contact between the mass of the wall and the interior of the building, within the insulation plane. Both an exterior-insulated concrete stem wall and a CIC wall offer this greater DBMS value, and given comparable total-wall R-value, they will therefore perform better than an ICF wall (Oak Ridge National Labs and Polish Academy of Sciences 2001).

However, when compared to the exterior-insulated stem wall mentioned above, another benefit of the CIC wall is that the insulation, which is susceptible to damage by termites, UV exposure, or impact (e.g., during construction or from lawn maintenance), is safely contained within the concrete wall and will not break down or be damaged over the course of time.

CIC, or "sandwich," walls are an effective way of including insulation in a durable placement within a foundation wall while maintaining the mass-effect benefits of the concrete. PHOTO COURTESY OF THERMOMASS.

CIC walls can be built in a variety of fashions, from precast or site-built tilt-up panels to being built in place within concrete forms. This leaves the owner or builder with a few options: to find a manufacturer to deliver wall panels to the site, hire an experienced crew to build tilt-up panels on-site, or work with a more conventional concrete-form company to install foam—using pre-manufactured proprietary through-ties—into conventional wall panels. Each has its benefits, and access to a manufacturing plant or skilled workforce will likely be the determining factor in making the decision; generally, pre-manufactured and tilt-up systems are found in commercial construction for whole-wall systems, while pour-in-place is more likely to be used in residential foundation construction.

Rubble-Trench Foundations

Pioneering architect Frank Lloyd Wright, who popularized rubble-trench foundations, began using them as early as 1902 (Velonis 1983). However, they have been in use for thousands of years in places such as the Middle East and Africa (Koko 2003). The rubble-trench foundation addresses the issue of frost protection by ensuring that no water can collect below the structural bearing points of the building; therefore, there is no water to freeze and heave. This is done with minimal excavation and very little use of concrete, making for a very low-impact foundation. Even reclaimed rubble, such as recycled concrete or "urbanite," can be used to further offset resource consumption.

To construct a rubble-trench foundation, the builder digs a trench mapping the footprint of the load paths of the structure (i.e., perimeter walls, interior load-bearing walls/posts) down below the frost line for that location.

The trench is then lined with filter fabric, to keep silt from migrating into the trench over time, clogging the pores, and impeding drainage. A small 1- to

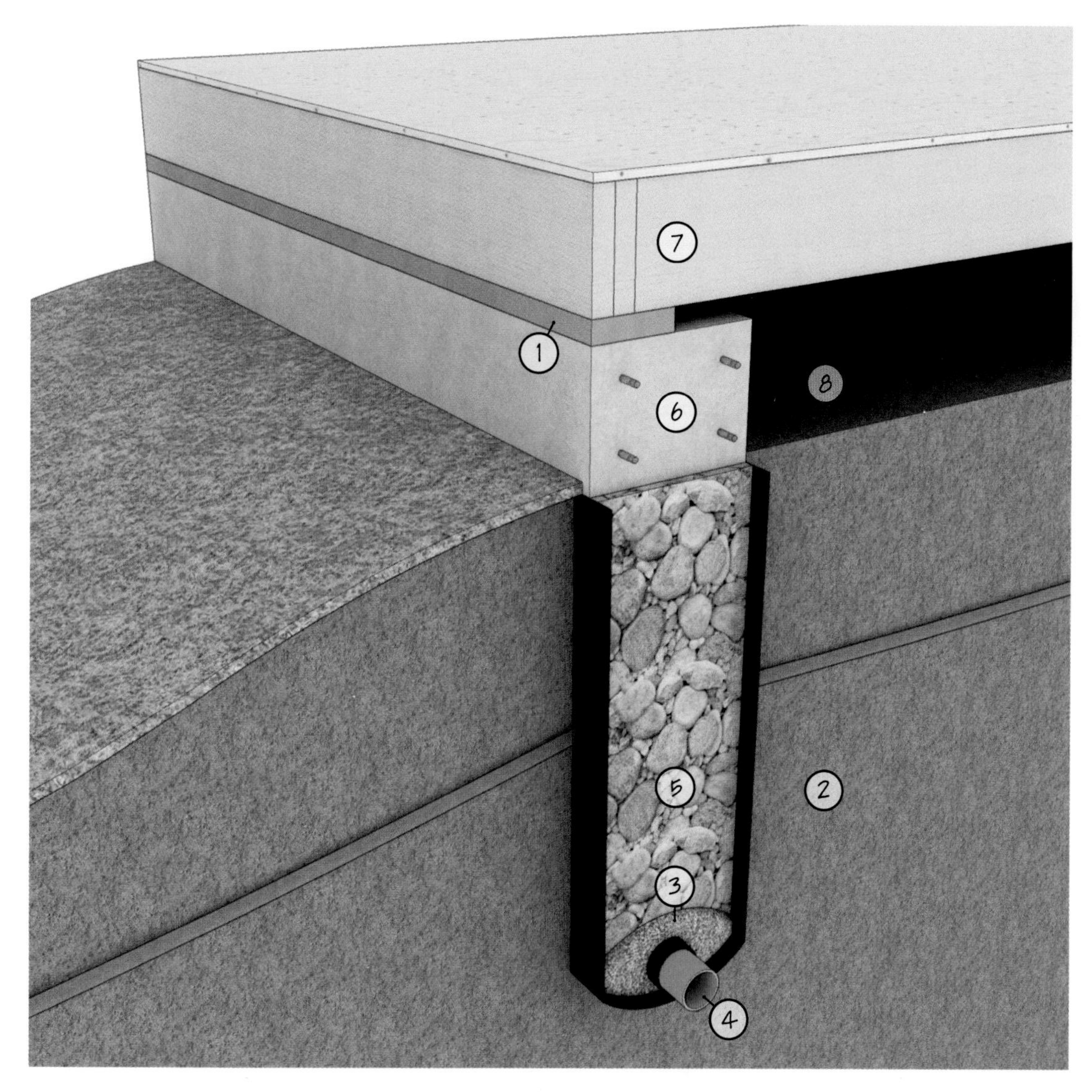

1. ROT-RESISTANT SILL PLATE (moisture break not shown)
2. UNDISTURBED/COMPACTED NATIVE SOIL
3. WASHED CRUSHED STONE
4. 4" DRAINPIPE (wrapped in filter fabric)
5. 2" CLEAN COMPACTED STONE (rubble)
6. CONCRETE GRADE BEAM
7. FLOOR FRAMING
8. VAPOR BARRIER (not shown)

Rubble-trench foundations address frost heaving by ensuring complete drainage of water below the frost line under the structure.
ILLUSTRATION BY AARON WESTGATE/FREEFLOW STUDIOS.

2-inch-deep bed of washed crushed stone (gravel) is laid on the bottom of the trench, followed by a fully interconnected run of 4-inch perforated drainpipe. Our preference is to place the holes at 4 o'clock and 8 o'clock, to allow for removal of water at as low a level within the trench as possible. It is good practice to wrap the pipe in an additional layer of filter fabric, as an added precaution against silt clogging the pipe over time. This pipe must be pitched at a slope of no less than 1 inch in 8 feet, which can be achieved by the excavation of the trench, by the thin bed of stone, by propping up the pipe, or a combination of the three. A high point in the trench must be established, and the pipe then slopes away from that point (usually in multiple directions, depending on the shape of the trench). A cleanout should be installed at this high point and at any other major interstices in the drain run, such as the corners; cleanouts generally consist of a 45-degree fitting running up to grade, and then capped above grade for future access.

Instead of using standard PVC for drainpipe, one might choose to use corrugated black ABS drainpipe. Unlike PVC, the ABS pipe is flexible, often comes pre-wrapped with a filter-fabric sock, and perhaps most importantly has a lower toxic footprint than PVC, of which dioxin is a component (and a major pollutant created in PVC's production). However, ABS is weaker than PVC, which can be an issue during installation, and its corrugated nature makes it much more difficult to clean out in the event of a clog; it is also more difficult to lay down pitched to the appropriate slope.

Once the drainpipe is fully connected, the builder then switches to solid pipe and runs the pipe to daylight, maintaining the pitch until the pipe breaks through grade. In the event of a very level site, a dry well—a large pit filled with crushed stone—may need to be built to receive the drain outlet. This dry well will need to be engineered respective to the soil type, water table, and other factors. Once the pipe is fully installed, the trench is then carefully filled with 2-inch clean stone, which is large enough to allow for easy drainage of water while avoiding clogging, yet small enough to be able to compact soundly. Care must be taken to avoid damage to the pipe and to ensure that full compaction, generally with the aid of a mechanical plate compactor, is achieved in 1- to 2-foot lifts until the trench is filled to grade. Failure to adequately compact the trench will result in uneven shifting and settling that can translate stresses throughout the structure over time.

Once the trench is filled, a number of different transitions can be used to support the structure; the most common are the grade beam and the slab-on-grade. In the case of a grade beam, a structural beam is poured out of concrete directly on top of the trench, on top of which a floor box can be constructed to support the structure of the building. Sizing of this beam is generally engineered by the load requirements of the building, and specifications and codes governing the crawl space that is created within the plane of the beam, below the floor box. The ground below the floor box within the beam should be covered with a layer of poly sheeting to reduce vapor migration into the structure, and allowance for ventilation of the crawl space should be designed into the grade beam as well. Alternatively, a slab-on-grade may be poured over the rubble trench, which is described in the next section of this chapter. This may well be considered redundant, but if there is a place for redundancy, it is in the foundation system of a building sited atop heavy clay soils in a cold wet climate. If you are planning on needing an excavator, leveling out a site, or installing a "curtain drain" (a shallow perimeter drain), for example, in the case of a site with a moderate degree of slope, it may not be much more expensive to turn that curtain drain into a foundation drain and dig a big ditch in concert with the site leveling.

On DVD, Chapter 12,
see RUBBLE-TRENCH FOUNDATION

Frost-Protected Shallow Foundations (FPSF)

FPSFs are exactly what they sound like: foundation systems that are protected from frost heaving, yet are shallow—meaning they are not dug down below

Rubble-Trench Foundations: A Builder's Perspective

We have built structures featuring both slab-on-grade and grade-beam installations over rubble trenches, with great results. In addition to all the performance and environmental benefits, nice features of a rubble trench are the control one has as a builder or project manager, as well as the ease of installation. As long as there is someone on hand with an excavator, no other specialty equipment is needed, with the potential exception, depending on the rubble material, of a plate compactor, which can be rented at any decent rental agency. Any time a job can be executed by you and not subbed out means less logistical hassle in scheduling (which, in a rainy spring, is no small matter!), no need for cumbersome supervision, and no anxiety about whether the concrete subs got all the details right. In the spirit of "if you want something done right, do it yourself," it is a great feeling to execute a well-installed drainage system, knowing you can put your full faith behind the longevity of the foundation. Because once it is in, it sure is hard to get back down there and fix it.

A grade beam is very straightforward to form and pour yourself, without the need for heavy specialty-form work. Builder Robert Riversong of Riversong HouseWright in Warren, Vermont, shared with us the great tip of using rough-cut 2 × 8 hemlock for forms, and lining them with 6-mil poly. After the grade beam set, we popped the forms and used them as floor joists! If pouring a slab-on-grade, again, the pre-pour work may well be something you would want to execute yourself; insulation, vapor barriers, form work, mechanical stub-outs—all these details need to be followed to a tee to ensure a satisfactory result, and as the contractor or owner/builder, you will want to ensure it is done correctly. If you've selected the rubble-trench option, you may be pouring only once, instead of the standard three times for the average basement (footings, walls, and floor slab), which means less hassle and more control.

Grade beams are frequently poured atop rubble-trench foundations to support a floor box or other framing. PHOTO BY BENJAMIN RAY GRIFFIN, GREENER LIVING VERMONT.

the frost line. The concept is that insulation is installed around the perimeter of the foundation—in cold climates, often a combination of vertical and horizontal, or "wing" installation—to capture geothermal heat, keeping the soil temperature above freezing and therefore eliminating any potential for frost heaving. So, whereas the frost wall beats frost by displacing water, and the rubble-trench foundation beats frost by draining water, the FPSF beats frost by keeping the soil, and any water present, from freezing. Used extensively throughout Scandinavia and other cold regions since the 1970s, FPSFs have become more prevalent in North America in the last twenty years and are now code-approved in

FROST-PROTECTED SHALLOW FOUNDATION (FPSF): Slab-on-Grade

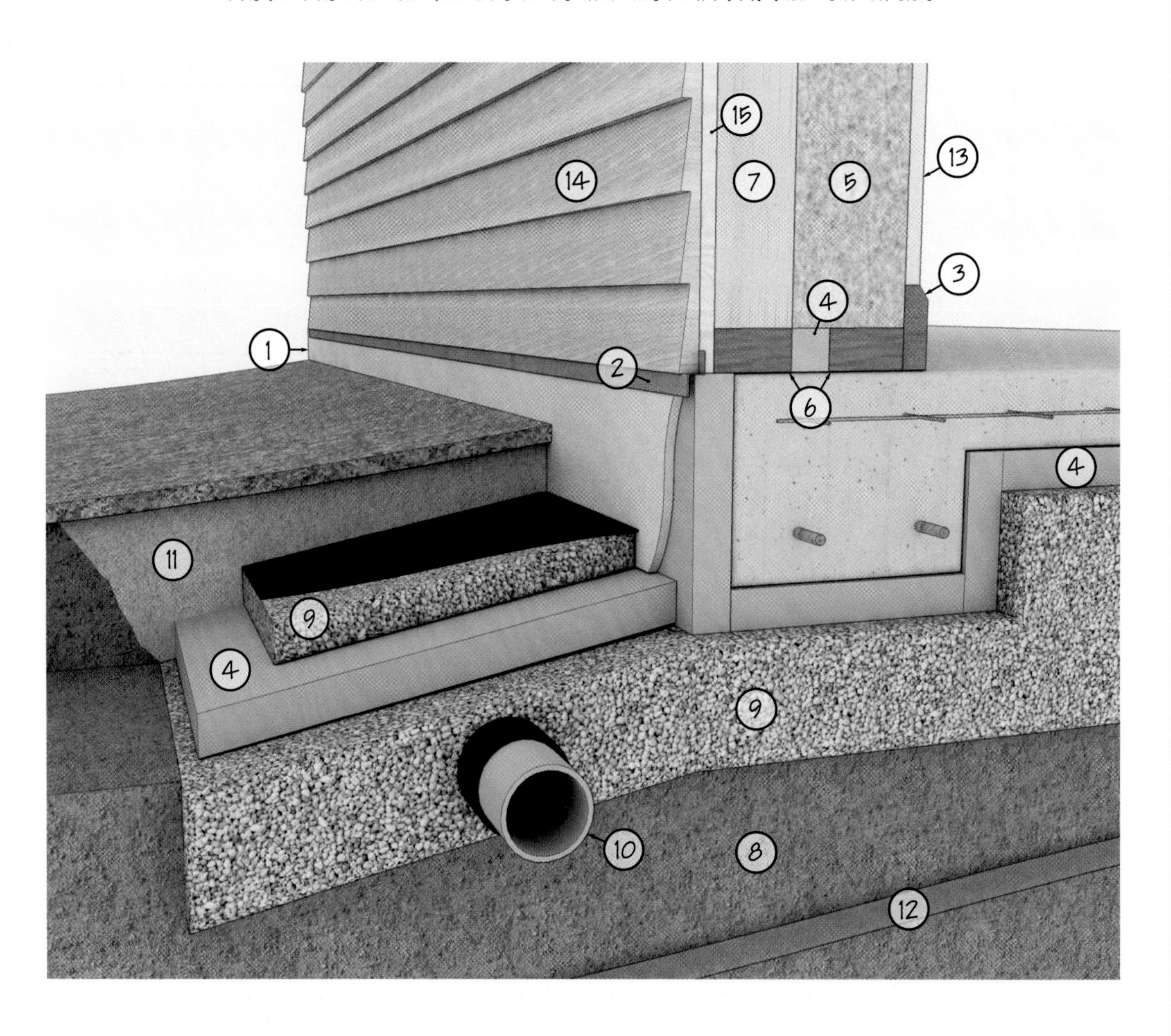

1. STUCCO
2. METAL FLASHING
3. BASEBOARD TRIM
4. 2" RIGID INSULATION
5. CELLULOSE INSULATION
6. ROT-RESISTANT SILL PLATE (moisture break not shown)
7. ROUGH FRAMING (staggered-stud wall)
8. UNDISTURBED/COMPACTED NATIVE SOIL
9. WASHED CRUSHED STONE
10. 4" DRAINPIPE, (wrapped in filter fabric)
11. POROUS BACKFILL (sand)
12. FROST DEPTH
13. ½" WALLBOARD
14. HORIZONTAL SIDING (clapboard, shiplap)
15. WALL SHEATHING

Frost-protected shallow foundations (FPSFs) can be poured as thickened-edge slabs . . . ILLUSTRATION BY AARON WESTGATE/FREEFLOW STUDIOS.

FROST-PROTECTED SHALLOW FOUNDATION (FPSF): Stem Wall

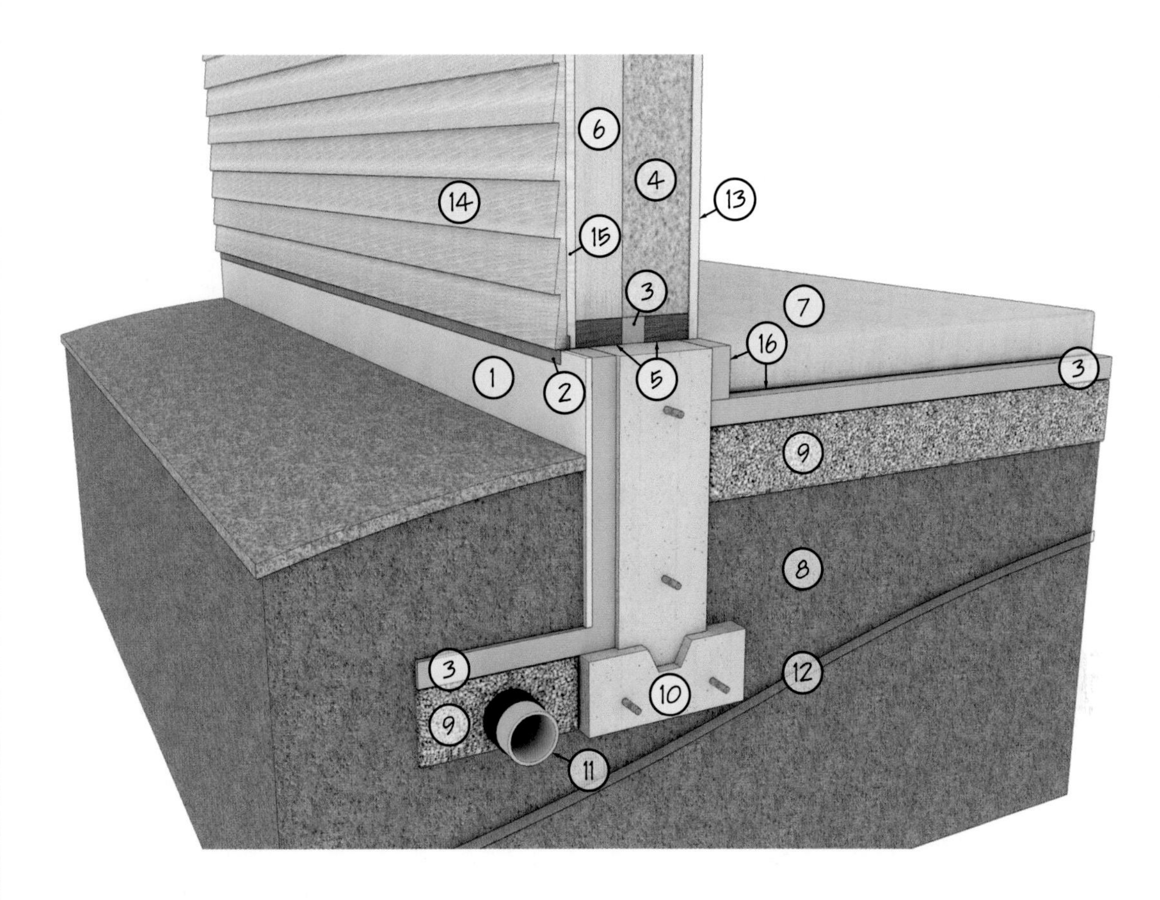

1. STUCCO
2. METAL FLASHING
3. 2" RIGID INSULATION
4. CELLULOSE INSULATION
5. ROT-RESISTANT SILL PLATE (moisture break not shown)
6. ROUGH FRAMING (staggered-stud wall)
7. 4" CONCRETE SLAB
8. UNDISTURBED/COMPACTED NATIVE SOIL
9. WASHED CRUSHED STONE
10. FOOTING
11. 4" DRAINPIPE (wrapped in filter fabric)
12. FROST DEPTH
13. ½" WALLBOARD
14. HORIZONTAL SIDING (clapboard, shiplap)
15. WALL SHEATHING
16. VAPOR BARRIER (not shown)

. . . or as stem walls. ILLUSTRATION BY AARON WESTGATE/FREEFLOW STUDIOS.

Different Approaches to FPSF Slab

There are a number of different approaches to the execution of a slab-on-grade, all of which are valid providing the details are all well attended to. Beginning from a level, cleared site of undisturbed or well-compacted soil, one approach to forming the pads and thickened edge is to simply dig out the relief of these elements, essentially earth-forming the cavity the pads and edge will fill. One must take into consideration any thicknesses of sub-slab insulation, gravel, and the detail of laying in a vapor barrier (for example, if the vapor barrier is wrapped around the pads, this may need to be done from separate pieces that are adhered to the main vapor barrier). A drawback to this situation is the difficulty in installing insulation to line the pads and thickened edge. Some designers and builders find this level of insulation to be unnecessary, given the increased depth of the concrete, as long as the slab itself and the exterior edge of the thickened edge are well insulated. If you are inclined to provide a continuous insulation plane below the structure—a strategy we strongly advocate—then digging out the formwork may ultimately prove to be much more trouble than building formwork. If you choose to build formwork for the thickened edge and pads to line with insulation, do not forget to consider the thickness of the insulation in sizing the form!

A built-form strategy was shown to us by Buzz Ferver of Overbrook Design in Worcester, Vermont, that is well detailed yet simple. Here, the forms for the thickened edge and pads are built on top of the prepped grade. Six inches (minimum) of gravel and 4-inch perforated drainage piping are installed under the entire slab and footing assembly on 10-foot centers to ensure the slab is always dry. The exteriors of the edge forms are built with 2 × 12s, whereas the interiors of the edge forms use 2 × 5s, to account for 5 inches of slab and 2 inches of insulation. The forms use a piece of 2-inch rigid insulation as the base, which runs below the interior form to ensure positive connection with the interior vertical insulation, which lines the form on the outside. Appropriate reinforcement is placed in the forms, pieces of rebar are installed to brace between the interior and exterior forms, and the exterior form is braced along the perimeter. Next, the interior of the foundation below the slab is filled with compacted 1½-inch crushed stone, level to the top of the interior edge form (and any interior pad forms, which themselves are lined with foam on the exterior). If mechanicals are to be run, they are done at this point and dug into the stone, or laid in as the stone is being installed, depending on elevations and the design. A durable vapor barrier can be installed, then sub-slab foam insulation that runs over the top of the interior vertical insulation and any pad insulation, ensuring continuous insulation. The slab and footing are then poured in one shot, and the project moves forward toward victory.

As is the case with many timber frame designs, there are frequently many pads spread throughout the interior of the structure. Add a masonry pad or two for chimneys or boilers, and the next thing you know you are tripping over formwork left, right, and center. If you

One approach to a slab-on-grade is an insulated, thickened-edge slab. PHOTO BY BUZZ FERVER.

Another approach for a slab-on-grade is a single 12-inch-deep pour, saving the difficulty of formwork. PHOTO BY MICHAEL BRODEUR.

are hiring out this formwork, the estimates may be surprisingly high. If you also mandate specific installation procedures, such as those mentioned above, you may not find anyone to submit an estimate. For this reason, it is not uncommon to find slabs poured at a level 10–12 inches across the field. While this undoubtedly increases concrete and reinforcement usage and materials cost—and the associated carbon footprint—it may ultimately come out comparable in cost as a result of labor reductions. Being much simpler in design, the chances of mistakes are reduced as well.

many jurisdictions and widely accepted as a viable foundation system in cold climates. One case study of a U.S. Army airport control tower built in Alaska near the Arctic Circle on an FPSF has proven its resiliency, even in the face of -60°F sustained winter temperatures and a 13-foot frost line. The obvious benefits of this system are evident in the reduced excavation and associated concrete use compared to pouring 13-foot frost walls in Alaska, for example, or even 5-foot frost walls in Vermont. Compared to a rubble-trench foundation, there is the potential for even less excavation and resource use, as not even a drainage trench need be built to frost line. It should be noted, however, that these foundation designs incorporate direct connection between the at-grade floor of the structure and the ground to ensure heat capture from the ground below the building, whereas a rubble-trench foundation allows for an elevated floor box and crawl space, allowing for more design flexibility. FPSF designs can be engineered for three-season or unheated structures, and although more involved than those below year-round heated structures, they are more resilient in the event that the building is left to freeze.

FPSFs generally take two forms: monolithic thickened-edge slabs-on-grade or grade beams with an internal floating slab or earthen floor.

In both cases, the topsoil is removed, the pad is graded, and a perimeter drain, similar to the drain at the base of the rubble trench, is installed. In the case of a monolithic thickened-edge slab-on-grade (also known as an Alaskan slab), both the concrete footings and the slab are poured at one time, directly on grade. Forms are built or dug out of the grade and lined with XPS foam, mineral wool board, or other suitable subgrade insulation material. (*Note*: insulation placed below footings must be of a density and engineered to receive a structural load.) Forming the footings

Stone

While not technically a foundation system, the use of native stone in building has a long history: foundations, walls, roof tiles, floor tiles, decorative accents, structural systems, you name it. In fact, every time we use sand, we are using tiny stones. Dry-stacked fieldstones were the original frost walls of the early New England settlers, and the use of stone, both structurally and aesthetically, is relevant to this day and keeps us connected not just to the lineage of builders in cold regions but to the earth itself. After all, as anyone who has tried their hand at gardening in New England can attest, if there is one crop we can reliably turn up each spring, it is fieldstone.

(as opposed to digging them out of grade) makes it easier to line the footing with insulation. These forms create a thickened edge on the slab to incorporate the footing, as well as thickened pads or ribs below the interior of the slab to support point loads, like posts or chimney foundations, or load-bearing interior walls. A thin gravel pad is laid to level the grade, and insulation is placed below the slab (this sub-slab insulation is less of a component of the FPSF and more of a consideration of the thermal efficiency of the structure). The insulation is designed respective to heat-loss calculations projected for the building; a single 2-inch sheet, achieving R-10 in the case of XPS foam, would be considered a standard minimum as a thermal break. Many Passive House strategies, however, employ much higher amounts of insulation below the slab; R-40 to R-60, or four to six sheets, is not uncommon. Depending on the quality and type of foam installation, a vapor barrier (commonly 6-mil poly) is then installed above or between layers of the insulation. As with all slabs-on-grade, provisions for electric and datacom (cable, DSL, etc.), plumbing, ventilation, and any other inter- or intra-slab utilities must be well considered.

The main difference between a grade beam with internal floating slab and monolithic thickened-edge floating slab FPSF is that the internal floating slab within the grade beam is poured separately from the footings. The grade beam would need to be only the depth required by structural load-carrying requirements of the building and insulation depth required by the FPSF design (a 2-foot vertical sheet, for example)—similar to the thickened-edge slab. Spread footings may not be needed, providing there is sufficient soil structure. Additionally, the grade beam can be designed to extend above grade to the desired height, which will provide additional protection to walls from splash-back and ground-source moisture. The slab can then be poured within the grade beam and atop any interior footing pads, with appropriate gravel pad, insulation, vapor barrier, and utility installation (National Association of Home Builders 1998). Alternatively, an earthen floor can be installed in lieu of a concrete slab, or as a finish topcoat over the concrete slab (see chapter 19 for more information).

On DVD, Chapter 12, see FROST-PROTECTED SHALLOW FOUNDATION

Pole and Pier Foundations

Pole foundations are generally found in agricultural buildings, in which structural posts are sunk deep into the ground to double as the foundation system. A very inexpensive, low-impact, light system, pole foundations excel in structures where high-performance, legacy-scale durability is not necessary. Poles were traditionally charred in a fire, and later tarred, to increase longevity; these days, pressure-treated poles are common, as are repurposing telephone poles, which have been treated for rot resistance. The liabilities of this system are first and foremost their vulnerability to rot (especially in locations with high water tables), difficulty in tight detailing, especially at the floor, and the danger of introducing rot-inhibiting chemicals to the building. While certainly not appropriate for your standard year-round residence, pole foundations earn a place as an affordable, light foundation system for the right structure.

Pier foundations are a step away from pole foundations. The piers in reference are generally poured-concrete columns, frequently formed with the ubiquitous cardboard Sonotube cylindrical form or site-built plywood forms, that sit on a poured footing (often poured singularly with the pier) positioned below frost line. These piers can be designed to come as high above grade as necessary and are located strategically to carry the load placed upon them—not unlike post-and-beam systems. In fact, piers can be designed to receive a frame directly, commonly fitted with a post base to structurally connect the frame to the pier and to isolate the end-grain wood from capillary wicking off the pier top. More frequently, a floor deck is designed to sit atop the piers, which can then receive any type of framing system. While it is far more common to see pier foundations than pole foundations in buildings, they offer similar benefits of light excavation requirements, minimal concrete usage, and low cost, with a more flexible design afforded by decoupling the framing system from the foundation system.

Liabilities still exist, however, for use of this foundation system. Achieving high-performance standards is much more difficult on a post or pier than for a building with a full connection with the earth. Although insulating and air sealing are easier with a pier foundation than with a pole foundation—since there are no interruptions of the posts through the floor insulation plane—significant heat can potentially be lost in either instance because buildings that have atmospheric conditions below the floor box are handicapped compared to those with full connection to grade. This also proves to be a problem when bringing running water into or draining wastewater from the structure. Whereas water lines are generally laid below frost line to ensure midwinter operation, with this type of foundation water lines must break above the frost line and even travel through feet of frigid air before entering the conditioned environment. A similar problem exists for draining wastewater from the building, particularly in plumbing traps that may hold still water in a freezing environment. This provides plenty of opportunity for failure, and potentially a pretty serious failure at that. Strategies to avoid this situation invariably involve energy usage or water loss, neither of which are desirable for a sustainable system. Accordingly, this type of foundation system is best applied to structures that are not intended for four-season permanent occupation and do not require plumbing, and their light footprint and easy installation often match well with the scale and use patterns of such buildings.

Both pole and pier foundations pose a problem with regard to frost protection, and to the structural function of the foundation in general. All the other foundation systems we discuss in this chapter are monolithic (of a single material) in nature, or at least contiguous—they are all physically connected throughout the entire foundation. In pole or pier foundations, for the sake of frost movement, there are essentially numerous independent foundations, and their connection to each other is not achieved until you reach the structure itself. If there is movement in any one of these little foundations, the motion differential is translated directly as a stress upon the building; rather than supporting the building by keeping it together, the building may now be trying to keep the foundation together! Therefore, uneven floors, sticky doors, and cracking wallboard joints should all be expected occurrences in the life of a pole- or pier-supported building, particularly in frost-prone regions.

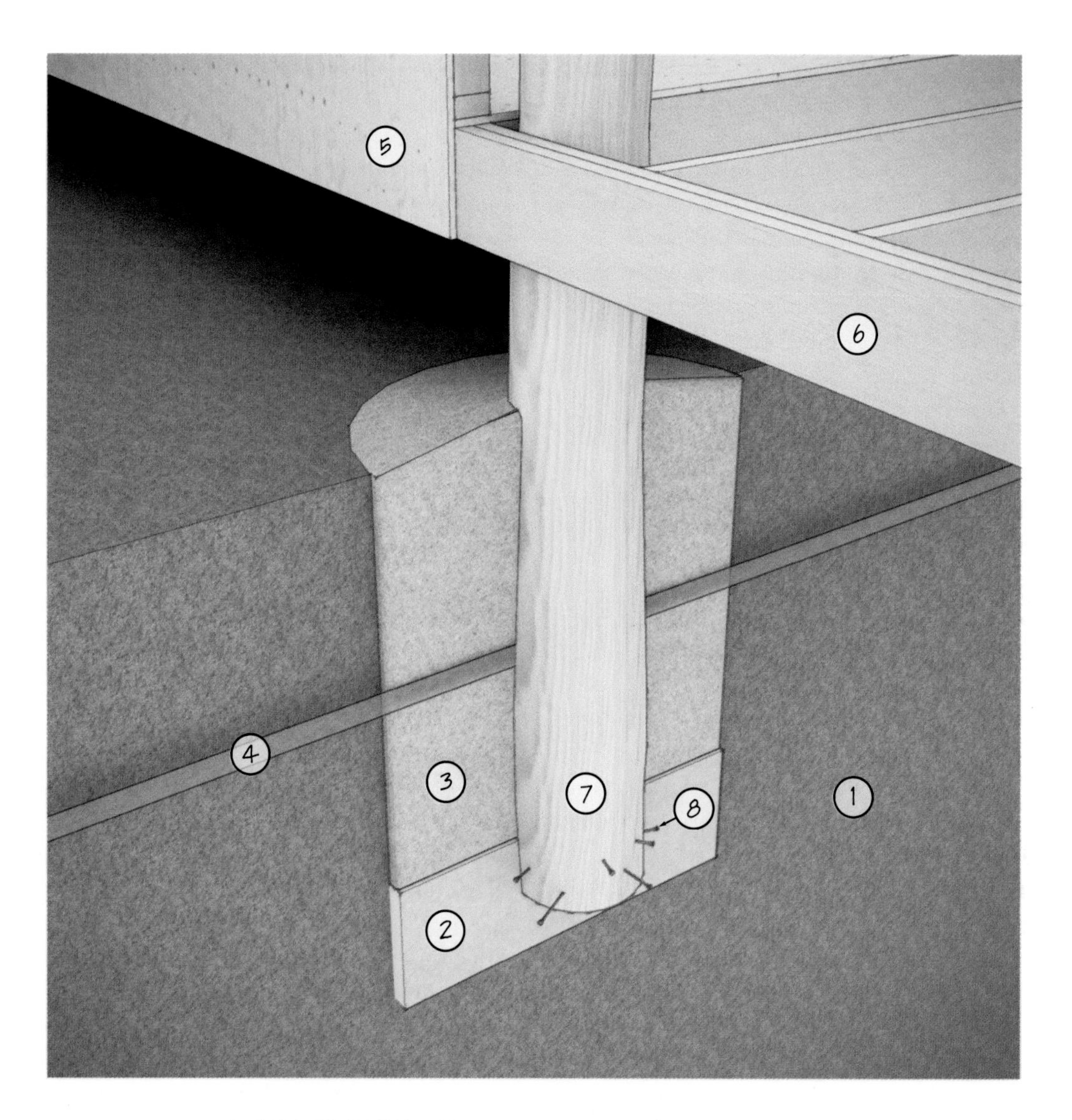

Pole buildings provide structure and foundation in a single system. ILLUSTRATION BY AARON WESTGATE/FREEFLOW STUDIOS.

Rammed Earth Foundations

It is popular among some factions of the natural building community to utilize rammed-tire or earth-bag construction techniques as a foundation solution. In fact, rammed earth is an incredibly strong material that is versatile, inexpensive, and abundant and has a small carbon footprint. Incorporating the use of inexpensive or reclaimed polypropylene bags—or even better, using up the glut of useless worn tires filling up rivers and landfills across the world—as formwork for the rammed earth is a logical extension of fulfilling the ecologically and financially friendly mandate of earthen construction. So what is not to love? Before you design your dream home to sit atop a tire foundation, there are a few concerns to consider.

The majority of the case studies for subgrade earth foundation systems are in relatively dry climates. Here in the soggy Northeast, we have a generally high water table, which, when coupled with subgrade frost, can make for some major instability in the ground. Therefore, if you choose to work with an earthen foundation system below grade, be sure that you consider how to deal with the water. This will most likely mean using an "earth" that is not frost-susceptible, such as gravel or other material with pore spaces large enough to encourage drainage and allow for water expansion within the structure of the soil. It may also involve use of earth that is stabilized with Portland cement to resist erosion. So while the thought of pounding tires full of loam may seem attractive, you may need to consider pounding tires full of gravel. It may be best advised, in fact, to keep the earth part of the foundation system above grade as a grade beam, used in concert with a rubble trench or frost-protected shallow foundation (see earlier in this chapter). If you are using polypropylene bags, you will need to make sure that they are covered quickly and permanently, as they will deteriorate rapidly with exposure to UV radiation.

Another major concern, specifically with the use of tires, is their toxicity. Laden with heavy metals and volatile petrochemicals, used tires are by no means a benign and safe building material. Placing them in direct contact with groundwater raises the concern of groundwater contamination, while having them in direct exposure to the interior of the building is a concern for both direct physical exposure as well as indoor air quality. While safe disposal of this rampant noxious waste product is necessary, incorporating tires into the building in this fashion may not ultimately be as safe as we would like, if not handled with care. Another way to incorporate used tires into your building—and to support progressive manufacturing companies—is to purchase and use roofing tiles made primarily of ground tires.

Finally, do not discount the tremendous labor involved in rammed earth. It is very difficult to mechanize the process of pounding tires or filling bags full of earth, and while there is an obvious and wonderful benefit to incorporating low-skill, labor-intensive processes to offset cost and build community, be sure to save some strength for the walls, the plaster, and the rest of the building (don't use up all your enthusiasm, or that of your community, in just getting the foundation built!). If you see that you may have a limited human-resource capacity, or a limited amount of time to build (i.e., in a New England summer), you may decide it is worth it to select a faster and easier foundation system, and invest your labor in parts of the building you will be able to see and appreciate every time you enter your home.

CHAPTER 13

Framing for Natural Buildings

In chapter 6 we learned about the fundamentals of what holds up the building and the forces at play that make appropriate structural design so important. In this chapter, we'll take a look at some of the different options for creating that structural support. The natural builder has many alternatives to choose from when considering a framing system. Different systems for creating a skeleton for the building accommodate different choices for fleshing out the body, so to speak. (Monolithic walls—such as load-bearing straw bale or cob—will be examined in later chapters.) As we evaluate our different framing options, we should keep in mind a few criteria to help evaluate the benefits and drawbacks of each option:

- *Resource use:* the total board footage of wood used, use of high-value timber, and embodied-energy profiles

- *Social impact:* support of the local logging industry, the wood manufacturing and processing industry, and the culture of sustainable silvicultural management practices

- *Cost:* the relative cost of the frame, and cost savings realized elsewhere in the project

- *Ease of construction:* the logistics, infrastructure, and skill required to execute the frame

- *Access to trade support:* the availability and cost of professional-trades support for frame construction, and the resulting impact on the budget and time frame of the project

- *Code approval:* the support and precedence in building codes

- *Longevity and durability:* the expected life span of the frame and its potential for adaptive reuse

- *Design flexibility in construction:* the ability to incorporate different wall forms, insulation materials, and exterior and interior finishes

- *Aesthetics and design goal:* how the frame supports the designed look and feel of the building, whether visible or hidden

Framing with Green Wood

Since the majority of the framing systems we will discuss are built of wood, before we review various framing options let's start by looking a bit more closely at the world of green wood, as it is used more frequently by natural builders than dimensional lumber. The term "green" can mean quite a few things: color, environmental impact, cost. But in this case, we are referring to the age and, by correlation, the moisture content of the wood as a result of not having sat around long enough—or undergone an active process—to dry out. Specifically, "green lumber" is freshly milled lumber that has a variable, yet generally high, moisture content. Practically, green lumber is also frequently "rough cut," meaning it has not been planed, or "dressed," and is still rough

Lumber from a local sawyer lacks the quality control of an industrialized process, putting the responsibility of grading into the builder's hands—when code allows its use. PHOTO BY BENJAMIN RAY GRIFFIN, GREENER LIVING VERMONT.

in texture and larger in dimension than dressed lumber (dressed lumber is referred to as "dimensional lumber," as it has been planed down to a consistent dimension); green timbers and lumber can also be planed at the mill by request, for a fee. Green lumber is also generally "ungraded," meaning it has not been evaluated and certified to be of a specific quality, although often mills will apply their own grading system, depending on the product.

The fact that green lumber is ungraded requires the builder to be more selective in choosing appropriate-quality material, which involves communication with the sawyer, more time spent evaluating material in the yard, and/or purchasing additional material for on-site grading. It also may factor into code requirements specifying the use of certified graded lumber, which may limit or even fully restrict the use of ungraded green lumber, at least in a structural capacity. In contrast, most lumber carried by building supply centers has been kiln-dried, planed, and graded for quality; it is all of uniform size and moisture content and can be predicted to behave all in the same way in the structure.

Green lumber has quite a few drawbacks: In addition to the potential material, time, and code liabilities of using ungraded material, rough lumber is much harder on the hands and is generous in its release of splinter material. Depending on the milling operation, it can also be quite dirty, which is tough on knife-edge tools such as planers, chisels, and saws; this is especially the case for site-milled lumber that does not have the luxury of the controlled environment of a permanent mill. The increased moisture content of green lumber, though, is the greatest obstacle to overcome; not only is the lumber significantly heavier as a result of the entrained water, but what happens when that moisture leaves can be unpredictable at best and damaging at worst. What happens as a piece of lumber dries depends on the species, how the wood is milled, drying conditions, forces imposed upon the wood as it dries, the size and shape of the wood, and surface treatment of the wood.

Wood twists, bows, cups, and shrinks as it dries, making its use while still green tricky. PHOTO BY KELLY GRIFFITH.

Since we can anticipate lateral shrinkage (wood is linearly stable upon drying), in addition to checking, warping, bowing, and cupping, the employment of appropriate design and construction technique will help either avoid these issues or keep them from interfering with the function of the wood. But the power of a drying timber is intense, and it can be difficult, if not impossible, to control. This can lead to beams that rock off posts, joists and rafters that hump and twist out of plane, studs warping, drywall screws or lath nails popping, floorboards shrinking and causing gaps to open, and window and door frames distorting and causing operable fenestrations to stick or lodge. All of these issues take more time to fix, and therefore cost more.

So with all of these liabilities, why bother using green wood at all? There are many benefits to using green wood, whether by virtue of its greenness or despite of it; quite often, it is either the path of least resistance or worth the resistance born of its limitations. Here's a look at why, in the world of natural building, we generally favor the use of green wood.

USE OF A LOCAL MATERIAL

As we are already well aware, the benefits of utilizing local materials are far-reaching. Unless you live in forested (oftentimes plantation-laden) regions of the northwestern United States, British Columbia, parts of eastern Canada, or parts of the southeastern United States (southern yellow pine country), where the majority of dimensional lumber is grown (Malin 1994), chances are your locally sourced lumber will only be green lumber, cut from surrounding woodlands and milled by a regional sawmill. In Vermont, there is generally still a small—frequently generations-old—sawmill every 100 square miles.

The social and ecological benefits of a local approach to lumber production, distribution, and use are many. For one, you have scale-appropriate woodlot management practices that are frequently performed by the very people who live within these forested regions. This does two things: it encourages sustainable silvicultural practices, as the loggers who manage the woods must consider their long-term health and production potential, and it keeps the majority of the profit generated by logging in the region where the industry is based, as opposed to being controlled by far-flung corporate headquarters and dispersed widely to groups of shareholders and "middlemen." Buying wood harvested from regionally or locally managed woodlots contributes to a viable regional economy and encourages a healthy, working landscape that is maintained for the benefit of all, human and nonhuman alike. The same goes for the processing industry; loggers would have no market were it not for regionally based sawmills and other wood manufacturers able to accept product at a smaller scale.

There are additional benefits for the consumer of using locally harvested wood, in addition to reaping the indirect benefits of healthy and prosperous natural and social ecologies. Use of local materials helps reduce the embodied energy of our building materials, as less carbon is used to transport timber across the country. This reduction is amplified when using material that is harvested directly from the building site. This not only reduces the footprint of the material even further but gets closer to the spirit of natural building, which includes developing deep and numerous relationships. The experience of cutting, milling, and ultimately crafting a shelter from the woods we call home is a way to illustrate how we are all supported by our environment in a direct and unique way, and it can empower us to be active members of, and participants in, that environment.

CONTROL OVER THE PRODUCT AND DESIGN ACCOMMODATION

When ordering dimensional framing material from the local lumber store, we are generally not given the option to request a specific species. Standard lumber is S-P-F, or spruce-pine-fir mix. There are generally a few species we can request by name—western red cedar, Douglas fir—in limited dimensions and at inflated prices, which perhaps more accurately

Multigenerational family-owned small-scale wood-product manufacturers, such as Hartshorn's Mill in Waitsfield, Vermont, are still alive and well in many forested regions in the United States. PHOTO BY MICHAEL BRODEUR.

reflects the true cost of production as a result of a smaller scale. But if you are looking for milled black locust as a rot-resistant sill plate, forget it—you won't find that at Acme Building Supply. Nor will you find 2 × 3 siding nailers. Your design is reliant on the pre-manufactured sizes of the material, often resulting in excessive material usage or inconvenient layout. While some mills are notorious for producing inconsistent or undersized lumber, a good mill (which is almost always also a mill that processes regionally harvested trees) will be able to deliver highly accurate and consistent results of any size requested. Whether you are looking for timbers of engineered dimension, lumber cut to the exact size necessary for the job, black locust plates, or curved cherry braces, the ability of local mills to mill species-specific custom orders of green lumber is a tremendous benefit to the designer and builder alike.

A portable wood mill facilitates use of site-cleared timber for building materials. PHOTO COURTESY OF BENJAMIN RAY GRIFFIN, GREENER LIVING VERMONT.

STRENGTH OF MATERIAL AND EASE OF USE

Rough-cut green lumber is stronger than its dimensional counterparts, presuming comparable quality of material (remember, your rough-cut

Log Cabins

Log cabins are a popular building style in North America, appreciated for being both rustic and elegant at the same time. Traditionally—meaning, when log construction was founded by need, rather than as a chosen architectural style—logs were sourced from the abundant forests of the settled region, either hewn, partially hewn, or left round, and then stacked on top of one another and notched together with any of a variety of different notching patterns, depending on the culture of the builder. Gaps between the logs were filled with saplings, moss, mud, or other similar materials. Log construction was brought to North America originally by Scandinavian settlers in the 1600s, and later by German and Swiss settlers; the style became an iconic form of settler construction and was valued for its ease and durability. Log cabins were abandoned in favor of more commercially viable structures of lumber or brick, until their popularity was revived as a rustic architectural style. Originally, log construction was a true natural building form, not unlike timber framing, that made use of locally harvested materials crafted to create durable, functional shelter.

Today, however, we see a host of issues with log construction that make it hard to classify as a natural building form, especially when constructed in what is now the conventional format. For one, the logs are rarely harvested sustainably from local woodlots. By far, the convention is for owners to buy and assemble kits—or hire contractors to assemble kits—sourced from the forests of western North America and trucked across the continent; it is quite common for these kits to be the products of industrial forest-management processes quite different from the selection practices of early settlers. Because rot and decay are a major concern when building with wood, most log homes—even those built of rot-resistant species—are treated heavily with chemicals to keep them from decomposing; this chemical treatment is repeated throughout the life span of the house as part of routine maintenance, exposing both workers and occupants to harmful chemicals.

Tree harvesting is another consideration. Logs for log home construction must be uniform and relatively large; the process of harvesting and procuring these large logs can lead to "high-grading" in a forest, which can result in a harvested area with only uniformly small and/or low-grade trees remaining, which negatively affects the health of the forest. Additionally, it takes more board footage to build a log building than to frame a conventional structure, since space taken up by the insulation that fills the open-cavity framing is otherwise filled by the depth of the logs themselves. This brings us to performance, another drawback. Given the relatively poor insulation value of wood—approximately R-1/inch—additional insulation is needed in colder climates, which leads frequently to the construction of insulated interior walls, further increasing wood consumption and essentially double-framing a building (in that the insulation wall could easily be made to be structural). Finally, the chinking between the walls must be done with great care to avoid massive convective losses; this is generally accomplished with either cement- or synthetic-based chinking materials, which increase embodied energy and/or petrochemical content of the wall system.

It is possible to build a natural log structure (a friend of ours chinked his log home with sheep's wool and lime plaster), but given the maintenance and durability issues, performance problems, and resource considerations, we feel there are better options for creating a natural building.

lumber is most likely to be ungraded). Considering that a dimensional 2 × 4 is actually only 1½ × 3½ inches, a rough-cut 2 × 4 will be the full dimension, or close to it, and thereby stronger simply by virtue of the fact that it is larger. This may help when making engineering decisions about the structure, and potentially lead to overall reduced wood use in the frame. In post-and-beam construction (see the next section), a choice may even be made to use a round beam instead of a squared beam to increase strength; maintaining the integrity of the grain pattern makes an unsawn timber stronger than a sawn timber of the same size. And despite all the liabilities of lumber drying in place, certain timber-framed joints can be designed and built to increase in strength as the timbers shrink, tightening the joint as they dry. Although not directly related to strength, an additional benefit in the world of timber framing is the ease of working with a green timber compared to a dried timber, especially when using hand tools to cut the joinery. In fact, this goes for framing lumber, too; the ease of driving a screw through green hemlock versus dried hemlock is quite substantial.

Framing Systems

Now that we have a good feel for the differences between working with green lumber and dimensional lumber, and the considerations at play in determining the best framing system for the job, let's explore our options.

POST-AND-BEAM

Post-and-beam framing systems are those in which larger, specifically engineered structural members (such as posts, beams, braces, and girts) are used in a custom-designed and engineered structural system. Large vertical members, or posts, carry large horizontal members, or beams. The sizing, spacing, and orientation of the members are all designed intentionally to address the different structural stresses identified in chapter 6. This broad category of construction has many subsets. Wood timber joined with metal fasteners is called heavy timber construction, or frequently just post-and-beam; we will be referring to this style, in this language, in this section. Timber framing (see the next section in this chapter) is a traditional form of post-and-beam construction featuring the use of wood-on-wood joinery as the fastening mechanism. Steel construction (also covered later in this chapter) is frequently built in the post-and-beam style as well, although steel studs are also commonplace.

Because post-and-beam frames have widely spaced posts, they offer little support and no substance for wall enclosures. They provide purely structure, which frees up wall enclosures to be just that—enclosure systems—with no need to serve a structural role unless designed to do so, in which case they are generally utilized for sheer strength. Occasionally a post-and-beam frame may be designed to be buried in a wall, but given the size of the frame, the corresponding increased value in the materials (larger timbers means more cost), and the opportunity for an architectural statement, post-and-beam frames are usually left exposed to the interior. This being the case, their construction generally requires a higher degree of attention and care—from the material selection and handling to the quality of construction and the selection and placement of fasteners—to ensure the frame is serving both structural and aesthetic mandates. Standard wall enclosures today are either large foam panels screwed to the outside of the frame or stud walls built surrounding the frame (see later in this chapter for more on stud-wall construction). While the latter option can be used successfully in a natural building—especially when double-stud-walled and insulated with a natural insulation—other options, such as straw bale or earthen wall construction, can eliminate the need for any further framing material while creating wall structure, insulation (in the case of straw), air barrier, and finish, all in one. That being said, if wood siding is desired over straw bale walls with an interior post-and-beam frame, essentially an entire second frame must be built to the exterior of the bales to support the siding, driving up the amount of materials used, complexity of detailing, and ultimately cost (see chapter 14 for more information on siding over straw bale walls).

Speaking of cost, generally post-and-beam frames are more expensive than stud-wall frames (see "Stud-Wall Framing" section of this chapter), for a few reasons. For one, the design costs are higher, especially if not using or modifying an existing design, because the engineering must be done more accurately and the structural design executed with more precision. For another, the materials costs are generally higher. Larger timbers are more expensive (cost per board foot) than lumber, and if not purchased from the local mill they can cost significantly more for custom milling and long-distance transportation. In the event a longer span is needed than can be accommodated by natural wood, engineered lumber—such as glued laminated timbers ("glulams"), laminated veneer lumber (LVL), or wooden I-beams—can be used, although this only increases cost and the liabilities of the use of an industrially intensive material. Additionally, and perhaps surprisingly, post-and-beam frames can often lead to increased wood use in a structure—especially when compared to light-framing stud wall techniques—unless specifically designed to reduce material use. This can be due to overengineering, oversizing members for aesthetic reasons, or complexities in the structure requiring additional framing members. Because of their heft, post-and-beam frames are also harder to build, requiring more heavy equipment (such as specialty saws or cranes) and/or human labor to craft and move than their light-member counterparts.

That said, if cost is not a critical issue for you, it can be argued that you get what you pay for with a good post-and-beam frame. For one, a post-and-beam frame is much more durable than a stud-wall frame. Not only is the framing system itself stronger, but when oriented to the interior of the structure, the frame is no longer placed in the most vulnerable position in the wall—the condensation plane, adjacent to the exterior skin of the structure. By keeping the frame on the "warm" side of the insulation in a cold climate, and exposed to inspection and drying, the frame will have very little chance of rotting or being damaged by termites. It could be argued that fire is of enhanced risk, yet timbers are far more likely than studs to char than burn. And reuse is much more likely with a post-and-beam frame than with a stud frame in the event that the structure need be reused in the future (see the next section in this chapter). But ultimately, one of the biggest incentives for choosing a post-and-beam frame is the visual strength and geometry an exposed heavy-timber frame lends to the interior of a building. In harmony with the aesthetics inherent to other natural materials such as plastered earth, straw, and native stone, timbers lend a natural, organic texture and offer a sense of security through the exposure of the massive structure they provide.

TIMBER FRAMING

As mentioned earlier in this chapter, timber framing is a style of post-and-beam framing. It is defined by its traditional, craft-based approach in which wood-on-wood joinery (such as the pegged mortise-and-tenon method) is used as the primary mechanism to fasten

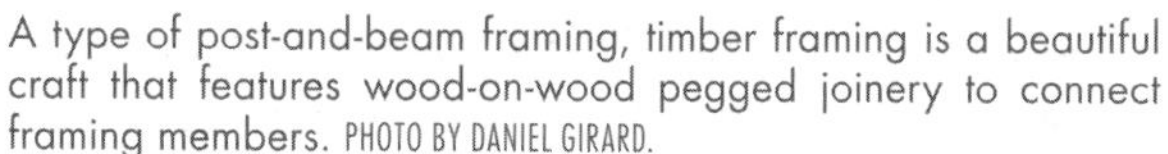

A type of post-and-beam framing, timber framing is a beautiful craft that features wood-on-wood pegged joinery to connect framing members. PHOTO BY DANIEL GIRARD.

Post-and-beam wood framing features heavy timbers connected together, in this case with metal brackets. PHOTO BY JOSH JACKSON.

the frame together, and is not unlike an enormous puzzle. There is a long and storied history of timber framing throughout the world and throughout the ages; most familiar to us in North America are the traditions of Europe and Japan.

Certainly, major elements of the attraction to timber framing—compared to mechanically fastened post-and-beam—are the rich tradition, the aesthetic of all-wood construction, and the enhanced craftspersonship invested in the project. However, this enhanced craftspersonship goes hand-in-hand with a higher required skill level, which invariably increases cost and the necessity of access to skilled timber framers. Another cost liability—and in some cases, even a feasibility issue—appears in code jurisdictions that demonstrate a distrust of timber frames. It is standard in some building codes for the shear strength of the diagonal shoulder and knee braces in timber frames to be discounted, requiring that additional (read: costly) shear support be added to the frame. It can be difficult to find an engineer that understands both timber engineering and mortise-and-tenon joint design well enough to advise, and ultimately sign off, on a timber frame design. And in some cases, the use of mechanical fasteners is simply mandated.

In regard to strength, timber joinery is generally stronger than mechanically fastened joinery, as the joints are tighter and designed to maximize strength and durability of the joint over time; in fact, they can even increase in strength as the timbers dry, as mentioned earlier in the chapter. This style of joinery arguably makes it even easier to reuse frames; there have been many houses built from reclaimed timber-framed barns across New England, and there is a brisk business in reclaimed timbers and timber frames. Therefore, it can well be argued that the higher up-front value of the timber frame will pay for itself over the life of the frame—providing the owner's progeny are around hundreds of years later to continue to reap the benefits! It would be fair to say that timber framing is at the far end of the spectrum in terms of both up-front cost, skill level required, strength, and durability, whereas a mechanically fastened post-and-beam frame could be considered something of a compromise in an effort to balance these different values.

Stud-wall construction allows for fast production with small crews using lower-value timbers than for most post-and-beam frames. PHOTO COURTESY OF YESTERMORROW DESIGN/BUILD SCHOOL.

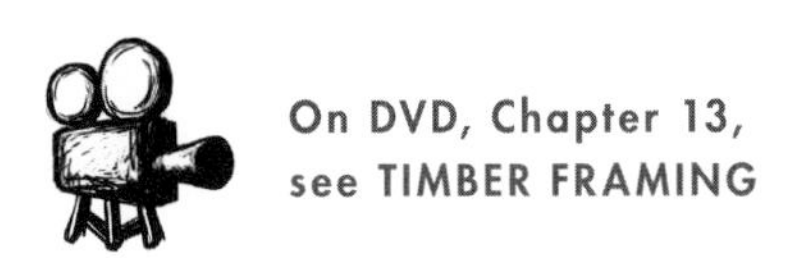

On DVD, Chapter 13, see TIMBER FRAMING

POLE FRAMING

In chapter 12 we introduced pole foundations as a combined foundation/framing system. While a more detailed description is presented in that chapter, it is worthwhile to say here that for the sake of comparison a pole frame can be considered in the class of post-and-beam frames, with the foundation poles acting as the primary posts off which the remaining framing (floors, walls, roof) can be constructed.

STUD-WALL FRAMING

As we learned in chapter 1, stud-wall construction has been the standard of modern construction practices in North America since the mid-nineteenth century following advances in building technology as a result, of the American Industrial Revolution, specifically in lumber-milling and nail-production technology. Stud framing (sometimes referred to as stick building) takes a different approach to structure than post-and-beam: rather than singular, intentionally placed framing members, a series of smaller framing members are joined together to create a repetitive framing system. The load is accepted and distributed more widely by the wall as a whole. In practice, the use of smaller, repetitive framing members plays out in a few ways. From a design standpoint, things are simplified. A pattern of framing is established, and as long as that pattern is followed—for example, stud placement 16 inches on center (OC)—no further custom engineering or design analysis is needed; the pattern is predictable and universally accepted. Using smaller repetitive members is also a lot faster and easier to construct. The framing members are easy to pick up, there is no fancy or complicated joinery, the walls can be built quickly on the ground with pneumatic and power tools and tilted up easily with a small crew, and the repetitive nature encourages efficiency of production in the field.

This ease of construction and its ubiquitous presence in modern building practice mean that trade support for this style of construction is abundant and significantly less expensive than for timber framing, which is a much more specialized skill. From an ecological standpoint, the use of smaller lumber can

Bamboo Framing

In many places in the world, bamboo grows with terrific abundance. We see bamboo crop up in buildings in many different ways such as wattle (in woven walls; see chapter 16), furniture, and flooring. (In chapter 19 we look a bit more extensively at bamboo through its use as a flooring product.) Bamboo has also been used in vernacular style across the globe as a structural material. Often associated with lower-class or temporary housing, bamboo is only now getting the credit it deserves as a high-performance structural material. Abundant, highly regenerative, and incredibly fast-growing, bamboo far outpaces wood in its growth time to maturity for harvest and its yield per acre, and it also proves to be more structural per unit size than wood; it is even stronger than concrete under certain loads. Given its light and flexible nature matched with this strength, bamboo is comparable to steel and is favorable for seismic designs. In fact, bamboo buildings have withstood magnitude 6.7 earthquakes without structural failure. This performance has earned some bamboo systems approval in ICC and other code organizations as a viable structural material.

When used for framing bamboo is most frequently used in a post-and-beam style, and infilled with any number of wall systems. While bamboo was traditionally lashed, advances in fastening technology are overcoming the hurdles of joining the round poles soundly together; metal fasteners, mortared joints, and other systems are all being used to make bamboo a more efficient and stronger building material. These systems are being employed to create incredible structures: a 28-foot cantilevered roof capable of supporting 11 tons, a 160-foot-span bridge, and a museum over 55,000 square feet in size. No longer the stuff of indigenous owner-builders utilizing a material growing abundantly in their locale, bamboo is becoming mainstream with the building industry at large; companies such as Lamboo are producing structural bamboo products, which they claim to be ten times stronger than wood in tension and 20% more stable with moisture and temperature conditions. Prominent architects are using bamboo as structure in their designs, including Colombian architect Símon Vélez, an innovator and leader in the field of bamboo architecture.

Bamboo does have its downsides. It is very labor-intensive, for one. Availability is difficult in many parts of the world, for another. A major consideration is bamboo's vulnerability to pests and decay; it must be protected from the elements and is often treated with borates or other preservatives. While we in cold climates do not have the benefit of regionally grown structural-grade bamboo, we can expect to see both its use and availability increasing as renewable resources continue to be sought for building solutions.

theoretically allow for more efficient yields from the woodlot—more "sticks" can be carved out of any given tree, and lower-quality and smaller trees can be harvested and effectively utilized. That said, two factors level this playing field. For one, the use of forked posts, curved braces, and other lower-valued timbers in custom timber-frame design can make use of trees that would not be suitable to mill even

Including forked trees in a frame makes use of low-value trees that might otherwise become firewood. PHOTO BY JACOB DEVA RACUSIN.

into studs. For another, the wood for studs is often procured through the conventional logging practice of clear-cutting large swaths of monocultured genetically engineered plantation trees, a practice that clearly has negative ecological ramifications. Only through selective cutting and sustainable silvicultural practices for stud-wood sources can the benefit of more efficient yields in stud-wall construction be argued when compared to heavy timber construction, the timbers for which are more likely to be harvested in a similarly responsible fashion.

Returning our focus to the building, there is an additional potential benefit of stud-wall framing in that the basic form of the wall is created by the frame, enabling a more direct process of enclosure in comparison to post-and-beam, in which separate walls need to be constructed around a structural frame. This is a potential benefit, realized only if the straight-walled form is desirable; as we will discuss in chapters 14 and 16, it is often divergence from this static form that lends character to and defines the design style of a building, as found in the curves, niches, and undulations of many natural wall systems.

Highlighted as benefits of a timber frame, durability and aesthetic form earn lower marks for stud-wall construction. Stud walls are not only weaker structures than well-designed post-and-beam frames, they are also generally more vulnerable to rot and decay as a result of their placement within the wall system, generally close to the exterior of the wall where interstitial condensation is most likely to form (see chapter 8) and susceptibility to termites is increased. While this does not inherently have to be the case, we tend not to see exposed stud walls held to the interior of a structure's insulation plane. Why? Because they aren't much to look at, which brings us to aesthetics. Again, if we are looking for a smooth wall without the interruption of framing, then this is no drawback; in fact, plastering over a stud-framed, straw- or wood-insulated wall can be worlds easier than working over a timber-framed wall, where the edge of the timber frame regularly interrupts the field of plaster. But presuming an appreciation for the form and contrast that a visible post-and-beam frame provides, a hidden frame can certainly pale in comparison. The presence of the frame within the wall plane also limits the options for wall materials; it would be difficult to cob or lay adobe block around a stud wall, and even straw-bale-wall installation is complicated by having to embed the frame within the insulation plane.

The modern stud wall as it is conventionally deployed in Western platform construction has many performance drawbacks, beyond its propensity to rot. Although less wood is used than with an overbuilt post-and-beam frame, stud-wall framing still requires quite a lot of trees to get the job done, and there is much waste in the practice that can be avoided through more thoughtful design. Additionally, thermal losses abound, especially when the structure is insulated in the conventional model of installing batts of insulation between the stud bays. There are significant conductive losses not only along the studs themselves, but also at the sill plates, midstory band joists, and top plates where rafters are received along the eave walls. And as we discussed in chapter 7, batts also open up the likelihood of convective losses through interstitial convective cycling. There are improvements over standard Western platform construction to address these problems. Regarding resource consumption, there is a value-engineered framing system called advanced framing technique (AFT) or optimum value engineering (OVE), in which strategic design decisions are made to reduce wood usage in the frame. Promoted by organizations such as the National Association of Home Builders and the United States Department of Energy, OVE techniques are code-compliant, thoroughly detailed, and well vetted for both structural performance and cost and resource savings in documented field testing.

In the OVE system, rather than a 16-inch OC layout, a 24-inch OC layout is used; instead of doubled top plates, single top plates are used, and rafters are laid out to bear directly over studs. These and other design modifications can potentially reduce wood consumption by 11% to 19% and overall framing costs as much as $1.20 per square foot. This reduced framing usage also reduces the conductive heat losses from thermal bridging. In one project, annual heating and cooling costs were calculated to be reduced by almost 30% when compared to a conventionally framed building (PATH/NAHB n.d., 2). It should be noted that, given the customization and restructuring of the framing pattern, greater investment in time is needed both in design and in the field, especially initially while designers and builders alike climb the learning curve of the modified system.

Even with the reduction of conductive losses through OVE, significant losses still occur by virtue of the framing and insulation strategy. Additionally, convective cycling and heat loss due to poorly installed batts are not addressed, nor is the fact that the amount of insulation is limited by the thickness of the studs themselves. A more proactive solution to the heat-loss issues involved with stud-wall construction is to use a double-stud-wall framing system, as described in chapter 7. Double-stud walls reduce or even eliminate any through-wall conductive loss, increase the insulation to the desired thickness, and, depending on the insulation used, reduce or eliminate interstitial convective cycling. Structurally, there are a few ways to approach the inclusion of double-stud walls. For new construction, either one or the other, or potentially both, of the walls can bear the structural load of the roof. Care must be taken here to clearly identify where the load paths are translating and to ensure that adequate structural support in all planes is engineered into the design.

On DVD, Chapter 13, see DOUBLE-STUD CONSTRUCTION

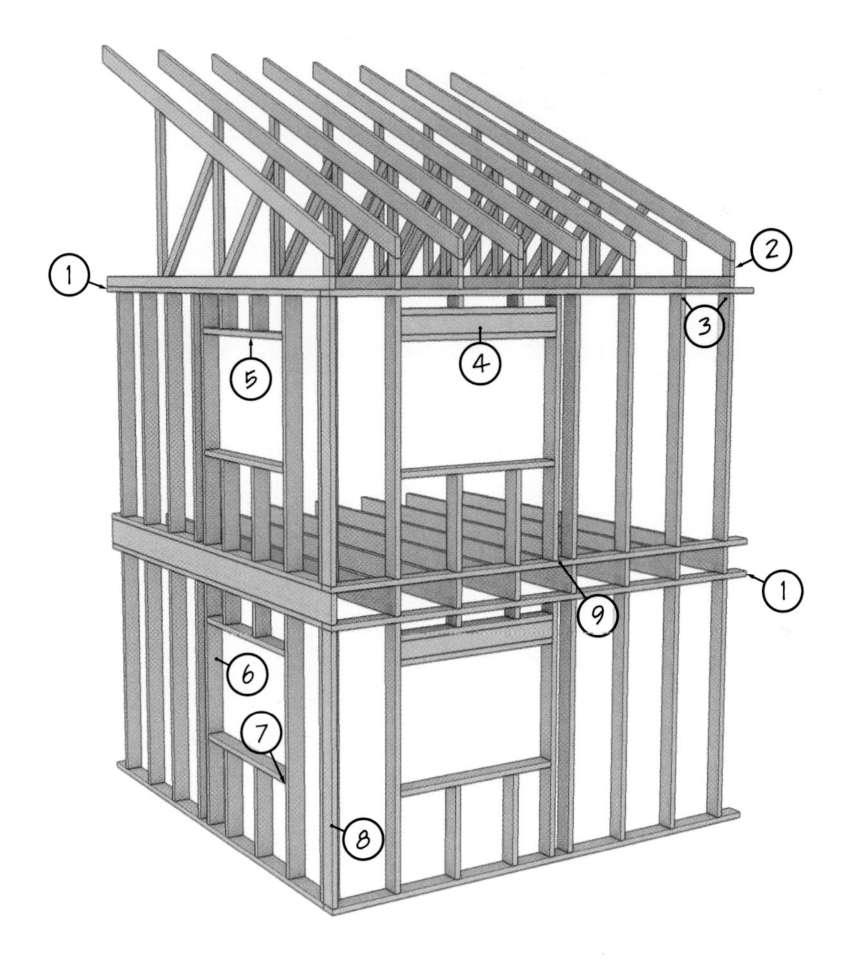

1. SINGLE TOP PLATE
2. RAISED HEEL
3. ROOF FRAME ALIGNED WITH WALL/FLOOR FRAME
4. SINGLE HEADER
5. NO HEADER IN NON-BEARING WALL
6. SINGLE STUD AT ROUGH OPENING
7. NO CRIPPLES UNDER WINDOW OPENING
8. TWO-STUD CORNERS
9. POINT LOAD TRANSFERRED BY RIM JOIST OR SOLID BLOCKING

Advanced framing technique (AFT) involves a series of framing design details to reduce wood consumption and minimize thermal bridging in single-stud frames. ILLUSTRATION BY AARON WESTGATE/FREEFLOW STUDIOS.

Steel Framing

Steel is used for framing material in a variety of styles, frequently in commercial and institutional construction, and increasingly in residential construction as well. Arguments can be made to favor the use of steel framing over wood, either as light-gauge steel studs or heavy-gauge post-and-beam framing. Steel is stronger and more dimensionally stable than wood and can be manufactured to spec, therefore reducing overall material use and simplifying the construction process overall. Steel is more durable in the presence of moisture. Steel manufacturers claim that steel is also more resistant to fire, although this claim is subject to debate as most steel loses strength quickly under high temperatures. Steel also incorporates high levels of recycled material—anywhere from 20% to over 90%, depending on the manufacturing process—and is itself highly recyclable at the end of its functional life. And in recent decades, the processes of both resource extraction and manufacturing for steel have improved dramatically in terms of environmental impact and energy consumption.

All that said, however, steel still has a significantly greater ecological impact when compared to wood, even considering industrial timber harvesting and wood manufacturing processes. Regardless of recycled content, steel production still relies heavily on open-pit mining and a very energy-intensive production and manufacturing process. The process of recycling steel in and of itself is an energy-intensive process, and although it does greatly decrease the embodied energy compared to virgin steel, even high-recycled-content steel has a higher embodied energy than sawn softwood. Another major drawback to the use of steel is its poor thermal performance—400 times worse than wood (Malin 1994). If steel is to be used in a building, it must be isolated thermally from conducting heat through the envelope. And despite the impacts of scale and the propensity for overuse of wood materials within a given structure, building with wood is still making use of a minimally processed, renewable, natural material, resplendent in its vagaries and inconsistencies. Steel—although sourced originally from the earth in the form of iron ore and combined with coal, limestone, and zinc—is so heavily processed and so greatly removed from an accessible manufacturing and production scale that it ceases to be recognizable to us as anything but an industrial material created in an industrial process. While there is great potential for small-scale steel remanufacturing plants to re-create steel products from existing scrap in a more accessible way, this is certainly the exception rather than the rule when evaluating the use of steel in today's natural building (Malin 1994).

In retrofitting an existing building, there are two options. One is to build an additional stud wall into the building; this keeps the shell of the structure largely intact but reduces interior floor space and can be very complicated in accommodating reinstallation of mechanicals and built-ins, especially in kitchens and bathrooms. The other is to build an additional stud wall to the exterior. Although pouring a foundation extension or structurally extending the floor deck is an option, it is costly and complicated. An easier solution is to use a Larsen truss system, developed by Canadian builder John Larsen in 1981, in which vertical ladder-trusses are affixed to the building, either to the sheathing or built off the studs themselves. The wall is tied together and back to the building through a plywood bottom plate and tied into the roof at the soffit or attached directly to the rafter tails. If the existing rafter overhangs are minimal to begin with, eave and/or gable extensions may be required, which will make the project more complicated and costly.

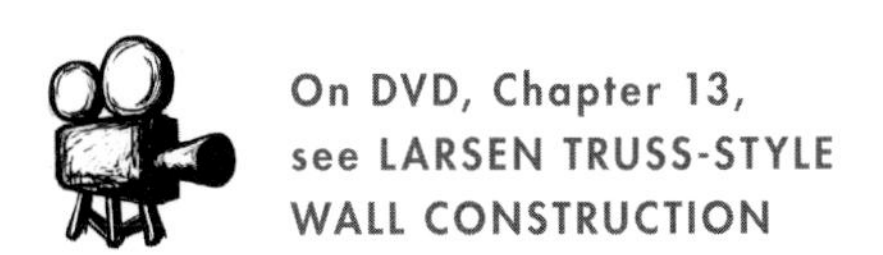

HYBRID/CUSTOM FRAMING

It is not uncommon at all for us to use a combination of framing systems in a project. For example, we will very often frame a stick-built roof system, and even stud-built gable-end walls, on top of a timber frame.

It is also possible to build a stud-framed shell, for the ease and speed of construction, yet incorporate timber-framed elements to the interior, such as exposed timber post-and-beam instead of a load-bearing interior stud wall, or a post-and-beam loft or second-story framing. This same strategy can also be employed with load-bearing straw-bale-wall systems; while the perimeter structure of the building may be built of bales, there may well be a need for additional structural support in the interior of the building, depending on the structural design. In this case, a hybrid structure of straw-and-stud or post-and-beam framing may be employed. If this is to be the case, special considerations for ensuring these two

Trusses

Trusses are structural components made up of a series of triangular forms created by fixed pieces, or "members," of wood, steel, or other materials. Using the strength inherent in the geometry of the triangle, and the control of forces at the joints of the truss members, trusses can be used to span incredible distances, support dramatic loads, and create interesting and unique forms in a structural capacity. Trusses are used in all sorts of ways: floors, walls, roofs—even bridges are built of trusses. Of greatest relevancy for us is the truss roof. Clear spans within a building envelope can be achieved with trusses that may not be achievable with standard rafter construction. Trusses can be built on-site on the ground or can be pre-manufactured to spec and delivered to the site; they are then installed by crane, greatly reducing labor and risk to building a roof in situ on top of the frame. They are also very uniform, if built with care, and can greatly simplify finishing tasks, from installing fascia trim to hanging a drywall ceiling. While trusses may be more expense or trouble than necessary for some buildings, they can be both time and cost savers for many others, and they certainly bear consideration when evaluating options for providing structural solutions for your next project.

structural systems work harmoniously must be taken, particularly concerning the pre-compression and settling considerations of straw. A similar condition exists when window and door framing in the plane of the straw bale wall is used in a structural capacity (see chapter 14 for more information).

In the case of a bale-wrapped post-and-beam frame, there are also cases in which we will want to insulate a portion of the wall with another material. A good example of this can be found in the south wall of many buildings, where there are larger amounts of glass—in either windows or doors—installed over the face of the wall. At a certain point, it may be unwieldy to stuff bales into strange areas between all of these framed openings, and a more flexible form of insulation may be desired. Double-stud-wall construction can then be used for this section of the wall to contain the insulation of choice (our preferences being straw-clay, woodchip-clay, or cellulose; see chapter 15 for more information). Quite often, what results is not actually a hybrid structure—the post-and-beam frame is still asked to do the heavy lifting, and the double-stud wall just holds the insulation. However, this approach can be designed for the stud wall to take up the load of the beam above it, perhaps freeing up a post or two that might otherwise be in the way of a window. To be successful, the design must be very thoroughly calculated to ensure that the structure is not weakened or compromised in this transition, and that it is in fact practical to build even if all structural requirements have been satisfied.

CHAPTER 14

Insulative Wall Systems: Straw Bale

Straw bale and plaster walls are superb options for an aesthetically beautiful, and ecologically beneficial, wall system. They are high performing in terms of R-value, durability, moisture resistance, and interior air quality and comfort. In the northeastern region of the United States, the materials needed to construct these walls are often locally sourced, either from the building site or just down the road, and are sustainable regarding their life cycle. They are enjoyable materials to work with and offer a connection to ancient ways of building while also allowing the builder to meet today's stringent performance standards.

But straw bale construction is not a silver bullet; it is not the best choice in all situations, and certainly, if poorly installed, it can make that choice even less well suited. As in conventional construction, the quality of the design, the material sourcing, and the construction method will determine the level of performance. In this chapter we will cover design principles that we have found to be crucial to quality straw-bale installation, then assess the material itself, and finally dig into construction methods.

One of the objectives of this book is to illustrate that straw bale construction merits a substantive, legitimate place in the lexicon of performance-oriented building techniques. Another is to increase the knowledge of those who practice this type of construction so we can collectively raise the bar by making good design, excellent detailing, meticulous execution, and clean finishing hallmarks of straw bale construction.

Straw bale wall construction proceeds in five steps: (1) design; (2) bale preparation, including air fins, stand-aways, and toe-ups; (3) bale construction and installation; (4) plaster preparation; and (5) plastering. We will cover plastering in chapter 17. The first four steps will be covered in this chapter.

Step 1: Design for Straw Bale Structures

We are advocating a design/build process that infers an iterative, dialogical progression between designing and building a structure, in which knowledge of each aspect informs the other (see chapter 10). One of the challenges of straw bale construction is that it requires knowing a lot about the whole process of building from beginning to end, during all parts of the project; the materials sourcing and quality are integral to the design entailed in putting

The Ford-Farlice-Rubio family in front of their straw bale home in Vermont. PHOTO BY KELLY GRIFFITH.

Bales on construction site. PHOTO BY JAN TAYLOR ALLEN.

those materials together, just as the design is critical to the actual execution of the construction. It is impossible to think through, for instance, the interior footprint—an important piece of information for the design stage—without understanding the anatomy of a bale wall in section, which entails knowing what makes up the layers of a bale wall and why. And it is difficult to understand the importance of including things like plaster supports in the design (both rough framing and finish) without having built a plastered straw-bale-wall system that does not have this designed in advance, and discovering too late the mistake. Following are some common recommended structural types allowing the incorporation of bales into a structure, followed by key design concepts and principles that are crucial to understand and utilize early on in the design/build process in order to achieve a quality, long-lasting straw bale structure.

TYPES OF STRAW BALE CONSTRUCTION

The first principle to understand is that there is not just one kind of straw bale structure, there are many. There are three broad categories of straw bale installation for exterior envelopes: bale wrap, where straw bales "wrap" around timber elements that are left visible on the interior; bale infill, where bales sit in between framing and framing members are either visible inside and outside or are buried inside the wall; and load-bearing straw walls, where the straw bales support the weight of the roof loads with no additional framing. Within these three broad categories, we will discuss the five most common building types for the Northeast: (1) bale wrap with exposed interior frame, (2) stud infill, (3) bale wrap with exterior rain screen, (4) bales with buried box columns, and (5) load bearing.

There is a commonality among the first four types: wood framing provides the structure of the building, serving to support the roof, thus allowing the roof to be fully on and "dried in" (providing waterproofing protection) before the bales are installed—particularly useful in the cold and wet climate of the northeastern United States. By comparison, the fifth type, load-bearing straw-bale construction (also called Nebraska-style construction), is challenging to execute well in this region owing to the frequency of precipitation. Load-bearing walls necessarily require the roof to be added on top of the pre-compressed straw bale walls after they are fully built, which leaves the bales vulnerable to rain for many days during construction. In the dry climates of places such as Colorado, Arizona, and California this is more conceivable; however, in wetter regions this presents a serious challenge.

Among the types involving structural wood framing, we have found either the bale wrap or the stud infill to be the preferred methods. A full wrapping of insulation around the building, such as both of these methods provide, yields a much tighter envelope that is more secure from air intrusions.

The term "infill" refers to the practice that derives from Europe, often called "half-timbered," wherein the timbers or framing members run inside to outside, and the "infill" material along with its coating of plaster sits in between those members. This method of installing straw bale, or indeed any infill material (see the section "Other Mass-Wall Types" in chapter 16), was initially tried in this region and discarded by European immigrants because they found that the cracks that opened up along every edge where the infill met framing members yielded a very drafty and cold building. Why did this work in Europe but not in the northeastern United States? The same reason that roses grow in Europe where at the same latitude in the United States they falter. There is enough of a climatic variation, due to ocean currents, that leaves the United States with a temperature swing of 100 degrees, while Europe experiences only a 60-degree swing. This climatic variance, which causes wood to shrink and swell and open cracks between itself and any abutting material, is one reason this approach was abandoned by the inhabitants of the Northeast in favor of more fully integrated wall systems. Additionally, wood framing members running inside to outside introduce a thermal bridge, which diminishes the thermal performance of any structure.

As discussed above, the load-bearing straw bale wall, although desirable from a thermal and plaster-continuity standpoint in that it is a continuous bale wrap without framing to break it, suffers from its vulnerability to moisture during the building process, making it less desirable in the Northeast than either the bale wrap or the infill method.

Bale Wrap with Exposed Interior Frame

In the first type, the bales and plaster wrap around the exterior of the structural timber frame, leaving the timber frame exposed to view on the interior of the structure. This is similar to how a structurally insulated panel (SIP) is used in combination with a timber frame, which is why this straw bale form is at times referred to as natural building's answer to the SIP panel. The SIP is a thick foam board that is sandwiched between plywood sheathing on one side and drywall for interior finishing on the other side. The foam acts as the insulation; the two skins provide the basis for exterior weather protection on one side and the substrate for interior finishing on the other. In the straw bale/plaster version, the straw is the insulator, and the plaster skins on either side provide both the interior finish and exterior protection.

Bale Infill

In the second type, the structural stud walls are framed such that a bale can be notched to fit cleanly into the cavity between the studs—18 inches on center for bales on-edge and installed upright—"on end." (For more on bale orientation, see "Know Thicknesses" in Key Design Concepts on page 179.) Bales laid on-flat can be notched into common framing patterns (16 or 24 inches on center). As a bale is much thicker than the depth of a 2 × 4 or 2 × 6 stud, the bale fills the cavity and its remaining thickness sits to the interior of the stud wall. After appropriate treatment and preparation of the embedded studs, plaster is applied to each side of the bale wall. This system is becoming quite popular as it combines the simplicity and familiarity of common stud-wall framing with the straw-bale and plaster-wall system. Also, it ends up being quite cost efficient when compared to the timber-frame/straw-bale method. Timber framing is a more expensive framing choice than conventional stud framing because it is more labor intensive. Also, in a stud-framing system, if the studs are spaced appropriately, the notches in the straw bales can be cut in an assembly-line manner, which saves time on the bale cutting and installation. Plastering with this system also amounts to a significant reduction of the amount of "edge" (i.e., all of the places where transitions occur between framing and bale/plaster), which results in a more streamlined and faster plastering process.

Bale Wrap with Exterior Rain Screen

In the third type, stud walls can be framed to whatever spacing and thickness desired, and the bales are placed against the interior face of the studs. The studs must first be prepared with air fins (see the "Air Fins" section later in this chapter). Then clay plaster is applied on the exterior between the stud bays, followed by siding, and the interior receives full plaster to finish. This approach is also gaining in popularity, owing to the ease of installation, if wood siding meets the design goals. The bales need

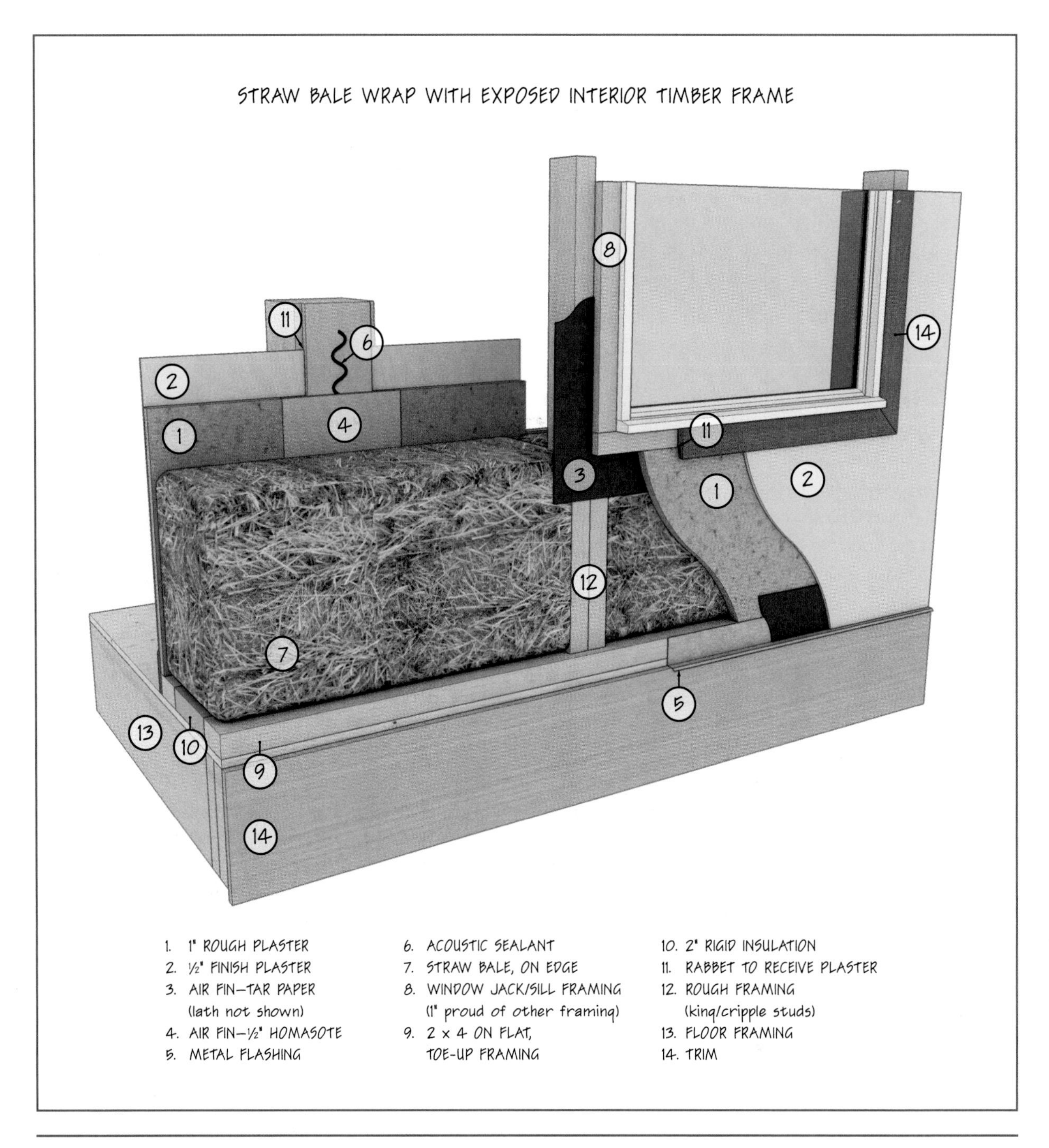

Section drawing of timber frame/bale wall, finished with plaster, showing planes, air fins, and the principle of "rough to rough, finish to finish." ILLUSTRATION BY AARON WESTGATE/FREEFLOW STUDIOS.

no notching at all for the framing, as they sit to the inside of the frame; however, each stud must be treated with an air fin, and plastering between the studs on the exterior before the siding is installed is more time-consuming. The plastering process on the interior is streamlined and faster because of the large, uninterrupted surface areas.

This approach is known in conventional construction as a vented rain screen, and vents must be added on the bottom of the stud wall and at the top to vent any condensation. It is also important in this assembly to resolve the discrepancy that exists between the window and door framing and the insulation plane. If the windows are framed all the way to the exterior—as they should be in a cold, wet climate—they will be floating outside of the insulation plane in the vented, uninsulated rain screen. A technique for addressing this is to create a framed insulated box around the window/door rough openings (ROs) using blown cellulose insulation, recycled foam board, wool insulation, or some other type of insulation. Fully insulating the stud cavity with cellulose insulation—an approach used by natural builders seeking to push the thermal-performance edge of straw bale wall systems—provides an opportunity to increase the insulation value of the wall system substantially, although further detailing concerning sheathing, siding venting, and the use/placement of exterior plaster must be addressed. Note that siding can also be incorporated into any of the other wall types through the use of additional framing, or in the case of the stud infill, by adding nailers to the face of the studs to support siding proud of the face of the exterior plaster.

Straw Bale with Buried Box Columns or Timbers

The fourth technique entails burying framed box columns—similar to posts, but made of plywood or less-desirable timber stock screwed together to the desired size and insulated—inside the plane of the straw bales. Like the stud-framed wall in the "straw bale notched into 2 × 4 studs" approach, they are surrounded by straw bales and encapsulated in the wall system. The benefits of this system are that box columns allow for larger spans in between structural posts than stud framing does, as timbers in a timber frame do; yet they are not meant to be attractive and left exposed, so they do not interfere with a large field coat of plaster. Buried box columns do not create a thermal bridge from inside to outside and therefore do not compromise the thermal envelope. Lastly, costs can be minimized, somewhat, with this system since the columns do not have to be of a finish-grade material. Our friend and colleague David Lanfear of Buffalo, New York, uses what he calls "superstuds," which are essentially site-built buried box columns, as one example of this system.

Load Bearing

In the final type, the straw bales support the weight of the roof; there is no vertical wood-framing support. Bales are stacked on the foundation walls or floor, with windows and doors framed in beefed-up "bucks"—rough window-and-door-framed openings—that "float" between straw bale courses. Walls are compressed using strapping, and the roof is then framed or set on top of the bales. Benefits of this system include use of less wood, and efficiency in the sense that bales do double duty as structural support and as insulation. Much has been written on this system, mostly for use in dry climates that have little chance of rain while the walls are being built (and the roof is not on yet). In the northeastern United States, where rain can happen any day of the building season, the most common way to practically achieve this method is to put up a giant tarp or circus tent to protect the straw before the roof can go on.

Another less-discussed but equally important reason why this method is not used more frequently in the Northeast is that the cement stucco frequently applied to load-bearing assemblies (because it adds greater compressive and shear strength to the walls) should be avoided with the high presence of moisture and vapor in buildings in our climate. In the wet and cold regions of the United States, the relatively impermeable cement stucco causes straw to rot,

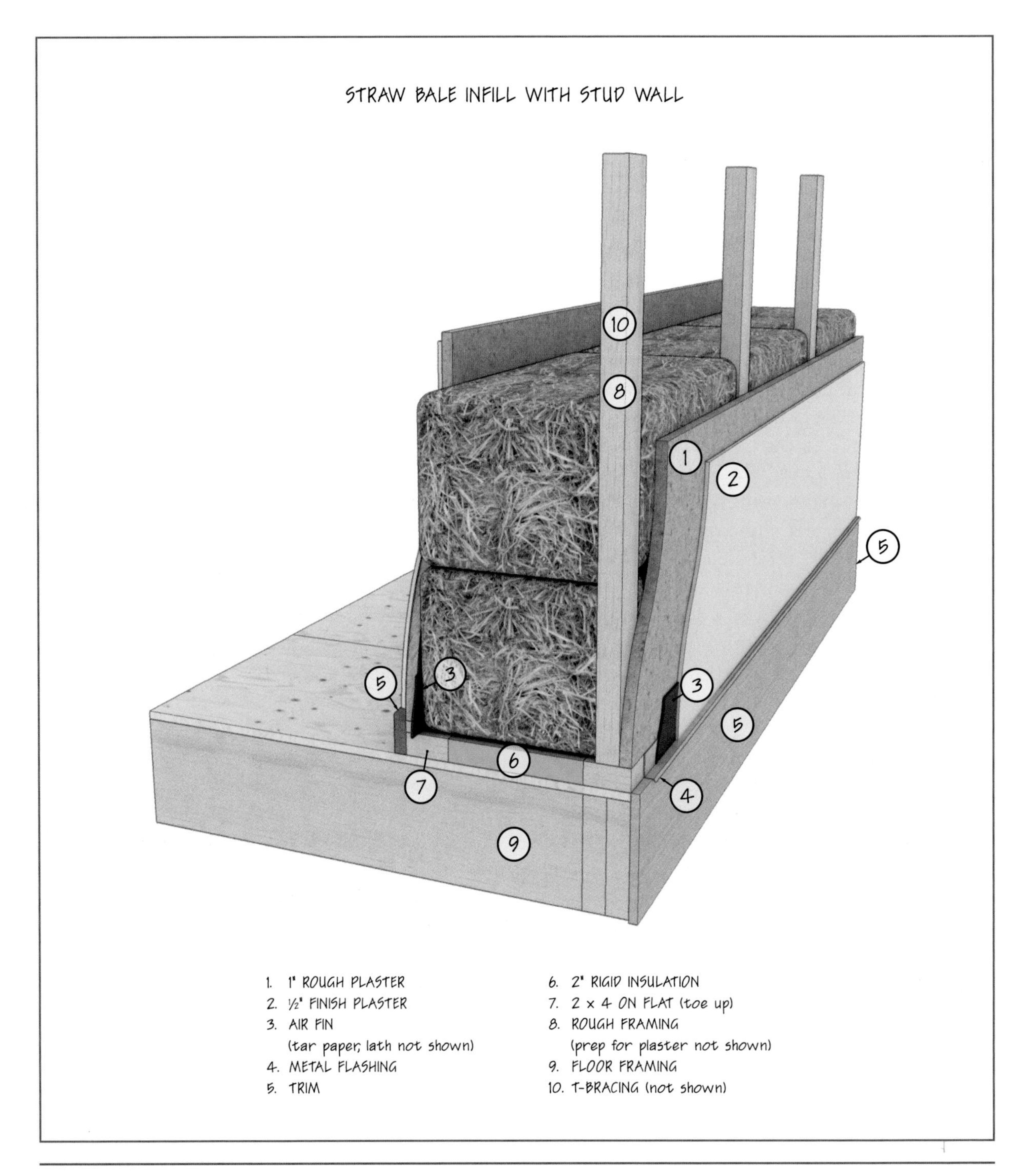

Section drawing of stud frame/bale wall, with plaster finish. ILLUSTRATION BY AARON WESTGATE/FREEFLOW STUDIOS.

hence the imperative for using clay plaster minimally as a base coat. See chapter 6 as well as Bruce King's *Design of Straw Bale Buildings* for more on load-bearing straw-bale construction. Note that clay and lime plasters can be used in load-bearing straw-bale construction, provided that the design accommodates the reduced structural performance of these plasters.

KEY DESIGN CONCEPTS

There are several design concepts to keep in mind that will increase efficiency and effectiveness (and reduce headaches) when you embark on the building process.

Know Thicknesses

The thickness of bale walls and the thickness of plaster must be taken into account from the beginning of the design process. When sizing the foundation and interior footprint, knowing the full thickness of the wall is essential information. The straw bale module itself measures 14 inches by 18 inches, and bales can be installed in either orientation—yielding different finish-wall thicknesses. The 14-inch direction is called "on edge," while the 18-inch direction is called "on flat." Straw bales are usually installed horizontally, in a running bond (staggering seams) if possible; however, bales can also be installed with their variable length running vertically. This orientation is called "on end" and can be done in either the 14-inch or 18-inch wall thicknesses. Each of the five types covered earlier in the chapter will have different effects on the wall thickness and must be decided upon early in the process.

Establish Planes

Another key concept related to wall thickness is the identification and location of the various pivotal planes in the wall section. These include, but are not limited to:

- the finished exterior plane of the straw bale/plaster assembly, be it siding or finish plaster
- the finished interior plane of the straw bale/plaster assembly
- the rough exterior plaster plane of the straw bale/plaster assembly
- the rough interior plaster plane of the straw bale/plaster assembly
- the location of the two vertical edges of the straw bale itself, which should equal either 14 or 18 inches, centered on the midpoint of the total thickness of the wall, pulled from the outside final plane
- the plane of the exterior side of the timber frame or post-and-beam, where it will meet the bale wall
- the plane of any rabbet (notch) or stand-away (attachment creating a notch) on the exterior side of the posts allowing the plaster to end behind the post rather than on it (this should equal the finished interior plaster plane); rabbets and stand-aways are discussed in depth later in this chapter

This information (on pivotal planes and bale and framing locations) will be essential to successful construction. Identifying and anticipating these key elements of placement and size in the design phase will allow the builder to accurately order materials and to affirm that the intersections of all materials work smoothly together in the front end of the project.

Rough to Rough, Finish to Finish

A specific concept useful in thinking through planes of the finished wall is the design concept that we like to call "rough to rough, finish to finish." This phrase describes an approach whereby the rough framing supports are located so that they capture and/or

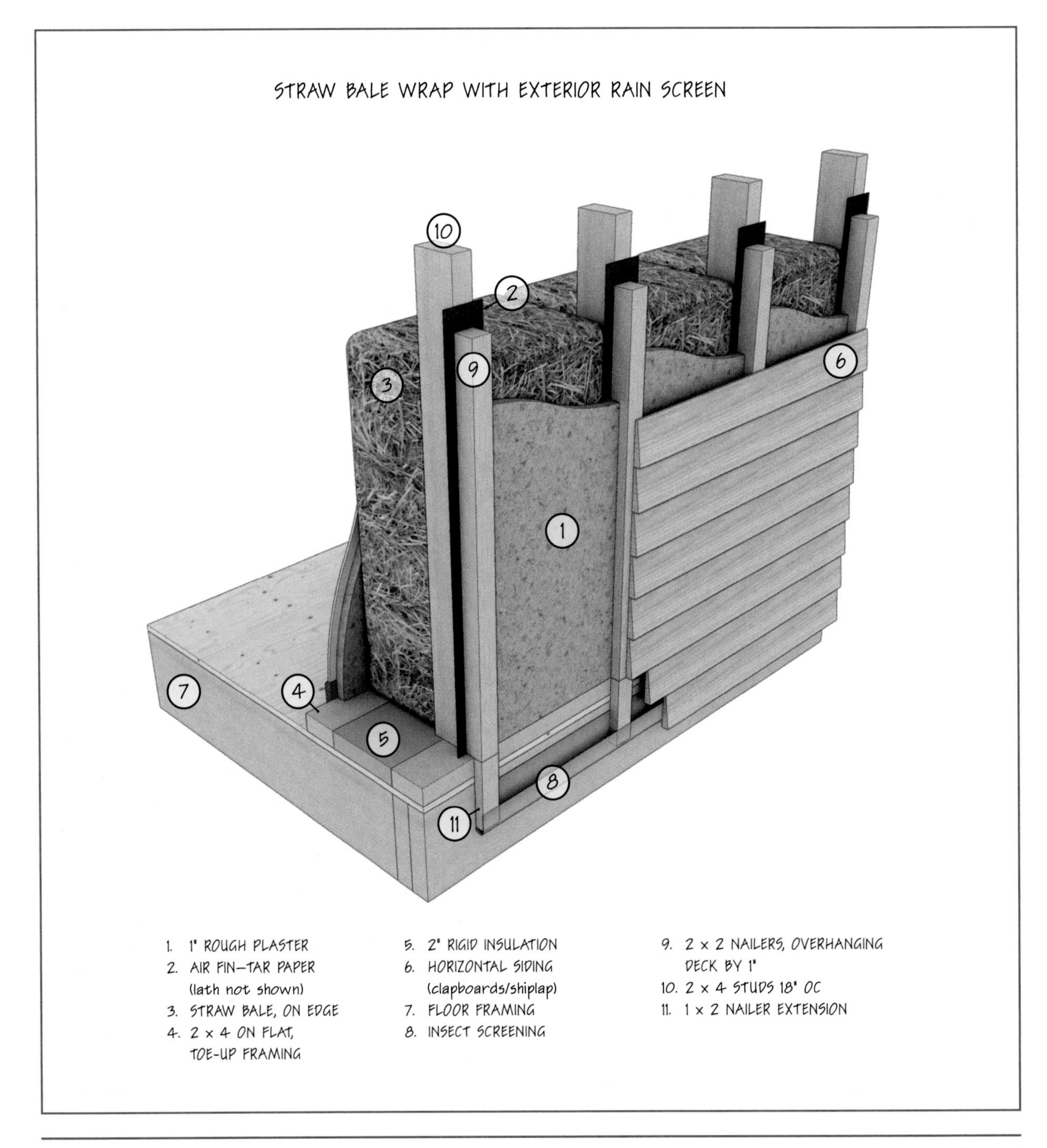

Section drawing of bale wall notched around structural stud framing, and finished with plaster and siding, showing air fins and rain-screen detailing. ILLUSTRATION BY AARON WESTGATE/FREEFLOW STUDIOS.

plane out with the rough coat of plaster (also known as the base coat). The rough framing provides a solid landing point for the plaster at the bottom of the wall and at the tops of windows and doors, and it also provides a solid, unwavering guide for locating planes and keeping the construction on point. "The finish to finish" part of the concept describes the process wherein the finish trim is applied (and carefully masked), and then the finish plaster abuts the finish trim. This avoids the finish trim having to be scribed to the undulating plaster, which is a time-consuming process and never as successful as the finish-to-finish approach. The finish trim also incorporates a rabbet for the plaster to tuck behind, a detail described later in this chapter. "Rough to rough, finish to finish" is the cleanest answer to plaster edges we have found yet, allowing for appropriate phasing, easy repair, efficient assembly, and crack mitigation.

Designing to the Bale Module

The concept of designing to the bale size is an important design principle that can save time, materials, and money or, if not done correctly, add costs in all of these areas. Straw bales are blocks of insulation of fixed size, which can be modified, but each modification adds complexity and labor time to your installation. Designing to the bale module means setting the window heights, and other enclosure elements, to an iteration of the bale module dimensions that makes sense. The simplest example of this is to factor in the height of the toe-up or pony wall (terms that are defined and discussed in the "Good Shoes" section later in this chapter) and to calculate where (either on-flat or on-edge; whichever the general installation orientation is) one or two bales (or as many bales as gets you close to where you would like the window situated) gets you. Also, installing the bales vertically (also called "on end") under a window can create a solid, tight support for the window because the bales compress and can tuck nicely under the window framing. This entails designing windows to a height that would work for common bale lengths (i.e., 32–37 inches).

This is also relevant for the tops of walls. For instance, if the walls are designed to come up under a flat soffit, sizing this distance so that the top bale course is done on-end or vertically (which will need to be forced in) will compress the wall some and ensure a tight and seamless installation. Much more detail can be found later, in the "Bale Construction and Installation" section of this chapter.

On DVD, Chapter 14,
see DESIGNING TO THE BALE MODULE

THE CORNERSTONE STRAW-BALE-DESIGN PRINCIPLE: GOOD OUTERWEAR

The old English proverb, "Good shoes, good coat, and good hat" provides a useful framework for how to detail to most effectively protect natural buildings. Because the northeastern United States is colder than England (and much of Europe, for that matter), and because of the increased focus on building tight, energy-efficient houses, we have added to this proverb an additional piece: "Good buttons, zippers, and belt," which emphasizes the need to attend to airtightness and detailing around all the transitions. Careful execution of the transitions that exist between walls and foundations, walls and roofs, and insulation and windows or doors will maximize thermal and moisture performance of natural wall systems. We address strategies to address such transitions in detail in the "Bale Preparation" section later in this chapter.

On DVD, Chapter 14,
see "GOOD SHOES, COAT, HAT, AND BUTTONS/ZIPPERS/BELT" STRATEGY

Good Shoes

For an in-depth discussion of foundations, see chapter 12. In this section, however, we will consider the connection between the foundation and straw bale walls, as well as good detailing practices.

As discussed in depth in chapters 8 and 12, water is the enemy of both good performance and longevity of natural wall systems. Liquid water and water vapor can enter, and threaten, walls from four places: ground moisture, precipitation, vapor condensation, and built-in moisture. When thinking about the "good shoes" of the structure, we must look for adequate protections from ground moisture (liquid and vapor) and ambient precipitation. The main guiding specification we use is to start the straw bale wall at least 18 to 24 inches off the finished exterior grade. This is so snow cannot pile up against the walls, and so rain running off the roof is less likely to splash back on the walls. It is akin to an exterior "toe kick" for the wall system.

In terms of design and execution, there are multiple options for how this can be achieved. What follows are several common foundation/bale-wall details we have been utilizing with success in the Northeast.

Slab-on-Grade—Toe-Up/Pony Wall

As discussed in chapter 12, Alaskan slab, also known as slab-on-grade or floating slab, can be utilized to good effect in conjunction with straw bale walls. The key to this—since the foundation is essentially on level with the exterior finished grade—is to create a "lift" for the bale wall. We have come to call this a "pony wall" because of its short height. This pony wall can be constructed and finished in many ways. Here are a few:

On DVD, Chapter 14, see PONY WALL EXAMPLES

Insulated Concrete Forms (ICFs) and Concrete Blocks

For specific information on ICFs, their installation and technical specifications see chapter 12. ICFs are one viable choice for the construction of this pony wall. Durisol and Faswall blocks, and concrete block, provide a stable pony wall for the straw bale walls to bear on.

Durisol-block frost walls used as pony walls to raise bales off grade. PHOTO BY ACE MCARLETON.

Durisol and Faswall, both made of woodchips coated in magnesium cement and formed into block shapes, are available in sizes almost compatible with the bale wall module. The thickest block module size is 14 inches, 3 or 7 inches thinner than an on-edge or on-flat bale with one inside and one outside coat of plaster. To account for the difference, one design option we have used is to bring a 14-inch block flush to the exterior plane of the ultimate wall-plaster finish. The exterior of the block pony wall is then finished with a durable finish of some kind—most commonly cement stucco, tinted to match the plaster color (although it can also easily be stacked stone veneer or exterior-grade stone or ceramic tile). The interior of the pony wall is then in need of something to make up the remaining difference. This can be constructed of lumber subframing. A thin pony wall of wood framing lends several benefits. First, it allows for electrical cable to be run, and electric boxes to be set, into a stud wall. This greatly simplifies the logistics and execution of electric—either for you as the builder or for an electrician. Second, regardless of the material chosen to finish the interior face of the pony wall, a transition between the plastered bale wall and the pony wall finish must be installed—and the top plate of the stud pony wall provides an attachment surface for this transition. A transition is important to provide some sort of a "plaster stop," or landing place, for the plaster and also is an opportunity to create a visual transition between the two wall types. Third, the framing provides a standard attachment for wood-finish boards, drywall, or cement board and tile. Fourth, it is a square material to which to bring flooring.

Of course, if masonry finish is desired on the interior, the block wall can be continued on the interior plane (for instance, two 10-inch blocks installed as a pony wall for a bale wall oriented on the flat). The bales would be 18 inches, sitting on a 20-inch block wall. Centering the bale would allow for 1 inch of a rough plaster stop on top of the block wall inside, and 1 inch outside. The finish plaster would then plane out in the same plane as the ½ inch of finish stucco on the block walls—with a drip-cap flashing transitioning between the two on the exterior and some form of transition (be it wood, masonry, or metal) on the interior.

Since the concrete of the block chosen may wick moisture up from the foundation or the outside, it is important to install a capillary break between the start of the bale wall and the top of the block wall. We use Grace Ice & Water Shield, foam board, or sill seal most commonly for this.

For an illustration of the important design concepts in these pony wall sections, see the image "Bale Wrap with Exposed Interior Timber Frame."

Framed Double-Stud Walls with Cellulose

Another option we commonly use is a double-stud pony wall blown with cellulose for insulation. This wall can be finished on the inside and outside with any material appropriate for wood framing. These include, on the exterior, siding, shingles, veneer stone, cement board, and stucco or tile or stone; and, on the interior, drywall, painted or plastered, or wood-board finish. A practical option is to inset shelving or storage into this pony wall, remembering that the deeper you go, the thinner the insulation at that point. Benefits of this method include many of the same ones as for the framed stud wall on the interior of the block pony wall: a square point of joining for flooring material, ease of electric installation, and attachment points

Stud-framed pony wall, finished with boards on the interior and stone facade on the exterior, with outlet boxes set in it, and filled with cellulose for insulation. PHOTO BY ACE MCARLETON.

for plaster-stop transition and for the chosen interior finish. An additional benefit on the exterior is that the top plate of the exterior framed pony wall is now available to use for attachment for drip flashing/ plaster stop at the bottom of the plastered bale wall.

The capillary break with the stud-framed pony wall blown with cellulose occurs at the bottom of the walls, between the bottom plate and the foundation material.

On DVD, Chapter 14, see INSTALLED DOUBLE-STUD CELLULOSE-INSULATED PONY WALL

Tall Stem Walls from Cement Basement

Of course, if the design calls for a full basement, it is a simple matter to design so that the concrete stem walls will be poured high enough off the intended finish grade to form the necessary lift for the bales. Again, 24 inches is ideal. A capillary break is essential between the cement and the bottom of the bale wall. It should be noted that older straw bale books recommended rebar to be set into the concrete stem walls, to pin the bales to the foundation. This is inadvisable in a cold, wet climate where the use of metal in bale walls is not recommended owing to metal's propensity to be a point of condensation for vapor at the place where the temperature drops within the wall system, when there is a significant temperature gradient from inside to outside.

In this system, the floor box is framed onto the top of the stem wall. On top of the floor, we often install two pieces of 2× framing stock, aligned with the interior and exterior rough-plaster plane. The resulting cavity in between is insulated with rigid foam, wool, or another appropriate material. This rough-framing transition between the bales, the floor box, and the stem wall serves as an attachment point, again, for the drip cap/flashing on the exterior and a plaster stop/trim board. Finishing options on the stem walls are many and varied. Often the interior floor will start above the stem wall, necessitating stairs or a ramp in from the exterior finish grade. In this case, the 2× material creates a lift for the bales from the finished floor, an essential detail because it has now become what we call a "toe-up" and will provide a multitude of functions—from moisture protection to electrical chase to providing a necessary place for baseboard and exterior drip-cap attachment.

Floors—Interior

The final aspect of the "good shoes" approach takes us inside the building to take a closer look at where the bottom of the wall meets the floor. Here, the concern is still to protect the base of the straw bale

Here the full basement stem walls are finished with stucco tinted to complement the natural plaster on the bale walls. The tile band acts as transition between the two. PHOTO BY ACE MCARLETON.

wall from damaging exposure to moisture. Inside, the risk is from an overflowing toilet or similar plumbing malfunction, or something as simple as a spilled bucket of mop water.

The design solution here involves a "toe-up," or a track, to act as an isolating transition between the finished floor and the start of the straw. Most commonly we use 2× material, either single or doubled or even tripled in height, one on the inside face of the wall and one on the outside face of the wall, which is insulated in between. The consideration of the height of the toe-up illustrates our design principles from earlier in the chapter; it involves simultaneously thinking through having attachment points for baseboard, as well as providing a plaster stop for the rough coat of plaster.

The design concept "rough to rough, finish to finish" is at play here as well. The rough plaster lands on the toe-up framing and provides an attachment point for the "finish to finish" of the baseboard/finish plaster.

Baseboard is recommended at the base of straw bale walls on the interior for three reasons: First, it can be designed to act as a plaster stop for the finish plaster. Here, rather than the finish plaster landing directly on top of the baseboard, we recommend a rabbet be cut out of the back of the baseboard where it meets the rough plaster plane, so the finish plaster can be tucked into that rabbet. Second, baseboard protects the wall finish from impact or damage—they are called "toe-kicks" for a reason! Third, toe-up framing provides a single, flat plane to which to cut and install flooring. The job of scribing flooring pieces, or trim of any kind, to follow the natural curving contours of straw bale and plaster is an unenviable one, so it is recommended to consider this in the design phase.

Good Hat

Precipitation is the enemy of structure longevity. Plastered straw-bale walls, like any wall system, require protection from rain and snow. In the "Good Shoes" section, we reviewed how the good straw bale designer and builder can minimize the stresses on these walls by attending to protective design details at the bottom of the wall. Now we turn our attention to the top of the walls and the roof. For an in-depth discussion of roofs in general, see chapter 18.

Overhangs

The most significant piece of advice about roofs in relation to straw bale construction is to ensure the roof has adequate overhangs to direct most of the rain and snow away from the walls. The minimum overhang sizing should be 2 feet on two-story buildings, and 3 feet on three-story buildings, for plaster-finished walls. Achieving these overhang lengths can bring up issues concerning the support of the overhang, as in the example of a timber-frame structure in which the last framing support for the overhang is situated almost 5 feet in (21-inch bale-wall thickness, plus 36-inch overhang) from the edge. This issue is addressed in chapter 18.

Sizing—Passive Solar Design

Here is where one of the design principles mentioned in chapter 10 shows up. Roof overhangs play an

The 2½-foot overhang pictured here is achieved with SIPs and supported by timber braces. PHOTO BY ACE MCARLETON.

important role in passive solar design, as they should be sized to shade the windows from absorbing the sun during months when the sun is higher in the sky (summer) and to allow the maximum amount of sunlight in during months when the sun is at a lower angle (winter). This does not protect the walls per se, but it does contribute vastly to the performance of the structure.

Timing

Minimizing risks of built-in moisture to straw bales includes sequencing the construction process so that the roof can be on and "dried-in" (meaning protecting the interior from rain and snow) before the bales are installed. This cannot be under-emphasized, as wet bales going in the wall or leaked on by an improperly prepared roof are that much closer to rot before the building is even inhabited.

On DVD, Chapter 14, see PREPARING A ROOF PRIOR TO INSTALLING STRAW BALES

Good Coat

The "good coat" can be plaster or siding. The function of the good coat is to protect the wall from the elements—including moisture, pests, and fire—on both the exterior and interior, as well as to support the entire wall system in creating a comfortable living environment within the house.

Siding

Siding, wood, metal, or other, is an excellent option for finishing the exterior. Rigid points of attachment—a stud wall or nailers on framing—must be provided as fastening locations for the siding. The stain or paint that the wood siding is ultimately finished with becomes the sacrificial layer to the elements, protecting the base coat of plaster and bales. These weathering impacts are then mitigated by renewing the coat of oil, stain, or paint regularly. Siding also provides options for the creation of rain screens, which are discussed in detail in chapter 8. Other than requiring attachment, venting, and air sealing details that work with a bale structure, the details

A beautiful pigmented lime-sand exterior plaster protects these straw bale walls from weather in Ithaca, New York. PHOTO BY ACE MCARLETON.

for installing siding are primarily the same as in a conventional structure—with the large exception that the bales still must receive a base coat, at minimum, of clay plaster with associated air fins to provide an air barrier and fire protection before the siding goes on. Hence we will focus here in the "Good Coat" section on describing the plaster systems that are necessary for protecting bale walls, whether underneath siding or as the finish in and of themselves.

Plaster as Air Barrier

First and foremost, plaster is the primary air barrier for the straw bale wall system. A quick refresher: an air barrier is essential in high-performance wall systems because convective heat loss—the heat loss due to air movement and air infiltration—is generally a greater contributor to poor performance than conductive heat loss. Conductive heat loss is what "R-value" and insulation address—straw bales installed in the wall block heat loss through conduction. Air barriers are what keep the air out, minimizing convective losses and affecting performance and comfort as significantly as the insulation choice. The base coat of clay plaster is applied at least an inch thick on the bales, and the goal is for it to be a continuous coat, with no breaks. Where breaks occur, as they inevitably will, such as at the points where framing meets the plaster, a secondary air barrier—an air fin—is employed to ensure the continuous coat. See the "Bale Preparation" section below for an in-depth discussion of air fins.

Sometimes breaks will occur in the plaster in the form of cracks. Cracking can be due to many things—from substrate instability and movement to wetting and drying cycles (from some unplanned introduction of liquid water to a plaster mix or a recipe that was too brittle) to improper drying conditions after application. Cracks, beyond being unsightly, constitute breaks in the primary air barrier—air leaks that compromise thermal performance and potentially carry water vapor into the wall—that must be addressed.

Plaster as Moisture Protection

An apprehension that is often raised regarding straw bale walls is whether the walls will just get wet and rot. This is a valid concern, and it is based on the fact that straw is vulnerable to water. It is also accurate, however, that wood, cellulose, and fiberglass are also vulnerable to water; in fact, most building materials are, with the possible exception of foams, stone, and some concretes, all of which have other downsides. Foam gives us good R-value that resists decomposition, but it is a human-made, bio-persistent compound with a significant ecological footprint. Concrete production also has a significant environmental price tag. And stone—which in actuality is immensely vulnerable to erosion through water exposure, although it occurs over a longer time frame than we generally are concerned about in building—has a negligable insulative value. The intelligent approach guides us to find ways to use materials with a small ecological and social footprint and good insulative properties, and to include drying mechanisms in the design that will protect the materials, allowing them to maximize their insulative properties and minimize their exposure to situations that will lead to rot. Clay plaster is one key part of this strategy.

A base coat of clay plaster will moisture-protect bales owing to clay's hydrophilic and vapor-permeable properties. PHOTO BY ACE MCARLETON.

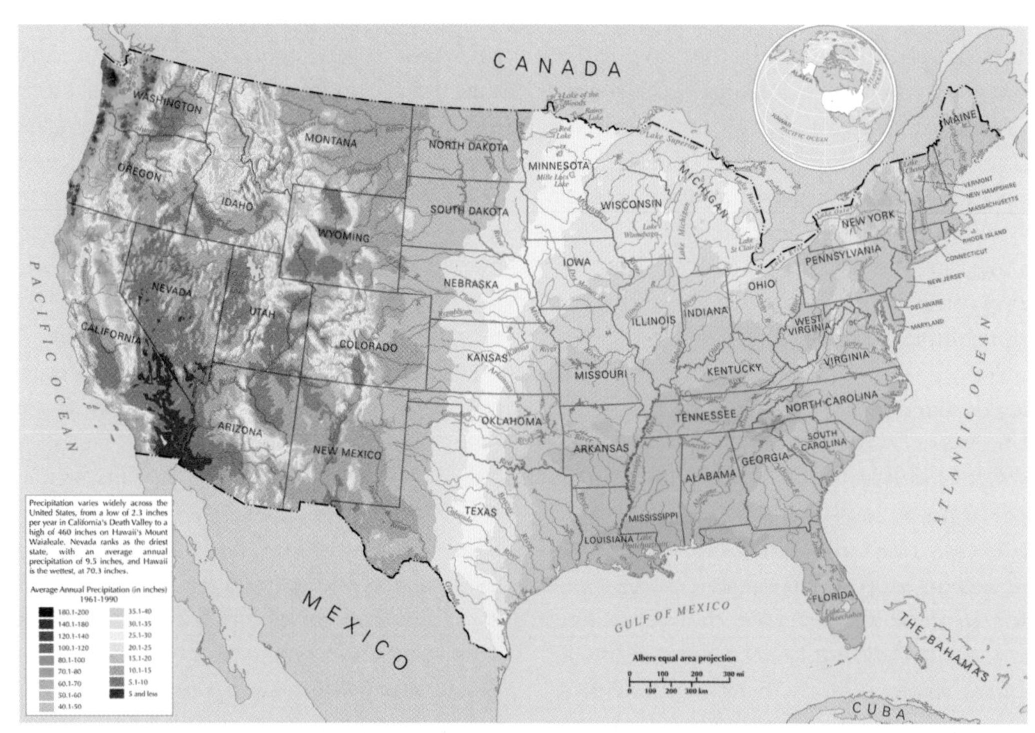

AVERAGE ANNUAL PRECIPITATION (in inches), 1961–1990

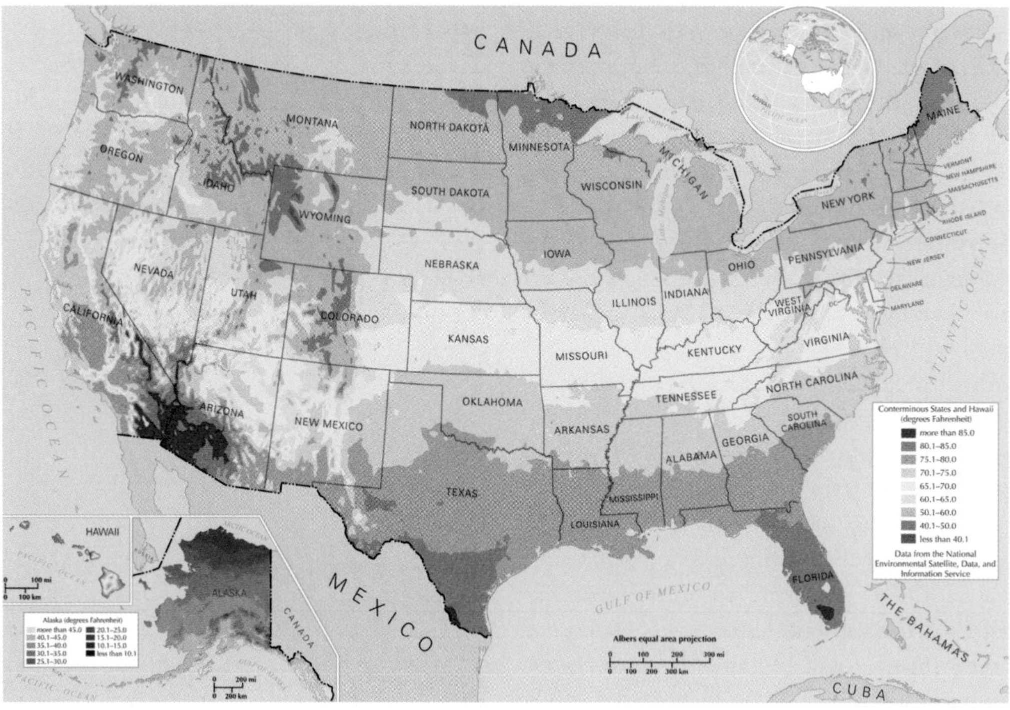

ANNUAL MEAN DAILY MAXIMUM TEMPERATURE

This map demonstrates the varied climatic conditions in the United States. Design that is climate-appropriate is essential to high-performance building. MAPS COURTESY OF NATIONAL ATLAS OF THE UNITED STATES.

Clay: Vapor Permeability and Hydrophilic Properties

The good coat of plaster is one of these essential mechanisms for protecting straw bale walls from moisture damage. But not just any plaster will do; using clay-based plaster as a base coat is vital. To debunk a common myth, cement-based plasters, long thought to be the best answer to straw bale protection, actually contribute to rot. This is because cement has a very low vapor-permeability coefficient, which means it traps water vapor inside the wall rather than allowing it to pass through and out. Cement plasters or stuccos have been used more successfully in dry places like California and the American Southwest. In fact they are sometimes used in seismic areas, given cement's high compressive strength. For more on this, see chapter 6.

Duplicating this model of building, which addresses important needs in arid and seismic climates, in a wet, cold climate such as the Northeast brings significant problems. We have seen widespread failure on straw bale structures plastered in cement; similarly, lime plaster is apparently utilized to good effect by our colleagues in drier Colorado, but when used without a base coat of clay here in the Northeast, moisture management is greatly reduced and the risk of moisture damage to bales increases. Clay's superiority results not only from its ability to allow vapor to pass through, but from its unique hydrophilic nature. Hydrophilic means "water loving," and the relevance of this for coating straw bales is that the clay plaster coat will actively draw moisture to itself, and away from the straw. Owing to its high vapor permeability, it will then allow the vapor to pass into the air. Neither cement- nor lime-based plasters, although certainly better than cement, will allow for this degree of transpiration. The emphasis here is on climatic variance and how it affects design choices and performance. We must go beyond thinking in a "one size fits all" approach. Rather, understanding *how* the systems need to work, what challenges they need to stand up to, and then the materials at hand and how to apply them successfully in each climate—this is the challenge of the natural builder. Paul Lacinski and Michel Bergeron tell a wonderful story in *Serious Straw Bale* about taking apart cob walls in Europe that were 500-plus years old and finding that the straw inside the clay was perfectly preserved and mummified. The clay kept the straw in an anaerobic environment and deprived it of exposure to any moisture. Learning from the past is an important part of understanding how to move forward with these materials and systems. Using clay-based plasters is critical in meeting the challenge of keeping straw bale walls dry in wet and cold places.

Durability

Another goal of the good coat is durability. Because the clay itself is vulnerable to erosive actions such as precipitation, we have found that applying a clay-based coat of plaster that adheres directly to the bales—the base coat—followed by a more durable yet vapor-permeable finish, such as a lime-sand plaster or wood siding, is a winning combination that addresses long-term durability for straw bale walls.

Flexibility

The good coat must also be flexible. Straw bales are not a rigid substrate, no matter how tightly installed. And given that the plaster coat also often spans over wood framing members that shrink and expand based on exposure to water vapor or liquid water, it is important to craft a plaster mix that will not crack with movement.

In chapter 17 we will more closely examine all of the potential plasters and their properties and applications.

Good Buttons, Zippers, and Belt

The big idea of the "good buttons, zippers, and belt" element of this design principle is the importance of evaluating and treating transitions appropriately for optimal thermal and moisture performance, and for longevity. Since our goal for the protective coat of the installation is for it to be continuous so it can best do its job, it is essential to look closely at transitions and edges, and to design in redundancies to minimize problems in the long run.

Step 2: Bale Preparation—Air Fins, Stand-Aways, Pony Wall/Toe-Ups, Window and Door Framing

Planning for air fins and rabbets or stand-aways must be included in the design phase. However, this is an excellent example of how the iterative design/build process works, in that it is impossible for designers to anticipate the finer points of air-fin design until they understand exactly what air fins are, how they are installed, and how they function. To promote this understanding, we will now describe what is involved with preparing the frame for the bales.

Homasote is a sheet air fin that works nicely on the back of a timber frame. PHOTO BY JAN ALLEN TYLER.

On DVD, Chapter 14, see INSTALLATION OF AIR FINS ON A TIMBER FRAME

AIR FINS

"Air fin" is a term used by straw bale builders to describe any material that acts as a control joint to block air movement at the edges of the wall system. An air fin can also be understood as a secondary air barrier that functions as a backup in the places where the plaster is compromised in its ability to act as the primary air barrier, namely, at the edges where the plaster meets wood trim, the floor, windows, or any other edge in the building structure. In any structure, the plaster comes to an end at the floor or foundation; where it meets the eaves and gables, ceilings, or top of walls; and where it meets the window trim, door trim, window seats, or interior walls. In a timber-frame/straw-bale design, additional plaster edges occur everywhere the plaster meets the frame—every linear inch of post, beam, brace, and girt. As discussed in the "Good Coat" section of this chapter, plaster is effective when it *is* a continuous coat (which is logistically impossible), or, more realistically, it is designed to perform *as if it were* a continuous coat, despite terminations. The mechanism for achieving this coverage is called an air fin.

As a preparatory step to installing the straw bales, the air fin material is attached to the back side of the frame (if it is a timber frame or post-and-beam) as well as every edge of the bale installation where it will meet window and door framing, floors, ceilings, eaves, or rakes of gables. Then the bales are installed. Care must be taken during installation to preserve the integrity of the air fin. After the walls are up and prepared for plaster, the air fins can be floated into the first or second coat of plaster. There they remain, embedded or floated into the plaster skin, creating an airtight barrier that spans between the field of the plastered bale wall and the framing member it is attached to—ready to spring into action when the wood framing member swells or expands and a crack opens up in the plaster right at that juncture.

To summarize, airtightness is critical in order to reap the full benefits of the potential R-value of a straw bale assembly. And because of the pressures on structures in cold, wet climates, the framing members tend to swell and shrink in accordance with the temperature gradient—and swings between hot/wet and cold/dry—more so than in more temperate or warmer climates. Attention to this detailing is akin to the in-depth measures necessary in active seismic areas. Again, the emphasis is on understanding what detailing is relevant given the climate in which you are building.

Tremco

Tremco is a rubberized acoustical sealant—favored by superinsulation builders for drywall and electrical-box connections—that we have begun using for our air-fin seals to wood members. It stays flexible the longest of any caulk, theoretically forever. However, it is the messiest stuff to use and is horrifically sticky and permanently gooey, and it will cover your body parts and clothing and any finish wood or other permanent, exposed construction elements. Care must be taken in its application for this reason.

We feel the benefits of Tremco outweigh its downsides. In this system, caulk acts as yet another control joint between the air fin and the framing member. The success of the system depends on the caulk not failing. In this situation, "not failing" means it being continuous, being thick enough to bond the two surfaces, and, more challengingly, remaining flexible enough to withstand the fluctuations of the wood member without cracking.

If Tremco is not available or not an option, the most flexible caulk should be used. Usually a silicone caulk is of higher quality than acrylic caulk and will not crack as readily under freeze/thaw conditions.

Air fins are made of any solid material that is also vapor permeable. We have used 30-pound roofing felt in combination with both metal and plastic lath, Masonite, drywall, and Homasote (a sound-deadening board). The air-fin material of choice is attached with a hand stapler, a pneumatic stapler, or roofing nails. Some air fins are pressed into a bed of caulk, some are not. The reasoning behind where to caulk—and where not to—is based on understanding where and how moisture-laden air will move. Caulk is required if the connection between the framing member and the air fin must remain airtight in order to keep air from moving between the framing member and the air fin into the bale wall.

Chapter 14, see AIR-FIN MATERIAL OPTIONS

Chapter 14, see INSTALLATION OF AIR FINS AROUND WINDOW AND DOOR OPENINGS

Tar paper and diamond lath mesh treat edges of the plaster around windows and at the transition to the pony wall on the exterior of the building. The dark patches in the field of the wall are light straw-clay knots filling voids. PHOTO BY ACE MCARLETON.

STAND-AWAYS

On DVD, Chapter 14, see INSTALLATION OF STAND-AWAYS ON THE TIMBER FRAME

The companion to the air fin is the stand-away. A term also created by the straw bale community, "stand-away" describes a notch that is created at the connection of the bale wall and a wood member, to hold the plane of the wall envelope away from the back of the wood frame. It functions in tandem with the air fin to tackle the problem of cracking plaster at the point where the plaster meets the wood member. To review, the air fin acts as a secondary air barrier to protect the bale wall from air intrusion where cracks can open up in the plaster where it meets a wood member. The air fin resolves a performance issue, but it does not address the secondary issue of the unsightly crack.

We have seen a lot of beautiful, skillfully milled and treated wood trim with gorgeous, velvety smooth plaster brought up to it that look amazing for the first several months but, as time passes (often just one heating season), ends up looking like hack jobs owing to large cracks running up the length of the trim. Caulking, installed to fill this gap, is unappealing visually and also cracks over time. Leaving the edge of the plaster, an inherently vulnerable part of the finish, exposed can lead to premature cracking, which is itself a performance issue. The best long-term, dependable, and aesthetically pleasing system we have discovered yet for joining trim or exposed wood members and plastered bale walls is the stand-away/rabbet and air-fin duo.

The main idea here is creating a hidden notch that will hold the straw plane back from the finished exposed timber edge—to allow space for the plaster coat to tuck into. The goal is to fill up that space with plaster until you can just pass the thickness of your plaster trowel between the plaster and the back edge of the notch. The seam, or edge of the plaster, will occur 1–1½ inches behind the timber on each edge. This means that when the unsightly crack inevitably occurs, it will be hidden from view—and the air fin will stop the air from getting into the wall.

Tremco rubberized sealant being spread on the stand-away to receive air fin material. PHOTO BY NICK SALMONS.

On DVD, Chapter 14, see INSTALLED AIR FINS AND STAND-AWAYS

Together, the air fin and the stand-away solve two aspects of the challenge of bringing a stable masonry substance like plaster into connection with wood. The air fin will protect the bale wall from air leaks, which can compromise the functioning of the bale wall both by lowering R-value and introducing moisture in the form of vapor. The stand-away will create a clean visual connection and protection for the plaster edge that will stand the test of time where two disparate materials are joined at the edges of the straw-bale/plaster installation.

Stand-Aways for Exposed Heavy Timber Frames

If there is an existing timber frame or post-and-beam frame that was not rabbeted, stand-aways should be added onto the back of members before the air fins and straw bales are installed. The process looks like this: Cut and attach thin-stock (we have found ¾ inch to be ideal) 4-, 6-, 8-, or 10-inch boards onto the back of the frame. Place the boards so they sit in from the edge of the member 1½ inches. Fit the joints between the boards so they fit fairly closely. The stand-aways themselves don't have to be airtight or to have airtight connections to each other if the wood member is in the middle of a wall; but they do provide the substrate to attach the air fins to—so in order for the air fins to be stable, supported, and continuous, the stand-aways should be consistently installed. If the wood member is at an edge, however, the stand-away should be caulked to the timber, as a wood-to-wood connection, created with fasteners, will not be airtight. A benefit of the rabbet—a type of stand-away—is that it eliminates the need for this additional caulking.

Rabbets

To streamline the process, we recommend rabbeting the back edges of the frame members before the frame is raised. It saves the additional ¾-inch boards from having to be added on to create the stand-away, and the labor of cutting and installing them when the frame is up, which can be awkward. It also provides the need to protect an additional seam with caulking.

To review one of the design points: know your planes. This choice becomes relevant for wall thickness, footprint, and foundation sizing. Rabbeting the back of the frame means an additional ¾ inch does not have to be added to the plane of the exterior side of the frame, before the straw bale starts. It doesn't increase the thickness of the wall section. It will take away a bit of the thickness of the timber members, but not in a way that is visually compromising.

All in all, the rabbet is just a cleaner, more efficient way to achieve the stand-away on the interior exposed frame. It is a great example of the value of design and forethought in the successful execution of the plastered bale walls.

Rabbets for Trim

Window trim, baseboards, exterior corner boards, and any wood trim intersection with plaster benefit from the rabbet version of the stand-away. Although trim versus "the trimless look" is very much an

Using a rabbeted stand-away is a clean and efficient way to bring plaster to the frame. PHOTO BY JAN TAYLOR ALLEN.

aesthetic decision, it can also be a functional one. We strongly encourage at least some trim. For example, baseboard is a really wise choice in a straw bale house because of its functionality. Running plaster down to the floor is not a durable alternative. Toes kick it, children run toy trucks into it, cracks open up as the floor moves, water that spills on the floor is directly wicked up into the plaster and weakens it. Flooring installed to a set plane of a baseboard or sub-baseboard is miles simpler and cleaner and will help preserve your investment.

The importance of the design phase, specifically the concepts of "rough to rough, finish to finish" and "know your planes," is illustrated here. Planning ahead to install the finish-trim baseboard means remembering to have framing set to attach it to before the bales go in. This fundamentally entails working backward from finish and planning your planes so that the line of the subframing is set so the rough coat will come flush with it—and so that when you attach the rabbeted trim board, the top coat neatly finishes out to the rabbet's dimensions.

PONY WALL AND TOE-UPS—BALE–FOUNDATION AND BALE–FLOOR TRANSITIONS

The most important things to consider at the foundation/bale-wall transition are capillary breaks and flashing. A capillary break must be installed between foundation walls, regardless of the system chosen (see the "Good Shoes" section earlier in this chapter as well as chapter 12), and at the bottom of the straw wall or pony wall. This will prevent the wicking up of any moisture from the foundation into the bottom of the wall. Flashing is important for two reasons: first, to build in mechanisms to shed water away from the wall; and second, to provide a place for the plaster to land on.

In terms of shedding water away from the wall, most often we use a metal drip cap, bent on a brake into an L-shape in profile, with a small bend at the tip of the bottom leg to discourage water from wicking back up the underside of the flashing. This flashing is attached to the toe-up, pony-wall framing, or some framing at the bottom of the straw bale wall that is partly in place for this purpose. Often we will design it so that the rough coat of plaster will land on the exterior 1 inch on top of the toe-up or pony wall or whatever framing; and then the finish coat will land on the drip-cap flashing installed between the base and finish coat. Alternatives to drip cap are stone or tile, angled on the top to shed water away, or a wood piece, preferably of some rot-resistant species, also beveled on top to shed water. These options can be weighed based on desired aesthetics, cost, efficiency of installation, and durability.

WINDOW AND DOOR FRAMING

Windows and doors are a huge topic of conversation in straw bale design and construction. How are they supported? What are they framed into, or are they not framed into anything? At what location in the wall section are they placed? Are they flashed like any conventional window? How are the bales supported over the window openings? What are the different finished "looks" that follow from different choices of lintels, windowsills, side reveals, and casings? Let's take a look at each of these considerations under both exposed-beam and stud-framing scenarios.

Exposed Timber Frame/Post-and-Beam

In the Northeast, we frame windows and doors typically on the far exterior plane of the wall, while the timber structure sits to the interior plane of the wall. The reasons for this are multiple, including:

Protection. The window penetration is a potential hazard for leaks, so pushing it all the way to the outside plane of the wall minimizes the depth and area of precipitation accessing the wall. A deep windowsill on the outside leaves exposed two horizontal edges where the sill material meets the bale and plaster. Snow and ice can build up and cause wear on the plaster. Rain will degrade it and stress whatever flashing has been placed there.

Deep exterior window returns, such as this one in New York State, have significant exposure to rain, ice, and snow and are not recommended unless well protected by a roof in the northeastern climate. PHOTO BY ACE MCARLETON.

Design and aesthetics. Deeper window wells on the interior of the structure create warm, conditioned spaces to extend the feeling and size of the rooms. They provide ample space to keep plants, art, snoozing cats, or other items. They can be used as bookshelves and storage benches. Using them to create a built-in seat is another popular option not only because it is a space-saving smart design element in the structure, but because it provides a place to really interact with and enjoy the thickness, depth, weight, and softness of the bale walls.

Attachment. The rough framing for the window and door openings on the exterior plane of the straw bale wall creates attachment points on both sides of the straw bale assembly. On the interior plane, there is the timber frame or post-and-beam to attach to, and on the exterior, the king and jack studs of the rough window framing are the attachment points.

Structural Stud Walls

The wall section of 2× stud-framed walls that support the roof looks quite different than the exposed heavy timber-frame wall section. The windows and doors are still framed on the exterior plane of the wall section; however, they are simply framed into the stud wall in a traditional manner.

In either structural stud walls or exposed heavy-timber structure, windows can also be "floated," or placed where the course of bales ends, which is accomplished by framing a strong box opening—big enough to hold the window and strong enough to withstand external pressures. This floating box can be placed at the top of a compressed bale row when it reaches the height desired, during the bale wall construction process. It is then pinned into place and remains there until it is time to put in the windows. This method was pioneered by the load-bearing straw-bale community, and more detail on it can be found in Bruce King's work, *Design of Straw Bale Buildings*. It is important to note that operable windows in floating bucks must only be used in bale walls that are well compressed. Otherwise, excessive creep and movement in the settling bale wall can distort the plumb and level of the window buck, causing failure in operable windows.

Step 3: Bale Construction and Installation

Straw bale construction utilizes grain straw, which has been cut and compressed by a baler in the fields into "bales," or blocks of straw.

THE MATERIAL

Straw is the cellulosic, woody part of the grain plant. It is hardier than hay, has less nutritional content (which would cause it to biodegrade), and most importantly has a tubular "straw" shape. This tubular shape allows for lots of air cavities in the bale block, giving straw bale its high R-value. In the Northeast we generally build with two-string bales. Three-string bales do exist and are used for buildings in other parts of the country, mostly out west. This chapter will focus on building with two-string bales, which measure, generally, 14 inches by 18 inches with a varying length; most common lengths are on either side of 36 inches (bale lengths can vary from 27 to 40 inches).

Ensuring Quality Materials for Successful Construction

The first step in straw bale construction is to understand how to ensure a quality building material. One of the most consistent challenges and potentials of straw bale construction is that the builder is responsible for the quality inspection and control of materials. (Currently, straw bales are not inspected for building quality at building supply yards.) Hence, understanding what "quality" means is an important part of fostering success with this construction method. Quality straw bales suitable for construction should:

- *Be dry.* As measured with a bale meter, or moisture meter, the moisture content of a straw bale that is to be used as building material should not exceed 15%, ideally. Straw starts to decompose at and above 23% moisture content. A musty odor, dull color, dark spotting, or a bale that feels damp to the touch are important sensory indicators.

- *Be tight.* Straw that is to be used for building purposes must be tightly bound. The best way to test this is to try to pick up the bale. If you can just barely get your fingers under the twine, that's a good sign! If the twine lifts up easily and can be pulled up several inches, this is too loose. Bales like this can be re-bound, if there are just a few in your delivery batch, but if the whole group is like this, it is really worth finding a better source. The quality of your walls will show a marked improvement. Loose bales will result in lower-performing walls (in terms of R-value), more difficulty shaving notches and shaping them during construction, and more slumping and cracking of plaster because the bale substrate is not as stable as it could be.

- *Have poly twine.* The use of poly twine is debated in the straw bale community. Minimizing our plastic consumption is one good reason. However, poly twine can withstand greater pressures than natural twine before breaking, and balers can be set to higher tensions if poly twine is used. We therefore recommend bales tied with poly twine.

- *Have long straws.* Many balers are being set to chop the straw short these days as it is better for every trade that uses straw except for natural builders. For gardeners and farmers, short straw is just fine and it is easier to make into bales. Short straw is not good for builders for several reasons. First, long straws will help the bale hold its shape even when cut into for notches or otherwise modified. Second, the R-value of the bale is due to the tubular structure of the straws, which long straws are more apt to maintain. Third, the light straw-clay we need as a complement to straw bale construction can only be made effectively with long straws.

- *Be pure straw.* A quality bale is one that is mainly straw, with minimal other content such as leaves, grass, hay, twigs, stones, or dirt. This is for many reasons, not the least of which is R-value; contaminants will detract from acheiving the desired value. Also, these "extra components" can add problems of their own: leaves, grass, and hay all have a higher nutritional content and can be more attractive to insects and quicker to decompose; soil is a very low insulator and unnecessary weight; unwelcome stones dull chain saw blades.

Sourcing

Sourcing bales is a process that varies depending on location and bale availability. Three likely sources are local farmers who grow grain, straw wholesale dealers, and farm/garden retailers. The first two are frequently the best options, as markup will be lower, larger amounts can be secured, quality can be verified closer to the source, and the material often can be delivered directly to the site. Farm and garden retailers carry straw bales, but often not in amounts needed for an entire structure, and the price will include their overhead and an additional markup (although this can be a good option in a pinch).

To locate grain producers and grain dealers, start with the state agriculture board. Remember, straw bales are made after the grain harvest, usually in late summer/early fall, so bales available in early spring will be from the previous year. If possible, secure the bales needed for spring/early summer construction the previous fall and ensure safe overwintering conditions, thereby making sure the material will be on hand when you need it, and at a reasonable price (all the more reason to work directly with a farmer or bale dealer early on in the design phase).

BUILDING METHODS

In this section we will look first at general construction methods of how to work with the material and then take a close look at design-specific methods, which follow the design principles discussed earlier in this chapter of "good shoes, good hat, good coat," and "good buttons, zippers, and belt."

Measuring a Straw Bale

The most important concept here is to understand how to accurately measure the bulbous, irregular building material you are working with. Straw bales are composed of five to eight "flakes," or sheafs of folded straws, about 2 to 4 inches wide, which are packed together like the pages of a book. If you cut the twine and allow the bale to fall apart, these "flakes" will be evident. The end of a straw bale presents several options for where to hold the end of the measuring tape. There is the tightest place—where the strapping or twine cuts into the flake and holds it tight. There are the two outside edges of the flake, which tend to "wing out" just past the two tight twine lines. Then there is the slightly mounded part of the end flake that is contained between the two pieces of twine where they cross the edge of the bale.

We have found that the most consistent place to measure from is the place in between the two lines of twine.

1. Push your hand steadily against the flake between the lines of twine.

2. Sight down from above, and line up the end of your measuring tape with the compressed bale end.

3. Stretch out the tape to the other end of the bale and sight down to note the measurement at the center of the flake on this end. If you need to mark the bale at a certain length, use spray paint or push a needle in to mark the spot.

Measuring straw bales is part science, part art. There is a level of precision that is possible and, in fact, often desirable, but it is a different type of precision than in conventional carpentry, where 1/16s are common tolerances, or in timber framing or finish trim, where precision can approach 1/32s and 1/64s. With straw bales, the number of variables present in the material makes it an impossible task to approach with the mind of the carpenter or timber framer. Straw bale builders are precise, because they understand what they want their outcome to look like, but in a different way than other builders because they must understand the limitations and possibilities of the material to achieve that outcome. There is a balance that exists in building the wall, and in making it tight, but not so tight that the window and door framing buckles, or the corners get pushed out. Hence we must find a measuring and sizing system for the bales that achieves that balance. This will take a bit of practice.

When a resized bale is needed (see the "How to Resize Straw Bales" sidebar), many people assume you want to upsize the bale relative to the measured space because the bale compresses; however, we have found this most often leads to problems. If you stay consistent and measure from the center of the end flake on every bale, always—both the one in the wall and the one on the ground you are resizing—you will more likely be successful working with this variable material. Another strategy to help tame the lumpy bales is to square them before measuring or installing.

Squaring Straw Bales

Owing to the tension they are under because of the twine or straps, straw bales bulge out in the center between the twine, and on either side. This may not initially seem like a big deal. However, the bulging shape of the bales makes it difficult to measure them precisely (as just discussed), and also, most importantly, if the bales are installed when they are this shape, there will be gaps at every bale intersection, obviously compromising the integrity of the thermal envelope. Though it is possible to stuff these gaps after the walls are up (indeed, stuffing gaps with light straw-clay is an essential part of the process that will be done regardless), not squaring the bales at the outset will necessitate an amount of stuffing that is less effective thermally, and more labor intensive. The time investment of squaring the bales in the front end will pay off as time savings later in the installation.

Straw bales fit together in the wall best when first squared on each end with a chain saw. PHOTO BY JAN TAYLOR ALLEN.

To square the straw bales:

1. Set up a station where the bale you are working on can be set at a comfortable working height for the cutter. Usually we achieve this by stacking two extra bales on top of each other as a "table" for the bale that is being worked on.

2. It's most efficient to square bale with two people—one person stands in one place with the chain saw in hand (which is safer; moving around a lot with a chain saw with lots of slippery straw on the ground is inadvisable), and the second person moves and places the bales for the cutter. The second person also maintains two piles—one of the uncut bales, one of the newly squared bales ready for building.

3. The cutter has the challenge of cutting the bale end so it is closer to flat and square and 90 degrees from the other two sides, while not popping or breaking the twine or strapping. We always advise the cutter to use "robot arm" (i.e., to resist the temptation to follow the curve of the bale). The goal is to drop the saw down at exactly 90 degrees from the top and two sides, and close enough to the twine to accomplish the job without cutting the strings.

If the bale construction is happening in the context of a bale raising or group work party, the squaring station can be a great place to plug in helpful folks who are comfortable and handy with chain saws—or those who have always wanted to learn.

On DVD, Chapter 14, see SQUARING THE ENDS OF STRAW BALES WITH A CHAIN SAW

Tools for Resizing

There are two main methods and sets of related tools and materials needed for resizing straw bales.

The straightforward and low-overhead option is twine. Make sure it is poly twine and that it is of a strong enough gauge to withstand tensioning.

You can make bale needles out of wooden dowels with a sharpened point and a hole drilled through the sharp end. This is a low-budget and simple option; however, the dowels will not stand up to repeated use.

For a more durable system that can handle lots of volume, we recommend using a setup from the packaging industry: poly strapping with tensioners, or ratchets, and metal clips or open seals for securing the strapping. The poly strapping is very strong and durable, and working with the tensioners (ratchets) and crimpers becomes quite smooth, efficient, and reliable once the user is accustomed to them.

The needles that we have found work best for this strapping are ¾-inch bar stock we had bent up and cut for us at a local metal shop. The needle we use is 26 inches long, with one end cut to a sharp point. Just above the point is an angled slot cut to receive the flat strapping. The opposite end is bent at 90 degrees to the main shaft, and that piece, the handle, is about 6 inches long. Why 26 inches? It gives us the ability to knock a needle behind timber posts on a corner, to the outside of the wall, so we can use the strapping to tighten up stray corners. For more on this, see the discussion of outside corners in the section "Special Conditions" later in this chapter.

Bale Stacking: The Process

It is an exciting moment when the preparation work is done and the first bales go in. This next section will take you through the installation of the bales themselves into the wall cavity.

Order of Operations

First, install the toe-up or pony wall (see step 2, "Bale Preparations," in this chapter for details). If it is a toe-up in your design, it is essential that any electric runs that are meant to be placed around the wall circumference for eventual outlets or switches get run in the toe-up. Once the toe-up is insulated and the bales are placed on it, that channel will be inaccessible from above. It is much simpler to do your electrical layout and rough in the electric before beginning bale installation.

Straw bales are best installed working from the corners to the center of the wall. The corners need to be strong and solid, so starting with whole bales on the corners is best. Other starting points should be windows or doors—working from edges to the center. This will ensure that the inevitable "cut," or smaller, resized bale, will be secured between whole bales, rather than ending up at an opening or corner.

Compression

Straw bales must be compressed as they are stacked, to ensure as little sag and settling as possible after the weight of the plaster is introduced. After each bale has been placed, but before fastening, make sure to use a sledgehammer or some other form of downward and lateral force (even jumping on them works well!). The bale will move in response to the force, and it will become evident when you are unable to move it any further. Then fasten the bale to framing. If there is no nearby framing, other creative methods can be used. See step 4, "Plaster Preparation," the final section of this chapter, for more information on

How to Resize Straw Bales

The process of resizing straw bales is an integral part of straw bale installations. Straw bales can be altered in length by resizing, or re-tying, which allows us to retain the tightness imparted to the bale from the baler. A description of how to resize a straw bale follows; or watch how it is done in the companion DVD.

Resizing straw bales with strapping and ratchet straps is efficient and accurate. PHOTO BY ACE MCARLETON.

1. Measure and mark the desired size on the bale with spray paint or some other means.

2. Thread a needle with twine, or strapping, and push it through the bale at the mark, next to the existing twine. Most often it is best to do this on the inside of the twine, rather than on the outside, because your needle will be going through something solid rather than the loose outer "wings" of the bale. Make an effort to keep your needle at 90 degrees to the top of the bale, so the end you are making will be as square as possible.

3. Reach underneath and grab the strap/twine, and bring it around and up to the top of the bale. Make an effort to line it up nicely, so it stays even and square when you tighten it.

4. If using twine, make a trucker's hitch, a loop, and thread the opposite end through the loop. Pull back on it until the desired tension is reached. Secure the loose end, retaining tension, with several knots.

   If using poly strapping, clip the end you just brought around into the ratchet, by lifting up the shoe and pinning the strap. Leave about 4 inches of the end of the strap exposed in front of the ratchet tool. Feed the other end of the strap through the front cutter and the side wheel. Pull out the slack and begin ratcheting to achieve the desired tension. Set an open seal or clip on the 4 inches in front of the ratchet where the two ends of the strap overlap, and use the crimping tool to secure the open seal. Finally, squeeze down on the handle of the ratchet tool so the cutter in the front cuts the excess strap.

Again, if doing the installation as a work party or bale raising, this phase of the work is a great place for helpful folks to plug in. Generally, whether it is a bale raising or not, we create a bale resizing station, where the strapping or twine, tools, extra measuring tapes, and the like are always located. This is helpful when the jobsite quickly fills up with straw and tool loss becomes a real possibility.

On DVD, Chapter 14, see RETYING STRAW BALES WITH STRAPPING AND RATCHET TOOLS

this. We have also taken to compressing the tops of walls, once bales are all installed, by using a car jack braced against roof framing to gain some mechanical compression.

Some systems of compression, such as using steel strapping and ratchets or Gripples, have been utilized to good effect. These have most often been used in the context of load-bearing construction, when compressing the bales before the roof goes on makes a stronger wall.

Fastening

There are a couple of methods for fastening straw bales. These are relevant for straw bale as an infill or wrap, but not for load-bearing construction.

Our favorite method, "vampire stakes" are stakes made out of 1 × 1, 1 × 2, or 2 × 2 rough wood stock, cut to about a foot long, with holes predrilled for fasteners to go through. These stakes are driven into the heart of the bale, usually behind a twine, so they are held and embedded into the bale, and pull the bale into the framing when the fastener is sunk. In terms of the fastener, TimberLoks are fantastic, but they can be costly. We like to have an assortment of lengths around, from 4 to 16 inches long. The shorter lengths will be available at any building supply; those 12 inches and up will need to be special-ordered. Pole-barn nails can be used as a more affordable alternative (the downside to these is that, once driven, they are much more difficult to remove and adjust, which is often useful to be able to do).

The second method involves toe-fastening bales with blocks or triangles with long screws or plastic washers. The fasteners are blocks made of 2½-inch plywood squares, or triangles for better "bite" into the bale, and are akin to homemade washers. They are most often predrilled with holes for the fasteners. These fasteners are positioned so they "toe-screw" the corners or edges of the bale to nearby framing. Plastic washers, also called locking plates, are very compact, can be ordered easily, come ready to be used, and are very effective and versatile.

On DVD, Chapter 14, see METHODS OF STRAW BALE INSTALLATION

Fastening a corner bale with a "vampire stake": a 1 × 2 stake driven halfway into a bale and fastened to the back side of the timber with a TimberLok screw. PHOTO BY ACE MCARLETON.

Caging

Another option for securing the straw bales is caging, and there are a few ways to do this.

Saplings

Cutting saplings from the surrounding woods is possible on many rural sites at which we work but obviously would be more difficult in suburban or urban sites. The system works this way: After the saplings are collected, drill pockets with a hole saw and a corded drill into the top plate and bottom sill. They can be placed ahead of time, much as the window and door framing is, and the bales notched to them as the walls go up; or the straw can first go in, and afterward a channel can be carved in the straw for the saplings to nest into the bale so they can be covered with plaster. The saplings can be installed in a staggered zig-zag pattern on the interior and exterior of the wall plane, or they can be placed directly opposite one another across the wall thickness. All of these options are good ones, and your choice will depend on the rest of your framing setup and how much linear footage you need to support with the caging.

Either of the fastening systems—the ratchet tools and poly strapping, or the twine and dowel rods—discussed earlier in the chapter can be used to attach the bales to the saplings as each bale goes in; or alternatively, strapping or twine can be used after the fact. Thread the resizing needles with twine or strapping, and push them through the wall next to a sapling. On the opposite side of the wall, pass the twine or strapping over the face of another sapling, and send it back through to the first side with a needle. Then tighten it. It will hold around the solid saplings, pulling them toward each other, and squeezing the straw bales tight.

Framing

If there are large spans in the wall where there is no framing to tie to, the opportunity exists to add in a 2 × 2 or 2 × 4 to provide more points of connection and stiffening to the wall. The process for doing this is exactly the same as for the saplings—you are just using planed, rectilinear lumber rather than natural rounds. Framing can be installed in advance and bales notched to it, or it can be retrofitted in after the straw bales are up.

Caging systems are not always necessary. Our usual rule of thumb is that if more than one bale is floating free, or otherwise not attached to anything, consider adding some stiffening to that area. After the walls are up, their solidity can be evaluated, and stiffening methods can be utilized during the plaster-preparation stage (see step 4 at the end of this chapter). One of the other options that exists is to send strapping or twine through the wall in the section that needs greater solidity and to strap it to adjacent bales that are secured to framing. This is one application where the poly strapping and ratchets are invaluable to attain the requisite tension.

Special Conditions

There are several instances that require special consideration, including stacking an outside corner, beveling at the sides of window openings, and utilizing lintels and baskets above window and door headers.

Outside Corners

One key design decision to be made regarding the structure is how the outside corners will be completed. There are several options.

Corners can be built by "bricking" the bales at the corners. This means, for example, that the end bale of one wall will go all the way to the edge of the foundation, and the wall that meets it to form the corner will abut that straw bale. Then, in the next course, the opposite will occur. This ensures solid construction of the corners and a staggering of joints. There are pros and cons to this method. It can be challenging to get bales tightly "held in" to the corner. Bales should be fastened to the timber frame or studs that are there, and this will go a long way toward stability, but still the tendency is for the corner to "wander" in terms of plane and to be loosely attached. There are several solutions to this. The first option is to temporarily place a stud as a guide to keep bales in plane during construction. This works well, if the folks doing the installation take care not to put force on this guiding member because doing so will cause the member to bow with the pressure of the bales, defeating its purpose.

A second option is to carefully track plumbness with a 4- or 6-foot level as the corner is being built. At the end of construction, when the corner is all the way up, two people can go back during the plaster preparation stage (see the final section of this chapter) and work together to "true" the corner. This involves one person outside, one person inside, and lots of shouting, whacking the bales with sledgehammers, and sometimes using the resizing strapping and ratchet tools to "suck in" or adjust parts of the corner until it looks good and reads plumb.

When nicely prepared, the bales make a really nice rounded corner when plastered. It is a fun challenge to plaster and can look quite sculptural and gentle, if that is the aesthetic look desired.

Alternatively, corners can also be framed. In this method, the straw bales will come up to, and stop at, rough framing members that are placed at the corners. The same care must be taken here as elsewhere: bales must not be shoved up against these framing members with great force, causing them to lose their shape. If its integrity is preserved, the rough framing acts as a plumb guide for the bales on the corner.

It is inadvisable to plaster over these framing members, however, because even if they are covered with roofing felt, lath, and a really strong plaster mix, the truth is that wood and plaster just don't play nicely together in the long term. They are too fundamentally different in their natures. So, an implication of choosing this corner option is choosing wood trim boards on the corners. This alternative can look great (depending on your taste)—"picture framing" the plaster walls really nicely at the edges and corners. Remember to use rabbets on the trim boards, and to use the concept of "rough to rough, finish to finish" for planing out all your materials.

Baling to the Window Opening/Beveling

Placing bales in the vertical orientation, "on end," at windows and doors allows for a more solid installation. The reason for orienting them vertically touches upon one of the key design principles highlighted in this chapter: think ahead to what the desired finish will be at the window openings and this will guide the installation of the bales. The sides will be beveled back with a chain saw after the bales are up for most openings. This is done to allow more light to enter the interior spaces, to create an easier transition from the wall plane to the inset window, and to create more space for benches or storage. Remembering to orient the bales vertically at the sides of windows and doors saves time and creates a more stable bale wall to work with.

The on-end orientation means the flakes are being compressed downward. This compression makes it easier to cut the straws. Before installing the vertical bale, center straps can be run around the bale, so when the twine closest to the window opening is cut, the flakes will still be held together by the one original twine and the new center twine. It also works well to angle the center twine, so that on the 18-inch face of the bale the center twine runs almost on top of or right next to the original second twine, and on the opposite 18-inch side, it is in the same position as the original first twine. This also serves to capture and compress the flakes more strongly in the absence of the first twine.

Conversely, the twine will be in the way of the chain saw during beveling for courses installed horizontally at openings. Approaches to resolve this do exist, such as diagonally threading twine or strapping in the shape of your desired bevel before you cut the twine, either before the bales go up into the installation, or after they are together in the wall. However, this approach is time-intensive—and more importantly, it does not leave a stable and strong bale to receive plaster (after the original strings are cut and the bevel is shaved back). Further, the flakes are oriented vertically so their tendency is to fall over and out into the open space.

Bales installed "on end" or "upright" on either side of a window or door allow for beveling without cutting strings, while maintaining the best structure in the bale. PHOTO BY ACE MCARLETON.

The ideal orientation for bales at window and door openings depends on design and aesthetics for finishing, however. For instance, if you are finishing the window well with trim boards, and therefore are installing rough framing to attach them, then you not only can but will have to make the diagonal bevel cut on the bales before fitting them against the angled framing you have installed. In this case, you only need to have the bale hold its trapezoidal shape. The process will entail threading two diagonal pieces of strapping through the bale and cutting the original strapping to reveal this trapezoidal bale. There are written instructions on how to make a trapezoidal bale below in the "How to Install Baskets" sidebar.

More about beveling can be found in the "Plaster Preparation" section at the end of this chapter.

Lintels and Baskets

There are two basic methods to support bales on top of openings: lintels and baskets.

A lintel is a solid ledge, usually made of wood although it can be made of other material, that is placed above the window opening to act as a support for the bale that sits above it. It can be attached to the header of the window rough framing and cantilever out some distance into the window well. In this case, the lintel must be engineered strongly enough to resist bending due to the weight of the bales above because often it is not just the single bale that is bearing on the lintel but all the bales above it as well. Lintels also can be wood members that span entirely across the opening, resting on either side of the window sides. This works only if the course break coincides with the top of the window height—which, in a non-load-bearing structure, can be planned if the height of the courses determines the window rough-opening header heights.

Baskets are usually made of metal lath, also known as plasterer's lath or blood lath—so called because it is very sharp and causes cuts if not handled

How to Install Baskets

Here are step-by-step directions on how to install baskets for window overhangs:

1. The installer cuts a piece of metal or plastic lath to the desired width and length and attaches it to the window header with roofing nails or finish staples of good length, so they hold firmly. Where on the header it is attached has everything to do with design decisions regarding where you want the rough bale plane at the top of the window to be. Remembering to factor for rough- and finish-plaster thicknesses, and where you want the trim to come to, will guide the placement of this lath on the header.

2. The installer then curves the lath up until the top cut edge rests against the back side of whatever framing it will ultimately be attached to—usually a girt, tie, floor joist, or some other member that is located horizontally above the window opening and in the interior plane of the wall.

3. Now the shape of the curve can be adjusted. The installer can either tack it in place, then getting down to stand and see how it looks, or work with a partner to set the shape from the ground.

4. The installer fastens the top edge of the lath to the back of the interior-plane framing.

5. A template of the shape of this curve can be made now, so that the installer can trace the shape of the basket onto the bale for cutting. Also, this template can be used to visually match the lintels, if that is desired.

The next phase is creating the curved bale to sit in this space. Usually we create a curved bale with a chain saw. There are many ways to do it, and here are just a few:

For a bale installed horizontally, either flat or on edge: (1) string a new piece of strapping or twine around the bale, so the strapping is out of the way of where you want to cut the curve; (2) take the template and use spray paint to trace the shape onto the bale; (3) cut the original one strap that is in the way now; your new strap should hold the straw mostly in place; (4) cut away the curved shape with your chain saw; and (5) place the bale in the basket to test its shape; trim if necessary.

It is also possible to create curved or trapezoidal bales (a bevel or triangular shape) through the two strings. The process looks like this for bales being put into the basket on end or upright:

1. Draw the line of the angle you are looking to cut the bale into down the two sides. Connect the top and bottom of where that angle starts and ends on the contiguous faces, so you know where to start and end the cut.

2. Stand facing the cut so you are sighting between the strings.

3. Cut down with the chain saw, between the strings, following the angle of the bevel. First cut close to the right-hand string. Then cut close to the left-hand string.

4. At the bottom of the cut, turn the blade of the saw horizontally, to cut across the face of the new angled plane. Clean out this notch you have created between the strings, and check that it is in the shape you ultimately want. Adjust if necessary.

5. Hook your fingers under the strings, and move them so they drop into the notch you just created; one on the left, one on the right. They will be looser than they originally were because they are holding less straw than they were (because you cut some away). We will fix this in a minute.

6. With the strings now in the angled notches, use the blade of your saw to carefully trim away the two remaining side "wings" of material, following your spray-paint line. Now you should have a trapezoidal bale.

7. The best way to tighten the straps is to add two new ones, adjusted to the new tension. You can also cut the straps, add on a piece, and retie them more tightly, but this method requires some finesse so that you do not lose the coherency and tension of your bale.

Diamond metal lath baskets for bales over window. PHOTO BY ACE MCARLETON.

carefully—but they can also be made from plastic lath. Baskets are the easiest way to create curved window overhangs, which is a beautiful and classic bale-structure look. Lath is a sculptural, flexible sheet that is rather like a hammock for the bales.

On DVD, Chapter 14, see INSTALLATION OF WINDOW AND DOOR BASKETS

Flashing

Are windows and doors in straw bale construction flashed like any conventional window or door? Yes, with the exception that there is generally no house wrap on an exterior straw bale wall to wrap around the window or door rough opening. Instead, we add a membrane such as #30 roofing felt wrapping the frame and terminating behind the exterior trim (when present), lapping over the rough plaster. At the top of the window, metal drip cap flashing above the head casing (when present) is designed to be wide enough to fasten all the way back to the rough framing behind the rough coat. For more information, see conventional current guides to window flashing. Some details that include flashing are provided in appendix B, or for other examples see Magwood et al.'s *Straw Bale Details*.

Window and Door Finish Options

Window and door openings allow for different aesthetic opportunities in a straw bale structure. These options for different finished "looks" follow from different choices in framing and construction of lintels, windowsills, side reveals, and casings. Certain things will remain the same regardless. For example, the thickness of the straw bale walls will remain evident at fenestrations, whether you finish the reveals in wood or in plaster.

Some construction options include:

- full-plastered, beveled, curved plastered basket for lintel, with curved plastered sill
- full-plastered, beveled, curved plastered basket for lintel, with wood or stone sill
- plastered, beveled sides, with partial lintel at top and windowsill options
- plastered, beveled sides, with full-finish lintel at top
- full wood-finished beveled sides, wood lintel at top, and wood or stone windowsill
- creative options like adding recessed lighting to framed lintel and window seats

Beveling window wells can be done at any angle or curve. The window openings can be prepared to receive wood lintels, wood sills, curved plaster reveals, tile mosaic, slate or other stone sills, and many other creative options. PHOTOS BY ACE MCARLETON.

Mechanical Vents

The rough openings for any vents or pipes that must go through straw bale walls are easiest to cut before bales are plastered. This requires knowing in advance the location and size of the rough openings for these vents or pipes. A chain saw is the best tool to make the hole. Carefully plunge-cut with a sharp saw, watching out for any strapping or twine, which will tangle in the body of the saw and can be dangerous. The exterior of this fenestration will need to be flashed and treated for air sealing and moisture protection. Combustion-exhaust vents that require clearances from combustible materials can be hung in plywood boxes—sealed for airtightness—that can float in the wall. The empty space in the box can then be insulated with a noncombustible insulation such as rock wool, where code allows.

Top of Wall— Bale-Eave and Bale-Gable Transitions

Another opportunity to detail for airtightness, clean transitions, and continuity of the thermal envelope is at the top of the wall, where the bales will meet the roof and roof insulation. As we saw in chapter 7 the greatest pressure in the structure exists at the top of the wall, and therefore vulnerabilities or sloppiness will cost more here than in other places in the structure. The guiding principles should always be to ensure continuity of the thermal envelope, airtightness (execute air fins with care here), and efficiency of building material use (fixed-form and flexible-form insulation should play the role that best fits the nature of their function).

The best example of efficient design is to bring the straw bales to meet the cellulose with which the rafter cavities are insulated. In this design, the straw bales come up to the point where they meet the exterior point of the rafters. This means that the level point—where we follow the top of the bale to that same point on the interior of the wall—is lower than the rafter plate, often by a foot or more. To address this, we install an extra-long air fin, pinned to the bale below. The cellulose, when it is blown into the rafter cavities, spills down on top of the straw and fully fills the transition from wall to roof angle. This is an example of using a fixed-form insulation type (in the form of straw bales) to do what makes the most sense for them, i.e., build up the bulk of the vertical wall, until they reach a flat point (in this case the rafters). Then the flexible-form insulation (dense-pack cellulose) fills the angled and bent shape at the top of the wall, providing a continuous thermal envelope.

The same concept applies at the top of gable walls, but with a different angle condition. The most efficient design for gables is to have the bales run up behind framing that has been set on the interior and exterior plane of the wall, to capture the tops of the bales, so a solid wall plane can be achieved underneath the framing members. Above that, the cavity is filled with cellulose insulation. Remember, as with any transition, to treat the wood members that the bales are tucking behind with air fins and "tooth" for plaster, as well as any relevant rabbets or stand-aways for a clean and durable plaster installation.

This transitional approach can, for the most part, replace the time-intensive process of making trapezoidal or triangular bales for gables. There still will be unavoidable moments where such cutting of bales will be necessary. Directions for reshaping bales for conditions like this can be found in the "How to Install Baskets" sidebar earlier in this section.

Step 4: Plaster Preparation

The final stage of the bale design/build process, "plaster prep" represents a group of activities that take the fully installed straw bale wall the whole distance to being plaster-ready. Plaster prep turns straw bale walls from a shaggy, lumpy, holey, occasionally shaky assemblage of individual bale blocks into a solid, cohesive wall unit that is adequately prepared for plaster application. In our classes we tell students to expect that plaster prep will often take *as long as the bale installation*. Students think we are exaggerating, until they experience it. One of the most basic necessities for plastering is a solid, stable substrate. Again, to return to the design principles covered in the beginning of this chapter, plaster prep means knowing what you will need to plaster successfully, and working backward from there.

TRUEING

Trueing refers to the process of evening up the walls and roughly plumbing them with a level. Usually two people are required for this process, one on the inside and one on the outside. Each person surveys the relevant planes they are seeking to bring the bale wall in line with. Most often this is defined by the toe-up, window and door framing, or edge of the foundation—it varies depending on the installation and how the planes are set (for more on this, see the section "Establish Planes" in step 1). What follows is a process of knocking the bales with sledgehammers or commanders, or kicking or hitting them with arms and legs until both people come to consensus about the "trueness" of the bale. It is wise to start at the bottom of the wall, and to move in a linear fashion in one direction, so both people know what the procedure is and can work together in a coordinated way. Often it is helpful for one person on one side to brace a part of the bale, or an adjacent bale, while the person on the other side applies force to the recalcitrant area.

Objections surface to this process at times based on peoples' desire for "sculptural," "lumpy," or "uneven" straw bale walls. While it's true that part of this process of trueing can be seen as an aesthetic choice to create cleaner lines, the majority of it is functional and performance based. Some main reasons not to leave the bales uneven or lumpy include: the compromise in R-value where the bales are not matching each other, edge for edge; the possibility for instability where bales are unevenly installed; creating more work to do later filling gaps with straw clay, the time involved in plastering protrusions from the plane of the wall; and the likelihood that rain or snow will gather on the protrusions on the exterior side, creating a vulnerable spot in your wall.

STUFFING

Gaps in the bale walls are stuffed with light straw-clay. Often called the natural builders' expansion foam, straw-clay is a material made from clay and straw that can be formed and pushed into any size and shape, to fill gaps or spaces. This is a critical step that minimizes air intrusion and thermal losses, helps protect the wall from fire, and creates a consistent, solid plane to which to apply plaster. (For more on straw-clay see chapter 15.)

On DVD, Chapter 14, see HOW TO INSTALL STRAW-CLAY KNOTS

Stuffing cavities with light straw-clay is an essential step for wall performance. PHOTO BY ACE MCARLETON.

To stuff gaps in straw bale walls, follow this procedure: Survey the wall. Look and, most importantly, feel for gaps between the corners of bales, at the edges of windows and doors—really everywhere. Use your hand to determine the depth of the gap. Take up a handful of straw-clay mixture. Hold it in both hands, aligning it so the long straws are oriented lengthwise with each other, spanning into both hands. Tug them gently, so they slide past and along each other, lengthening the mass of them but not pulling them all the way apart. With both hands, twist the elongated mass of straws. Fold the twisted group of straws in the center so that you are folding it in half.

Twist the straws below the fold with both hands. This will make what we call a straw-clay "knot." Push this knot, knotted side first, into the gap in the wall you have identified. If it is a vertical gap: Insert your first knot, then push it down hard. Insert your next knot, and push that down too. Continue in this way until the last one is tight to put in. This creates a tension-fit—a very solid surface of knots. If it is a horizontal gap, insert your first knot, then push it to the far right or left side of the gap. Continue this process until the last one is tight to get in. If a gap can be filled with a single knot, then choose your handful of straw-clay to match the size knot you need for the hole. Resist the temptation to loosely place shallow clumps of straw-clay into holes. They will look good initially but then fall out under the weight of the plaster.

It is imperative that you do not make more problems while attempting to fix a problem. It is easy to get carried away with the knot stuffing and to actually displace the bales, causing extra time later to pull out the work that was done and have to redo it. The concept here is similar to fitting and sizing the straw bales into the wall space: tightness is desirable, but cramming or oversizing them will merely cause problems by pushing out framing or corners. Make the plug well fitting, not gigantic. Additionally, use good judgment about where the plugs are necessary. For instance, do not fill the stand-aways or rabbets that are intended as plaster notches.

There are very important reasons to do a careful, thorough stuffing job, including fire protection, thermal performance, and setup for success in the plastering stage. Plastering onto a solid, even surface is many times easier than plastering onto a surface that is uneven and full of gaps, which can eat up more cubic yards of plastering material than may have been anticipated. Also, huge thick areas of plaster filling large gaps will take forever to dry and are more likely to crack due to thickness differential between the gaps and the plane of the wall. Having plaster in the gaps rather than straw-clay will also diminish thermal performance, since straw-clay has a significantly higher R-value than plaster. The plaster will also stick nicely to the straw-clay plugs owing to their clay content.

TRIMMING

Once the walls are up, they are shaggy, with loose straw hanging off their faces: time for a haircut and a shave. Again, this is a time-saver for plastering. Plaster spreads smoothly onto surfaces that are stable. Loose straw gets in the way of plastering, as the plasterer ends up having to pull it off and deal with it in the midst of doing a very different activity.

We use an electric weed whacker, or string trimmer to trim the walls. It's light, quiet, and effective. Gas weed whackers are loud and smelly, and chain saws work but are slower, heavier, more awkward, and more dangerous. Hand-held angle grinders, fitted with chain-saw wheels called "Lancelots" or "Squires," are good for hard-to-reach small areas.

BEVELING

Beveling refers to shaving back, or relieving, the square edges of the bales that frame window and door openings. This is most commonly done to allow more light into interior spaces but has the added benefits of creating more usable space in the window well and an easier visual transition between the wall plane and the inset window.

Use spray paint to mark the point to which the bevel should come to, on both the inside of the window opening and the face of the wall. Usually we begin our bevels 6 inches (or so) inside the window buck, and 9 to 12 inches from the corner of the bale out into the plane of the wall (but this is really an aesthetics-based choice). The desired shape of the cut is also subjective; you can make the cut sharp and angular or curvy.

Usually we make a template, or a jig, that we can fit onto the window buck and test to see if we've created the shape we want. This is a trick learned from Paul Lacinski, coauthor of *Serious Straw Bale* and charter member of Natural Builders NorthEast (NBNE) and GreenSpace Collaborative. Almost anything lying around the jobsite can be used; plywood and wood scraps work great. The benefit of the jig is that bevels on openings around the structure can be more closely matched as it provides a consistent guide.

Careful and intelligent setup will yield safe, accurate, and efficient cutting. Survey the bales that make up the sides of the window or door, and make sure all the twine or straps are cut out of the way.

The chain saw is useful for removing the bulk of the material, and the angle grinder with the Lancelot or Squire attachment is useful for more detailed fine-tuning of shape. It also can be useful to have the same person cut the window bevels, for efficiency and consistency. Depending on how many fenestrations exist in the structure, it can be a substantial task to cut and shape them. Remember that during the plaster stage, some tweaking to the shape is possible by building up a bit of plaster here or there—although it's wise to create the desired finish shape as much as possible in the construction phase.

FINAL PLASTER PREP DETAILS

A few final details will bring the plaster preparation to a close and move the process toward plaster application. Among them are the setting of electrical boxes and prepping for cabinets, interior walls, niches, tiling, and other decorative features. Finally, the spraying of clay slip will speed the plastering process.

Setting Electric Boxes

Most major electrical runs are best routed through interior walls, if possible. However, setting electric boxes—either switches or outlets—into straw bale walls is possible and very commonly done. The "plaster prep" phase is the time to install these runs and boxes.

Outlets and switches are set to standard height in a straw bale structure, just as in any structure. If the bottom of the wall is a pony wall (see the "Good Shoes" section earlier in this chapter), the outlets can be done whenever the pony wall finish is about to be installed and are unrelated to plaster preparation. If, however, the outlets will be set in the bale wall itself, the runs of Romex or conduit should be roughed into the toe-up before the bale installation begins (see the "Building Methods" section of this chapter).

On DVD, Chapter 14, see RUNNING ELECTRICAL SERVICE IN WALLS

Carving a niche for an electric box with a chain saw attachment for an angle grinder. PHOTO BY ACE MCARLETON.

Setting Electrical Boxes

1. After the walls are up, use spray paint to mark the location of the outlets and switches.

2. Using a chain saw or Lancelot tool, carve a notch that will fit the box size specified for that location. The chain saw is preferred, as a square cut is more easily achieved.

3. Often, we will pre-fasten the box onto a scrap of wood, usually a 1 × 4 × 12-inch piece, and then carve a slot for this wood scrap as well. This wood piece serves to stabilize the box in the straw and also, in the case of outlets, to provide a much larger surface area behind the plaster to diminish the likelihood that the box will, if tugged on, pull out of the wall.

4. With the Lancelot or chain saw, make a groove in the bale, tracing up from where the roughed-in run is at the toe-up to where the box is located.

5. Thread the Romex or conduit through the box, as you would in any electric setting process.

6. Set the box/wood piece assembly, with the electrical wires or conduit threaded through and poking out, into the niche you have just created.

7. Stabilize, affix, and level the box/wood piece assembly by pinning it with long staples (10 inches); we special-order these from garden supply houses. Often long TimberLoks can also be used, shooting off at different and opposing angles into the bale through the back of the box.

8. Be certain to caulk any penetrations in the electrical box for airtightness. This is a strategy used in any structure where good thermal performance is a goal.

9. Using the long staples, secure the electric cable into the groove that traces from the toe-up to the box.

Preparing for Cabinets, Interior Walls, or Heavy Wall Hangings

If the design calls for cabinets to be mounted high on the wall, commonly called "uppers" or "upper cabinets," or if the counter is going to be attached to the wall for security or to allow for a backsplash, this horizontal framing must be inset into the bale wall before the base coat of plaster goes on. Again, here it is relevant to think through planes and structure: is a 2 × 4 or a 2 × 6 necessary to support the weight and accommodate what will ultimately be fastened to it? How deeply should it be inset to the bale wall (so that it is flush or so it protrudes ½ inch, or 1 inch)? The answer to these questions is related to the design and the chosen finishes, and it can be thought through using the informational tools and critical thinking processes presented in this chapter and this book.

The same process is used for walls or interior walls that will be connected to the exterior straw bale walls. At the connection point of these walls, placing a stud, or a sapling (if the walls are going to be wattle

Tiles can be inset into plaster for a decorative effect. PHOTO BY ACE MCARLETON.

and daub, for instance; see chapter 16), fitted with an air fin, is an important step to accomplish in the plaster-prep stage.

Another common question relates to hanging art or other things on the plastered straw bale walls. Most art can be hung on the plaster using typical drywall anchors, or sometimes just small sharp nails, in the same way you would hang anything in any type of standard plaster wall.

If the item in question is anticipated to be quite heavy, such as larger art, a coat hook rack, or some other item, the plaster prep stage is the time to embed and attach some sort of framing in the place the item will go. In cob construction, this is often called a "dead man." Vertical, horizontal, free-floating, or attached to the rest of the frame—it's up to the design. Another option that is important to mention is to run copper pipe (or some other strong rod) at the top of the wall between posts, up near the tie beams, and dropping fishing line down to hold art, which is an attractive way to solve this problem as well.

Niches, Truth Windows, Tiles, and Other Decorative Additions

Another step of shaping the walls before plastering is to include niches or other decorative additions to the wall. The Lancelot or Squire tool that is used for wood carving, discussed earlier in this chapter, is remarkably efficient for carving into straw. After being plastered, niches can be used as a place for keys, art or sculpture, a cup of coffee, or even recessed bookshelves. Similar to the choices about windowsills, these niches can be installed with wood, stone, or tile shelves or sills for items to sit on, or they can be fully plastered. It is important to keep in mind that when straw is removed from the wall, we reduce the overall R-value. However, in useful or interesting places on the walls, niches are a wonderful addition and add character and uniqueness to a structure.

"Armatures," or horizontal shelves made of sticks inserted into the bale wall and woven together with supple branches, then filled with cob, can also be added. Either inside niches or as a protruding element, these details can add interesting aesthetic or functional features to the walls.

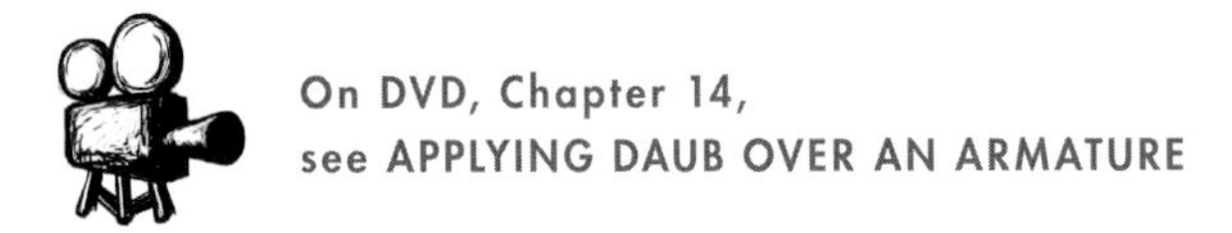
On DVD, Chapter 14, see APPLYING DAUB OVER AN ARMATURE

Truth windows have become a very popular and interesting hallmark of straw bale structures. They come in many and varied types, and new versions are constantly being innovated. We've installed anything from old wrought-iron heating registers modified with glass to stained-glass star and fully framed wooden doors that when opened reveal the straw that makes up the wall. The truth-window installation process depends very much on what it is, but the basic themes apply. It should be stable and solidly affixed so it cannot fall out or be knocked out.

Sometimes the strongest way to install a truth window is to embed it in the base coat of plaster and then bring the finish plaster to it last, which would work well in the case of a simple pane of glass as a truth window. In the case of something like the old heating register treated with glass, its weight necessitated that we secure it before the base coat of plaster by carving a shallow niche to rest it in, then pin it with long staples and/or long TimberLoks, and later bring the plaster to it. Regardless of what phase of the process the truth window is installed in, if you desire one, think early about where to put it so you can make certain not to stuff that area with straw-clay or spray it with slip.

Truth windows do represent a break in the plaster and a potential way into the wall assembly for air and water vapor. Air fins and gaskets should be used to ensure a tight installation whenever possible.

Clay Slip Spraying

The final step in plaster preparation is to spray the walls with clay slip. We do this as a necessary pre-step to plastering on-edge bales, owing to the lack of "tooth" present in the orientation of the straw. We have also begun spraying slip on flat-oriented walls as well, given the noticeable time-saving aspects this lends to the plastering process.

Spraying clay slip with a drywall texture gun and air compressor. PHOTO BY ACE MCARLETON.

Clay slip is made by blending clay, either dry bagged clay or wet site soil, into suspension in water. The desired consistency resembles a loose milkshake. Slip can be made in a mortar mixer, a bucket or trash can with a drill and paddle mixer, or a pit in the ground. For this application, the slip should be screened through ⅛-inch hardware cloth to remove any foreign matter or chunks. The slip is then poured into a drywall texture gun, which is attached to an air compressor. The spraying is messy, so be sure to cover anything that cannot be coated in liquid clay or easily cleaned.

Now you are ready to plaster—see chapter 17 for an in-depth look at natural plastering for straw bale walls.

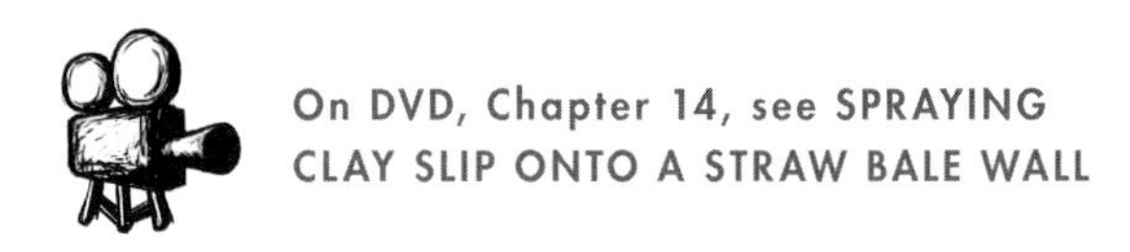
On DVD, Chapter 14, see SPRAYING CLAY SLIP ONTO A STRAW BALE WALL

CHAPTER 15

Insulative Wall Systems: Straw-Clay, Woodchip-Clay, and Cellulose

In this chapter we will examine three alternative building assemblies that fall on the insulative end of the insulation-mass continuum of common natural building systems: straw-clay, woodchip-clay, and dense-pack cellulose. All three employ basic materials that insulate—straw, woodchips, and recycled paper. Each is essentially a different form of cellulose, which, as was discussed chapter 5, is the woody plant fiber found in trees and stems of grasses.

Insulative Walls: An Overview

Straw bale insulation is an example of fixed-form insulation, where the bales are block modules set to a standard size that can be modified only minimally and are contained by twine. Unlike straw bales, straw-clay, woodchip-clay, and cellulose need a framed wall to contain them. They are flexible-form insulation types, meaning they fill and mold into the shape of the space in which they are installed. Chapter 7 discusses fixed-form and flexible-form insulation and the comparative thermal benefits and drawbacks of each. One of the major benefits of flexible-form insulation is that it can be installed in a wall assembly in a way that mitigates or eliminates interstitial convective cycling. This kind of convection is a major source of heat loss in a wall assembly, so using natural, flexible-form insulations like straw-clay and woodchip-clay is a strong, performance-based solution.

Also, recall from chapter 7 that it is air pockets that insulate by blocking heat from conducting through a material. The cellulose-based building materials we cover in this chapter contain air pockets that provide insulative properties. Insulation blocks conduction. This stands in contrast to mass, which thermally refers to a dense solid that accepts heat into itself like a battery, releasing it over time when the source heat goes away. For an exterior wall in a cold climate, a mass wall performs poorly as it does not stop conductive heat flow as long as it is colder outside than inside, and if it is allowed to get cold, it will take a tremendous amount of heat input for it to warm up, or it will more likely remain cold. Think of stone castles. This is the foremost reason that in a climate that gets very cold and remains so for five months consecutively, insulative wall systems such as

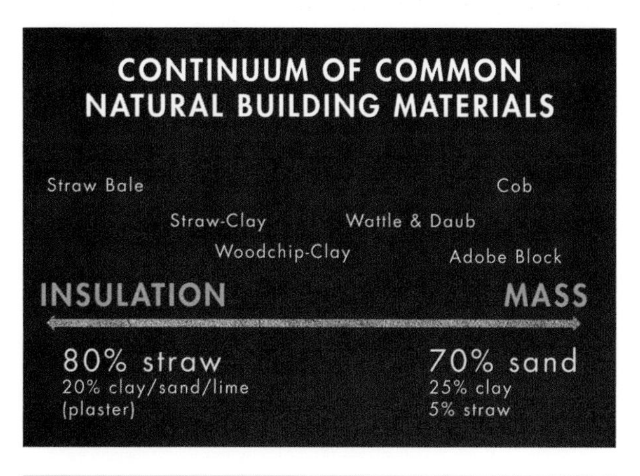

Natural building materials and methods fall on a continuum in terms of their performance. IMAGE BY ACE MCARLETON.

straw bale, straw-clay, woodchip-clay, and cellulose are a more thermally effective choice than cob or adobe block for exterior shell walls (see chapter 16 for more information on mass walls and heat).

Clay is the binder for light straw-clay and woodchip-clay, both of which are referred to as "light clay" flexible-form insulation materials. It is from the German tradition that we get the term "light clay." It is a translation of *Leichtlehm*, which literally means "light loam," and referred historically to straw-clay. The "light" also refers to the amount of clay that is in the mixture—which is just enough liquefied clay to "glue" the mixture together, not more, not less. To create these materials straw and woodchips are mixed with clay slip and placed wet into forms to dry in place. In the case of woodchip-clay, it must be contained by wood lath or some other caging system, as the woodchips do not hold their form with the infill method; finishes are then applied to the wood lath, harkening back to the old "lath and plaster" walls found in many old houses in the Northeast. When using slip-form straw-clay, the finish is placed directly onto the exposed face of the straw-clay panels, without any additional caging.

Cellulose, the third "other insulative wall system," is very insulative, the highest insulator of this group in fact, owing to the wood fiber having been processed into paper with lots of tiny air gaps among the fluffy paper pieces. Cellulose—shredded romance novels and newspapers—is mixed only with a dry borate for flame retardancy and mold protection, although some inferior products use urea formaldehyde for this purpose and are best avoided. Cellulose can be installed into vertical open cavities by mixing with a small amount of water at the tip of the blower as it is sprayed into the cavities, a technique called "damp-spray"; just enough water is used to cause the material to stick against the cavity wall, frequently the exterior sheathing. "Dense-pack" cellulose is installed dry into vertical or sloped cavities at a high density to avoid settling. The cavities must be fully enclosed with materials such as gypsum board, sheathing, or see-through netting designed specifically for the purpose. Finishes are then placed onto the sheet material rather than directly on the cellulose. Larger installation equipment is needed to ensure proper density can be reached, and greater operator expertise is needed to ensure a complete fill in closed cavities, where inspection may be difficult or impossible. "Loose fill" cellulose can be achieved with smaller equipment that is common to rent; while it is suitable for horizontal applications such as attic floors, it is not relevant to the discussion in this chapter as it is not appropriate for use in vertical installations such as walls, where it is prone to settling.

Interior partition wall: woodchip-clay, one coat of clay base coat, and a finish coat of lime-sand plaster. PHOTO COURTESY OF NEW FRAMEWORKS NATURAL BUILDING.

All of these systems can be installed in an airtight manner, maximizing performance. Attention should be paid to air-sealing transitions and to application of an air barrier. In the case of the light-clay assemblies, the plaster coats serve as the air barrier. None of these systems utilize vapor barriers, as all these materials are hygroscopic and allow for vapor to diffuse through them quite readily. These systems, when installed with air fins, are airtight but vapor permeable, which creates a high-performing envelope in cold climates (see chapter 8 for more on vapor permeable wall assemblies).

EVALUATING USAGE AND PERFORMANCE OF INSULATIVE NATURAL WALLS

Each of these assemblies has its own attributes that bear scrutiny in order to determine which is the best fit for a given project. These systems can be compared based on R-value per inch, common thickness, wall R-value, vapor permeability, materials used, method of manufacture of the material, method of application of the material, whether the material is fixed-form or flexible-form, what the drying time is, and how the material is finished. We can also consider mass effect of the wall systems, the tightness of installation that limits convection in the wall, the quality of the experience of working with the material, accessibility, and cost. Following our proposed metrics for decision making and ethical determinations within natural building as outlined in the introduction and chapter 2, many factors must be considered and weighed in deciding which system is right for meeting a given set of project demands.

It is important to note here that R-value is only part of the measure of the performance of a wall assembly. R-values are measured in the laboratory, and only the material itself is tested. The numbers in table 15.1 therefore do not give us any information on total R, which is the full picture of how a structure's systems and the transitions between them perform in resisting the movement of heat. This is how it happens in the real world in an actual building. Total R—the R-value of the windows, doors, walls, framing, roof, and all of their connections and transitions—gets us closer to the true picture of performance in a structure.

But beyond R-value, which measures conductive losses, there are two more types of heat loss to pay attention to with regard to any wall system: convection and radiation. Of these, convective losses are the greatest problem. Air sealing will greatly affect the building's performance, regardless of wall-system type or how high the R-value is rated in a lab. An installation of a material that includes lots of air gaps that will create convective currents inside the wall cavity itself will drastically compromise the thermal performance of that wall system (see chapter 7 for more on this).

Wall systems that combine thermal mass and insulation will perform somewhere between systems that are primarily one or the other. This is also a way in which R-value is only one predictor of how a wall will perform. The mass effect, measured as the "dynamic benefit for massive systems" (DBMS), has a significant impact on how a space will feel and behave in terms of comfort. Mass will help a space avoid sharp temperature swings, creating greater comfort for inhabitants. Both straw-clay and woodchip-clay, and straw bale for that matter, include the combined thermal effects that insulation (cellulose in the form of straw or woodchips) and mass (clay or sand and clay in the case of plaster) together bring to the thermal envelope, creating a very comfortable, high-performance wall system (the topics of mass and heat are covered in chapters 7 and 16).

Embodied energy is another metric for consideration. How can we calculate or understand the difference in embodied energy between woodchip-clay, straw-clay, straw bale, and cellulose? It is difficult to do this with any degree of universality or accuracy as there are no data nor is there a scientific study that we are aware of that analyzes the differences. However, we can use common sense to consider the inputs of production, transportation, application, life span, and end-of-life/decomposition characteristics of each type. See chapter 2 for an in-depth look at ecological costs and embodied energy of building materials. Financial cost is another metric. For an in-depth look at costing for natural building systems, see chapters 10 and 11.

TABLE 15.1. COMPARING INSULATIVE NATURAL WALL ASSEMBLIES

Attribute	Straw bale	Straw-clay	Woodchip-clay	Cellulose
R-value (per inch) at set density	1.45 (for 2-string bales)	1.6 (for medium-density applications)	untested, presumed between 1 and 2	3.45 (for dense pack only)
Common thickness	14–18 inches	12 inches	12 inches	12 inches
R-value for wall thickness	27	19.2	12–24 (presumed, untested)	47
Vapor permeability	excellent	excellent	presumed similar to straw-clay, untested	excellent
Materials	straw bales	clay, straw	clay, woodchips	cellulose (recycled paper), borate, mineral oil
Method of manufacture	baler in field, delivered to site	tossing by hand or pitchfork	mortar mixer or by hand	factory
Method of application	stacked and modified by hand	by hand or bucket, hand-tamped	by hand or bucket, hand-tamped	with truck-based compressor and blower, industrial hose
Potential for DIY	medium	high	high	low
Fixed-form vs. flexible-form	fixed-form (block)	flexible-form, unless made into blocks in advance	flexible-form	flexible-form
Drying time until covering is OK	ready immediately	week/s	week/s	24 hours for damp spray, otherwise ready immediately
Embodied energy	low	low	low	medium-low

Source: Created by Ace McArleton for slideshow lecture for Yestermorrow "Insulative Walls" class. Information compiled from Proposed International Code for Straw-Clay and Canadian Mortgage and Housing Corporation 2005. See references for complete citations.

Note: This table is a summary of current testing and known factors regarding these materials and systems. More testing is encouraged to get more data on woodchip-clay's performance.

STRAW-CLAY AND WOODCHIP-CLAY

Human beings have coated straw with clay and packed the resulting material in or around forms to create walls or cavities for a very long time. In colonial times in the United States, we have recorded evidence of straw-clay being used as infill between timber posts in early European settler homes and in chimney walls (see chapter 1 for more information). In Europe, the combination of clay and straw goes back to twelfth-century half-timbered houses, where clay and chopped straw were pressed around a light formwork of saplings and woven branches, allowed to dry, then shaved with a knife and additional material then added, in layers, until the wall was complete (Anderson 2002).

Comparison of Light-Clay Systems: Straw-Clay and Woodchip-Clay

Today, straw-clay is used for both exterior walls and interior partition walls, as insulation layers in earthen ovens, and as stuffing for straw bale walls to fill any gaps. Our crew likes to refer to straw-clay as the natural builder's expansion foam when we are installing straw bale walls or woodchip-clay walls, as it fills those awkward spaces and blocks air gaps. Straw-clay is a versatile composite whose materials are easily obtained in most places. It is easily mixed by the novice, is very flexible in its use, and can be molded to the size and shape of the place needing filling. Some people have experimented with making straw-clay blocks and used them for ceiling insulation (Anderson 2002).

As with many natural building methods, with light straw-clay the builder oversees the factory on-site and is in charge of combining raw materials to create a quality product. As noted earlier in this chapter, benefits of straw-clay include its use of readily available, inexpensive, low-embodied-energy materials. Significant savings are possible if the installer is also the owner because with light straw-clay it is the labor that is time-consuming and expensive if hired out. The mixing procedure, while time-intensive, is also quite simple and easy to learn, creating the possibility for the participation of many types of people, old and young. Lower cost, both financial and ecological, comes into play here because the builders do the manufacturing work rather than paying a company to make the material and then package, market, and ship it to the site.

Potential liabilities of using straw-clay include that it takes more time and work and places more risk on the builder to ensure quality without much current available industry support or trades knowledge. Another downside is that the mixing of straw-clay is labor-intensive and the material requires significant drying time, which can preclude using this wall method if drying conditions on-site are not sufficient or time is short. The drying needs also limit the thickness of this installation to 12 inches for exterior walls, which levels out its insulation maximum value at R-19.2 (1.6 × 12). Although this is certainly a limitation, remember that the R-19.2 value can actually be achieved with a straw-clay assembly that is in a double-stud wall—unlike a common insulation strategy, fiberglass placed between studs, which does not achieve its stated R-value in practice. However, the thickness limitation of straw-clay walls means straw-clay cannot be entered, by itself, into the pantheon of superinsulation strategies.

Arch made of straw-clay. PHOTO COURTESY OF NEW FRAMEWORKS NATURAL BUILDING.

Woodchip-clay is a more recent German innovation from the 1980s. This method resulted from a revival in clay building that the country experienced in the wake of a rash of cement repairs that had been made post–World War II, which caused moisture damage to historic structures that had survived perfectly well with clay for centuries. It experienced an upswing of interest in the 1990s, and builders in the United States also started using this technique.

One of the most significant benefits of woodchip-clay balances out one of straw-clay's most limiting factors: time. Woodchip-clay is fast. It is easy to mix, especially in a mortar mixer, and it is a breeze to install into the wall forms, if the site has been set up intelligently. It also utilizes what had often been

Woodchip-clay held by wood lath. PHOTO COURTESY OF NEW FRAMEWORKS NATURAL BUILDING.

considered a "waste product," and is now more of a by-product, in wood-producing regions or anywhere trees grow somewhat abundantly: woodchips. It is worth noting that it is possible to rent an industrial chipper and to chip up clean lumber and timber construction waste and turn it into the walls rather than burning or otherwise disposing of it. We experienced this process of creating walls from woodchips made on-site for the first time at a project in Deering, New Hampshire, that was headed up by Ben Graham, of Natural Design/Build in Plainfield, Vermont.

DESIGN FOR STRAW-CLAY AND WOODCHIP-CLAY INSTALLATIONS

Many of the design concepts for straw-clay and woodchip-clay (as noted earlier, both can be referred to as "light clay") structures are the same as those for straw bale structures. They entail thinking through interlocking parts of the system—such as framing and how it will interact with the infill, or the plaster and how it will interact with the framing—and hence the placement of these elements and transitions between them. Design for straw-clay and woodchip-clay installations must include planned vapor-permeable air barriers for protection from air infiltration and moisture infiltration from ground, outside air, inside air, interior floor water (as in a burst pipe or overflowing toilet). The design needs also to anticipate protection from impact, wear, and use—such as planning for a toe-kick or baseboard, or including corner boards in the design.

The four-pronged approach outlined in chapter 14—"good shoes, good coat, good hat" and "good buttons, zippers, and belt"—is also essential for a long-lasting, clean design for all light-clay walls. See that chapter for more details on foundation transition detailing, siding or plaster finish detailing, roof design and detailing, and air sealing transitions.

The process starts with identifying the building type. What will the foundation be? What is the framing system? For more on selecting and distinguishing foundation, framing, and roof systems see chapters 12, 13, and 18, respectively. Recommended designs for light-clay walls in the Northeast include:

- *Timber frame with double-stud wall for light-clay infill.* In this design the timber frame is the structural support for the roof loads, and the stud walls act as caging, reinforcement, and vertical ties from floor to ceiling of the straw-clay or woodchip-clay mass. This system is preferential for those who wish to see a timber frame on the interior and to pair this handsome, visible building support system with a natural wall "wrap" that serves as an alternative to foam panels. Downsides include cost, since a timber frame is a more expensive framing option if contracted out, and redundancy, since building the full stud wall on the exterior just to hold the insulation can be seen as duplicative.

- *Double-stud wall (structural) with straw-clay slip-form installation.* This option results in walls that do triple duty: the stud caging required for the straw-clay infill is the roof support, and the straw-clay is the insulation and also the substrate for plaster. Benefits of this design are that the walls that cage the infill material are also the roof structural support, hence saving materials and cost.

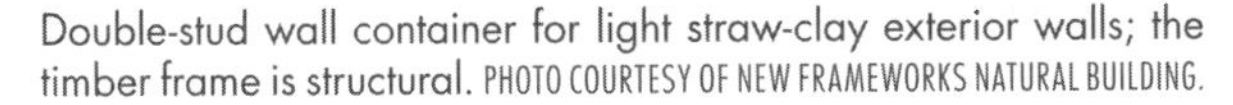

Double-stud wall container for light straw-clay exterior walls; the timber frame is structural. PHOTO COURTESY OF NEW FRAMEWORKS NATURAL BUILDING.

Light straw-clay walls as interior partition walls for a bathroom in an open floor plan. PHOTO COURTESY OF NEW FRAMEWORKS NATURAL BUILDING.

- *Double-stud wall (structural) with woodchip-clay infill.* The infill is caged in this system with wood lath. The wood lath becomes the primary substrate for the plaster coat/s. The primary benefit of this system is that woodchip-clay is a much faster infill method than straw-clay, in general.
- *Single-stud wall (structural, or nonstructural when paired with timber frame) with straw-clay or woodchip-clay.* This system would most often be chosen for a room or structure where neither heating nor cooling was needed, as it would not be thick enough to have much insulative value. This design is beneficial in that it is simple to build and install and quick to complete because it has less material, and it can create a great minimally conditioned space.
- *Single-stud wall (interior) with straw-clay or woodchip-clay infill.* In the Northeast we do a fair number of light-clay interior partition walls in bale houses because the walls provide sound deadening, we utilize what is often a waste material from the primary construction of the rest of the house, we are able to create a continuity by using similar materials and a recurring aesthetic, and the walls are beautiful and functional. Given the lesser total R-value of light-clay walls compared to straw bale, utilizing this material for interior walls, rather than exterior walls, is a sensible choice if a more highly insulative envelope is a concern.
- *In any of these systems, woven-wattle walls can replace planed rectilinear lumber stud walls.* A less-industrialized option, a wattle wall, which refers to uprights of round saplings interwoven with supple horizontal lengths of (generally) smaller branches, also serves as a caging system for any of the light clay systems, or damp-spray cellulose, explored in this chapter. The wattle wall should be pre-plastered for cellulose installations, and plastered after installation in the case of light clay infill.

The wall section should be considered next. In designing for light-clay exterior walls, establishing the thicknesses and locations of various wall components is essential. This is best done with a section drawing. Where will the stud wall (if doing a single-stud wall) or walls (if doing a double-stud wall) be framed to avoid thermal bridging? Where will this establish the plane for the base coat of plaster, how thick will that coat be, and where will the plaster stop be attached for the finish coat? Mapping it all out in the design stage will streamline the building process—and the results will be cleaner and better executed.

Specific Design Considerations

Specific design considerations for straw-clay and woodchip-clay installations include the following factors.

Framing for Windows and Doors—Beveling, Lintel Support

Planning for the wall framing must include designing the reveals, or edges, of the fenestrations. All the windows and doors must be framed to provide the setup for whatever the finished outcome will be. We often recommend that most windows and doors have beveled reveals—meaning that the straw-clay or woodchip-clay walls open up at an angle from the edge of the window or door. This allows for more light at the window openings, provides a more generous entry space for a person entering or exiting carrying a load at the door openings, and creates an easier visual transition for the wall thickness to resolve against both. There are times when bevels are not the right choice: for a window designated for a window seat, for example, it will likely be more comfortable to lean back against a reveal that is perpendicular to the window. It also is an aesthetic question. The window and door reveals provide an opportunity to think about the feeling you would like the space to have, and the way the shapes and finishes will together communicate that feeling. The bevels can be flat and angular, or more rounded and gentle. Finishes of wood, plaster, or something more unusual like tiles, stone, or sheet metal help articulate "contemporary," "cottage-y," or "classic." Openings treated with finish-wood paneling on the sides, a wood windowsill, and wood lintel create solid, consistent openings that feel utilitarian yet refined—reminiscent of Cape Cod. Openings treated with thick, rounded plaster feel organic and sculptural. These sorts of choices are the basis for creating the look you want in a structure, and it is important to show that natural materials do not all have one aesthetic—that the materials themselves are versatile if we understand their constraints and opportunities and work with them.

The term "lintel" specifically refers to a member that carries the weight of the wall above a fenestration by spanning that opening. In natural walls that have some significant thickness to them, the lintel strategy must be considered not just as a structural necessity but, similarly to the reveals, also as an aesthetic choice. The desired finished look must be accommodated in the design and construction stages by providing rough framing to attach finish trim to, or by setting the desired angle or curve in the rough-framing stage. In double-stud-wall construction, it is really just the interior framing that will have to be set for the bevel, since the exterior framing creates the rough opening for the window or door.

Air Fins

Air fins (described in detail in chapter 14) are also necessary for airtight light-clay-wall installations. Unlike in straw bale installations, however, air fins in most locations in light-clay walls can be installed after the walls are up and filled. Just as with straw bale walls, planning ahead with air fins will make them successful. The best guideline to follow is to attend to the edges of the installation—bottom of wall, top of wall, window and door edges, edges of posts, braces, and girts—and plan how the plaster will meet that edge. The plaster is the primary air barrier for the wall system. It is compromised everywhere it ends, and especially where it ends against wood, which moves with moisture and will cause a crack to open up where the two meet. It is essential to make sure the design includes an air fin that bridges this location and that can be embedded into the plaster coat to protect walls from moisture and air leaks. The air fins can be added in one efficient installation by pre-attaching them to the back side of the trim, baseboard, or cornerboard—and then installing the whole assembly onto the rough framing. The rough or finish coat of plaster then can be brought onto the air fin and tucked into the stand-away or rabbet cut onto the back side of the trim piece.

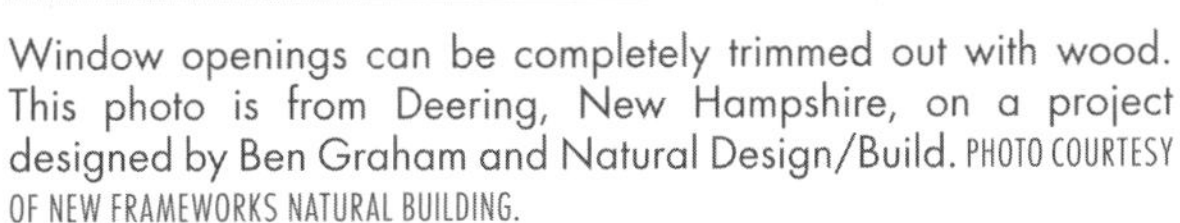

Window openings can be completely trimmed out with wood. This photo is from Deering, New Hampshire, on a project designed by Ben Graham and Natural Design/Build. PHOTO COURTESY OF NEW FRAMEWORKS NATURAL BUILDING.

Picture-framing light-clay walls offers protection and a nice aesthetic. PHOTO COURTESY OF KELLY GRIFFITH.

Trim

The stud walls that form the support and caging for the light-clay infill also provide a ready point of attachment for wood trim at the edges of the installation. These transition pieces can add to the look of the installation by "picture framing" the walls. They can also create extra protection on corners and at the bottom of the wall or around doors to deflect impact from the plaster coat.

Electric and Plumbing Runs and Electrical Boxes

The runs and boxes must be set into the stud cavity before the straw-clay or woodchip-clay goes in. Planning and designing the electrical plan and mapping plumbing runs (if in interior walls) is an essential first step to planning the light-clay walls. Care should be taken in the installation to pack light-clay infill around these intrusions in the wall. It will be necessary to check with the local code enforcer and a licensed electrician to ascertain what will be required in terms of your electric installation. Most often Romex is all that's needed, but at times direct-burial cable, BX or MC cable, or conduit is required. It is absolutely possible to run plumbing through these walls; it is perhaps not the best setup, however, if repairs are needed or the plumbing fails. In this case, it may be advisable to design for the plumbing runs to be in a chase that is accessible without demolishing the plaster and light-clay infill.

Decorative Touches

Special touches can be added to light-clay walls, including portals, windows, truth windows, tiles,

Portals are openings in interior walls that provide interior views. PHOTO COURTESY OF NEW FRAMEWORKS NATURAL BUILDING.

and bottles. Planning for these in the design stage will more easily bring the artistic vision to fruition. Portals are most useful on interior partition walls, where an opening from one side to the other would not compromise the purpose of the wall. Colored bottles, often seen in cob walls, can be set into the light-clay wall as it goes up, for light and a whimsical touch. Tiles can be set into the plaster coats, with additional pinning support; and truth windows are done similarly to how they are done in straw bale walls (see chapter 14 for more details on tiles and truth windows).

Finishes

Like straw bale walls, light-clay walls can be finished with wood or other siding or with finish plaster. Also similar to straw bale, they require a base coat of clay-based plaster as a primary air barrier, fire retardant, moisture manager, and insect and rodent deterrent. After this coat is completed, either a final coat of lime plaster on the exterior or an air space and siding (rain screen) can be installed. The lime plaster, as is covered extensively in chapter 17, will require several coats of high-quality limewash as the ultimate finish to the structure. Selecting what finish your project will ultimately have is an important part of the design stage, as the earlier stages of construction will set up for the successful finish.

CLAY SLIP FOR STRAW-CLAY AND WOODCHIP-CLAY

Clay slip is a mixture of clay and water, made to a particular liquid consistency. The consistency varies based on the purpose of the slip. For straw-clay or woodchip-clay, the desired consistency is similar to that of a thin milkshake. We often test the consistency with our students by having them immerse their hand in the slip; when it is removed, the hand should be evenly coated, without the clay slip "webbing" in between their fingers—a glove, not a mitten. The "light" in "light straw-clay" or "woodchip–light clay" refers to the slip being used lightly as a binder—as light an amount as can be used without losing binding potential. The heavier the slip is, the more the infill

Slip is a "milkshake" of clay suspended in water. PHOTO BY KELLY GRIFFITH.

mixture becomes like the heavier, more massive wall types, such as adobe and cob. Our goal here is to maximize the cellulosic components—straw and woodchips—and minimize the clay slip in a perfect balance that will hold together well and have density, yet still be insulative.

Clay slip can be made from clay-rich subsoil, either found on-site or delivered by excavator from a local clay source. Clay slip can also be made from dry, powdered, bagged clay, such as OM-4 ball clay or other ball clays. For building, we tend to prefer the ball clays (which are secondary clays) rather than the kaolins (which are primary clays), because they are stronger. See chapter 4 for more background on clays. A detailed look at making clay slip from clay-rich subsoil and from bagged clay can be found in chapter 17.

Constructing Straw-Clay Exterior Walls

FRAMING AND SLIP FORMS

Wall framing for straw-clay double-stud walls is accomplished following conventional framing guidelines. The International Building Code for light-straw-clay construction specifies no wider spacing between studs than 32 inches on center. The code also calls for horizontal reinforcement—bamboo 1 × 2s that run horizontally through the wall panels and are embedded in the infill like rebar in concrete, and serve much the same function. See earlier in this chapter in the design section for more information on framing for window and door reveals.

Most exterior straw-clay walls will be wraps around a timber frame or structural stud walls in themselves. As discussed above, walls can be either double-stud walls or single-stud walls, but single-stud walls are most commonly done as interior walls in our climate, due to the reduction in R-value. It is also possible to build double-wattle walls in lieu of sawn lumber and wood lath. Wattle and daub walls are, of course, the technological precursors to stud walls with wood lath and plaster. This would be a time-consuming yet interesting option if it met other project goals.

Slip forms for straw-clay can be made of any flat, strong, planar material that can be tightly affixed to the face of the studs. Most often we use plywood, as it has all of these attributes. However, 1× boards can also be used. Typically they are sized to fit so they are 2 feet high, but almost any size will do so long as whoever is doing the installation can comfortably detach and reattach them. For attachment, screws are the best—simple drywall screws or deck screws will work—as they will need to be backed out when the top of the slip form is reached, moved up, and reattached.

On DVD, Chapter 15, see ATTACHING FORM BOARDS TO FRAME

STRAW: THE MATERIAL

Straw used for straw-clay should be loose, long (more than 8 inches), and dry: loose, so each strand can be tossed and effectively coated with clay slip; long, so that the strands really knit and interweave with one another to create a strong matrix; dry, so that uncontrolled water and potential decomposition are not introduced into the wall infill. The ideal is to have the straw dry enough that it will absorb the clay slip, in the appropriate consistency and amount that can be adjusted during the mixing process.

Sourcing the straw is usually quite straightforward. For tips on sourcing, see chapter 14.

Hand mixing by tub. PHOTO COURTESY OF NEW FRAMEWORKS NATURAL BUILDING.

CALCULATING AMOUNTS

When making straw-clay, the straw is freed from the bale form in order to mix it with slip. In a loose form the straw does not have the density it has when in a tightly packed bale, but this density is re-created (potentially at an even higher density) by tamping the new mixture into the forms. It is difficult to calculate precisely how much straw will be used in the wall because there are too many variables, such as how tightly the wall will be tamped. Calculate how many bales would be used, volume-wise, to fill the size of the cavity as a starting point then get 10% to 20% more.

Calculating the amount of clay that will be used is similarly elusive. Often, since a base coat of clay plaster will be applied as the primary protection for the wall, getting "a lot" of clay is a good idea. See the section below, "Constructing Woodchip-Clay Exterior Walls," for information on estimating clay amounts for straw-clay infill.

STRAW-CLAY MIXING

Straw-clay mixing is potentially the most arduous and time-consuming part of the process, possibly rivaled only by the actual installation and tamping. This process, done in any volume, is not for the faint of heart or impatient. The method that is most straightforward is mixing by hand with pitchforks. Two people stand opposite one another. The straw is scattered loosely on a sheet placed between them. Clay slip is drizzled evenly over the straw piles. The two people simply toss the straw with the clay slip, working around in a circle, staying opposite each other, until the straw and clay are well mixed. Each straw should be evenly coated with the slip for the ideal consistency. The balance is then adjusted by adding more clay or more straw. The well-mixed material is then transported via wheelbarrow or bucket to the wall cavity, and wall building begins. Some inventive folks have built straw-clay mixing machines, to try to replicate the tossing action that coats and mixes each strand of straw in hand mixing, and that is essential for a quality material. Mechanizing the mixing process shows a great deal of promise in making straw-clay walls a more viable building strategy by addressing efficiency in mixing.

Machine for mixing straw-clay, built by Sarah Highland. PHOTO COURTESY OF NEW FRAMEWORKS NATURAL BUILDING.

TAMPING

Ensure that the form boards are tightly affixed to both sides of the stud wall. If the form board is loose, the action of the tamping will push the straw-clay out of the plane of the wall and into the gap that is created. Later, this will create a difficult substrate for plaster. For more on this, see chapter 17.

Lift the straw-clay and place in the wall cavity by hand or by pitchfork. Reach in and distribute the layer of straw-clay evenly around the space. Especially ensure that the material is added to the corners of the cavity, as tight corners make the installation firm and strong. Apply downward force with a tamper. Most simply, a tamper is a piece of 2 × 4, 2 × 6, or 4 × 4. Or even more simply, someone can get inside the cavity and jump on the material, which can be very effective. The amount of tamping required is a variable that each installer must determine; after trial and error, the installer will determine the right balance. Too little tamping, and the material will not hold itself together and loose straw will be evident—yet the highest R-values come from straw-clay that is lightly tamped. This balance is difficult to strike. More tamping yields straw-clay of a high density, which diminishes the insulative capabilities, is exhausting, and reaches a point of diminishing returns in terms of effort on the part of the installer. Testing conducted by the U.S. Forest Products Laboratory and Design Coalition (2004) corroborates the assertion that the more the straw-clay is tamped, the greater density it will have and the lower the R-value will be. Finding an ideal balance is crucial so that the material is loosely tamped yet solid.

On DVD, Chapter 15, see INSTALLATION AND TAMPING OF LIGHT STRAW-CLAY

Tamping straw-clay into double-stud wall with a 2 × 4. PHOTO BY ACE MCARLETON.

When the space created by the form boards is filled and tamped, remove the form boards and move them up. Set the lowest board so that its bottom edge overlaps the straw-clay below by at least 2 inches. This ensures that the new material, when it is tamped, will not push out the top of the original installation owing to lack of containment.

Keep going until the top of the wall is reached. The last few inches should be filled with straw-clay knots, as described in chapter 14 in the section on stuffing to fill horizontal cavities. Straw-clay has the potential to settle, and walls should be checked after significant drying has occurred. Fill any gaps that have opened up with straw-clay knots. Straw-clay walls dry at a rate of 1 inch per week, given average drying conditions. Straw-clay walls must be coated with a base coat of clay or clay-lime plaster. The finish can be either lime-sand plaster or siding on the exterior. For more on plastering, see chapter 17.

Reinforcing Rods

Reinforcing rods are often required for straw-clay installations. The International Green Construction Code (section 508.6.2) calls for stabilizing bars: "Non-structural horizontal bars to stabilize the straw-clay infill shall be installed at 24 inches on center and secured to vertical members. Stabilizing bars may be any of the following: ¾-inch bamboo, ½-inch fiberglass rod, 1-inch wood dowels, wood 1 × 2."

Holes can be drilled with a Forstner bit through each stud on a level line, spaced several feet vertically up the plane of the wall. Through these holes saplings or dowels can be threaded. These horizontal reinforcing rods act as rebar does in concrete: to stabilize and provide backbone for the mass of the material and to lock it into a location. In the case of single-stud straw-clay slip-form construction, panels that have been installed without this have been known to fall out after they shrink away from studs after drying.

On DVD, Chapter 15, see REINFORCING RODS: THEIR IMPORTANCE, HARVEST, AND INSTALLATION

Constructing Straw-Clay Interior Walls

Straw-clay can be a sensible choice for a complementary wall area in a straw bale project—such as an exterior wall section for a bump-out or dormer, or interior walls—because loose straw is an abundant by-product of straw bale construction. Each day of straw bale construction can produce up to twenty contractor-size bags of loose straw from the resizing process. One challenge of harvesting it for making straw-clay is to attempt to segregate the long straws from the short chop that results from shaving (or from floor sweepings), yet it can be done if the intention is held during cleanup. When deciding how to construct interior walls for a straw bale house, it makes a lot of sense to take all those bags of loose straw you will have (there's only so much garden mulch one can reasonably use!) and do something constructive with them. Straw-clay interior walls are wonderful because they are made from what is essentially a waste product of another process. They are also easily constructed by lesser-skilled people on the crew, and they are thinner than exterior walls (usually 4–6 inches thick). They provide sound deadening, interior mass, and moisture management of interior air, due to the large amount of clay in the mix. Because they are finished with the same plaster that the exterior bale walls are finished with, a seamless aesthetic can be created at the transitions.

Case Study: Building an Interior Straw-Clay Wall Step-by-Step

Mary Ellen Blakey's home is an excellent example of combining multiple natural building methods. It's a high-posted cape, with an owner-designed open floor plan. New Frameworks Natural Building consulted in the design stage, installed and plastered the straw bale walls, and returned to help with interior walls, floors, and other finishes. The straw-clay walls were designed to be the walls for a first-floor bathroom. The design program called for sound deadening for privacy and to separate the spaces. A similar look and feel to the exterior straw bale walls was also desired, as was utilizing waste products from the bale installation. Mary Ellen did all of the mixing, installation, and tamping of the walls over a two-month period. She went at her own pace and, though she said it was hard work, loved the results and the feeling of having done all that work herself in her structure.

The starting point for the straw-clay walls was an earth floor on crushed gravel within frost walls. The first step to prepare for the walls was to pour a cement pad for the bathroom. Note the Durisol block stem walls that create the necessary separation from grade for protection of the straw bale envelope.

Cement pads are poured for the bathroom. PHOTO BY ACE MCARLETON.

Walls are framed on the concrete pads. PHOTO BY ACE MCARLETON.

Reinforcement rods—saplings—are threaded through holes drilled in the studs to support the walls. PHOTO BY ACE MCARLETON.

The next step was to frame the walls. We built the walls on the ground and tipped them up into place, with a capillary break of 30-pound roofing felt between bottom plate and concrete pad. Framing was set at 2 feet on center. This was done to minimize wood use and works great with straw-clay panels.

Note that where the walls connect to the exterior envelope walls, there were two posts. Although this was the case in this design, it is not always the case. When interior partition walls meet exterior straw bale walls, it is advisable to inset a vertical framing member (a 2 × 4) into the straw in anticipation of receiving the interior wall framing. This can also be retrofitted after the bales are in—a notch carved with a chain saw or Lancelot—but that is harder. Good attachment of that anchor framing member to the girt or plate above, and the sill or floor or toe-up below, is desired.

Next was mixing the straw-clay. Clay slip was prepared in a mortar mixer with either clay-rich site soil (over 30% clay content) or a bagged clay such as ball clay. We added a little lime (about 10%) to cut down on biologic growth in the weeks while it was drying, and to help it dry faster. The clay slip was then tossed with loose, long, dry straw left over from the bale installation. Mixing can also be done by hand on a tarp or sheet of plywood (or similar) with pitchforks.

Next we drilled holes with a Forstner bit through each stud on a level line, spaced several feet vertically up the plane of the wall. Through these holes we threaded saplings that were harvested from Mary Ellen's land to act as horizontal reinforcement.

Next, we added horizontal 2× blocking on the interior plane of the stud wall, in anticipation of later sink attachment. In the left of the photo are the electrical boxes, which were set with wire runs before installation of straw-clay began. It was then time to attach the form boards. Form boards were sized to be 2 feet tall, wide enough to span several bays, and to land halfway on the end studs. They were attached with screws for easy movement. However, when placing the form board it is quite important to ensure that the form is attached tightly against the studs. If it is not well attached, straw-clay that is being tamped from above will work its way into that gap, creating annoying lumps in your wall that are

The first panel is filled. PHOTO BY ACE MCARLETON.

hard to shave back. It's easier to just ensure a tight seal with the form board and the face of the studs in the first place.

We brought straw-clay to the wall in a bucket or wheelbarrow from the storage tarp where the bulk mix was kept. It was then transferred to the wall cavity and dropped, with attention paid by the installer to which places in the cavity needed more or less. It is especially important to attend carefully to the corners, where a solid and full application will make a big difference in the quality of the installation. Tamping should be done with a long block; usually we use scrap lumber. Of course the heavier the tamper, the more force you as the person operating the tamper get out of each lift-and-drop, but the heavier it is each time to lift. We tend to go light and go often to save our bodies. This is one place in the installation where the density of light straw-clay comes into play. Looking at the U.S. Forest Products Laboratory's chart, which shows performance data from different densities of light straw-clay, we can see that density has an impact on the wall's performance. The amount of tamping done during installation is one factor in the wall's ultimate density.

Interior partition walls made of light straw-clay drying. Note the patterns of drying in the exposed straw-clay: lighter gray is drier, darker is more recent and wetter. Straw-clay dries, on average, an inch a week. PHOTO BY NEW FRAMEWORKS NATURAL BUILDING.

There are a few quirks to the straw-clay install. When going around embedded items in the wall, such as plumbing or electrical boxes, the installer must take care to fill and tamp fully around the intrusion, which can be tricky with the confined area of the form boards. When the top of the wall is reached, one form board must be placed all the way to the top of the cavity, to fully seal it, while the form board on the opposite face can be brought up so there is still a narrow window through which to add the last few inches of material. Filling this area should proceed as is described for stuffing straw-clay knots into deep, horizontal cavities in chapter 14, in the "Plaster Preparation" section.

The walls are totally filled and prepared to receive the base coat of plaster. PHOTO BY ACE MCARLETON.

Base coat on the light straw-clay interior walls. PHOTO BY ACE MCARLETON.

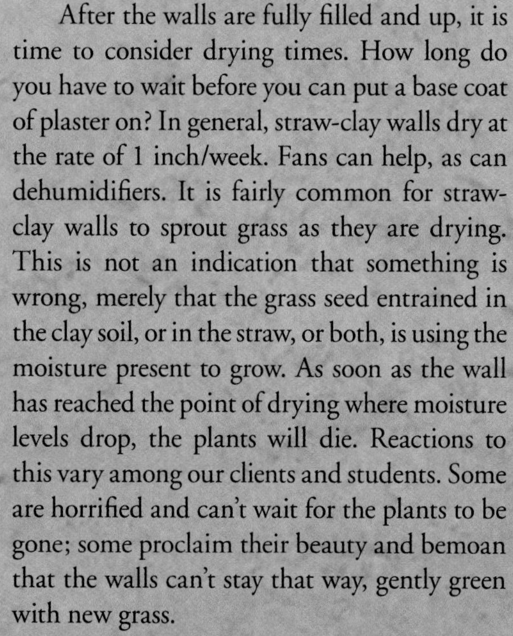

After the walls are fully filled and up, it is time to consider drying times. How long do you have to wait before you can put a base coat of plaster on? In general, straw-clay walls dry at the rate of 1 inch/week. Fans can help, as can dehumidifiers. It is fairly common for straw-clay walls to sprout grass as they are drying. This is not an indication that something is wrong, merely that the grass seed entrained in the clay soil, or in the straw, or both, is using the moisture present to grow. As soon as the wall has reached the point of drying where moisture levels drop, the plants will die. Reactions to this vary among our clients and students. Some are horrified and can't wait for the plants to be gone; some proclaim their beauty and bemoan that the walls can't stay that way, gently green with new grass.

After the walls were sufficiently dry, we returned to apply a base coat of clay-lime plaster to the walls. Note the blue tape to protect the trim boards, which in this case went on before the base coat. Ideally, the finish wood trim would be installed after the base coat but before the finish coat, and it would have a rabbet on the back side that the finish plaster could tuck into for a clean finish. For much detail on plastering on straw-clay walls, see chapter 17.

The finished straw-clay bathroom partition walls, finished with a lime plaster and clay paint tinted with ochre pigment. PHOTO BY ACE MCARLETON.

The straw-clay walls meet the straw bale exterior walls. PHOTO BY ACE MCARLETON.

Constructing Woodchip-Clay Exterior Walls

This section focuses on mixing and constructing woodchip-clay exterior walls, and options for finishes.

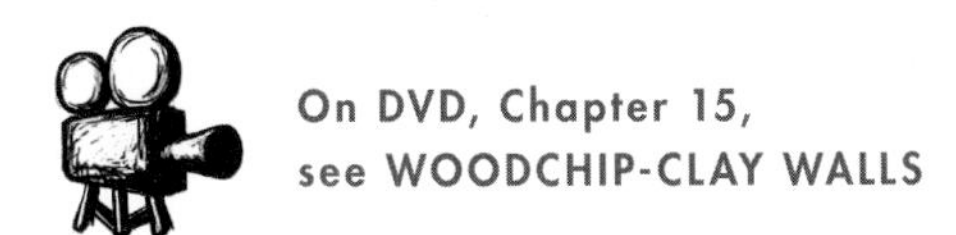
On DVD, Chapter 15,
see WOODCHIP-CLAY WALLS

Wood lath is set up as caging to contain the chip slip. PHOTO BY ACE MCARLETON.

FRAMING AND CAGING

Stud framing for woodchip-clay walls follows conventional framing guidelines, with the exception of spacing, which can often be 2 feet on center rather than the traditional 16 inches on center. Double-stud walls are gusseted together periodically by a plywood or other gusset, though the use of these should be minimized to reduce thermal bridging. For more on design specification considerations, see the "Design" section earlier in this chapter and chapter 13.

Wood lath, strips of ¼ × 1½-inch rough-sawn wood often 4 feet long, are attached to vertical studs with a set spacing that accommodates plaster, and in this case, also serves as a cage or container for the woodchip-clay infill. The goal is for the woodchips to be contained and for the lath to provide a good substrate for the plaster. Setting the spaces between the lath so the plaster can move in between and bulge out behind the lath creates what is called a "key." For more on this, see chapter 17, specifically "Step 2: Evaluate the Substrate." Generally speaking, the lath should not be set wider apart than ¾ inch. This specification comes from the conventional plaster trade, where lath and plaster is one of the most recent precursors to drywall. However, in that application, there is usually not any infill for the plaster to attach to. In our case, the clay-coated infill material is going to provide a good connection to the base coat of clay plaster. At times, wider spacing (such as 1–1¼ inches) between the lath is also successful.

Lath can be crosscut with a hand saw, with a circular saw, or on a chop saw. Lay out your cuts so the lath pieces break in the center of a stud. Attaching the lath is most quickly and effectively done with a pneumatic finish stapler attached to an air compressor. This tool makes attaching the lath a breeze.

Tip: turn the stapler sideways when you staple the lath to the stud, so the wood grain is captured by the staple, rather than the staple running lengthwise with the grain, where it much more easily pulls off.

One entire side of the wall can be caged in advance of any filling. However, choose carefully which side is completed, otherwise you may find yourself unable to access the side left open for installation due to some impediment—the timber frame, the grade dropping away, a staircase—that was hard to really know would be in the way until installation starts and access is difficult. Survey the space, imagine yourself standing to dump and fill, and ensure you have good access on the side left open. The open side can also receive lath up to 2½ feet, a good working height for the first pour. Doing more than this will make it difficult to make adjustments to ensure an even installation, and less will just be annoying, as the woodchip-clay dumping delivers a good volume quickly—one of the great benefits of this method.

WOODCHIPS: THE MATERIAL

Woodchips should be dry, clean, and well graded in terms of size. The largest pieces should be around 3 inches, while the smallest can go down to sawdust. Ideally most of them will be 1–2 inches. Dryness is important to start with, so that rather than beginning with wood pores that are filled with moisture from who knows what source and for how long, they are instead filled with clay slip during the mixing process. This ensures that uncontrolled moisture and potential rot are not introduced into your walls. "Clean" refers to a uniform product, not sullied with pine needles, leaves, bark, grass, or any other plant material besides wood. Finding chips without bark can be difficult but is worth the effort, as bark can serve as a vector for insects and moisture and is more likely to decompose in the wall. Any species of wood will do, although in general softwood, being less dense, is more insulative than hardwood.

Sourcing Woodchips

Finding a good source of chips can at times be the trickiest part of building with woodchip-clay. The best sources include forest-product suppliers, landscaping suppliers, and tree workers who chip up fallen trees. The first two are most desirable, as they will be more likely to have uniform product that can be examined, evaluated, and easily arranged for pickup or delivery to the site. Tree workers will often throw branches and other miscellaneous tree waste into the chipper with the bulk wood, which will yield a less desirable product for woodchip-clay walls. Sometimes waste from a wood construction process can be utilized, such as from a woodshop or the framing process of a structure. Most often with these sources you will have to arrange for the chipping yourself. It is possible to rent an industrial chipper and chip up the wood waste. It is a loud machine, and expensive to rent, but at times this can be the path of least resistance, and a satisfying process to engage in—using the waste-wood scraps from a wood-building project that would otherwise be burned or sent to a landfill to make the walls.

Calculating Amounts

The amount of woodchips you will need is a simple volume calculation. Multiply the thickness of the infill cavity by the height and length of the walls to determine the amount needed. Woodchips will most often be measured in cubic yards, similar to clay or sand.

Calculating the amount of clay that will be needed is trickier, but you will need enough to turn into slip and coat all sides of each woodchip. Most projects will be using clay for other things besides just the infill material, such as for the clay base coat of plaster. Since site clay (delivered from an excavator or found on-site) will be extremely inexpensive (for the material itself), and it is only the cost of delivery that is relevant, it makes good sense to just get "a lot." An average house will have at least a 10-cubic-yard pile of clay to work from, and this pile will service the infill and the base coat of plaster with some left over; and of course it depends entirely on the size of the project.

Tip: If you are using bagged clay, it is more important to calculate the estimated amount needed. A good starting point is to think through how much slip a bag of clay makes, and how much of that slip will make one "batch"—as defined by the size of the mixer. Our mixer, "Little Green," is a 6-cubic-foot mixer, which actually can mix only 2–3 cubic feet at a time of material. Half a bag of clay will make enough slip for one 2- to 3-cubic-foot batch of woodchip-clay, so we use this to calculate enough for the space.

Mixing

The simplest, quickest way to mix woodchip-clay is in a mortar mixer. Unlike straw-clay, which must be tossed by hand or in a specialized machine, the dry woodchips are simply shoveled into the mixer, where in a few short minutes they combine readily with the clay slip. It is also technically possible to mix woodchip-clay by hand or by pitchfork or garden hoe, much like straw-clay. However, this process would be immensely more time-consuming and should be chosen only if good reasons exist to eschew the use of a mortar mixer.

Woodchip-clay is quickly and efficiently mixed in a mortar mixer. PHOTO COURTESY OF NEW FRAMEWORKS NATURAL BUILDING.

The first step is to make the clay slip. If possible, this can and should be done ahead of time, as clay-site soil is more difficult and time-consuming to turn into slip, and hence is more efficiently made by itself prior to making woodchip-clay. If using bagged clay, using the mortar mixer to make the slip just prior to adding the woodchips will make this first step in making woodchip-clay easy.

Then, add the dry woodchips. It will take some experimentation to find the perfect balance of slip and chips. Each woodchip piece should be evenly coated with slip, and the slip should not be so thick that the chip mixture is sloppy and there are globs of unmixed clay in the mixture. The chip-clay mixture should fall and tumble readily, and not be too sticky. Water content will affect this also.

Add water to the mix as needed. The mix should not be either too dry or too wet. A mix that is too wet will be soupy, and the clay will not hold to the chips. If it is too dry the clay-coated chips will not bond to one another when they are in the wall cavity. Water amounts should be adjusted in the mixer before removing the mixture. Empty the mixer into a wheelbarrow or similar transportable container, and bring it to the wall cavity.

Installation

Next, the chip-slip mixture is taken from the wheelbarrow and poured into the wall cavity. There are many ways to do this. The most effective is to simply use a 5-gallon bucket. This works best if the cavity is wide enough to accommodate the size of the bucket opening when upside down in the dumping position. Otherwise, the bucket edge will bump the opposite lath cage and make it more challenging to empty the bucket. In this case, we have modified a common kitchen-size plastic rectangular trash can by cutting away three sides of the opening at an angle that mirrors the angle at which the trash can is held in order to dump the chips. This ensures that the far side of the container will not hit the lath on the opposite side of the wall as the material is dumped. The mixture can also be shoveled in or transferred by hand. Since one of the best things about woodchip-clay is how fast it is to make and install, the most efficient process of transfer is the best.

As discussed above, the consistency of the mixture should be such that the chips easily fall, fill, and move to evenly occupy the space. Tamping should be minimal. The only thing that should be monitored is that there are not large spaces or air holes in the installation. This usually occurs because a larger woodchip has fallen in such a way that it is caught sideways and is preventing other chips from falling into a space. Look through the lath where you have just put a load, and sight any significant holes. A simple tool such as a stick is all that is needed to dislodge any stuck chips. After removing the blockage, the material should easily tumble into the space.

This process is repeated until the walls have been filled as high as is possible. The top of wall is

The lath that cages woodchip-clay bends easily for curved shapes. PHOTO BY ACE MCARLETON.

a challenging spot, given that the preferred method of putting in the woodchip-clay is to let it fall into place, and at the top of wall it is often not possible to get above the top plate to fill any more from above. Use your hands to push as much woodchip-clay as possible into the top-of-wall cavity and then fill the top remaining inches with straw-clay knots, following the process described in chapter 14 for filling large horizontal gaps.

On DVD, Chapter 15, see INSTALLING WOODCHIP-CLAY INTO A WALL CAVITY

Drying Time

There is no stipulated drying time for woodchip-clay walls. As with other methods, drying time depends on ambient conditions. Accepted rates of drying for straw-clay are 1 inch per week, and we can assume that due to its lower density, woodchip-clay would dry faster than that. It must be emphasized that every ounce of water that is added in the mixing stage must come out of the installation by evaporation. This is many gallons of water. It is essential to ensure that a mechanism for drying is well considered and implemented, especially if winter is imminent. Often the use of dehumidifiers and fans is required, if ambient conditions of wind and temperature are not conducive to drying potential.

Air Fins and Finishing

Woodchip-clay walls should receive at least one coat of clay or clay-lime plaster for protection against moisture, fire, insects and rodents, and air infiltration. See chapter 17 for more details on the making, application, and function of clay and clay-lime base coats on natural walls. In woodchip-clay exterior walls, the base coat of plaster serves as the primary air barrier to seal the house and maximize performance.

Just as with straw bale walls or any other natural wall system that is sealed with plaster, treating the edges and transitions of the plaster with a flexible gasket (air fin) becomes essential to maximize performance. But unlike in straw bale walls, most air-fin situations in woodchip-clay walls can be dealt with after the walls are filled. This is because the air fins are placed on the face of the studs at the edges of the installation, to be floated into the base or finish coat of plaster. It is actually easiest to apply them before the finish coat. The rough-to-rough, finish-to-finish technique applies here: provide rough framing that is proud of the plane of the wood lath, so that the thickness of the rough plaster (about ¾ inch) comes into line with it. Then attach the air fin either to that rough framing or to the back side of the trim piece, and install that whole assembly. This is done by rabbeting the back of the trim board, caulking and attaching the air fin, then fastening the assembly to the edges of the wall as a baseboard, cornerboard trim, or top-of-wall crown molding. The trim piece is then protected with blue tape or some similar strategy, and the finish coat of plaster is applied and ends up being tucked into the rabbet on the back of the trim piece. This system neatly takes care of air sealing, creates clean plaster edges, protects corners for longevity, and allows for effective sequencing and setup for each stage of the work.

Constructing Woodchip-Clay Interior Walls

Generally an interior wall is a single-stud wall. It is framed, and any electric, plumbing, or blocking for cabinet or sink attachments is installed first. Then the lath is attached, as with exterior walls (covered in an earlier section in this chapter), and the woodchip-clay is poured in. Thinking through the plaster support at the bottom of the wall for the base coat and finish coat is important, as is planning for the transition around any doors or windows and at the top of wall. The design concept of "rough to rough, finish to finish" is as relevant here as it is in bale walls.

Dylan Ford and Bobby Farlice-Rubio chose woodchip-clay walls as their interior partition walls for their timber-frame/straw-bale house. Dylan and Bobby were committed to building a high-performing, ecologically sound, locally made house, so building interior walls that reflected that commitment made sense. As parents of three exuberant small children, sound isolation became a priority for the interior spaces. Because moisture balancing and interior air quality are also benefits of clay, woodchip-clay walls were the right choice for them. Note that bathroom walls may need to contain vent pipes; during installation the woodchip-clay fills in and around them with a minimum of effort, but if the pipes ever need to be accessed for repair the walls will need to be deconstructed.

Plastered woodchip clay walls help to soundproof the children's bedroom. PHOTO BY KELLY GRIFFITH.

Cellulose-Insulated Walls

Cellulose is an excellent recycled product that is flexible-form insulation with high R-value. It is used to good effect paired with natural walls, such as in pony walls, roofs, and floors. In one of our favorite details for connecting the roof insulation cavities to the walls, we design so the cellulose in the roof rafter cavities spills down onto the top of the bale walls, creating a continuous insulative envelope. This ensures that there will not be a gap between the roof and wall insulation at the most vulnerable point for thermal losses.

Cellulose also can be used as wall insulation: a popular high-performance strategy involves building double-stud walls to a 12-inch width, installing sheathing on the exterior and airtight drywall on the interior, and blowing in dense-pack cellulose to fill the cavity. Cellulose provides a high-performing wall insulation that is mold, moisture, and fire resistant, reduces airflow in cavities, can safely store moisture without degradation, is vapor permeable, is relatively nontoxic, and has light embodied energy and toxic footprints. It has the distinct advantage over the other insulation forms mentioned in this chapter of not building high volumes of moisture into the wall during its installation. This is a significant advantage that allows for great flexibility in its installation. For example, it can be installed into a closed cavity without the need to release gallons of moisture, and during times of poor drying conditions. It is also, as was noted earlier in the chapter, a good option for remodeling and renovation.

Is cellulose a natural building material? Cellulose belongs in the category of recycled building materials, a category unto itself (another material in this category is cotton blue jeans insulation). Cellulose could be seen as the wild card in this chapter. It is not traditionally considered a "natural building material," in that its creation requires factory processing and its application frequently involves use of petrochemical-driven machines (although smaller, electric-powered units can be used in some applications), thus removing it further from the immediately accessible and low-impact realm of sand or clay or straw. Yet it has a very low carbon footprint owing to its recycled content, and it is the perfect fit for renovations in urban areas, especially where high R-value is needed with a minimum of mess and added thickness to existing walls, and in conditions in which building high quantities of moisture into the walls during construction is unacceptable. It is important for natural building practitioners not to simply focus on new construction, but also to emphasize ecologically and socially responsible renovation—and cellulose insulation is a perfect tool for this. Despite its drawbacks, we natural builders in the Northeast use it in almost every project: as a fast, dependable, flexible-form insulation that will insulate our roofs, floor boxes, and pony walls; easily connect to straw-bale or straw-clay wall assemblies; and handle moisture similarly to, or better than, its natural insulation counterparts—and has a lower ecological impact than foam.

Yet even after removing these preconceptions, some argue that notable distinctions persist. Natural building takes cellulose in its unprocessed or minimally processed forms, such as straw or wood, and combines it with other unprocessed or minimally processed binders and minerals like clay and sand. At their best, these technologies encourage us to return to our roots, to fully use our bodies through the participatory process of building these assemblies. They connect us to human history through rediscovery of knowledge about materials and methods our ancestors used for many centuries, and they encourage us to merge these ancient crafts with contemporary needs. They invite us to know our homes through a sense of place. They persuade us to smell the mineral smell of clay mixed with sand, manure, and lime and show us that a field of cereal grain we have watched our neighbor grow will become the walls that will protect us from cold, heat, rain, and wind.

Currently, dense-pack and damp-spray cellulose insulation—the forms appropriate for filling vertical or sloped cavities—is installed by and large by skilled professionals with specialty equipment, and the material itself is made in a factory and sold plastic-wrapped in building-supply stores. For many who seek to work directly with the material that insulates their walls, or to be a part of its harvest and mixing, this can be a downside of cellulose when compared to other natural building methods. This exists in balance with the fact that cellulose is a recycled, repurposed paper product—long ago harvested from trees by a paper company somewhere, turned into newsprint or romance novels, and re-born as an incredibly efficient insulation—that has its own quality of wonder.

We will allow this articulated tension to stand unresolved. Is cellulose insulation a natural building product? We argue it belongs on the spectrum of options, for which it is important to weigh costs and benefits, as with any building system, and in some cases it will win out over clay and straw, or it will be a perfect complement to them, as in the hybrid straw/cellulose walls mentioned in chapter 7. This will remain an open question, up to the constantly evolving dialogue among all of us who engage these important questions through building to answer. Perhaps at times our answer will be "yes," and at times "no," and perhaps this is the perfect thing as we strive to implement integrative design/build practices that merge green and natural building methods to improve our built environment.

CHAPTER 16

Natural Mass-Wall Systems: Earth and Stone

Mass-Wall Basics

One of the major distinguishing characteristics of natural building in the Northeast is the predominance of insulated-wall systems over massive-wall systems in residential and light commercial construction. The reason for this is simple: the short warm season, cold winter temperatures, and high year-round atmospheric humidity levels preclude widespread adoption of mass-wall systems in this climate. Massive-wall (also referred to as mass-wall) systems are buildings whose walls are built of materials that are of very high mass, such as stone or earth, as opposed to less-dense walls built of wood, straw, or cellulose. Note that there is an inverse relationship between the mass of a material and its capacity as an insulator; therefore, mass walls offer very little insulation, hence their limited use as exterior walls for heated spaces in cold climates. Massive-wall systems can be quite attractive for seasonal-use and smaller-scale buildings, however, and mass building techniques can be applied in many creative, functional, and aesthetic ways within insulated buildings to pleasing effect, taking advantage of their beauty, the abundance of their source materials, and the benefits they can lend in the thermal performance of a well-insulated building. Although we do not recommend mass walls as the first choice for cold-climate enclosure systems, they have much to offer the world of northern natural building and can be effective as efficient enclosure walls in moderate to warm climates.

It is worth noting that there are indeed many mass-walled structures throughout the colder parts of North America, including cob, adobe, stone, rammed earth, and brick. In some cases, these buildings do suffer performance losses that go so far as to compromise comfort. But there are a few strategies, if you want to include mass walls in your building envelope, that will improve your experience:

- *Build tight.* The most comfortable mass-walled structures are airtight and minimize convective heat losses.

- *Build small.* While indeed building small is a good recommendation for all buildings, it is all the more so for mass-walled structures; if the conductive losses of a structure are high (as is the case for a mass-walled building), it will be much easier to keep the building warm if there is less building to heat.

- *Insulate the wall.* There is no reason why a massive wall cannot be insulated. While it may create for some redundancies in form or structure, there are many ways massive walls can be insulated, from building double walls that are insulated in the center to wrapping the walls in insulation or insulating within the wall system. It should be noted here that isolating the mass from the interior of the building will eliminate the "mass effect" of heat conditioning of the space (as described in chapter 7); it is this mass effect that allows massive walls to perform beyond their limitations as insulators. The most efficient forms of mass/insulation walls invariably involve exposing at least portions of the mass to the conditioned space.

As we will discover, the combined values of availability, affordability, access (low-skill and low-technology), adaptability (form and style), and in many cases function (thermal mass) are what make earth construction the most ubiquitous building style in the world.

BENEFITS OF MASS WALLS

Working with massive walls built of natural materials—primarily earth and stone—offers many of the same benefits we have been discussing for other types of insulative walls, which makes sense considering that all of these wall systems involve the use of naturally sourced materials using high-labor (and ideally interactive and fun) application techniques and can be similarly finished with clay- or lime-based plasters. Accordingly, they share similar attributes of allowing for very tangible, process-oriented experiences—in the materials acquisition, on-site manufacturing, and application. They also share a similar range of aesthetics—from simple and rustic to detailed and complex—with those of other naturally built structures, all of which are widely adaptable in style, form, texture, color, and feel.

The use of mass for the wall form is particularly favorable when evaluating the impact of resource extraction and material production, the abundance of feedstock materials, and the relative material cost. Building with earth and/or stone is, as is often said, "dirt cheap," and if you have a good supply of building-quality soil, it costs only what it takes to pull the soil from the ground and put it into a form or set it into a wall. Extraction simply involves removing topsoil—which, if on your own property, can be piled for later use in agriculture or landscaping—and using a backhoe or excavator to remove the subsoil for use. Transportation of a part (i.e., sand) or all of the soil may be necessary, and transportation can certainly add to both the cost and the embodied energy of the material (although there is generally favorable soil or stone for use within any given region, and it will still be cheaper when compared to other materials).

The production of raw earth and stone into suitable building material has potentially the lightest impact of any other material we explore. While mixing of cob, adobe, or rammed earth can be easily mechanized by any range of technology—from hand-powered ram forms to small tractors and compactors to massive custom-developed machinery—quite often the materials are manipulated lightly by hand (or foot), using simple tools such as a hammer and chisel. There is a greater potential for enthusiastic builders to execute their structures utilizing "low-tech" methodologies with massive materials. The same cannot be said for insulated walls that rely on more extensive agricultural or, in the case of cellulose or woodchips, manufacturing processes to accrue an acceptable quantity or form of quality building material.

One of the most exciting aspects to building with natural mass materials is the tremendous diversity of form and style available to the designer or builder. Brick is an essential element of Federal-style architecture, while stone can be used in a range of contexts, from the humble cottage to the imposing castle. Earth is particularly valued for its flexibility of form; it is as adaptive as concrete but without the impacts of embodied energy and at a lower price point. Earth can be used to create contemporary flat walls as rammed earth, traditional thick-walled structures as adobe, sculptural organic forms as cob, serpentine domed coils as earth bags, and much more. Even fired earth—ceramic—is used as both a building material and a full enclosure form, fired in place.

The structural capacity of massive walls, both in compression and in shear (horizontal forces in the same plane or against the plane of the wall) as a monolithic wall, is the greatest of the natural wall systems available to us. To so readily have both wall form and structure provided to us in a single process is unique; straw bale is the only other such natural insulation material to also provide that benefit, but it does so only after being coated in plaster, which bears the lion's share of the structural forces. For more information about the structure of massive walls, see chapter 6. Additionally, massive walls are also exceedingly durable. Since they are primarily geologic in their makeup (see chapter 4), massive walls are highly resistant to mold, rot, and

The many types of mass walls can be readily integrated to create a variety of texture and form in a building. PHOTO BY TIM REITH.

other forms of decay. Massive walls are also largely resistant to macrobiological damage, resisting insects, rodents, and other common pests that can plague cellulosic or other biologic materials. Massive walls are durable in the face of erosion. Stone, brick, and limestone are all quite resilient, requiring minimal maintenance when appropriately detailed to avoid excessive exposure to moisture. While earth-based materials are more prone to damage from moisture than other mass materials, they are certainly more resilient than their biological-based counterparts and can readily be protected by other masonry-based finishes, such as lime plaster, when necessary. In the case of earthen walls, which have a high clay content in their composition, moisture management is another strong benefit. Testing conducted on a cob structure in British Columbia, Canada, clearly displayed the walls' ability to manage dramatic spikes in atmospheric humidity, maintaining consistent and appropriate interior humidity levels. Furthermore, the walls were shown by virtue of their construction to be largely immune to the dangers of condensation that plague many other wall forms, particularly insulated walls (Goodvin et al. 2011).

As we discussed in chapter 7, utilizing the effect of mass in regulating temperature is a principal consideration not only in passive solar design, but in all forms of good climate-responsive design strategies. The ability of mass walls to absorb and contain heat is of particular advantage in moderate to hot climates in regulating interior temperatures. While this strategy is less effective in colder climates, as mentioned above it can still be applied in certain contexts; the positive effects of mass can readily be realized in colder climates when situated within an insulated envelope, where rather than buffering heat transfer from the hot outdoors to the cool indoors, passive (solar) or active heat can be stored and retained in the interior

Mass walls are very effective when used as interior walls in cold climates, where their use as exterior walls is limited. PHOTO BY KELLY GRIFFITH.

conditioned space to be released slowly over an extended period of time (e.g., through a cold night). This can be achieved as a massive wall insulated from the outside, as a component of an insulating wall (as in the interior plaster of a straw bale wall), or as an interior or other feature within the home (such as a floor or chimney). Again, in moderate to warm climates the effects of mass on moderating temperature fluctuation and storing heat help compensate for poor insulative values and can therefore allow these walls to be used as enclosure walls exhibiting good performance levels. Using perlite, a "puffed stone" aggregate that is a good insulator, as part of the aggregate of a mass wall formulation can also somewhat improve the thermal resistance of the wall.

Finally, massive construction can be highly supportive of human health on many levels, when well executed. Because earth, stone, and brick are inherently nontoxic materials and free of VOCs, many of the hazards of contemporary building can be avoided with their use, both during manufacturing and construction and throughout the useful life of the structure. The comfort and beauty provided by these materials cannot be discounted either when considering their benefits. Indeed, one's well-being is certainly increased when leaning up against a sun- or fire-warmed earthen wall on a cold winter's day, basking in the subtle glow radiating from the earthen plaster.

DRAWBACKS OF MASS WALLS

Building with mass, of course, does have a number of disadvantages, some of which are particular to their use in colder climates, and some of which are endemic to their nature. When compared to materials that are generally produced with some industrial-scale assistance (such as sawn lumber, straw, and cellulose insulation, which require major machinery to produce, process, and/or install them), earthen materials are even more labor-intensive than other forms of natural building. As discussed above, this high-labor form of building can be an asset in terms of access, affordability, and limited environmental impact. It can, however, also be a liability if this labor is not readily provided, or is provided at a cost. In those cases, if a lot of effort is required to produce large quantities of adobe blocks, for example, project costs will increase dramatically if the labor for the wall installation is hired out. While this balance exists with all natural building techniques, it is perhaps even more extreme in the world of massive materials.

Perhaps the biggest drawback to working with mass in a cold climate is mass's poor performance as an insulator, as discussed in chapter 7. In fact, the R-value per inch of a straw bale wall is nearly ten times greater than that of an earth wall, depending on the orientation of the straw. Again, while the mass effect can help offset the practical implications of this poor conductive resistance performance, it is simply not enough to overcome the heat losses associated with the cold temperatures in the northern latitudes to achieve comparable performance to a well-insulated wall. This is all the more true in cold climates that are especially cloudy in the winter. Whereas Colorado will still realize significant heating potential from the sun during the coldest months, which makes the choice of mass as a wall system a sensible one owing to thermal performance, the lack of visible sun in the Northeast during much of the winter does not allow mass to realize its beneficial contributions as part of a passive-solar-heating design strategy. While insulation can always be added to an earthen wall system to reduce its conductive heat loss—and in some cases formulated into the mix itself, such as using perlite as a component of the aggregate for cob or adobe—this invariably adds time, cost, and complexity to the project that all might otherwise be avoided by selecting a more insulative wall system, and in many cases it will result in compromises in the ecological impact of material choices (such as utilizing rigid foam board as an exterior insulation material to keep the mass in direct exposure to the interior of the building). Given that natural builders hold thermal performance as a high standard in evaluating the selection of different wall systems, it is no surprise that the inherent limitations of mass walls as enclosure systems inhibit wider use in colder regions.

Another limitation of utilizing many forms of massive-wall technology in cold climates is the short length of the building season (without springing for the added cost and logistics of tenting and heating). Although some mass-wall building styles, such as dry-stack stone, are not weather-dependent, it is unwise for still-drying bodies of earth to be allowed to freeze, as the subsequent expansion of frozen water within the wall system can lead to weakness at best or failure at worst, depending on the moisture content of the wall and extremity of the temperature.

One last drawback of utilizing mass walls in cold, wet climates such as the Northeast is that the short summers are often rainy and humid. As if it isn't hard enough having a short building season, for that season to be frequently humid and rainy makes it that much harder to successfully dry out a wet earthen building, particularly one built on a larger scale.

Applications for Mass in Cold Climates

Although it can be challenging to find successful applications for mass-based enclosure systems in cold climates, there are other ways in which mass can be effectively designed into a structure. While some of these applications may not be in wall form, the materials, their characteristics, and design parameters all remain the same.

INTERIOR WALLS

Interior walls are great places to utilize mass in a structure. Smaller in scale, easier to integrate flexibly into a construction schedule, and able to be executed in a controlled climate with increased drying potential, interior wall applications serve to take advantage of most of mass materials' benefits while avoiding many of their liabilities. Some of these advantages may even be more readily realized as interior walls; for one, the superior acoustical isolation performance of mass is highly valued, especially for bedrooms, music rooms, entertainment rooms, and children's rooms where it is desired to either contain or prevent sound from exiting or entering the space (see chapter 9 for more on acoustics).

Another reason for using mass walls on the interior of a building is the flexibility of form a nonstructural interior partition wall allows. In nearly all cases, an increase in the complexity of the form of the structure—circular, polygonal, amorphic—is going to increase the cost, time, and complexity of construction, from the foundation to the roof, and all of the transitions therein. If such freedom of form is desired but need not be manifested in the shell of the building, then interior walls can offer incredible opportunity to define spaces and create forms in unique and expressive ways, and even enhance their effect in a room in contrast to the rectilinear form of the exterior walls or exposed frame (in the case of a timber frame).

Mass walls provide the potential for thermal heat storage in sun-starved regions of northern climates. If it can be expected that the potential solar heat gain will be limited, especially during the heating season, then placing the mass where it can be directly associated with an active heat source allows it to realize its benefits as a temperature moderator. One example of this is surrounding a flue with a mass wall, thereby storing wasted heat from exhaust gases before they leave the chimney. (*Note*: please be sure to consider the effect that this may have on both back drafting and creosote or other deposits in your flue; consult a professional, if needed, to ensure a safe design!) Another example could be building a hydronic radiant wall. Although more commonly seen in floors, there are also benefits to running radiant heat tubing in a massive wall. These benefits include flexibility of design or construction where in-floor tubing may be more difficult to run; application in a remodel situation; or use of radiant heat while avoiding the health implications of a heated floor, as identified by the Bau-Biologie community.

Of course, structural considerations for safely bearing the dead load of these heavy massive walls need to be addressed, but again this may be easier when decoupled from the primary structure of the shell and frame.

Enclosing a chimney in a mass wall is a terrific way to retain heat that might otherwise exhaust out of the building. PHOTO BY KELLY GRIFFITH.

BUILT-INS AND CHIMNEYS

Built-ins, chimneys, and other large permanent features within a structure offer additional opportunities for the introduction of mass, and being smaller, they may be even better suited for direct positioning in concert with the heat source. For example, if running a flue through a massive interior wall is unfeasible, then building a massive free-standing chimney may be an easier possibility. Many chimneys are still built of brick or plastered block; when placed toward the interior of the structure—as opposed to adjacent to or, even worse, outside of an exterior wall—then the retained heat can radiate into the room before exhausting to the outdoors. Using a massive heater, such as an earthen, brick, or block rocket stove, or a masonry heater, will serve this goal with even greater efficacy. See chapter 21 for more information on heat sources.

Considering that placing an interior wall in direct exposure to the sun may not be compatible with the floor plan, an easier way of bringing mass to the sun could be through a built-in bench, daybed, chaise, or chair; not only will this provide an inviting place to sit on a chilly yet sunny day—warming you through conduction as you sit on the sun-heated couch—but the heat will continue to be stored up throughout the day and then be radiated into the house as you sleep, offsetting nighttime heat losses.

On DVD, Chapter 16, see RUMFORD FIREPLACE: FUNCTION, CONSTRUCTION, AND FINISHED PRODUCT

FLOORS

Floors are a very efficient application for utilizing mass heat storage. Unlike the other options, there is no disruption to the floor plan and no planning around heating or utility systems, and they are an integral part of the structure, rather than an accessory or luxury. Floors are often in the most direct exposure to solar heat when sunlight enters the building and are often integral parts of passive solar design. Additionally, many massive floors, such as those made from concrete or earth, are designed to work as part of an active radiant heat system, as discussed in chapter 21. Perhaps one of the only drawbacks of a massive floor is its hardness, which can be a detraction for those who prefer a softer floor, such as wood or cork. There is the option, however, to optimize positioning of thermal mass storage in floors near southern windows and to utilize a softer floor in other, more heavily used areas in the house. This strategy may not only offer a suitable compromise but can also create a unique and expressive design feature, encourage traffic patterns throughout the building, or even define use patterns and impose "rooms" in an open floor plan. Other pros and cons of mass floors are discussed in chapter 19.

SMALL STRUCTURES

Small structures and uninsulated buildings, as discussed earlier, are the best applications for massive walls as enclosure systems in cold, wet climates. While this may be of limited use in an urban environment, in suburban and rural environments there are many such opportunities, as many properties host, in addition to a primary residence, at least one garage, garden shed, wood shed, studio, cabin, sauna, barn, or other similar accessory or agricultural structure. In fact, the durability, strength, potential affordability, and wealth of potential forms of massive walls may prove to be superior to other styles of natural building in many of these applications.

Considering that high-performance enclosure walls are but one part of the many different facets of a building—and may not need to be high performance depending on the type of building—a careful evaluation of the options of massive natural building materials and techniques will guide us toward how best to include them in cold-climate designs.

Most Common Mass-Wall Types

There are many different types of mass-wall systems available to the natural builder. In this section, we will explore a few of the varieties that are used in warm and cold climates alike. While any number of different mass-wall systems could potentially be used, the following types tend to be the most supportive of both design goals and practical logistics.

COB

Cob construction is often one of the first forms of natural building mentioned by neophytes when asked "What is natural building?" Throughout the development of the modern natural-building movement, cob has maintained an iconic status, built upon the material's inherent qualities of simplicity in construction, affordability and ubiquity in sourcing materials, and sculptural, expressive aesthetic style. This is all for good reason, as indeed there is a lot of precedence of unique and beautiful cob buildings built extraordinarily inexpensively by otherwise unskilled owner-builders. Historically, cob may be most closely associated with the U.K., where more than 100,000 cob buildings were constructed in the last five centuries, many of which still exist; the earliest cob structures in the U.K. date all the way back to the thirteenth century. Like other simple earthen technologies we will explore in this chapter, the use of cob appears across the globe; there are examples of ancient cob construction throughout the Middle East and Far East, as well as Native American cob buildings dating back 1,000 years in the southwestern United States (Weismann and

Cob buildings have been built for thousands of years across the globe in a range of styles representative of the culture of their creation. PHOTO BY ACE MCARLETON.

Bryce 2006). Cob is a flexible material that utilizes the same ingredients already on hand for our earthen plasters, and it can be used for interior walls, heaters and/or chimneys, as well as built-in benches, shelves, or other features.

What Is Cob?

Cob is a type of earthen wall that is constructed of clay, sand, straw, and water; the materials are mixed together and are hand-formed (the term "sculpted" is frequently used to describe these walls) in situ, one lump, or "cob"—as the English called the earthen loaves—at a time. The resulting aesthetic can be somewhat diverse, although common themes persist. The walls are invariably massive—full-height exterior walls are often upwards of 24 inches at the base, tapering to a narrower width at the top of wall. Common characteristics include deep-set fenestrations (from either the interior or the exterior) and soft, undulating plastered wall surfaces. Styles tend to diverge from that point, however. Many of the hundreds-of-years-old English cob barns and houses were built with relatively clean lines and simple forms. Examples of cob construction from North Africa and the Middle East (where it is more commonly referred to as "monolithic adobe"), such as the multistory apartments in Yemen and mosques in Mali, not only feature larger-scale and more complex building forms but are embossed with intricate moldings and symbols representative of their cultures. Modern American cob buildings tend to be much more free-form in style, sporting amorphic or fluid shapes that may feature any number of creative additions, from bottles laid into the wall for a stained-glass effect to niches, carvings, and arches to bas reliefs and moldings gracing the walls. Any of these traditions can be drawn upon to inspire the cob elements brought into the contemporary natural home.

It is important to note that cob as a material can be used not only for walls but for other uses in a building as well; this is of particular relevance for cold-climate builders attracted to cob but unwilling to use it because of its poor insulating performance. For non-wall uses such as for built-ins or heaters, a variety of different techniques can be used. Some forms may be built similarly to the way walls are built. In other cases, for example, creating an arch over an area (also helpful in wall building) or making a durable shelf, a corbelling technique can be used, in which loaves of cob are successively cantilevered out to achieve the desired shape. In still other cases, a light formwork of sticks, called an armature, is used, such as for creating a thin shelf. In the case of an oven or heater, there are many different designs in which cob can be used in many different ways, from corbelling to using a sand or stick form to temporarily create the shape of the oven to using cob as a finish or veneer over a masonry-built heater or oven.

The source materials for cob are very similar to those of a site-soil-based earthen plaster; in fact, the basic proportions between the materials, mechanics, and troubleshooting approaches are the same as well. A clay-rich site soil is selected, with additional sand added as necessary to achieve an appropriate ratio. Straws are then added to provide tensile strength to the mix, making up approximately 15% to 20% of the overall composition. The largest distinction between earth plaster and cob is that because cob walls and features generally occur on a much larger scale than what is required for an earth-plaster application, the scale of the materials for cob tends to be larger as well—specifically the sand and straw needed to make the cob. In their book *Building with Cob* (2011), Adam Weismann and Katy Bryce call for aggregate that is "4 inches minus," meaning that the largest particles are 4 inches in diameter, and the overall composition of the mix represents a range of sizes from this largest size down to the finest particles (known as "fines"). If mixing by foot, a smaller aggregate size may be required. Note that some cob builders in moderate or cold climates use perlite as a component of the aggregate to improve insulation values. The straws are preferred to be full-length out of the bale. More information can be found in chapter 17, or from the listings in appendix A.

How to Build with Cob

Cob can be built entirely by hand, or machinery can be used to help with the process. The simplest method (from a technological standpoint) is to mix cob by foot, on a tarp. First the site soil, with any additional clay or sand to achieve a proper ratio, is laid in the center of a tarp and lightly moistened. The material is ground together by a twisting motion of the foot. Once the material is spread across the tarp, the tarp is pulled up from a couple of sides to re-consolidate the material, then stomping resumes. After the soils are sufficiently well mixed, the straw is added and the process continues, with water added as necessary, until a material of consistency similar to cookie dough is produced.

It is good practice to allow this mix to sit overnight or for a few hours, to allow it to adhere better to itself and increase in plasticity, not unlike an earthen plaster. A step up in technology would be to follow the old method of mixing cob, which is to tether a horse or cow to a stake and drive it in a circular pattern while the appropriate materials are added to the track. Manure from the animals assists in durability of the material (see chapter 17 for more information on manure as an additive). Further along the technological continuum, some have had good luck with mortar mixers, although we find that in Vermont the material needs to be mixed too wet to be well incorporated by the mixer. A more efficient means of mechanical production is to use a tractor with a bucket loader to mix cob in large volumes.

Mixing cob is a laborious process that can be very enjoyable when done as a community. PHOTO BY JESS AHLEMEIER.

Cob can be sculpted into form in a few different ways, but it is most commonly done by forming the cob into manageable loaves that are tossed or passed along to the wall. The cobs are then formed onto the foundation material, floor, or previous layers (or "lifts") of cob. It is critical for the cob to be fully integrated into the existing cob on the wall, to allow it to perform as a monolithic wall and avoid weakness and joints in the wall. This takes considerable effort, and a "cobber's thumb"—a hand-held thumb-like wooden device, or perhaps merely a stick—is frequently used to aid in this process; stomping on the wall (with feet) is another successful strategy. It is also important to ensure that the wall is well-moistened between lifts, to allow the new lifts of cob to be fully integrated into the existing wall. Generally, lifts need to be kept to about 12 inches added a day to allow for drying (depending on the consistency of the material and weather conditions); if more than this is added the wall begins to slump and deform. At the beginning of the following day, the excess material is trimmed using a machete, spade shovel, or hand saw (note that any tools used for cob should not be expected to be of use for cutting wood again, as the cob will most likely irreparably damage the edge of the blade or teeth of the saw). The trimmings can then be reincorporated into the new cob mixture. Care must be taken when forming up a wall to ensure that it is plumb, or appropriately sloped as it tapers from a wider base to a narrower top; this variation in thickness must also be considered in design.

Hand-sculpting a wall is a unique and powerful experience for those interested in not only the product of the building but the process of its creation. PHOTO BY JESS AHLEMEIER.

Cob is traditionally finished with plaster, either lime (for the exterior or interior) or clay (for the interior). In the case of lime, bonding between the dissimilar materials must be considered and attended to. Chapter 17 gives full information on creating a variety of plastered finishes over many different substrates, including cob.

ADOBE

The history of adobe is a long one. It was one of the first manufactured building materials, with evidence of adobe construction going back to 8000 B.C. in the ancient city of Jericho, and 7000 B.C. in India. With the conquering of Spain by Arabs and Berbers, the technology was brought to Europe; as Spain invaded and conquered the societies of South America, so was adobe brought into widespread use throughout the Western Hemisphere. As with music, food, and art, the dissemination of architectural styles into new realms is a beneficial by-product of the history of conquest that formed the world as we see it today. Although the Spanish missionaries and conquistadors popularized the use of adobe as they moved throughout the New World, there is evidence of earthen technology, such as mortar and plaster, and even of early adobe-construction practices among precolonial South American and southern North American communities (Chiras 2000).

What Is Adobe?

Adobe has similar properties as cob by virtue of its ingredients, which are nearly the same—clay and sand, and occasionally straw. Adobe buildings are constructed of unfired earthen bricks, also called adobe. These adobes can be manufactured in much the same way as cob—on-site, in small batches, by hand—though a more mechanized process can also be utilized. Unlike cob, however, adobe buildings are built in the same fashion as "brick and mortar" buildings seen throughout industrialized North America. Whereas cob is molded and formed in place in the building, adobes are made using wooden or metal forms to create preformed blocks. This combination of ease and speed of construction and flexibility of manufacturing technology (i.e., no industrial kilns required for firing bricks allows for on-site small-scale production) has made adobe construction one of the most ubiquitous on the planet. Indeed, adobe can be found not only in the arid and semiarid regions with which it is most closely associated but also throughout the world, including the northern United States and Canada (certainly not arid regions!), China, Japan, throughout South America, and beyond.

Adobe brings many advantages as a building material. Built-in moisture can be reduced in the building as the adobes are pre-dried before installation. Adobes are small, light, easy to move, and easy to store. Since they are unfired, they are also easy to rework and shape, being both softer than fired brick and able to be sculpted when remoistened. As bricks, adobes follow common patterns of construction and enjoy the support of a well-established masonry industry. For those looking for low-impact, inexpensive, and design-flexible solutions to integrate mass into a new or existing structure, adobe might just be the answer. Perhaps one of the greatest assets of adobe construction, as for other earthen forms, is the low embodied energy of the material.

TABLE 16.1. EMBODIED-ENERGY COMPARISON OF WALL COMPONENTS AND SYSTEMS

Material	Energy (as gasoline)	Energy (as BTUs)
8 clay bricks	1 gallon	125K BTUs
1 bag Portland cement	4 gallons	500K BTUs
10 × 10 ft. wood-frame wall	6 gallons	750K BTUs
10 × 10 ft. steel-frame wall	22 gallons	2,750K BTUs
10 × 10 ft. adobe wall	0.2 gallon	24K BTUs (6K calories for two person-days)

Source: Chiras 2000, 132.

Note: The energy usage for each of these different building systems clearly illustrates the low embodied energy of adobe construction.

How to Build with Adobe

When we build with adobe, we connect to a cultural lineage thousands of years old. The advent of modern, faster-to-build-with construction materials and techniques, coupled with an increasing association between adobe shelter and poverty, has led to the decline of adobe's use. However, a broad community of preservationists, natural building enthusiasts, and practical-minded folks who are following generations' worth of footsteps in seeking affordable, durable, and beautiful shelter have been able to keep the tradition of adobe construction alive and well.

Adobe bricks are made primarily from earth containing clay and sand; some recipes call for the use of straw, manure, blood, cactus juice, lime, or even small quantities of cement (cement-stabilized earthen blocks). Asphalt emulsion, a petroleum by-product, is favored by some to make the adobes water-resistant or even waterproof, although this is not a practice supported by traditional adobe builders. The balance of ingredients and inclusion of additives—as well as common troubleshooting scenarios—are very similar as for earthen plasters; see chapter 17 for more information. While adobe blocks are often greater in sand content—20% clay to 80% sand is a common recipe—some U.S. building codes require 25% to 45% clay, which more closely resembles an earthen plaster in composition, or even exceeds its clay content.

Mechanical adobe production is a fascinating evolution of an ancient craft in the modern age. Particularly in the southwestern United States, there is a burgeoning industry of machine-produced adobe block production. While some are making unstabilized blocks mentioned above, most use cement stabilization to address a few of adobe's limitations. For one, cement stabilization meets the demands of many building codes. For another, the adobes cure very quickly and can be turned around into production at a fraction of the time—this may be of particular note for cold-climate builders for whom drying time is scarce. Stabilized blocks are stronger and can store more easily without degrading. While not all regions have access to local industrial adobe production, you may be in a region that does. For a more accessible technology, look to hand-powered block presses, such as the CINVA Ram. These are single-block, easy-to-manufacture presses that, while initially created for the developing world, can be efficient and practical tools for any owner-builder.

If you can't find locally manufactured adobes, or if you'd simply prefer to steer clear of mechanical methodologies, forming adobes on-site is gratifying work. The soil can be readily mixed by foot (see the section on cob earlier in this chapter), by hand with a mortar hoe in a mortar tub, by machine in a mortar mixer, with a tractor fitted with a loader, or even with a rototiller. The mud is often left to sit for a day to allow the material to integrate and break up lumps of clay. At this point, the mud is shoveled and packed firmly into wooden forms. These forms are built slightly larger than the desired finished product to account for shrinkage (which will vary depending on clay type and percentage) and can range in size from single-block forms to larger multi-block forms requiring two or more people to manage. Predictably, the forms are built to a modular size; this varies regionally and is best "spec'd" to respond to the design, local building codes, and the proclivities of the builder. In the United States, this is often 10 × 14 × 4 inches, weighing approximately 35 to 40 pounds (slightly lighter than a two-string straw bale); while in Iran, for instance, 8 × 10 × 2 is the convention. An important detail in building a form is to include handles that make setting and removing the form easy. Quite often, the exposed tops of the adobes are scratched to increases bonding potential between the block and the mortar, not unlike between coats of plaster. Fingers or various scratch tools can be used for this purpose.

After the mud has dried sufficiently to hold shape, the forms are removed, and the adobes continue to dry; as the process continues, they are successively flipped on edge, then stacked, to both ensure even drying and create more space. Drying can take anywhere from two to four weeks under the best drying circumstances (warm, sunny, dry, and breezy); in colder and damper locales, the time can take much longer. It should be no surprise, therefore, that there is a significant logistical liability to cold- and/or wet-climate builders seeking to build and dry 10,000 bricks for a modest building in a rainy summer, whereas the semiarid-climate builder enjoys the benefits of long stretches of high temperatures, uninterrupted sunshine, and dry atmospheric conditions. Adobes can also be fired in kilns, which increases their hardness. This also increases not only their embodied energy but the demands of more particular formulation, as well as their hygroscopicity, leading to spalling (delamination or separation of the material exacerbated by moisture intrusion) or other moisture-borne problems if not properly waterproofed. If adobes are fired they are quite similar to common fired bricks, which are made primarily of clay, sand, and other minerals and are generally not considered to be a natural building material.

Upon drying—confirmed by breaking into an adobe to ensure consistent lightening of the soil, an indicator of dryness—the adobes are ready to use. Preliminary testing can be accomplished at this point. In *The Barefoot Architect* (2008), Johan Van Lengen lays out a few simple tests for structural adobes (which are less relevant if using the blocks for interior, nonstructural purposes). For the first test, place an adobe on two other adobes spaced apart, so only an inch or so of the test adobe is bearing on the ones below. Step firmly on the block; it should not break under your weight. A second test involves soaking an adobe for four hours; when the block is broken apart, the dampened area should not be greater than 1 centimeter. In a third, a soaked adobe is placed similarly to the block in the first test, spanning two other adobes. Stack an additional half dozen blocks atop the soaked block; the test block should stand for at least a minute before breaking.

Adobes are stacked up in traditional "one-over-two" (staggered positioning) pattern, also known as brick-laid (similar to brick structures) or running-bond (broken vertical seams) pattern, and glued together by mortar. The mortar in adobe buildings is most frequently earthen, although gypsum and lime mortars are used as well. The composition and integrity of the mortar is as important as that of the blocks themselves; walls must be built at a rate that is slow enough to prevent the mortar from slumping or deforming, and the structural integrity of the wall relies on sound mortared bonds between sound blocks. To aid in seismic structural integrity, rebar is frequently placed throughout an adobe wall, to stabilize the wall itself and to help tie the components of the structure (foundation, walls, roof) together in a solid fashion.

Another aid in structural stability is the bond beam, a solid beam that caps the top of the adobe wall and safely distributes the compressive load of the roof evenly over the wall; the bond beam can also be used to tie the roof to the walls and foundation in seismic structural systems. These beams are frequently poured-in-place reinforced concrete, although wood can be used as well (provided that code allows this). Windows and doors are achieved through the use of lintels, again frequently poured concrete or wood, that span the opening in which a door or window is placed; chunks of wood, called "deadmen," are often mortared and pinned into the wall as fastening points to hang fenestrations. Vigas, or peeled logs running horizontally across the tops of the walls, are a hallmark of the traditional adobe aesthetic and are frequently used in part or whole to support roof framing; they can also transform the aesthetic of a building, and they are frequently included in the design even if not used structurally.

Yet another approach is one similar to cob (which itself is sometimes referred to as "monolithic adobe"), called "poured adobe." In this technique, a much wetter-consistency mud mix is poured into forms directly on the wall; as the mix hardens, the forms are removed or slid farther up the wall. It is clear with this approach how the distinctions between adobe, cob, and rammed earth can blur as their production processes overlap and advantages of one system are combined with another to arrive at a comprehensive methodology that works for a given situation and set of goals.

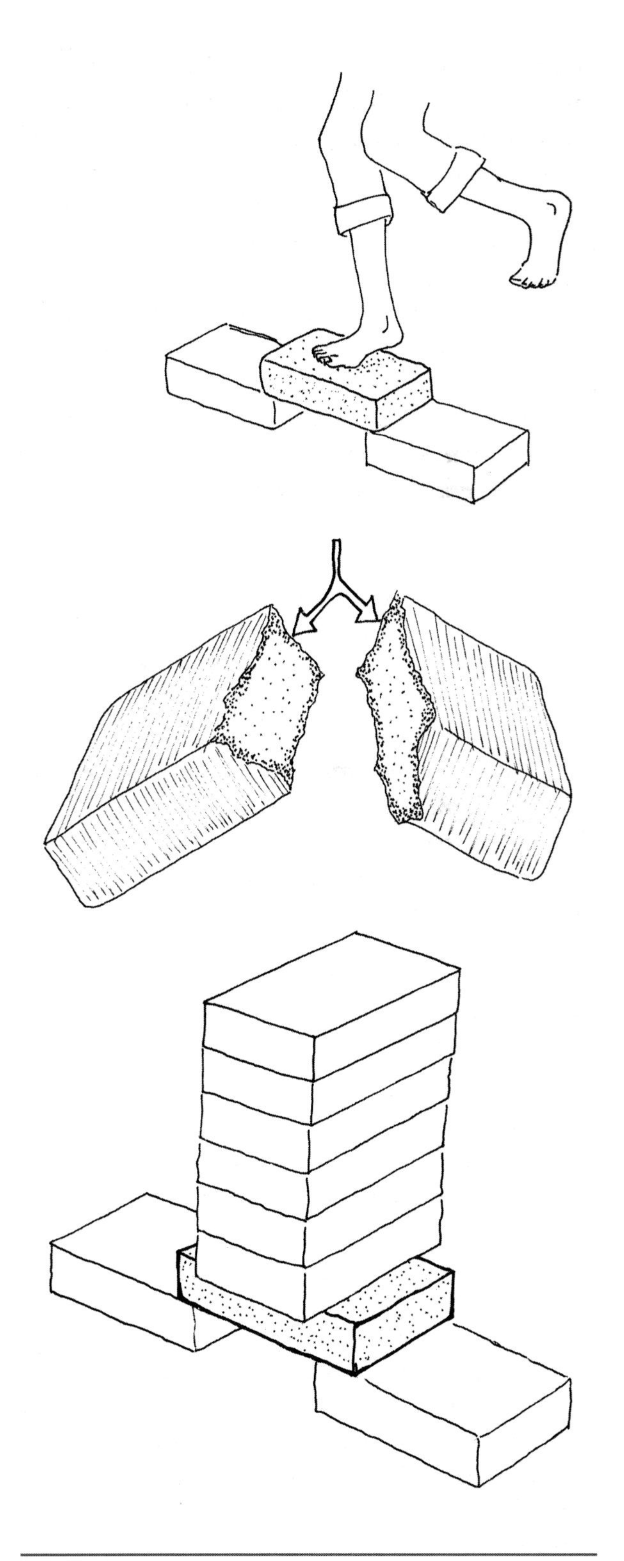

Three simple tests for evaluating the structural capacity of an adobe block. ILLUSTRATION BY BEN GRAHAM.

And as with most other natural wall systems, plaster is the finish of choice. Many adobe buildings have been harmed through the use of cement stuccos, which trap moisture within the wall, as discussed in chapter 8; therefore, the information presented in chapter 17 regarding appropriate application of plaster applies as well for adobe walls. If one chooses to use adobe for an external wall system in a cold climate, a few options are available. In addition to making the walls as thick as possible to help encourage heat concentration toward the interior of the structure, insulation can be placed surrounding the exterior of the structure, either as foam board, insulated stud walls, or even straw bale walls. Another approach is to build a double adobe wall, with an insulated cavity.

WATTLE AND DAUB

Wattle walls are some of the earliest wall forms used in forested areas of the world; as noted in chapter 1, many of the indigenous peoples of North America, and later European settlers, used wattle walls to construct their shelters. Eight thousand years ago, early European settlements on the Aegean coast were built of wattle and daub. In fact, the earliest recorded human settlements in the Nile Delta region of northern Africa were built of woven walls covered in clay. In Japan, historically bamboo was used for wattle material.

What Is Wattle and Daub?

It is a bit hard to categorize the wattle-and-daub wall system. On the one hand, the wattle is woven of biologic material and is certainly not massive. On the other, the daub (a heavy earthen plaster, great for added stability) is a material very closely related to others mentioned in this chapter. For this reason, we include wattle and daub here, although it stands out somewhat from its other earthen-material cousins.

This raw technology was steadily refined until major changes came with the advent of milling technology, which replaced wooden poles with studs, wattle with wood lath, and daub with plaster. After gypsum came on the scene in the building industry, gypsum lath replaced wood lath, which was in turn quickly replaced by gypsum wallboard. However, as is the case for other traditional techniques displaced by modern conveniences, there is still a place for wattle-and-daub walls. They are incredibly inexpensive and simple to build from a wide variety of materials. Low-value thin-diameter tree thinnings and understory saplings are ideal for wattles, and in cultures where coppice forestry is still practiced, wattles can be a product of the management system (see sidebar "Harvesting Wattle" in this chapter). Bamboo is also frequently used, for those in more temperate or tropical regions.

Wattle-and-daub walls incorporate many materials and processes common to the world of natural building—if you can make plaster, you can make daub—and sticks, branches, and bamboo rods find their way frequently into a variety of wall systems. Wattle and daub is highly flexible in form, readily adapting to curved floor plans, rough framing, or other nonlineal forms. Wattle and daub is very lightweight and thin in profile; thinner walls could be as thin as 4 inches, only half of which is heavy daub (the other half being wattle). And—if you enjoy it—the undulating, textural form of the wall can be quite desirable, either to match an existing aesthetic created by a sculpted cob wall or to create contrast to a flatter straw-clay wall.

An important detail to mention is that wattle-and-daub walls are generally nonstructural walls, unless the poles have been designed into a structural system; we use wattle-and-daub walls occasionally as interior partition walls to great effect. Wattle-and-daub walls can be great vehicles for including mass in a structure without the commitment of a full mass wall. Next to a heater, in line of sight with the sun, surrounding a bedroom—the benefits of the massive daub not only are inherent to the form but can be modulated easily by increasing the thickness of daub as desired. Beautiful and functional effect can also be achieved by "skip-daubing," or leaving portions of the wall without daub, exposing the beauty of the woven wattle. This can be desirable where ventilation, light, or sound transfer between interior walls may be desired (or controlled via curtains); the shape of the skip and changes in texture, color, or peeling of the wattles can create bold statements or subtle elements in a wall. Taking it a step further, skip-wattling will leave sections of a wall fully visible between poles, or leave portions of the poles themselves visible in wider skips.

On DVD, Chapter 16, see WATTLE-AND-DAUB WALLS

How to Build with Wattle and Daub

Building a wattle-and-daub wall is quite simple in basic practice, and subject to a wide variety of interpretations. The standard form is to use strong, straight, vertical poles approximately 2–4 inches in diameter (be careful the tapered end does not reduce too greatly in thickness), spaced approximately 12–18 inches apart; tune the spacing to both the length of the wattles and to place the uprights evenly across a wall. Be sure to have poles capping both ends of the wall, to secure the ends of the wattles. Poles can be secured to the building in a variety of fashions; the easiest way is to use top and bottom plates to help secure them to floor and ceiling. Our favored method is to bore a 2-inch hole into the plates with a Forstner or paddle bit, approximately halfway through a rough-cut 2 × 4, and tension-fit poles with their ends cut into tenons to fit into the bored "mortises" on the plates; often these poles can be tension-fit into place, although we always follow up with a screw or two to ensure the poles stay put upon drying. Another

Wattle-and-daub walls are some of the oldest walls known to humankind, and they are still relevant for use today. PHOTO BY BEN GRAHAM.

Harvesting Wattle

There are a few important characteristics to consider when selecting a sapling for harvest to be used as a wattle: the sapling should be straight, it should be of even thickness with little taper, and there should be little or no side branching. According to Mark Krawczyk, founder and owner of Keyline Vermont, a permaculture design and education company, a forest management practice known as "coppicing" encourages this type of growth. With coppicing, hardwood tree species are cut to encourage regrowth, which comes in the form of multiple, smaller stems sprouting from the same root system. These stems are more tightly clustered than the original mature tree spacing, which concentrates the growth in a clustered area for ease of harvesting and increased yields; this growth pattern also encourages competition between the stems, causing rapid regrowth of straight, branch-less stems. Harvesting of the wattles encourages regrowth and support of the management practice; regrowth takes just four to six years for most smaller-diameter wattle production. Upon harvesting, larger wattles can be split; the wattle becomes more flexible, and this increases the total volume of weaving material. This is easily done with a machete, or even better a billhook, which is similar to a machete but with a curved tip.

Krawczyk notes that, traditionally, hazel has been used as a coppice species for wattle fences and walls, for a few reasons: "its growth pattern as a multi-stem understory plant, coupled with its great flexibility and strength when twisted and bent, naturally lend it to be used in a wattle system." That said, many other species can be used. Willow is a common and ideal species. We've also done well in walls that are easy to wattle (lots of access, well-spaced poles) using hardwood saplings and branches, such as red maple, that are still green, allowing for maximum flexibility.

method is to use a divided plate: 2 × 2s are installed top and bottom, and poles are nailed or screwed to the face of the plate. A second 2 × 2 caps the other side of the poles, and can be further fastened for stability. This tends to lead to a more wavering line, but this may be desirable, or at least acceptable, in certain conditions. A bottom plate could be doubled, if receiving baseboard trim, for increased nailing surface. Straight-grained hardwood poles or bamboo are ideal, although softwoods can also work well; predrilling to receive attachment screws will greatly reduce splitting of the ends.

The wattles are then woven in any desired pattern; the simplest is a horizontal weave, alternating courses to ensure an even substrate, as well as to provide "key," or surface area, between wattles to receive a firm attachment of the daub. However, the sky is the limit, and we have seen gorgeous decorative weavings that one would be loath to slather in daub. The weavings do not need to be screwed, nailed, or otherwise fastened, which is good, as their desired thickness is only a half inch to an inch. Using wattle from different species can make for some lovely visual effects. The wattles should be woven tightly enough to maintain stability (remember that drying will be taking place, shrinking the wattles and potentially loosening the weave) yet loosely enough to still allow for daub to key into the spaces between the wattles.

The daub follows much the same protocol as used for cob, discussed earlier in this chapter, or earthen plaster in chapter 17. Slightly more malleable than cob, yet thicker with more and longer straw than plaster, daub serves an intermediary purpose of partially forming the wall and partially forming the covering. The daub is complete when no wattle is exposed and the greatest dips in the wall (formed in the centers between poles where the wattles cross) are filled out to the extent of the daub's capacity. Subsequent coats of plaster can then be applied to reduce wave in the wall and refine the quality of the surface.

We have also installed vertical wattles, in which horizontal poles are set (often within a stud wall) and vertical weaving is daubed. The process is the same, and can be preferable depending on access into the wall (in some cases, it may be difficult or impossible to weave horizontally, whereas a demi-wall might allow easy access from above). The vertical wattles are not as effective in resisting the pull of gravity on the daub, but a sturdy wall can easily be built nonetheless.

While wattle-and-daub walls have not been favored as enclosure walls since the dawn of the sawmill, there is potential for lightweight double-wattle walls with an insulation cavity, in the same fashion as described for other mass-wall insulation techniques. Although we have not ourselves attempted this system, it holds promise as a low-impact, easy-to-build, well-performing wall system suitable even for cold climates.

STONE

Stone buildings are among the earliest buildings known to humankind. In fact, caves are a form of stone architecture—naturally formed stone shelters, built in the negative (material removed to create space). Since emerging from cave dwellings, stone buildings have been and continue to be primarily built in the positive—stones added together to create durable, functional, beautiful space—and such a tremendous diversity of space can be created by using stone! On every continent (save perhaps Antarctica), in every culture, stone has played a major role in the development of structure, from the humble beehive huts of early Irish missionaries to the staggering temples of Mesoamerican societies; from simple and strong sandstone houses along the Himalayan Plateau to the monuments of the Roman Empire. Even today, stone is still a highly valued building material used in a wide range of applications, including but not limited to free-standing walls, structural foundation walls, enclosure walls, veneer walls, heaters and hearths, floors, roofs, counters, roads, patios, and walkways. If we include sand and gravel in our definition of "stone" used in building, then we quickly realize just how heavily dependent we are on this most basic of materials.

Stone as a Building Material

In what is now a familiar theme we see the use of stone dramatically reduced with the advent of industrial materials—such as the concrete masonry unit (CMU)—that prove to be cheaper, easier to use, and more structurally predictable, and therefore easier to apply in the larger-scale development and production that is the hallmark of our modern industrialized built environment. And this is indeed understandable when one considers the liabilities of working with natural stone. Although, like soil, wood, and grass, different types of stone provide ease of use in different applications (for example, the use of easy-to-cleave slate for shingles), stones are inherently non-uniform in their nature and require a great deal of skill and attention on the part of the builder to ensure not only the proper technique of application but the knowledgeable evaluation and selection of quality stones. Herein lies yet another example of the conflict of natural building in an industrialized society: use of natural products requires the knowledge and skill of a designer, manufacturer, and builder all at once, in a society that favors industrialized production and specialization of labor. Although there are "manufactured" stone products available, these products are still variable in their

Stone buildings are still relevant today, as shown by this stunning home built by Clark Sanders of New York. PHOTO BY ACE MCARLETON.

nature and fall closer to "wild" stone in the spectrum of developed masonry products. As a result, the stone buildings of both yesterday and today are regarded as master works of craft and are valued not only for their functional durability and aesthetic power but for the reflection of quality displayed on the part of the craftsperson who created form from the stone itself.

How to Build with Stone

Selecting stone is of course the most important place to begin. There are dramatic differences between the strength, appearance, working properties, availability, and cost of different stones; as we explored in chapter 4, "stones" can have dramatically different properties. Schist will cleave well and can be readily fashioned into an easy-to-build form. Sandstones and limestones are softer and easier to work, although they will erode faster than harder stones such as granite, which although superior in durability are much harder to shape. As mentioned above, developing a relationship with the native materials available to you will greatly assist in both process and product.

You may have luck salvaging stone from an old building with a stone foundation; in this case, much of the selection work of quality stone has already been done (provided it is a good foundation). In some areas, ribbons of old stone walls snake through the landscape and may provide a good source of stone, although a thorough investigation of the quality and type of stones available in these walls and their suitability for your project might indicate that it may be more appropriate to simply leave the walls where they are. It should be noted that landowners must absolutely grant permission before you salvage old stone from private property.

New stone can frequently be secured from local quarries; here, you may have the benefit of selection, light processing (washing, grading for size), and even transportation to the site. It is certainly worth considering that the farther you must go to source stone, the greater the expense, embodied energy, and logistical hassle. If importing stone from a distance, be sure to carefully evaluate the material, not only for quality but also for aesthetics; having a grand stone façade that does not match the native stone elsewhere in the landscape can certainly detract from the desired aesthetic of blending into the surrounding ecosystem.

Finally, stones found around the property or in the area can also be used to good effect, with a careful eye toward quality. Helen and Scott Nearing, in their book *Living the Good Life*, describe taking daily walks during which they would collect stones and return them to their property, organized by position or use in a future building (corner stone, keystone).

There are three primary approaches to building with stone: dry stacked, mortared, and slip-formed. The first two are quite similar. Stones are laid up atop a solid foundation and placed carefully to ensure that they are stable and firmly seated. Like bricks, a running bond is necessary to avoid vertical seams, which greatly weaken a wall by compounding the vulnerability of the joints. Stone walls are generally two stones deep, with the occasional "tie stone" placed through the depth of the wall, to avoid an internal vertical seam by tying the two halves of the wall together. Stones frequently need to be cut or shaped, most often using a hammer (which can range in size from 20 ounces up to full sledgehammers) and a stone chisel. Obviously, the harder the stone and more rounded the shape, the more laborious and difficult this process becomes. Smaller stone wedges or shims can also be used to help stabilize a stone as it is being laid into a course, although care must be taken to ensure that the wedge is not bearing structural load.

Laying up a dry-stacked or mortared stone wall is often done in courses—even layers a foot or two in height—to help maintain stability in the wall. Different masons have different patterns for different stone; part of the magic of stone walls is the unique expression of the mason's relationship with the material. Indeed the process, although seemingly simple, can be quite time-consuming at first as one learns how to understand stone, how to sort efficiently through a pile for the right-size piece, how to lay it firmly into a wall, and how to evaluate for the next piece. Many masons describe this as a process of listening to the stone, which speaks well to the patience and attentiveness required for efficient, quality workpersonship.

Building with stone requires patience and attunement to the stone. PHOTO OF STONEMASON THEA ALVIN BY AIMEE FLANDERS.

Moving Stones

The use of rollers, levers, barrows, and other tools—even loaders and skid-steers—to aid in proper ergonomic lifting technique is critical for anyone undertaking a stone building project. Most injury is a result of rushing and distraction, not strength or agility. Rob Roy, a builder of stone megaliths who is well versed in moving multi-ton stones around by hands and simple tools, puts it this way: "You must move at the speed of stone."

Mortared stone walls are quite similar in fashion to dry-stacked stone walls. In this case, a cement-lime-sand mortar is used to help secure the stones to each other and make the wall airtight. A mortar mix should be well tuned to the stone it is supporting; if the mortar is harder than the stone, it will encourage premature wearing of the stone, whereas a softer mortar will wear before the stone and can be maintained easily for greater longevity of the wall. It is important to note that mortar is not a glue that holds the stones together; experienced masons often "dry-fit" their stones before laying in a mortar bed. One cannot build a good mortared wall without understanding how to build a good dry-stacked wall. As a lime- and cement-based material, mortar is weakened when it dries too quickly, and it can set up rapidly on a wall. Therefore, a few tips are helpful when working with mortar. Begin by cleaning the foundation or existing course with a brush, and wet down the surface to reduce suction of the mortar. Spread out a thin bed of mortar only a few feet at a time, and as the wall is being built occasionally mist down the wall to keep the mortar from drying too quickly. This should be done periodically over the course of a few days, and in particularly hot and dry weather protecting the walls with soaked burlap or tarps will aid in keeping the mortar from drying too quickly.

When setting a stone into a mortar bed, rock the stone gently to seat it fully in the mortar bed and work out any air bubbles. Spread more mortar on the sides of the rock, or on the sides of the next rock, to achieve a solid mortared joint running vertically between stones. Occasionally as the wall is being built, or at the end of the day if drying conditions allow, go back over the wall and "point" the face joints, removing excess mortar and toweling or fingering additional mortar into underfilled joints. Some aesthetics will dictate recessed joints, whereas others will encourage the mortar to fill out to the face of the stones. Walls should be built only a few feet high at a time, before allowing the mortar to cure, to avoid damage to the mortar as it cures—not unlike adobe block, mentioned earlier.

Slip-forming stone walls has been referred to as a "democratic" approach to building stone walls, in that it requires less skill and time than stacked stone walls, so that a wider population of people have access to the technology. In fact, the Nearings helped re-popularize the form in their buildings, constructed of their found stones. In this approach, temporary wood forms are built on the interior and exterior of the wall and wired and braced together. These forms are plumbed up from the foundation, a bed of mortar is set, and the stones are placed within. Unlike stacked walls, however, flat stones are favored for their aesthetics in slip-forming stone walls, and they are often placed vertically, against the exterior of the wall. Additionally, 1 to 2 inches of space is left between the stones for the mortar joints, so bear this in mind when setting the stones. After they are set, the rest of the space is filled with mortar, and the process is repeated until the top of the form is

The mortar joints in this stone wall were carefully pointed to give the appearance of a dry-laid wall. PHOTO BY KELLY GRIFFITH.

reached; this makes for a wall that has much more mortar than in other stone wall approaches—in some cases more than half the wall may be mortar. After the walls have cured for a couple of days, the forms are removed, slid up the wall, and secured to receive the next course. Although this technique makes for faster stone laying, it does take time to build, set, plumb, and secure the frames throughout the process. It also makes for a particular aesthetic to the wall, particularly on the interior, where much of the wall surface is likely to be mortar.

To attach wooden elements such as door or window frames, sill plates for floors, or top plates for roofs, anchor bolts (often called J-bolts or L-bolts in reference to their shape) are mortared into the stone walls. The wood members are then predrilled to receive the bolt, and a washer and nut will securely hold the wood in place. Care must be taken to protect the wood from potential moisture damage from being in contact with the stone; a moisture-resistant membrane can be placed between the wood and the stone, or treated wood can be used to reduce rot.

Stone is similar to other mass forms in regard to its structural vulnerability in case of seismic activity, and it must be reinforced in seismic areas. As has been noted earlier in this book, the thermal performance of stone is also lacking. The options for insulating stone walls are similar to those for insulating adobe walls—double walls with insulating cavities or adjoining insulated stud walls are both common solutions.

Other Mass-Wall Types

There are many more materials and techniques that can be utilized to create massive-wall systems that are not found as frequently in natural buildings in cold climates for the reasons explained at the beginning of the chapter, or because of their more limited use for interior features, or in some cases because of their relative obscurity. That said, these additional methods are highly relevant to the field of natural building and bear inclusion in this chapter. For further reference on these technologies, please see appendix A.

RAMMED EARTH

Like other forms of earthen construction, rammed earth has a very long history of use throughout the world. In Africa, Europe, China, and across the Americas rammed earth has been used by various civilizations for many thousands of years to build all types of buildings from modest homes to palaces and temples; even parts of the Great Wall of China are built of rammed earth. In North America, rammed earth came into favor at different points throughout the last five hundred years. First, Spanish settlers used the technique using a mix of soil and ground seashells, called *taipa*. A resurgence of rammed earth use came about in the 1840s, and then again in the 1930s, when Frank Lloyd Wright helped to popularize the technology as an affordable option for Depression-era owner-builders. In the 1970s, another resurgence was experienced as a part of a more environmentally conscious approach to building, spurred on by the "back-to-the-land" movement. Rammed earth buildings are still constructed today, by both owner-builders and professional contractors, with the majority of the activity in North America centered in the southwestern United States.

Rammed earth is in many ways similar to conventional concrete construction. Essentially, temporary forms are erected, and a soil mix—generally clay and sand, sometimes with cement stabilization, although some builders withhold clay entirely and use only cement for binding—is slightly moistened, poured into the forms in sections, and

Rammed earth construction uses formwork to construct an earthen wall, not unlike a poured concrete wall. PHOTO COURTESY OF WIKIMEDIA COMMONS.

tamped with either hand-powered or mechanized compactors. This is repeated until the form is filled. Rammed earth walls are incredibly durable, akin to stone—in fact, the process is not unlike creating walls of sedimentary stone, and given the right soil mix and proper detailing to avoid moisture, rammed earthen walls will last a very, very long time. While many rammed earth walls are plastered for their final look, many are left exposed, especially on the interior (the exterior can be left exposed in arid climates); many builders have created incredible beauty with their walls by changing the composition of the soil mix to create different visual effects. For more on rammed earth foundations, see the "Rammed Earth Foundations" sidebar on page 155.

RAMMED TIRE

Rammed tire buildings are, like rammed earth, cob, and adobe, earthen walls at their core, but in this case utilizing waste automobile tires as permanent, stay-in-place forms. Given the dramatic disposal issues related to car tires, this method was devised as a means of removing materials from the waste stream and creating a low-cost, ecological building system at the same time. The use of rammed tire retaining walls was popularized by Michael Reynolds of Taos, New Mexico, in his "earthship" design, in which a curved, bermed northern wall is built of tires, compacted in place with an adobe soil mix; variations on this theme include gravel for areas with higher water tables or cement-stabilized soil mixes. Given their toxic and unsightly nature, these walls are almost always plastered. For more on rammed tire foundations, see the "Rammed Earth Foundations" sidebar on page 155.

EARTHBAG

Another container for earth can be found in polypropylene sandbags or feedbags. The use of a technique commonly used by the military for fortification to create ecological and affordable housing is a powerful example of a "swords into plowshares" approach to appropriate technology. This technology was initially developed by the late architect Nader Khalili, director of the California Institute of Earth Art and Architecture (Cal-Earth), for NASA as an extraterrestrial building solution, and Cal-Earth has gone on to make it available for communities in need of accessible and affordable housing around the world. The technique, often referred to as "earthbag" building or more pejoratively "dirtbag" building, has been popularized in the modern natural building movement by professional builders as well, such as Doni and Kaki Kiffmeyer of Moab, Utah, authors of the book *Earthbag Building*, among others.

The earthbag building prominently features curvilinear forms, such as domes and arches, but any variety of forms can be created. As with other earthen wall technologies, the structure and substance of the system is a clay-sand soil mix, although many other fill materials have been used by experimental builders such as perlite, pumice, or even rice hulls. The bags—sometimes individual feedbags, other times a continuous tube bag—are filled in place on the wall, either by hand or by machine. The bags are tamped into place, and strands of 4-strand barbed wire are laid between courses to provide tensile strength and to keep the bags from shifting. The bags are very UV-unstable and will decompose in the presence of sunlight, and so they are frequently plastered to protect their form.

Earthbags are a very versatile and inexpensive building medium, allowing for relative speed and flexibility for an earthen-wall system. PHOTO BY JACOB DEVA RACUSIN.

Green Mass Walls

In chapter 12 we discussed a number of foundation options utilized by the green building community. These same foundation-wall systems may also be used effectively as enclosure-wall systems. While decidedly not "natural building," they are worth mentioning here because they are relevant to the discussion of mass walls. The fact that these walls all offer varying degrees of insulation integrated into the system make them all the more relevant in cold climates, as similar insulation strategies could also, in some cases, be applied in natural mass-wall systems.

Concrete-insulation-concrete (CIC) walls, or "sandwich walls," feature a layer of insulation inserted within the form boards of a poured concrete wall. This not only serves to insulate the concrete wall but permanently protects the insulation from UV, insects, impact, or other damage. Additionally, the interior layer of concrete is both insulated and exposed to the interior of the structure, providing for thermal mass benefit to the building. This same strategy can be employed with rammed earth construction, or with any double-wall masonry construction (e.g., two adobe or stone walls with a center layer of insulation).

Insulated concrete form (ICF) walls come in any number of different styles of stay-in-place, permanent modular forms that either are insulated themselves or come equipped with insulation inserts; the cores of the forms are poured with concrete and reinforced with metal rod. Although insulating, these walls provide little mass effect, as the concrete core is thermally isolated from the interior of the building.

Aerated autoclaved concrete (AAC) block is a "puffed block" that offers insulation at a value of approximately R-1.25/inch. It is constructed in the same way as with standard concrete masonry units. Additional insulation may be necessary in colder climates.

That said, especially in the case of cement-stabilized soil, the bags do not serve a structural role and are merely forms for the adobe-like soil mix. We have seen earthbags used as grade beams for cob walls to good effect in the northern latitudes; use of insulating fill materials such as those listed above can make this system even more applicable in cold climates. For more on earthbag foundations, see the "Rammed Earth Foundations" sidebar on page 155.

CERAMIC WALLS

Nader Khalili also pioneered the development and use of ceramic buildings, known as the "geltaftan process." Inspired by the longevity and durability of the ancient village kilns in his native Iran, Khalili set about devising ways in which earthen homes in rural Iran could be fired, to improve the structural performance, maintenance, and durability of the buildings. After constructing an adobe building, the doors and windows are closed in, and a fire is lit inside the building, firing the structure as if in a kiln to a temperature of over 1,000°F. While this is by no means a system that can expect widespread adoption, it serves well to expand our ideas on how we can work with earth in many ways to create durable and effective shelter. Not unlike traditional "fired adobe" or brick, ceramic houses offer many of the same benefits through a different, unique mechanism.

Cordwood Walls

Cordwood masonry is a distinctive and increasingly popular natural building form found in northern forested regions in North America. In this style, logs cut to a set length (usually 12 to 16 inches, though walls can be built much wider) are stacked in a bed of mortar to form the wall, either structurally or as infill walls for a post-and-beam frame. The abundant wood resource makes this technique relevant in these regions; the history of cordwood building, although vague, can be traced to Siberia and parts of Europe dating back close to a thousand years ago, and in North America in Wisconsin, Ottawa, and parts of the northeastern United States going back to the late 1800s—all forested regions.

This technique is dubbed "cordwood masonry," and the phrase refers more to the brick-like application of the logs, mortared into a wall, than to the logs' classification as a masonry material; most wood species have an R-value on either side of 1 per inch, making them perform poorly as mass, yet not terribly insulative either, when compared to common insulation materials that range from R-1.5 to R-5. The best species to use is eastern white cedar, which is abundant, rot-resistant, more stable (shrinking and expanding with drying and wetting), and better insulating than hardwood species.

Accordingly, to provide insulation in the wall system, the log is laid on strips of mortar that support each end of the log, leaving a cavity between the two mortar beds (for example, a 12-inch log might have two 4-inch mortar beds with a 4-inch cavity in the middle). This cavity is then filled with insulation—often sawdust treated with lime—creating a thermal break between the mortar beds on either side of the wall and improving the overall thermal performance of the wall system. The wall's mass effect is a result of the copious amounts of exposed interior mortar, often cement-based, but sometimes earthen (referred to then as "cobwood masonry").

The benefits of cordwood are chiefly found in the aesthetic appeal (which is subjective, of course; some are enamored by the walls, while others find them distracting), the affordability (presuming the cost of wood is low or free), the accessibility to lesser-skilled builders and nonindustrial nature shared by most natural building forms, and the combination of mass and insulation provided in the manner described above.

The drawbacks, however, are numerous, particularly concerning performance and durability. Siting, design, and material selection are all critical to ensure longevity of the wood. As we know, wood is highly susceptible to decay, so placing wood

Cordwood building is a unique style that uses low-value wood blocks as bricks in a mortared wall. PHOTO BY JACOB DEVA RACUSIN.

repeatedly in direct contact with cement—which is all too happy to transfer its moisture to the wood it surrounds—and running the logs perpendicularly through the building, leaving the most porous butt-ends of the logs exposed to the elements, is tempting fate indeed, particularly in wet climates. Any presence of existing rot or decay in the logs will only seal the fate of that portion of the wall. There are performance issues as well: If the logs are too wet when installed and shrink when they dry, gaps and spaces will form between the logs and the mortar. Checking and cracking will also occur, creating voids running through the lengths of the log, and therefore through the wall. If the logs are too dry when installed and prone to expansion upon wetting, the swelling logs can crush and damage the mortar, which creates a potentially serious structural issue. The general approach to dealing with these issues is to either plaster over the wall (losing the aesthetic), or to repeatedly caulk the cracks and seal the log ends with any number of commercial caulk and sealant products. We find that the potential compromises of both durability and thermal performance make this wall system inferior when compared to the many other options listed in both this and the preceding chapters. That said, if all other factors—including resource availability, maintenance support, and personal inclination—lead toward its use, then comfort can be taken in the existing body of information and case studies available to support best-construction practices; see appendix A.

CHAPTER 17

Natural Plastering

Plasters have long history as a critical component of building, but much of the knowledge accumulated over the centuries about how to detail walls for a long-term, resilient plaster finish has been lost. In this chapter we will present what we have learned in our harsh climate, as well as how colleagues across the globe in their respective climates are applying finishes of clay, lime, gypsum, and manure. We will look at the spectrum of finishes, from rough plasters for straw bale walls to finish plasters that go over any wall surface.

Our approach in this chapter is to provide the tools necessary for you to better understand plaster and plastering, so that you can make good choices in your own projects. We will provide some recipes, but the emphasis of this chapter is that a recipe is useless unless you understand the characteristics of the materials and are able to problem-solve based on the materials and project at hand. There is no one answer for natural plasters, and this is a good thing. The strength of natural plasters comes in the user's flexibility to respond to variation in substrate, ambient conditions, locally available materials, desired performance capabilities, preferred aesthetic, and problems such as cracking or dusting that may come up. We will share with you our strategies for plaster mixing, application, and problem solving, and also provide you with information to help you make your own mixes and your own decisions and choices, which will yield beautiful walls in your building project.

CONTINUUM OF COMMON NATURAL BUILDING MATERIALS

Straw Bale
Cob
Straw-Clay
Wattle & Daub
Woodchip-Clay
Adobe Block
INSULATION
MASS
80% straw
20% clay/sand/lime (plaster)
70% sand
25% clay
5% straw

Natural building materials and methods fall on a continuum in terms of their thermal performance and relative ratios of the basic elements of straw, sand, and clay. IMAGE BY ACE MCARLETON.

What Is Plaster?

Plaster is wonderful, almost magical stuff, and a critical component of almost every natural wall system. If we think of natural materials and methods as a spectrum, or continuum, of different formulations, sizes, and applications, plaster would be found on one far end. On that end are natural building methods that use mostly clay, lime, and sand with just a bit of straw; and on the opposite end are methods that use mostly straw with a bit of clay and sand.

Next to where plaster would fall on the continuum is cob (see chapter 16). Plaster is similar to cob in that it is a mixture of water and materials like clay, sand, straw, and manure—except it is on a finer scale, and used as a finish on an existing surface rather than as a structural or free-standing wall, structure, or surface itself. We can think of plaster as a thin, non-freestanding wall made up of materials in such fine amounts as to mix with one another, and water, into a spreadable, frosting-like consistency that will coat an existing wall surface. Even paints, to follow our thought experiment all the way through, fall on this continuum on the extreme end (they would fall *beyond* plaster on the far end of the continuum). Paints are simply extremely fine aggregates suspended in a liquefied binding agent with color added (see chapter 20 for more on natural paints).

Of course, plaster is not just a covering for natural wall systems. It is also a finish regularly used in conventional building. Here, "plaster" most commonly refers to gypsum plaster. Products such as Red Top and Imperial are used as finishes on drywall. A system referred to as "modern plaster," a mixture of gypsum plaster and joint compound, is also coming into favor as it is easy to use, easily available, and finishes clean.

Lime plaster is now known in the more mainstream trades as a restoration plaster. In Europe and the United States, many buildings exist that were originally plastered with lime and require upkeep and maintenance. Lime plastering carries a strong tradition and craft history. *See Using Natural Finishes* in appendix A for excellent interviews with traditional lime plasterers. Lime also has been used for fine finishes—so fine, and treated with soap and oil, that they resist water, as in the *tadelakt* tradition of Morocco. In our projects in the northeastern United States, we often use lime plaster as a finish on drywall board in renovations or new construction.

In terms of clay plasters, the straw bale movement has done much of the work to set the groundwork for builders and occupants to see the beauty and functionality of clay. The definitive work of Athena and Bill Steen of the Arizona-based Canelo Project has done much to remind us of the vast and valuable history of native American and Mexican clay plasters, and how they are being used today in ways that meet modern needs and celebrate the best of traditional methods and aesthetics. Most recently, clay plasters have come into the mainstream building consciousness through prepackaged products such as American Clay, based out of the southwestern United States. Also, people working to address sick building syndrome and find alternatives to toxic building products are touting clay plasters as a healthy choice for structure finishes.

In the field of natural building, especially here in the Northeast, our plasters tend to be clay- and lime-based. There are good reasons for this, which are explored in-depth further in the chapter and can be summarized as follows: of the four binders—clay, lime, gypsum, and cement—clay and lime have the most positive attributes, and the fewest negative ones, for the building challenges we face in our climate. In other climates, or where seismic activity or dry conditions exist, other binders may win out. If we understand natural building to follow an ethics-based logic that guides our choices for material selection, rather than a dedicated adherence to certain materials, then choices like binder selection can flow from looking at the facts of each binder's characteristics, and evaluating their appropriateness in each situation.

Finish clay-lime plaster on drywall in bedroom, finished with lime paint. PHOTO BY ACE MCARLETON.

HISTORY

The evolution of plaster is fascinating to investigate and helps us understand the context of how it is used today. Plastering is an ancient building practice: making a spreadable mixture of fine aggregates (sands) and binders such as lime, clay, or gypsum has been practiced the world over for upwards of 4,000 years. The oldest example of plaster was most likely daub in wattle-and-daub walls made by our prehistoric ancestors, where the mixture of clay, sand, and fibers was pressed into a formwork of woven reeds or branches to make a wall (see chapter 16 for more information on wattle and daub). Next we see plaster on stone, mud brick, or cob in sacred and secular buildings in many cultures such as the pyramids in Egypt, Mayan temples in Central America, and ancient temples in Asia. Much later, in the United States, split or riven wooden lath was attached to studs, spaced so the plaster would press into the spaces and "key" to hold the coat on. This was followed by milled lath and plaster with horsehair in it mixed and applied in situ. Gypsum board, the ubiquitous standby of conventional building, is the next step in the evolution in which the gypsum plaster is mixed and formed into sheets of consistent size (4 × 8) in factories and faced with paper on two sides, and chemical hardeners are added for durability. Gypsum board is a standardized, pre-hardened plaster wall that requires mudding, taping, and sanding the seams, and then applying some sort of finish to its surface such as paint or other plaster.

Tadelakt is very fine lime plaster finished with oil soap and burnished with stones to a waterproof finish. PHOTO BY ACE MCARLETON.

Lime kilns were once numerous and supplied local areas with building lime. PHOTO COURTESY OF WIKIMEDIA COMMONS.

Lime has been used in Europe for 4,000 years, and this is actually what people mean when they say the Romans invented "cement." Roman cement was actually lime plaster mixed with *pozzolans*, or fired clays, which increased the strength, decreased the set time, and allowed for lime plaster to set up in the presence of water. This modification to its properties allowed the Greeks and Romans to build aqueducts and highways, and to support the many engineering innovations that expanded the Roman Empire. This type of lime is called hydraulic lime.

An old law in Great Britain dictated that lime pits where lime was being slaked had to have been soaked for a generation before the lime could be used. This type of lime was hydrated lime. Hydrated lime has been fired at high temperatures and carefully slaked with water to make an active putty that, when mixed with sand and perhaps fiber and spread onto a surface, reacts with the air to turn back into limestone, or calcium carbonate. Most buildings in the United Kingdom, until after World War II, were finished with these lime (as well as gypsum) plasters. Most towns had a lime kiln, and lime production was local and specific to the regional geology. Even today, driving across New England it is easy to spot road signs that read Lime Kiln Road. The transition from lime to cement as a common building material was part of postwar booms in many countries and dovetailed with the increases in large-scale industrial production. Cement was touted as the stronger, better binder (over lime), and its production as being more efficient and centralized, which illustrates an aspect of why lime has been a part of the natural building revival: the potential exists to re-create control over the production of building materials through rebuilding the knowledge and infrastructure to harvest, burn, slake, and use lime on a local scale.

FUNCTION

"Plaster" comes from the Greek word *plassein*, meaning "to mold" or "to daub on." Plaster is a wall covering that creates a solid coat of consistent material over all fields and relevant edges and corners and has a number of performance-related functions. In a straw bale, straw-clay, or woodchip-clay wall, for instance, the base coat of clay plaster acts as an air barrier, fire protectant, rodent and insect screen, and mold and mildew inhibitor and adds long-term moisture protection to the straw owing to its hydrophilic properties (and also "short-term" or immediate protection on the exterior from rain and snow). Cement stucco also can play a key structural role in load-bearing straw bale construction. On a gypsum board or wooden lath wall, clay or lime plaster can serve moisture-balancing, antistatic, and/or hygienic purposes. Plaster can also act as a "finish," or a final, decorative or aesthetic wall coating. Whether it serves as a textural substrate for an alis, or clay paint, or another natural wash (see chapter 20 for more details on paint or wash finishes), or whether the plaster itself will be the final finish, plaster plays an essential role in many wall systems.

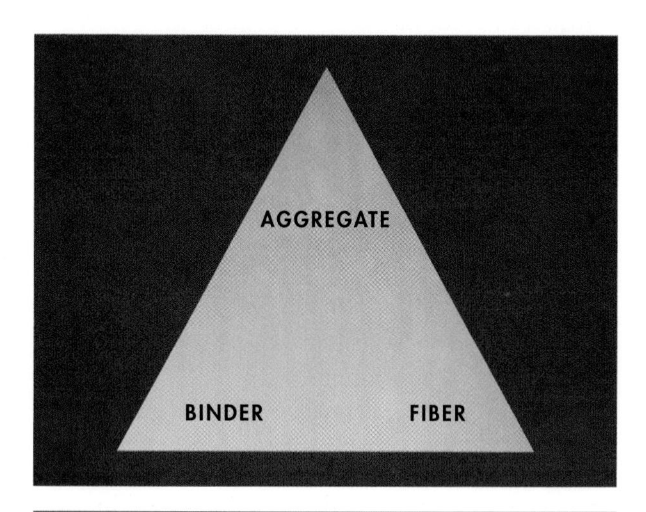

Basic structure of plaster. IMAGE BY ACE MCARLETON.

General Plaster Structure

Plaster is composed of three main ingredients; a fourth ingredient is optional. Aggregate and binder are the main components that create the basic structure of plaster. The third element is fiber, which is most often included to add tensile strength to reduce or eliminate cracking. The fourth component is the additives, a potentially endless list of optional ingredients chosen for the particular characteristic they contribute. Each plays a role, and it is up to the mixer to discern what elements, and in what amounts, will be present in the mix.

On DVD, Chapter 17, see PLASTER: STRUCTURE, COMPONENTS, AND THEIR FUNCTION

AGGREGATE

Aggregate, in this context, refers to a collection of tiny stones of varying sizes, which together make up the building blocks of plaster. An often-used metaphor is that the aggregate is like the bricks in the "wall" of our plaster. Aggregates are sand, gravel, fine sand, and stone powders. Aggregates are one of the most underemphasized elements of plaster, but the most important for performance. Aggregate, generally speaking, should make up approximately 85% of the mix (Weismann and Bryce 2008)—although this depends greatly on thickness of the coat, as some very fine and thinly applied lime plasters can push the binder:aggregate ratio quite far past the guiding principles we present here (Chivers 2011). The structure, strength, and stability of the plaster are all provided by the aggregate owing to the fact that the small stones lock and are held together by the binder, and are also chemically inert. Sand used for plasters must be well graded because the range of particle sizes makes for a stronger structure, as the different-size particles fill different-size niches in the structure of the plaster. They should also be sharp, because round sand just falls apart and is weak since it does not lock together as jagged edges and irregular shapes will. Understanding the weathering and raw material of an aggregate is helpful in learning what aggregates are going to be strongest and most useful for plaster.

A well-graded aggregate includes a variation of sizes and shapes. PHOTO BY KELLY GRIFFITH.

Fiber is often thought of as the most significant crack-reduction tool in the plasterer's toolkit, but actually appropriate aggregate size is also significant in terms of resisting cracking. Plaster made with aggregate that is too small for the thickness of the application is often the cause of cracking and/or failure. The best way to test whether a sand is sharp enough and has the appropriate mix of sizes and shapes is to spread a thin layer of it out in the palm of your hand and inspect it. Choose a size and grading range of aggregate that fits the plaster's thickness. The general rule of thumb is that the largest aggregate size should equal one-fifth of the total thickness of the plaster coat (Weismann and Bryce 2008).

Some lime plasterers claim that although sand is inert chemically, the chemical nature of the sand can make a difference in the strength of the plaster. In a fine-finish lime plaster, limestone sand—which is sand derived exclusively from limestone—is used to maximize the chemical relationship between aggregate and binder (Chivers 2011).

Another less-considered attribute of aggregate is that it lends color to the plaster. A grayish cast, often seen in cured lime-sand coats in our area, is due not to the lime, which is bright white, but to the sand. This can be used to our advantage in places where a muting effect of the bright white of lime is desired. White and other colored sands are also available for purchase, although confirming their size and shape for the thickness of the application is still up to you as the fabricator.

Sands

For base coats and plaster coats intended to be 1 inch or thicker, the use of a washed, well-graded, sharp sand with particles sized 3⁄16–1⁄8 inch and below, depending on thickness, is recommended. Sand can be secured in bulk from a local excavator, who can deliver it to your site. If that option is not available or bagged product is for some reason easier for your project, often play sand—found in bags at building supply or landscape supply yards—is suitable. Mason's sand is often available from building supply yards, but it is usually

Limestone sand is bright white and so lends that color to the plaster. It also has a chemical relationship with the active lime that is the binder, making a stronger plaster. PHOTO COURTESY OF NEW FRAMEWORKS NATURAL BUILDING.

too uniform and too small to be strong and stable in most natural base-coat plaster applications, though may be perfect for a finer finish mix. The size and shape of the particles should be assessed and matched to the application, as with any sand, no matter the source.

Fine Sands

Silica sand is the most common example of a fine sand, although limestone sand and other sands can also be found in this size. Sand, as you may recall from chapter 4, describes a size category for particles, and thus can be found in different chemical compositions and within a range of sizes. Most sand is predominantly silica, but silica sand has been industrially processed to be composed of only quartz, or silica, with no additional rocks or minerals present. Silica sand is a very fine sand, sized ⅛–¼ mm in diameter and smaller; it comes in mesh sizes down in the one-thousandth of a millimeter. Particles this size are grouped with the stone dust category.

Stone Dusts

Stone dusts, such as granite powder, marble powder, mica, fine silica, or limestone sand, are used when very fine finishes are desired. Because of their very fine size (less than 0.2 mm), stone dusts, used as the only aggregate, are most suited to plaster applications that are 3⁄16–⅛ inch or less. In New England, granite mining and production is common so granite dust is both an abundant and free material. Marble powder and limestone sand are used in *tadelakt*, which as noted earlier is a fine lime plaster polished with oils and soaps to be near-waterproof that hails from Morocco. Stone dusts can also be a beneficial auxiliary aggregate in a plaster that is intended to go on thickly and be more multi-purpose than a very thin, very fine finish coat. In this case, the stone dust would be added to the larger aggregate to include a "finely packed" finish surface of smaller aggregate, in addition to the bulk function of the larger aggregate to support the thickness of the plaster body in general.

BINDERS

Plasters are classified by their binders. Though we speak of "lime plaster," "clay plaster," "gypsum plaster," and "stucco" (cement plaster), in each case we are referring to the binder acting as the glue in the wall, pulling and holding all the elements together in the plaster structure. A binder works by surrounding the aggregate, fiber, and other additives and sticking to them and to itself.

Clay

Clay has many wonderful attributes for the natural builder. First, clay is naturally occurring and can be found in deposits in most continents on Earth, making it one of the most ubiquitous, minimally processed, and democratic building materials the world over. Clay is found in the subsoil, the layer of minerals that lies underneath topsoil. It is a composition of microscopic phyllosilicate minerals that have an ionic charge. These particles are so small that they have a lot of spaces in and among them for water to hang out. This is the second useful attribute: clay has a massive capacity to hold water. The third useful aspect relates

TABLE 17.1. BINDER COMPARISONS

Binder	Chemical composition	Vapor permeability (for 1 inch of material)	Weather resistance	Structural strength	Availability in North America	Social-ecological impact
Clay	microscopic, platelike, phyllo-silicate minerals, impart plasticity and harden when fired or dried	17–18 perms	low (interior finish mostly, exterior if treated or protected)	low	high in most places; clay-rich soils, bagged clay	low
Lime	$Ca(OH)_2$, calcium hydroxide, derived from limestone	11–13 perms	medium-high; exterior or interior	medium-high; hydraulic stronger	high; available in powdered, bagged form—hydrated Type S	medium
Gypsum	$CaSO_4 \cdot 2H_2O$, calcium sulfate dihydrate	18 perms	very low; interior only	low	high; bagged	medium
Cement	CaO, SiO, Al_2O_3, Fe_2O_3, MgO, ⅔ calcium silicates, ⅓ alum & iron-containing clinks	1–3.2 perms	high; exterior or interior	high; especially if fiber reinforced	high; bagged	high

Source: Ace McArleton of New Frameworks Natural Building, LLC; content from slides on "plaster theory" lecture at Yestermorrow Design/Build School.

Note: This table compares, at a glance, some of the most relevant characteristics of different binders for natural plaster, in an effort to provide information with which to potentially choose between them for a project.

to the ionic charge of the clay particles. Water, also being ionic, is attracted to the clay particles. This phenomenon is what categorizes clay as hydrophilic, which means "water-loving." For natural plasters, this has tremendous benefit to indoor air quality and longevity of building materials protected by that clay coat. It means that clay will draw moisture *into itself* in large volumes, far greater than any other material it is in contact with. It can be said therefore that clay has a higher moisture equilibrium content—the point at which the substance will reach equilibrium with water, and hence stop the diffusion drive of water coming into it—than other substances around it. (For more information on clay, see chapter 4.)

The fourth principle of clay useful to the natural builder is that clay is also vapor permeable. In table 17.1, we see that clay has a permeability rating of 17–18 perms, which we can compare to cement at 1–3 perms, and lime at 11–13 perms. A "perm" is a unit used to describe the rate of permeance, or the amount of water vapor that moves through a material during a set period of time. A lower perm rating indicates that the material is less permeable than a material with a higher perm rating.

What we can understand from this, in combination with the three previously noted principles, is that clay will attract water, it will hold a great deal of it, and it will also allow that water vapor to pass into the air. The most important implication of these three properties for us as natural builders is that clay will protect moisture-vulnerable wall-building materials, like straw bale, from long- or short-term moisture degradation. This is a lesson that was learned in the United Kingdom when many structures built of earthen or natural materials that had been originally plastered with clay or lime plaster were stuccoed with cement stucco after World War II bomb damage. Many of these structures showed moisture damage rapidly, demonstrating clearly the differences between vapor-permeable and vapor-impermeable finishes (Mitchell 2007). Clay acts as a moisture battery, collecting and storing water vapor into itself, while slowly releasing it into the ambient air to disperse it. This principle illustrates why clay is so useful as a wall finish in bathrooms, kitchens, or other humid places—which are distinct from wet places. Clay will take on liquid water as well and will absorb it until its absorption capacity is reached, but at this point the water will overcome the clay's binding structure, the clay will liquefy, and the plaster will wash off the wall. Clay does not stand up well to liquid water for any significant length of time.

This leads to the fifth principle of clay: it does not undergo a chemical change when it sets up into a plaster but goes through a physical one. It simply dries out. Water merely leaves, and in that way clay "sets up" into its binding structure. This is in contrast to lime and cement, which set up via a chemical reaction. Because clay's transformation is a physical one, it can be wetted again and returned to its original state. The benefit to the natural builder therefore is its ease of use. Perhaps more importantly, it is also easy to repair, as there is no permanently changed, chemically transformed plaster structure, but rather a plaster body that can be rewetted and reworked perhaps infinitely.

As mentioned in chapter 8, the best moisture-control strategies involve designing problems out of the structure rather than solving them after they have been needlessly designed into the enclosure. This line of thinking stands in opposition to the mainstream (and sometimes green) building principles pertaining to vapor barriers and vapor-impermeable wall systems where the intention is to design water out of structures, often resulting in moisture damage failures due to faulty installation or tiny errors. Instead, natural building proponents acknowledge that water is a fact of our built environment, and our world in general, and understand our buildings must deal with it actively. We embrace the integrative design/build approach that challenges us to work with ecological realities and to use natural materials that will break down in the presence of too much water (as ease of breakdown and reintegration into the natural ecology is a key consideration in using these materials), yet to use them intelligently and set them up for long-term success and performance by savvy design. Understanding how to appropriately use clay is one important tool in the natural builder's toolkit for designing moisture-management strategies into the enclosure system.

Site Soil

Depending on the geology of the area clay can be often readily found close to the building site, or even on it. Clay feels both slippery and incredibly sticky; it holds water and will hold a boot that steps in it. Often local excavators will know if there are clay deposits around, although take care as they will often recommend subsoil that is predominantly silt because they are unaware of the difference. We have learned the hard way that many people in the Northeast—including established pottery manufacturers who supply straw bale builders—will claim they have "pure clay," and yet what they think is clay is actually silt. All subsoil is some percentage clay, some percentage silt, and some percentage sand. For construction, we do not want silt to be greater than 25% of the total clay content, and most importantly, we want clay to be at least 25% of the subsoil content. The presence of actual clay in amounts needed for building can be determined by empirical field tests, such as the ribbon test, leg test, dropping-the-dry-ball test, or jar test (see the sidebar "Evaluating Site Soil"). However, a strong note of caution: the ribbon test is the only one used by soil scientists. We have found many samples thought to be clay that did not pass the ribbon test, though they passed other tests you will find in natural building books. Since clay is so essential to the performance of our structures, we highly recommend the ribbon test. If the sample passes, you can be reasonably confident that the soil has at least 10% to 25% clay or above.

Evaluating a pile of raw site soil with low clay (10%) and high silt content. PHOTO COURTESY OF NEW FRAMEWORKS NATURAL BUILDING.

For building, we look for 20% to 30%, but this can be hard to come by. Another good test is to make a plaster sample with the site soil in question, and after drying, evaluate its durability. Enough clay in the soil, and not too much silt, will make a strong plaster.

The benefits of using clay-rich site soil are many. It is usually free or very inexpensive. It is usually available locally, so financial and ecological costs of transport are few, and the joy of using something that comes from the ground where you live to build your house is a powerful benefit. The negatives include that the responsibility for assessing the quality and clay content is on the user—the negative effect of using a mix with too much silt and too little clay is plaster failure, so proceed with care. We have found many natural building books to be much more lax on the quality of clay required for building, and that is one important emphasis needed for durability in our demanding climate. It is also a more challenging process to get the soil into a workable state for building. See step 7, "Mix Methodology," later in this chapter for more on processing site soil.

Bagged Clay

Clay can also be purchased pre-processed in bagged form. In the Northeast, this material is most commonly used in the pottery trades, and the best bet for finding it is to call around to a pottery supply business in the area. Colleagues out west apparently find bagged "fire clay" or "mortar clay" readily and inexpensively at gravel yards and building supply houses. There are many bagged clays available. Our favorite for both base-coat plasters and finish plasters is ball clay. Ball clay is found in a very few geographic places where the geology is just right. Ball clay has a strong plasticity and strong holding capacity as compared to the kaolin clays, so it is excellent for base-coat work, which tends to be thicker (1–3 inches thick) and in need of greater strength. In our region, it is a very pale gray-brown color, which is not an issue in terms of base coats that will be covered but does at times limit the use of ball clay for finish work. It is very difficult to get a light or bright color of paint or finish plaster when the base binder is already a gray-brown color. In these cases, we tend to go to the kaolin clays. Most commonly we use EPK, which is a lighter white clay that is quite fine but does not seem to exhibit the strength and plasticity of the ball clays, and hence is more suited to fine finishes or paints.

The benefits of purchasing bagged ball clay for a project include guaranteed quality clay content, verified by an outside monitored commercial source; easily moved and handled bags as opposed to a huge pile on a tarp of site soil; and a dry powdered form that is predictable and easy to mix and measure.

Lime

Lime is not naturally occurring in the state for which it is used for building. It is most frequently made by quarrying limestone, but it also may be made from crushed shells, coral rocks, or chalk. This raw material is heated in kilns, usually to 1,650°F (900°C)—and as heat is added, carbon dioxide gas is given off while calcium oxide remains, which is also called quicklime. This quicklime is combined with water, slowly and carefully because it is quite reactive, and what is left is calcium hydroxide. Calcium hydroxide is known as hydrated lime. In the northeastern United States, the lime readily available for building can be purchased in bags labeled "Type S Hydrated Mason's Lime." This is calcium hydroxide that has been partially

Evaluating Site Soil: Sand vs. Silt vs. Clay

BY BEN WATERMAN, SOIL SCIENTIST

Soils in their natural state are a mix of particles of different sizes, namely sand, silt, and clay. Each has its own chemical and physical characteristics. Clay, owing to its microscopic particle size and its platelike structure, makes it sticky. Clay particles are also electrically charged, which accounts for their ability to attract water. Silt particles are too big to stick together, yet they are small enough to be easily carried by wind and water. Silty soils have very little structural stability. Sand particles, on the other hand, are not as easily moved because they are hefty enough and irregularly shaped enough to lock tight with other substrates in the soil-particle matrix.

The degree of influence of sand vs. silt vs. clay determines a site soil's overall suitability for plastering. Clay is the plasterer's best friend, but only if not overshadowed by too much silt and sand. Sand adds structural stability to any plaster, but only if there is enough clay to act as a binder. Silt, well, silt is just plainly not welcome when it comes to plaster! A good plaster has at least about 25% clay by volume or more if very little sand is present, and there is a risk of silt dominating the mix. Clay's influence is very strong relative to the traits of sand and silt, hence we can get away with only one-quarter of the sample being clay yet it still being a very effective plaster.

Soil scientists typically use the "ribbon test" in the field to estimate the ratio of sand, silt, and clay in soil in its natural state. A soil sample the size of a golf ball is simply moistened to a "not too wet, but not too dry" consistency, and then pressed by the thumb over the forefinger in the attempt to flatten out a "ribbon." In subsequent pressings of the sample by the thumb over the forefinger, the ribbon can be extended. The more clay the sample has, the longer the ribbon can grow, and sometimes it will even curl over the hand and stay intact. While the ribbon is being formed and pressed (or failing to because there is not enough clay), the thumb can also get a good feel for the sample's grittiness or its sand content. The best use of the ribbon test is to judge one sample relative to another, but for making absolute "pass" or "fail" calls for plaster suitability, there is a good chance the sample has at least 25% clay if the ribbon can be formed very easily, can be pressed to greater than 4 inches, and holds its form after you are done pressing it.

On DVD, Chapter 17, see RIBBON TEST FOR EVALUATING SITE SOIL

hydrated, then dried out, so the process of combining with water can be completed by the craftsperson. The lime cycle is completed when calcium hydroxide (or hydrated lime) is mixed with sand and other ingredients and spread on a wall or floor—and as it contacts the air, it reabsorbs carbon dioxide to form calcium carbonate. This is chemically the same as the raw material it started as, however, now it is in the new shape that has been created—a bright, smooth, soft wall covering.

Lime has many properties that are desirable for us as natural builders. It is soft, flexible, and more vapor permeable than cement. Lime is stronger and more stable when compared to clay. It can tolerate

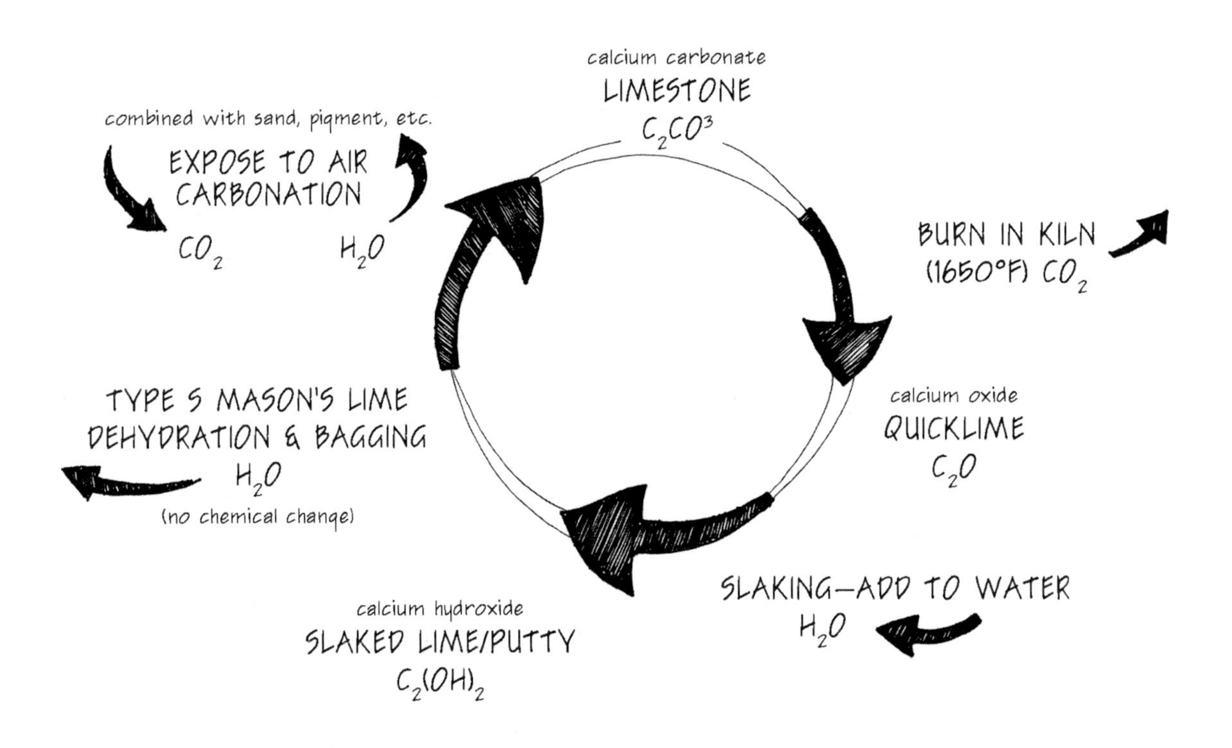

The hydrated lime cycle explained. ILLUSTRATION BY BEN GRAHAM.

the softness and flexibility of lime-stabilized clay as an undercoat, while providing a more weather- and impact-resistant finish. High-quality lime exhibits autogenous healing, which means that even if small cracks open up, it tends to fill them itself over time as the particles continue to carbonate (whereas cement will tend to crack owing to its extreme strength and rigidity and allow liquid water into the wall). Lime will still allow vapor to pass through the assembly, with a permeability of 11–13 perms; whereas cement, with a permeability of 1–3 perms, will not (see table 17.1). Lime also carries a somewhat smaller carbon footprint than cement, in that it burns at a slightly lower temperature, has fewer impurities, and uses fewer industrial by-products in its production. If lime were produced locally again in more regions—which would remove the environmental and ecological costs of large-scale, industrial lime production and distribution—small-scale industry (and the associated skills) could be added to the local economy and to the list of lime's benefits.

Gypsum

In the Northeast, we tend to see gypsum as having all the negatives of clay, with few of the benefits. Its general availability in bagged form from building supply stores and familiarity to mainstream tradespeople, along with its stickiness and vapor permeability, are some of its advantages. Due to its stickiness and the lightness of the available bagged mixes, it lends itself well to places like ceilings. It is worth mentioning that it is predominantly the sand type and size that make lime and clay plaster mixes heavy, and not the binders themselves. Combining a lighter aggregate with these binders would yield similar lightness to gypsum mixes; however, manufactured gypsum plasters are formulated with additives to allow for the use of smaller aggregate in lower amounts than in most clay or lime plasters.

However, if gypsum is desired as a plaster on straw bale walls, do remember that it will not protect the bales in the way clay will. Gypsum is not hydrophilic to the same degree as clay. Further, its set time is quite short, giving the installer less working time. It does not have the "softness" and "aliveness" of either lime or clay but instead has a flat, neutral look and feel. The crystalline structure of lime allows for a "double-refractive index" so it sends out double the light it takes in, which gives lime plaster its glowing, alive look. Clay also is described as having a softness and depth. Gypsum generally does not exhibit these traits.

Cement

Cement is the strongest of all the binders. It is, at the basic materials level, fired limestone just like lime. However, the lime has been changed by the addition of pozzolans and industrial waste additives of many sorts: fly ash, clays, blast furnace slag, all heated together in a kiln to 1,450°F. Early in bale construction in the northeastern United States, well-meaning builders thought that cement was the best choice for plastering straw bale structures. It was the most durable plaster available and had been used to good effect in the warm, relatively dry regions like the Southwest and California. Cement stucco sometimes is an integral part of the load-bearing, earthquake-responsive straw-bale designs in warm, dry, seismic regions. It seemed to have the durability and weather resistance to really protect the bale walls. Unfortunately, it did precisely the opposite. The low permeability of cement (1–3 perms; see table 17.1) trapped whatever moisture was in the wall system and kept the straw soaked, ultimately rotting it in many cases. It did not exhibit the hydrophilic characteristics of clay; rather than wicking moisture away from the straw and allowing it to evaporate in the air, it cracked, as cement is wont to do because of its rigid nature, and allowed bulk liquid water into the wall. The straw inside many experimental early bale structures that were plastered with cement stucco has entirely rotted away, which is why knocking on these stucco buildings yields a hollow sound (those that have been torn down were found to contain a pile of composted former straw on the ground!).

Clark Sanders, a natural and straw bale builder in Meredith, New York, always plastered his houses with cement stucco, and recently he has shifted to clay- and lime-based systems due to empirical research in the field. Some of Clark's early experiments on bales plastered with clay, lime-clay, and cement were featured in *Serious Straw Bale* and helped lay the early groundwork for understanding the importance of vapor permeability and of hydrophilic qualities in bale coverings (Lacinski and Bergeron 2000).

Hence, cement has developed a negative reputation in natural building in wet, cold regions. As mentioned earlier, however, in warm, dry climates or cold, dry climates cement stucco can sometimes be appropriate and essential for both load-bearing straw bale construction and structural support in general, and for seismic areas in particular. Once again we return to the principle of natural building that emphasizes locally specific, regionally appropriate choices for structure. Our hope is that the lessons of plastering with clay, lime, gypsum, and cement that we have learned here in the Northeast can be combined with the knowledge and experience of these materials in other regions, and by doing that, we will know more about the context-dependent strengths and limitations of these materials.

FIBER

Fiber is the element of the plaster that lends tensile reinforcement. It is one of the most important anticrack elements in the plaster. Concrete walls contain rebar, metal reinforcing rods that aid in lending tensile strength to the composite. Fiber serves a similar role in the plaster. It supports the coat's ability to resist stresses caused by movement in the building, which inevitably occurs both during and after construction. Examples of fibers include chopped straw, animal hair, cellulose, other fibrous plant material like cattail fibers or hemp fibers, and manure. The size and amount of the fiber needed is related to the thickness of the desired coat. Thick coats, like clay base coats, which can run up to 2-plus inches in one go, need longer and more substantial fibers to resist cracking. Thin, fine-finish coats, such as a plaster using stone dust for aggregate and going on in a thickness of ⅛ inch or less, will require a very fine fiber, such as a manure tea—which at that point is almost more of an additive rather than a fiber (see "Additives" section on page 277)—or no fiber at all.

We can see how the straw works by seeing it trying to span this crack. PHOTO BY KELLY GRIFFITH.

Chopped Straw

Chopped straw is one of the most common fibrous additives to natural plaster. It is especially useful in base coats, where its varying diameters and lengths lend strength to this relatively thick application. It also is one of the more visible fibers, and so it is chosen in applications where either that look is desired in a finish or it will not be seen—as in a base coat. Often, if plaster is going on top of natural wall systems where straw has played a part in the construction, such as straw-bale, straw-clay, woodchip-clay, or cob, there is straw left over as a by-product of that construction process, and it makes good sense to use it if possible. If the plaster project is a renovation, or going on a substrate where straw did not play a part—such as gypsum board or other substrates—acquiring straw and chopping it can be less immediately efficient. In these cases, we will often look to other fibers such as cellulose or manure, also because they are smaller and less visible, and gypsum board requires a less thick application.

Athena and Bill Steen, who run the Canelo Project, make a plaster consisting solely of chopped straw and clay called straw-clay plaster. This is a good adhesion coat to straw bale walls and can function as a protective coat for a short time. We have come to call this the "Steen coat" in honor of their invention.

Animal Manure

Animal manure is a substance that results from the animal consuming fibrous material, like grasses or straw, and processing it in its intestinal tract to extract the nutritional value. Since what we are looking for is not the nutritive aspects of the grasses but rather the fibrous, cellulosic structure to enhance the plaster, manure is well suited to our purposes. Horse manure is often used simply for the fiber itself, while cow manure is useful for the cellulosic fibers that have been digested, fermented, and softened by the rumen, as well as its proteins and saliva, which create a slippery, gelatinous substance in the plaster that increases durability. This is further described next in the "Additives" section.

Animal Hair

Animal hair, like horse and oxen hair, was used frequently in old lime and gypsum plasters found on wooden lath systems in homes in the United States that date back to the early part of the 1900s or earlier. It was the short body hair that was used, not the tail or mane hair. When considering using animal hair as plaster fiber, the most important qualities to be present are that it is rough and strong (human hair, for instance, is too smooth and weak). It is also important not to "wet-store" the plaster in a lime plaster with the hair premixed in it, as the action of the wet lime will attack and break down the proteins in the hair, resulting in failing plaster. When using animal hair in lime plaster, adding it just before applying the plaster is important (Holmes and Wingate 1997/2002).

Cellulose

Cellulose fiber is the same cellulose that is used for insulation. It is small, shredded bits of recycled newspaper, or, as newspapers are becoming less common, pulp fiction novels. It is a useful fiber additive for finer finishes, as it is a finer fiber, fairly inexpensive, and readily available at most lumberyards. One of the common liabilities of using cellulose is that it tends to clump in the mix. The best way to minimize this is to soak it in water overnight, and blend it up with a paddle mixer attached to a corded drill—yet still the plaster finish may reveal small dimples and drag marks where the intact cellulose piece survived even diligent mixing. This can be an interesting effect or a problem, depending on the desired look.

Cattail and Hemp Fibers

Plant fibers such as cattail fibers or hemp fibers can also be used to good effect. The main drawback for cattail fibers is the inefficiency of harvesting and processing enough to use for a project. But if you have time, it can be a great addition. In terms of hemp fibers, we are sadly unable to access many hemp-based products in the United States owing to drug legislation, and we have yet to try them. For good information on hemp building products, see work from our Canadian and United Kingdom colleagues: Weismann and Bryce's *Using Natural Finishes* or Magwood et al.'s *More Straw Bale Building*.

Poly Fibers

Poly fibers are used in the cement industry as an additive to aid against cracking in cement applications. We tend to steer away from them because they are plastic, but they can be a good tool in the plasterer's toolkit. They are very fine and almost translucent. We had good luck repairing a persistent, fairly wide (¼ inch) crack in lime-sand plaster by mixing poly fibers with lime and silica sand, and they performed quite well. Straw would have been too big and awkward, cellulose seemed too small and weak for an area with stress, and manure too difficult and cumbersome to obtain and process for such a small repair.

ADDITIVES

Additives are a category that includes any other elements that are added to the plaster mix to provide attributes to the plaster beyond those of the binder, aggregate, or fiber. Wheat paste, animal glues, oils, casein, and manure are some examples of additives for natural plasters. Cow manure, as mentioned above, has the additional benefit of having passed through the rumen, commonly known as the "fourth stomach" in ruminants such as cows and sheep. The rumen provides an anaerobic environment where fermentation and multiple mastications of ingested cellulosic matter break down this material. The upshot of this is that the straw or hay in the cow manure has been chewed, fermented, and digested until it is small and softened. Additionally, there is a sticky, slippery substance added in the ruminant digestion process, most likely a combination of saliva and lipids. For plaster, this additional substance does several important things. It adds to the plasticity and workability of the mix; a manure plaster is just so silky and pleasant to work with. It has a gluelike effect, which serves to bind the plaster together more tightly and to make it more weather resistant. We have also noticed that manure-clay-lime coats have a pretty significant improvement in durability and strength in their resistance to rubbing or exposure over several winters. Finally, the gluelike substance found in ruminant manure coats over the lime molecules, restricting their direct exposure to the air and thus slowing carbonation. This is favorable for lime, as it is strongest if it sets slowly and in the presence of water, rather than drying out quickly, similar to the "curing" process of cement.

Pigments can be added to natural plasters, though we do not actually recommend their use in this way in most applications. Pigments are expensive and resource-intensive, generally speaking, and the amount that will need to be added to the plaster to achieve desired tinting strength is significant enough to be inadvisable. In our view, a better use of resources would be to make a pigmented paint or a wash, and apply that to the plaster once it is dry. It is also easily repaired. If the color wears or gets a nick, it is simpler and less expensive to rehydrate some of the tinted limewash or clay paint you saved and touch up the spot than it would be to get an accurate color match in a plaster and then have to re-plaster.

Pigments can be added to plaster coats, however, and this can be effective if tinting is controlled and repeatable for each batch, and if the coat is not so thick that an inordinate amount of pigment is necessary. Achieving full color density through the plaster also makes nicks in the finish less obvious. If choosing this option, add wet pigments to wet mix, and dry pigments to dry mix. See chapter 20 for details on natural finishes and on working with pigments.

Plastering Methods

In the second half of this chapter, we seek to give you, the reader, the tools to feel confident working with natural plasters. An important part of this is to demystify the materials and processes so they can be easily understood; another part is to explore action steps and critical-thinking tools that can be applied to any building situation. What follows is a nine-step process that can be used to do almost any plaster project. We proceed through the process from start to finish—from materials to mix ratios, preparation, and aftercare.

STEP 1: IDENTIFY PERFORMANCE REQUIREMENTS

Is the plaster coat intended as an air barrier, moisture protection, and base coat in a two- or three-coat finish assembly on straw bale walls? Will the coat be the finish over existing painted drywall in a crowded local art-house movie theater, meant to capture both an old-world and a contemporary feel? Is the coat wrapping the exterior of a woodchip-clay structure that will be exposed to lots of wind and rain? Is the plaster coat outside, but shielded from weather by an ample porch? Will it serve as a finish in a busy ADA-compliant bathroom, where heavy traffic and moisture are concerns? Natural plasters can serve all these functions and more, and understanding what is needed out of the plaster coat is the first essential step to successful application.

Taking the time to assess the context the walls exist within is worth the investment. Identifying the goals of the application, what stresses it must be able to withstand, and the desired aesthetic will guide you as you formulate your recipe and set up for a successful application. Important parameters to

The rain-screen gable needs just one coat of clay plaster for air and moisture protection, whereas the lower story will receive a final coat of lime-sand plaster and be exposed to weather. PHOTO BY JAN ALLEN TYLER.

determine include: Will this plaster coat be on the exterior of a building or on its interior? Will the coat function as a base coat for another finish or is it in and of itself intended as finish? What are the moisture-performance requirements—is it in a bathroom, kitchen, or wet area? What is the weathering potential of the intended application site—does it receive strong wind or precipitation or sun? What are the desired aesthetics of the coat—should it look soft, or textured, or very smooth and clean? What is the plan for and availability to do maintenance of the coat—will it be easy to do touch-ups each year, or more like every five to ten years? What then are the related longevity and durability requirements, and the use pattern of the area of application? Will the coat receive another topcoat, such as wood siding, metal rain screen, limewash, or a translucent topcoat such as starch paste or sodium silicate?

The list of the performance requirements for the installation is similar to the design program in the design/build process: it will serve to guide you in your decisions as you move forward. For example, if resistance to liquid water (precipitation) is a key requirement, clay will probably not be the first binder you consider. However, if clay is wanted for other reasons—availability (on-site for free), ease of repair, or aesthetics, for instance—you might look to other ways to protect the clay and mitigate the harm from liquid water. Depending on your other requirements, you may look to a porch on the exterior of the structure to protect the wall from direct water. You may consider a lime-clay mix to increase the durability and strength while still retaining clay as a strong element. This is one example of thinking like a natural builder, interacting with the materials and the process of creating structure and finishes creatively, using knowledge of materials' properties and balancing those with the needs of the program requirements.

STEP 2: EVALUATE THE SUBSTRATE

The second step in preparing to plaster is to evaluate the substrate you intend to plaster on. As we will see, this process is key, and considering the wide range of potential variables, quite involved. Therefore, developing an organized approach to this process is key. Here are several variables to keep in mind.

Variables

Substrate solidity: test and maximize

"Natural plasters," which for our purposes are clay and lime plasters, can be applied successfully to painted or unpainted gypsum board, wooden lath, straw-clay, cob, straw bales, other plaster coats, or Durisol block. Substrates that don't work well are wood that doesn't have a key, cement, and anything that has no chemical relationship (or "bond") with the plaster (or moves with moisture). Natural plasters, to be successful in most cases, must have a "key" that provides "tooth" for the coat. Old-style wood lath exemplifies this quite well. In this system, if we were to remove the lath, the plaster would be dried into a sheet with horizontal rows of "key" where the lath was. This illustrates that it is not the bond between the wooden lath and the plaster that holds the coat from slumping or falling, but rather the keys of the plaster that mechanically hold it. This is also true for straw bales. It is a common misconception that straw bales must be wrapped in a wire or other mesh for the plaster to adhere. The jagged ends of the individual straws and the spaces between them provide for a very "toothy" or textural substrate for the plaster to dry into and then hang on.

It is when the substrate has no suction, no chemical relationship with the plaster, and might move with moisture—like wood or wood lath—that a key becomes especially important. One of the biggest causes of plaster failure is an unstable substrate. Any bouncing, shaking, crumbling, or moving of any kind will stress the plaster, and cracking will be likely. Test your substrate (be it drywall, wooden lath, metal lath, straw bales, straw-clay, or cob) and try to move it. If you can shore it up in any way to stabilize it, such as by adding more fasteners, do. Some substrates, like wood, will always move with moisture as they uptake and release it. In this case, adding "tooth" to the wood is necessary; we most commonly do this by attaching

The box beam at the top of the wall is treated with roofing felt and diamond lath to receive plaster. PHOTO COURTESY OF NEW FRAMEWORKS NATURAL BUILDING.

Pony wall to bale wall transition prepared for plaster with aluminum drip cap treated with metal lath for plaster tooth. PHOTO COURTESY OF NEW FRAMEWORKS NATURAL BUILDING.

metal or plastic lath. The material used must have good tooth, and we have found that hardware cloth, owing to its flatness, is not as useful and durable as its cousin, diamond lath, which has a thickness to it that allows for better long-term connection.

Some authors (Cedar Rose Guelberth, Adam Weismann and Katy Bryce) have recommended other strategies, such as making "natural mastic" of sand and wheat paste to apply to the wood for the plaster to grab onto, another version of creating "tooth." Nails can be hammered into the wood, a process called "spragging" (Holmes and Wingate 1997/2002). Finally, scores or grooves can be carved into the wood, which will create a minimal kind of tooth. This method is the most time-consuming and least effective of all of them.

For smaller wood elements, such as a single 2 × 4, often burlap can be floated into the plaster as it is being plastered. This is discussed later in this chapter.

Evenness/shape:
know your desired aesthetic and tolerances

A plaster coat will only look as good as the substrate. The finest, most beautiful plaster applied on top of a lumpy or uneven substrate will not be able to have its fineness realized in the overall aesthetic. This is especially relevant for straw bale walls, which have the potential to be quite variable in their shape. Before applying the plaster is the time to make certain you love the basic shape of the substrate, and to adjust it as needed.

Transitions and edges:
trim, drip edge, plaster stops, rabbets

Functionally and structurally, at the bottom of the wall or at the top of a window, the plaster will need a place to land—what we call a "plaster stop"—and these supports should be in place to receive the plaster coat before plastering begins. This can be drip cap, a beveled wood piece, stone, or a similar feature. Many options exist, and benefits or drawbacks depend on location and stresses on that transition. For exterior plaster stops we commonly use aluminum drip cap, to shed rain away from the wall and to provide a landing place for the plaster. On the interior, we commonly use baseboard or wood trim that is treated with a rabbet on the back side of the trim that meets the plastered edge, for the plaster to tuck into. This allows for cleaner edges, by designing out the visible crack potential where the plaster ends against the trim. Creativity is encouraged here; use your design vision to choose the plaster stops or transitional elements that work best—copper, other metal trim, plastic, and rope have all been used where they fit the look or ecological or social priority.

Rabbeted corner boards, window casings, and top-of-wall trim make for a clean look for plaster. PHOTO COURTESY OF NEW FRAMEWORKS NATURAL BUILDING.

Timber-frame members protected with blue tape and paper prior to plastering. Note timber knee braces framed to the interior for less edge against the plaster. PHOTO BY ACE MCARLETON.

Protection of surrounding edges: taping, papering off, etc.

Taping off and protecting surrounding floors, exposed framing members, windows, or anything adjacent to the plaster coat itself or the plastering process is essential. Plastering is an incredibly messy process. The fact that blue painter's tape is expensive and wasteful, and masking materials such as plastic or kraft paper are also waste products that cannot be recycled, does not outweigh the need for preserving the surrounds. Creative masking solutions can be used, such as scrap newspaper or cardboard, but do not use masking tape or other cheap tape; you will regret it, as it permanently sticks to whatever it is on. Also important is to mask and tape before pre-wetting; if reversed, the tape will sadly not stick, a frustrating situation that has happened to us hilariously often. Oiling the timbers so they do not absorb any plaster and stain is possible as an alternative, however, it must be closely tested if staining will occur anyway; cherry is notorious for permanent discoloration due to clay or lime plasters, despite being oiled. Oiling must also be done well in advance of taping, to ensure the tape will stick to the dried wood surface. Resist the thought, "I will just be careful." You are mixing, hauling, staging, and pushing goopy mud against your walls, and often up against edges that will only look clean and well executed if they are masked and attended to. Taking the time and utilizing the necessary resources to attend to this in advance will greatly pay off after the tape has been pulled.

Pre-wetting the substrate

Pre-wetting is the process by which the substrate is dampened before the plaster application commences, to control for suction. "Suction" is the term that describes the rate at which the substrate withdraws the water from the newly applied plaster. If there is too much suction, the plaster will dry out prematurely and crack. If there is not enough suction, the plaster will slide off the wall. Whether or not to pre-wet, and how much, how far in advance of plastering, and how often to pre-wet, varies with each substrate and each plaster application. If the substrate is a green (or not fully dry) clay plaster, on the interior of the structure where the air is moist and still, less pre-wetting will be necessary, because there is still water in the substrate and the plaster is not going to dry too quickly in the ambient conditions. If the substrate is a fully dry cob wall being plastered on a windy, sunny day, pre-wetting should occur for at least one full day before plastering, as often as possible, and should continue ahead of the plaster application. At times this is impractical, and a wetter mix is applied instead.

A good hose will be essential for your plastering setup regardless; have either a splitter at the source, with one hose at the mixing station and one for the wall, or some setup where it is efficient to spray the walls with water, and everyone on your site is not fighting for the hose. A backpack sprayer can also be incredibly useful. It is excellent for lower-volume and less-frequent spraying, and also for hard-to-reach areas. Hand-held spritzers are useful for spot or incidental dampening, but they are incredibly inefficient for large areas or large volume.

Substrates

Next are brief descriptions of the types of wall systems (and therefore substrates) to which the plaster can be applied.

Gypsum Board/Drywall

To prepare for clay and lime plasters on drywall, the gypsum board should be primed with sanded primer. We have had quite good luck plastering both lime-sand and clay plasters directly onto unpainted raw drywall without a sanded primer, but applying sanded primer is always a good safeguard measure to take if you are unsure. It is necessary to "plank" or float over the joints with something beforehand. We have had best luck using a small amount of the plaster we are applying, with finer aggregate in it. Larger pebbles or sands, even if they are desired for texture or strength in the plaster coat, will drag on the paper of the sheetrock and will not flatten on the edges of the "plank." This process is akin to mudding and taping: attach self-adhesive mesh tape to the joints before planking with your natural plaster. We have found unprimed gypsum-based joint compounds to be unreliable as substrates for natural plasters. Make sure to coat over any joint compound or Durabond with sanded primer. Our colleague Ryan Chivers of Artesano in Colorado utilizes Elmer's glue with sand regularly in place of a sanded primer as a coating on drywall to ensure good stability of the final lime-sand coats (Chivers 2011).

Also, think through transitions. At the bottom of the wall, what will support the plaster? We recommend baseboard or toe-kick of some kind. If you are doing a base coat, you want to set things up so the base-coat finish plane lines up with whatever rough framing supports are in place to receive finish trim. Then the finish coat will plane out to just tuck behind the rabbet in the finish trim. This system is clean and works really well, and it entails thinking through your planes and thicknesses in advance. In terms of longevity and integrity of the installation, providing transition elements for the plaster to come to as the plane changes from the wall to ceiling, for instance, discourages the plaster from cracking. Aesthetically, we have also found the plaster will look best if it is "picture framed" or edged on all sides with trim or transitional elements of some kind. The precise lines of the elements at the edges play off the soft-textured depth of the plaster in the fields. Cove molding is a great option for the ceiling/wall transition as well.

Finish lime plaster can be applied to drywall, and cove molding makes a nice transition at the top of a wall or between dissimilar materials. PHOTO BY ACE MCARLETON.

In terms of pre-wetting, we have found that it is not necessary with gypsum board since the suction is minimal and the added water causes the plaster to slither to the floor. The paper of the drywall is quite slippery and does not give the plaster much resistance anyway, and the additional water makes the situation worse.

Straw Bale

In terms of preparing straw bale walls for plaster, much was covered at the end of chapter 14 in the "Plaster Preparation" section. To briefly reiterate, we recommend the walls be trued, beveled, stuffed with light straw-clay, trimmed with a trimming tool, and in many cases sprayed with clay slip before plastering can begin. In terms of minimizing the movement in straw bale walls to provide a stable substrate for plaster, one of the most important things to do is to find loose or low places in the straw wall and build them out with lath and light straw-clay. Also, any edges, where cracking is likely, should be treated with an air fin that not only functions as a secondary air barrier and control joint, for airtightness, but also provides tooth for plaster. Air fins should span the edge of the installation and extend and be fastened to the plane of the wall. This provides a stable substrate for the plaster to come up to at an edge, to handle the stresses that occur at edges and be less likely to crack. For more on air fins and their installation, see chapter 14.

Foreground shows electric boxes set, loose areas shored up with diamond lath, and top-of-wall transition treated with an air fin. PHOTO BY ACE MCARLETON.

The nature of the connection of the plaster to straw bale walls is through the excellent "tooth" the straw provides. The plasterer's first job for the base coat will be to really work the plaster into the bales, so the link between plaster and bale becomes integral. Loose straw is the enemy of a strong plaster coat. The system's success depends on the integration and bond between the clay coat and the solid straw.

Let's take a moment to talk about aesthetics and straw bale walls. This phase of the process, preparing and evaluating your substrate for plaster, is the phase where you will have a great deal of control over the finished look of your walls. Many who are drawn to straw bale as a building method are attracted to the curve, contour, and texture of the walls. It is partly the move away from rigidly flat, industrial planes that gives straw bale walls their appeal; however, a clean, tight, relatively true installation does not deprive straw bale of its unique beauty. Rather, it allows for a high-performing, long-lasting marriage of the plaster and wall that sets you up as the plasterer to control for shape, curve, and texture in the plaster application process. The alternative, a lumpy wall, leaves the plasterer with very little choice as to how the plaster will look. Taking the time to at least get a clean surface on the bales with gentle curves (never a flat plane) highlights the natural beauty and character of straw bale walls, while not overly distracting the eye with bulges or lumps.

Functionally, it is better for the longevity of the structure that there not be bulges on the walls where snow or rain can build up. Significant changes in the depth of the plaster can also cause cracking, which detracts from the thermal and moisture performance of the plaster coat and compromises the bales. If the plasterer were to go thick in a low spot and thin over a high spot while plastering, for instance, because there was a great deal of variation in the bale wall substrate, cracking at that place where the change in thickness occurs is more likely. It is also a lot more difficult to plaster a variable substrate because it takes more time, and the plasterer cannot make a clean pass over a field of a consistent thickness. When plastering, it is the edges and details that eat up the hours.

If lumpy, wavering planes on walls is the aesthetic you are striving for, make sure that performance criteria are being met: that the straw bales are tight, and solid and all gaps stuffed; that whatever variations in the plane exist, they do not provide a welcoming place for rain, snow, or ice to build up; and that you have the time and patience to plaster the plane variations with a relatively even thickness of plaster.

In chapter 14 we emphasized the importance of design and knowing your planes. Here is where the chickens come home to roost: do you have rough framing set for the plaster to land on and plane out to meet, ready for your finish trim to attach to? Trace around all edges of your installation, and imagine where the plaster will land. It is important to have plaster stops installed at termination edges of the plaster. This includes drip cap over windows and doors and at the bottom of walls, treated with metal or plastic lath for plaster tooth.

Pre-wetting is also important. Perhaps ironically, after spending all the time keeping the straw dry during installation, spraying down the straw bale walls before applying the clay plaster is necessary to promote adhesion of the plaster and to control suction. This is true regardless of whether the walls have been sprayed with clay slip or not. Straw bale walls will not need repeated or frequent dampenings; usually one good spray just before plastering will do the job.

Straw-Clay

Straw-clay can be installed as walls using the slip-form method or by being tamped into cavities caged with wooden lath. We will cover the first here, and the second below with woodchip-clay, since both can be caged with wood lath, and the plaster preparation needs are similar.

Slip-form straw-clay walls are panels of straw-clay stuffed between framing members, stabilized during construction with lateral reinforcement of some kind to tie the panels to the framing. For more on the construction of straw-clay walls, see chapter 15. The panels should be quite stable, but now is the time to check. Perform a visual and tactile inspection. If there are any loose or unstable areas, the loose material should be gently excavated and patched. Approaches to patching could include stuffing fresh straw-clay into the gap, following the method discussed in chapter 14 in the "Plaster Preparation" section. Driving nails, gently, into the surrounding straw clay, so that they stud out and provide a pinning connection between the patch and the wall section, could also be a useful strategy to ensure connection.

The strength of the plaster connection to straw-clay slip-form walls comes from a chemical bond of like to like—between the clay plaster and the clay substrate—and from suction. The texture of the straw will also provide some tooth for the plaster to grab on to. Slip-form straw-clay walls will often have empty pockets, places where the straw-clay did not tamp down evenly and fill, which could not be seen until the forms were removed. Depending on the size of these, it may be advisable to fill them with a pre-coat of plaster. Think of this as spackling the holes in the drywall before painting. It is not necessary to pretreat every pocket, just the ones that are 1½ inches by 1½ inches or more. The other pockets, especially the ones that run horizontally and are essentially a groove in the straw-clay, are not necessary to pre-treat and will act as "keys" for the plaster as long as they are not too deep, do not run through the wall, or do not indicate a weak spot in the substrate.

Another common substrate condition for slip-form straw-clay walls is that there will be places where the straw-clay pushed out the form and dried proud of the plane of the wall. These areas will cause lumps in the plaster coat and make plastering an even coat difficult—so knock them down or shave them with a sharp trowel and/or snips to cut the straw.

Regarding aesthetics, straw-clay slip-form walls will be flatter than straw bale walls, because of the stud walls they are packed into. They are not as flat as drywall, and their fairly uniform appearance allows for a substrate that takes plaster easily, and some texture or curve can be added in the plastering process.

Edges of straw-clay panelized wall treated with rabbeted trim to accommodate clean plaster lines. PHOTO BY ACE MCARLETON.

Base coat being applied to woodchip-clay walls. Note the corner trim boards for long-term durability. PHOTO BY ACE MCARLETON.

In terms of plaster stops for straw-clay slip-form walls, use the same guidelines as outlined for drywall installations (outlined earlier in this section under "Gypsum Board"). Protection of surrounding edges is also the same as for straw bale or drywall.

In regard to pre-wetting straw-clay slip-form walls before plastering, they will need to be sprayed down. If they have been dry for a while, you may need to spray them up to a day in advance of plastering, and several times before beginning plastering. Dry clay is incredibly hydrophilic and has immense water-storage capacity. So dry straw-clay walls will suck the water out of any plaster immediately and disastrously, causing cracking and difficult application conditions. If the straw-clay is green—not fully dry—less pre-wetting may be necessary; however, the dry time of straw-clay is already a challenge, so adding more layers of wet material may be inadvisable.

Woodchip-Clay and Straw-Clay (Wooden Lath)
Woodchip-clay and straw-clay that have been tamped into stud cavities and caged with wooden lath are some of the most stable and predictable substrates we cover in this book. In this system, the plaster is supported primarily by the "keys" hanging on the lath. There will also be a like-to-like chemical bond between the clay-coated infill if the plaster touches it. In general, however, the most important thing to ensure is that all laths are fastened solidly to the framing. For more on installation and specifications for lath installation, see chapter 15. However, it is possible that gaps will exist in the infill, which were missed in the filling process. These should be stuffed with light straw-clay. If you can squeeze the patch material through the lath, do that. If not, remove lath in the area you need to repair and fill the gap, replacing the lath when you are done.

The shape of the substrate is fairly well established in a wood-lath caged system. There may be some places where stray woodchips or straw are protruding proud of the plane of the lath. A once-over on the wall surfaces, trowel in hand, knocking off or back the stray infill, should set you up well for a smooth plaster application. As in any lath and plaster system, the strength of the plaster comes from its ability to form really good keys. Ideally this means the plaster not only comes in between the lath but dries into bulges behind the laths. Hence, tamping of the woodchip-clay not only will reduce its thermal performance (see chapter 15 for more on this) but will also reduce the spaces for the plaster to key. Installing support and/or trim pieces at the edges of the installation is similar for wood-lath caged systems as for other systems. Note in the photo of plastered woodchip-clay walls that corner trim boards were chosen to preserve the vulnerable corners and also to add an aesthetic, structured look to the walls. The need to protect floors, trim, ceiling, and exposed framing elements is the same here as for other systems.

Dampening the lath is recommended. One pass of the hose right before plastering should be effective in controlling suction and preparing the substrate to receive the plaster. But as in all these substrate conditions, your evaluation of the level of pre-wetting required is going to be the best guide. Ambient conditions and the wetness of the plaster are factors that must be considered.

Earthen (Cob, Adobe Block)

Well-installed, well-compacted, and dry earthen substrates are among the more stable substrates to apply plaster to. They are masonry materials, made of clay and sand, are generally stable, and will provide good suction and "like to like" for plaster materials such as clay and clay-lime. In our region we see quite a few cob walls that are cracking due to poor detailing on foundations. We also see quite a few cob structures that are exposed to weathering from rain and snow due to poor detailing on roofs or other weatherproofing strategies. The cob or adobe block must be stable as an installation itself (not cracking or moving) before it can be a good substrate for plaster.

Cob and adobe are "thirsty" substrates. They will extract the water from the plaster that is being applied quite rapidly if they have not been thoroughly pre-wetted to control for suction. Despite the excellent suction and like-to-like interaction between the earthen substrate and the clay-based plaster, it is still ideal to provide some key. If lime is being applied with no clay component, it becomes even more essential to ensure a mechanical key. In their book *Building with Lime* (1997/2002), Holmes and Wingate describe the process for preparing earthen walls to accept lime-sand plasters. They recommend indenting the surface of the cob with a sharp tool to create "pockets" for the plaster to push into and hang on. They also recommend painting on a limewash before plastering with lime, which dovetails with the concept we have noted several times that "like bonds to like"—a lime plaster put onto a smooth clay wall will potentially shear off. Holmes and Wingate provide strategies for avoiding this challenge. If these conditions are met, clay and lime plasters are beautiful, appropriate, and durable finishes for earthen walls.

Finish lime plaster on cob walls. PHOTO BY ACE MCARLETON.

STEP 3: CHOOSE YOUR WEAPONS

Just as in a *Choose Your Own Adventure* book or video game, the weapons you choose determine to some extent how well you play in the game; selecting the elements that best match your goals is an important step to successful plastering.

On DVD, Chapter 17, see WORKING WITH A PLASTER MIX AND ITS INGREDIENTS

What Kind of Plaster?

Here are some rules of thumb we consult for deciding what kind of plaster to use in which applications.

When Clay?

Clay plaster is the right choice when:

- its particular hydrophilic and vapor-permeable qualities are called for
- its high water-storage capacity, nontoxic, and antistatic properties are desired
- its low ecological and social impact is desired

TABLE 17.2. A NATURAL PLASTERER'S MATERIALS KIT

Binders	Aggregates	Fibers	Additives	Topcoats
clay • site • bagged • OM-4 • EPK • red art or other colored clays	sand • washed and screened • mason's sand	chopped straw	wheat paste	limewash
	silica sand	longer straw	casein	clay paint
	mica	manure • horse, • cow, • other?	manure	milk paint
lime • hydrated Type S, dry or soaked • hydraulic	stone dust	cellulose	oil	oil
		animal hair		wax
		cattails		sodium/potassium silicate (water glass)

Source: Ace McArleton of New Frameworks Natural Building, created for Yestermorrow Design/Build lecture slides on plaster theory, 2009.

- the installation will be protected from liquid water and heavy impact
- ease of repair and ease of maintenance are desired
- a wall finish that feels soft and gentle (yet is surprisingly durable to everyday wear) is called for
- it is a regionally or locally accessible material

When Lime-Stabilized Earth?

Lime-stabilized clay coats are appropriate when:

- either more durability of the plaster coat is desired
- the goal is to increase the bond between the base coat of lime-stabilized clay and subsequent coats of lime-sand plaster
- the right amount of lime can be determined

Adding lime to the mix of clay, sand, and chopped straw/manure increases durability of the plaster. Lime molecules bond with clay particles, causing the clay particles to be larger. This process is called flocculation; it strengthens the mix and causes it to be more water resistant when sufficient lime is added to the mix. If too little lime is added, the clay flocculates only to the point of turning into silt, dramatically weakening the mix. The amount of lime that should be added varies widely depending on the type of clay being used, but it ranges between 15% and 50%. Empirical testing should be conducted when stabilizing unknown clays. The downside of lime-stabilizing a clay plaster is that the larger sizes of the flocculated molecules create less space to hold water and fewer charged clay ions to attract water molecules, resulting in a reduced vapor permeability due to lime being introduced to the mix. Lime-sand plaster is still vapor permeable at around 10–13 perms (see table 17.1), certainly more permeable than cement stucco at 1–3 perms, but it cannot compare to clay at 18 perms. While the permeability of lime-stabilized clay plasters is variable and largely untested, it can be presumed to fall between that of pure lime and pure clay plasters. To sum up, lime stabilization will slightly decrease clay's vapor permeability and hydrophilic properties, two of the most important functions of clay in natural walls. It is a tool to be used wisely, and its liabilities as well as benefits must be understood.

We use a lime-stabilized clay coat as our base coat for straw bale walls. Inspired by Paul Lacinski and Andy Mueller's work with GreenSpace Collaborative, based out of Massachusetts, we have used the strengthened-earth base coat as a good

Case Study: Savoy Theater, Montpelier, Vermont

The most common place we use clay in our region is as a base coat on straw bale, straw-clay, or woodchip-clay walls. As a part of a vapor-permeable, airtight natural wall system, clay is essential for the longevity and performance of these installations. However, clay is also an excellent choice for a finish on conventional gypsum board walls. A recent case study of ours where clay was selected as the wall finish of choice was in downtown Montpelier, at the Savoy Theater, a local art-house movie theater in need of major renovations. A new owner solicited the help of the community tradespeople, as well as volunteers of all kinds, to renovate this important community resource. After a consultation with the owner, we decided finish clay plaster was the perfect addition to the tired, flat, painted wallboard. Clay's soft and earthy feel would add, we decided, a unique feeling to the very angular storefront lobby space as well as the huge long wall in the theater. It would also provide moisture regulation and air quality for lots of people respirating in a small space. The owner, Terry Youk, was interested in the clay adding mass and sound-deadening in the theater to improve acoustics. Since it is a fairly high-traffic area, he was also drawn to clay's ease of repair if it were to get bashed by a sharp object. Terry was especially inspired by the local nature of the materials and was interested in supporting our local crafts-knowledge of working with the materials. He also wanted warm, inviting colors, and lime brightens every pigment with its bright whiteness. Together, we came up with a plan of using natural pigments to tint the clay plaster several different colors for opposing walls in the lobby space, and a brick red to play off an existing brick wall in the theater space.

In terms of site setup, we primed the existing grimy latex-painted drywall walls with a regular flat primer with some handfuls of sand thrown in. We worked with the carpenters, Walter Hergt and Harry Strand from Montpelier Construction, who were milling and installing the wainscoting to set us up for a successful transition by asking them to include a rabbet on the top board for the plaster to tuck into.

A finish clay plaster in a community art-house theater.
PHOTO COURTESY OF NEW FRAMEWORKS NATURAL BUILDING.

We started with a basic mix of OM-4 ball clay and sand from a local excavator that we knew was sharp and well sized for our application over painted drywall (about ¼ inch), in a ratio of 1:3. We chose cellulose as our fiber, for ease of acquisition and because its small, short fibers were adequate for the relative thinness and evenness of the application. Soaking cellulose well in advance in water, to loosen and separate the fibers, helps minimize clumping. We added wheat paste to the plaster to increase its durability and cohesiveness, and to minimize dusting. We also used a local waste product: granite powder, to add some fine aggregate to get a smoother finish and to ensure a warm color to the plaster.

The Savoy Theater is enjoyed anew in its new trimmings by an enthusiastic Montpelier community. The clay plasters dress out the new lobby and theater space with a soft, inviting, old-world feel.

When in Doubt, Prime!

A colleague of ours learned the hard way about putting clay plaster over existing painted walls without priming. He plastered a stairwell wall with clay plaster in a renovation. The handrail was mounted to this wall. The plaster went on fine, and dried fine, until parts of it near the handrail started cracking and falling off. David excavated and discovered that years of people going up and down and brushing hands against the wall near the handrail had built up oils on the existing wall. So, if in doubt about the state of existing walls, primer with a handful of sand thrown in is always a good option to ensure success.

balance of beneficial attributes for straw bale walls in cold climates. The clay protects the straw by wicking moisture and allowing it to pass into the air, while creating an airtight covering. In Paul Lacinski and Michel Bergeron's book *Serious Straw Bale*, the authors refer to straw found in an old cob wall that was taken down in Britain as being "mummified." Deprived of moisture and exposure to air, the straw, encased in clay, is perfectly preserved. This principle is at the root of our assertion that clay is essential as a base for straw walls in this climate where moisture protection and airtightness are so necessary, and it has been confirmed by field moisture testing. (See moisture studies referenced in chapter 8 and appendix D for more information.) However, we also have a harsh climate where often the base coat will sit exposed for a season or more, in which case clay's softness and nonchemical set are a liability. Adding lime to the clay coat has proven to harden it up, make it more weather resistant, and also make it more compatible with the lime-sand coat we apply as a finish coat.

Lime-stabilized clay coats are also appropriate when the clay content of the site soil being used is lower than is desirable. If you are using site soil rather than bagged clay, it is essential to determine the clay content of the soil. As noted earlier, lime stabilization is a process whereby the clay is flocculated into larger particles, so it can increase the strength of the clay that is present in a particular subsoil.

Lime stabilization also improves the bond between the base coat of lime-stabilized clay and subsequent coats of lime-sand plaster. We have found this to be true firsthand: when taking apart plastered straw-bale walls, a lime-sand coat will shear away from the pure clay coat beneath it, particularly if the clay coat was not properly scratched. Clay coats with at least 15% lime—sufficient for proper stabilization in many cases—seem to refrain from shearing. This again illustrates the chemistry adage "like bonds to like." Harry Francis, former technical director of the National Lime Association, confirms that a pure lime coat applied to a clay coat will turn the exterior layer of clay to silt as a result of partial, inadequate stabilization, causing delamination (Francis 2011). It is important to note that proper preparation of the clay base coat by scratching is critical to ensure a strong mechanical bond, as lime stabilization alone is not enough to guarantee a durable transition from a clay plaster to a lime plaster.

Clay-lime plaster also makes a nice finish plaster on drywall or as a final coat on natural walls, with the same properties described here: greater durability with many of the benefits of clay.

When Lime?

Lime is called for when:

- greater durability and weather resistance are needed
- some vapor permeability is also required
- antiseptic qualities are desired in the finish

- the plasterer has some skill in working with lime
- restoration work and an old-world feel is the nature of the work

We most often use lime-sand-manure plaster as a finish plaster, or second plaster coat, over a lime-stabilized, clay-based base coat on straw bale, woodchip-clay, and straw-clay wall systems. It provides a still vapor-permeable yet more weather-resistant and durable coat for exteriors or interiors than a clay coat.

Lime-sand can also be an excellent choice for finishing existing walls in renovations. In a recent bathroom renovation, we used a lime-sand plaster and limewash for topcoating and color on the drywall. The bathroom suffered from being too dark with no window before the owners, Samuel and Eli, decided to remodel. So in addition to a new window letting in the light from the southwest, the lime finishes seem to glow in the light because of lime's "double-refracting index." The house is 200-plus years old with existing lath and plaster walls, so lime-sand plaster also was in keeping with the house's aesthetic and historic feel. Lime has hygienic properties, so it is appropriate for a bathroom. The limewash that we crafted has a bit of yellow pigment in it, so lime's natural bright whiteness is tempered to a soft buttery glow and plays well with the locally crafted tile and wood shelving from local sawmills.

Finish lime plaster on drywall in bathroom remodel. PHOTO BY ACE MCARLETON.

When Gypsum?
Gypsum is called for when:

- the easy access of the material is desirable
- skilled tradespeople exist who are more comfortable with it than lime or clay
- the lightness and stickiness of the material are wanted, e.g., for a ceiling
- it is in a highly protected environment (interior only)
- a quick set is desired or required

Gypsum is one of the least-used binders for us as natural builders, because it exhibits all the downsides of clay—susceptibility to water, low durability—as well as not enough of the benefits of lime or clay. Although gypsum is vapor permeable in equal measure with clay, it is less hydrophilic than clay, and hence is a poor choice in our climate for coating natural wall systems. Gypsum is a useful option for certain occasions—usually on drywall, ceilings, and more conventional building situations when a bagged product is appealing—and thus is a good tool for the toolkit. It is also very helpful when a quick set is required, allowing for multiple coats plus paint in a single day. This is especially helpful when doing a repair or working in a inhabited space when multiple days of drying time is a large inconvenience.

When Cement?
Cement plaster, or stucco, is called for when:

- its greater compressive strength is called for, as in load-bearing straw-bale applications
- its strength functions as a component in a seismic engineering strategy
- its low vapor permeability is not going to pose a threat to the underlying substrate
- the craft is well established in the local trades

- it might match the vernacular architecture of the area
- its higher ecological cost is balanced out by other factors
- the easy access of the material is desirable
- its durability is required, such as in an exterior grade-level environment

Cement stucco was used often in the early days of straw building in the Northeast, following precedent established in the relatively drier Southwest and western regions of the United States. Its low vapor permeability and tendency to crack due to its rigidity and allowing bulk water inside the wall led to many wholesale failures of stucco-coated straw walls. Cement stucco has its place, such as on the outside of foundation walls where higher durability and strength is called for. It is a finish that is more suitable for exterior applications. There are areas where cement stucco is very much the vernacular style, and these places tend to be dry and relatively warm, such as California, New Mexico, and Arizona. It is more frequently in these places, too, where seismic conditions are a concern in construction, and stronger finishes such as cement stucco are required by local codes.

When Extra Fiber?
Extra fiber is warranted when:

- there is a chance of movement in the substrate
- the plaster will be thick, and hence will be prone to cracking
- artistic elements in the plaster are desired, such as bas-relief sculptures and shelves

Extra fiber, or tensile strength, is desirable when you anticipate that the plaster will be under greater-than-usual movement stresses. When a client of ours insisted on plastering over a 16-inch band joist that ran around the perimeter of the house between the first and second floors, the plaster was going to have to be on straw bale above and below it, and then span onto and over the wood member without cracking. The different substrates and the two different edges on the exterior of the structure presented concerns. First we treated the band joist with 30-pound roofing felt and plastic lath that extended across the joint on either side of the wood member and onto the plane of the straw, where we pinned it. We wanted to maximize the flexibility of the coat we applied, so we did a special mix with extra manure and straw in it, to supercharge the mix with tensile strength and the ability to resist cracking if there was movement.

This was a two-pronged approach of treating the substrate appropriately with felt paper to provide a barrier so the wood would not up-take moisture from the plaster coat getting soaked with rain, and fiber mesh sheet to give it a flexible web to cling to, and crafting a mix that would be engineered to tolerate movement. This is an example of how to use the knowledge of plaster's liabilities and strengths to solve building problems. Adding extra fiber also helps when building out bas reliefs or shelves in plaster, in other words, anytime depth and thickness are called for.

When Manure?
Manure is great when:

- processed fiber is needed to minimize cracking (especially when larger straws would make texture that's unwanted in the finish)
- it's readily available
- additional weather resistance and strength are needed
- it's cow manure, which creates a gel that helps lime cure more slowly

Many classic, artisanal plaster techniques prescribe manure as an additive to lime, clay, and cement, as the animal-processed fibers add extra crack resistance and a flexible strength to the mix (Holmes and Wingate 1997/2002). At times, manure is easy to source just down the road from the job. Other times,

Evaluating the strength and durability of a manure-lime-clay base coat on straw bale by scraping, pushing on it, and generally trying to damage it to see how well it resists. PHOTO BY NICK SALMONS.

it is a fiasco to acquire in its fresh and unsullied state and to transport it to the worksite. It is certainly in no way necessary for a good coat of plaster, but it can be a beneficial additive if conditions are right.

When Additives?

Additives are a good idea when:

- extra strength is required
- stain resistance is desired
- added hardness will help achieve performance goals

Wheat paste can help control dusting and provide a gluelike cohesiveness to the plaster. Casein can add strength, particularly with lime plasters and paints. In a clay plaster, it will add a stickiness and additional binding agent for cohesiveness. Oil can add water resistance and stain resistance and should only be added to a plaster with lots of care; too much oil will cause the plaster to fail. We tend to use oil as a topcoat to the plaster after it has already dried.

Many additives exist and are dependent on local resources and needs. Ox blood was used for a long time for plasters and earth and lime floors as a hardener. Cactus juice is used in areas where cacti grow abundantly and adds a workability and durability to the plaster.

STEP 4: CRAFTING THE MIX

On DVD, Chapter 17, see THE PLASTER MIXING PROCESS

Now that you are familiar with some of the reasons to choose one binder, aggregate, fiber, or additive over another, it is time to either find an existing recipe or create your own. In appendix C we have included our most commonly used recipes for natural plasters and paints. *Using Natural Finishes* by Adam Weismann and Katy Bryce is also an excellent resource for recipes and methods.

No recipe can replace your local knowledge of your ingredients, however. Clay is different. Sands are different. The minerals in your water may cause slightly different effects than another water source in a different place. Some rules of thumb for either tweaking or creating your own recipes are as follows.

Binder-to-Aggregate Ratio

The binder-to-aggregate ratio must be not greater than 1:2 and not less than 1:4. This is a good starting point and rule of thumb—as mentioned earlier, our colleague Ryan Chivers will sometimes make very fine lime plasters where the binder is in a 1:1 ratio to the aggregate, but this is definitely the exception rather than the rule. We will make test samples in advance of a job that explore these ranges. In our plaster classes at Yestermorrow Design/Build School, we create a "plaster lab" where students try making a clay base coat at 1:2, 1:3, and 1:4 ratios. We talk about the different experiences of the students as they mixed them—*Was it hard or easy? What did you notice?* Then we travel to an existing wall surface and try to apply the three different ratios. The experience of working with the mixes supplies another necessary piece of information—*Was it difficult or easy to trowel? What happened when you added more water?* The upshot of this is that usually every plaster will end up somewhere in this range of relative amounts. Where it lands on that spectrum depends greatly on many factors: the properties of the materials themselves, which can vary depending on their source; the type and state of the substrate (bone-dry cob walls, straw bale walls, and drywall walls will all have different needs in relation to the mix); and the application—*How workable is it? How experienced are the applicators?* And, of course, water content is crucial.

Percentage of Fiber in the Mix

Fiber generally should not be more than 30% of the mix. Experimenting with the amount of fiber in a mix is an important factor in a successful plaster. Too much will make the mix unworkable, while too little will not give it adequate crack resistance. A plaster like the "Steen coat," as has been discussed previously in this chapter, is a very straw-heavy mix with just clay (no sand or lime or anything else) and represents an exception to this guideline, as straw is probably more than 50% of this coat. An application of a ⅛-inch finish coat on a drywall ceiling of a lime-sand coat, however, will require the tiniest fiber available (finely processed manure perhaps? or well-soaked and blended cellulose?) in the lowest amounts possible to prevent cracking but not be visible in the finish (unless that look is desired). Again, fiber amounts will be determined by the needs of your substrate, the thickness of the coat, the stress on the plaster, the desired finish look of the plaster, and the workability of the mix.

Percentage of Additives in the Mix

Additives usually should not be more than 2% to 15% of the mix. Any more than this and the basic strength and structure of your plaster will be compromised. Remember that plaster is, in its essence, that triangle of binder-aggregate-fiber, and additives are included to add properties or elements to this basic architecture, not to replace them.

Whatever you decide your recipe should be, remember to first document well what you have decided to do, then document any changes you make while mixing (i.e., "added another half yogurt container of lime"). In evaluating the results of step 5, knowing the ratio and quality of the sample you are looking at will be essential.

Applying test patches of the mix to a scrap of drywall and letting it dry allows you to evaluate its attributes, and make changes if necessary. PHOTO BY JAN TAYLOR ALLEN.

STEP 5: TESTING

On DVD, Chapter 17, see TESTING PLASTER PATCHES

Test patches are important, unless you are going over a familiar substrate with a tried-and-true recipe with similar ingredients (and even then sometimes we will do test patches). Take the recipe you have decided on during the previous step, or from a common recipe made with your local materials, and express that recipe in "parts." This allows for the mix to be made in any amount, as long as the ratios remain the same. So for test patches, make a small batch and test out your ideas before you bust out a whole 30-gallon tub of the stuff in your mixer, only to realize something isn't quite right.

The substrate for your test patches should ideally be the one you will ultimately plaster on, to achieve

Casein is the additive in this clay plaster. PHOTO BY ACE MCARLETON.

the best sense of how it will work. Make sure to do the patches large enough that you can get a sense of how they will work; usually we do patches of 5 × 5 inches. The test patches should mimic all the stresses you anticipate may happen to the actual coat of plaster, such as going from thick to thin in one patch, to see how the plaster handles the change. Also ensure good suction control by pre-wetting if the substrate warrants this, as well as replicating the curing and drying conditions you anticipate the plaster coat will be exposed to. If it is a clay coat, waiting until it is dry will be essential to seeing how it will behave. If it is a lime coat, one to two days will be minimum to start to see its dry character.

STEP 6: EVALUATION

How are the test patches looking? As mentioned at the end of step 4, good documentation of what the test patches are composed of is essential to replication or adjustment of the plaster. Begin to use your problem-solving tools to figure out why a test patch with a higher binder-to-aggregate ratio might be exhibiting hairline cracking, while another might not—and most importantly, assess how to adjust the mix to minimize undesirable affects. Common things to look for while troubleshooting:

Cracking:

- mix was too wet, substrate too dry
- not enough fiber
- needs more aggregate/too much binder
- needs bigger aggregate

Dusting:

- too much binder
- not a high-quality binder (too much silt in your clay? lime inactive?)
- maybe add a gluey additive (manure, wheat paste, casein)

Weak and grainy:

- not enough binder (sand content too high)
- went on too dry, or substrate too dry

Too rough-looking, not smooth:

- use smaller aggregate for a tighter pack to the aggregate on the surface where you trowel
- use smaller fiber

If the test patches are not showing the properties you want, return to step 1 and work through the steps again, choosing new materials from the natural plasterer's chart.

STEP 7: MIX METHODOLOGY

The order in which materials are added and combined matters quite a bit in plaster mixing, and just like in baking it can make the difference between a successful mix or a sloppy, messy substance that refuses to integrate into a uniform batter.

Clay-Rich Site Soil

Working with clay-rich site soil is one of the rewarding, and time-consuming, challenges of natural building. Being able to take rough, thick chunks of ground and turn them into smooth, creamy plaster that will coat the walls is incredibly gratifying—or frustrating, depending on your perspective. There are definitely some tricks. Clay-rich site soil that is in a big pile, left there by the excavator or dump truck that brought it, or that you are digging out of a nearby bank by hand is most commonly in chunks of various sizes. The clay is acting as a binder, holding the subsoil particles that are not clay (that are sand or silt) together into chunks. These chunks are hard to break up, or should be, if the clay is any good and present in any quantity.

There are three ways to process clay-rich site soil. First, you can process it dry. This means spreading the clay out on tarps in the sun for a long time until it is dry enough to pulverize. For pulverizing, a hammer mill can be used, and you could of course also do this slowly by hand using a tamper or mallet. Pulverizing will yield powdered clay that can be stored in a dry place until it is ready to be used. The powder will more readily combine with water and other ingredients. This method works best where there is enough dry weather to dry the clay for powdering.

The second method is to process it wet. This means making some form of vessel or vessels to soak the clay chunks in water until they start to soften and can be broken apart, then blended into a smooth consistency for use in plaster. We have used pits in the ground lined with plastic, a method wherein people stand knee-deep in clay-slip water and stomp the chunks with their bare feet. Obviously, there are times of year when this is more viable and enjoyable, and there are efficiency sacrifices with this approach. A more mechanized method is to use 30-gallon trash cans, filling them with water and raw site soil. Alternatively, 5-gallon buckets can be great. Either of these, if left to sit for two days to a week, should be blended with a paddle mixer attached to a corded drill. The blended mixture of clay and water is called clay slip—and clay slip is what is incorporated into the mix. The slip can be poured through a screen—hardware cloth works well for this—to isolate and remove any remaining chunks or stones.

Processing site clay can be rewarding. PHOTO BY ACE MCARLETON.

The third strategy is a tricky one, which colleague David Ludt pioneered. Ludt uses the mortar mixer to make the clay clods "open up" into slip and then just adds the rest of the plaster ingredients to it, combining or closely sequencing the two steps so the slip does not have to be made separately, in advance of plaster mixing. The trick is, he uses lime to help do this, so it is essential if you attempt this method that you choose a mix with lime-stabilized clay. The system we use for plastering straw bale, straw-clay, or woodchip-clay wall systems is a base coat of lime-stabilized clay plaster, often with manure added, followed by a finish coat of lime-sand plaster. This method is most often the one we'll choose if we are working with clay-rich site soil. Shovel clay clods into the mixer, in the amount needed for the plaster batch, as the mixer is running. The clods tumble around, tossed by the paddles. Add lime (powdered or lime putty); the amount should be approximately 15% to 20% of the clay amount. Add some water, but not too much.

The idea here is to get the clay to do what potters call "wedging." The clay will immediately physically change appearance when the lime is added and should begin appearing more plastic and creamy. The correct amount of water at this point is the amount that allows for the clay to sound like one big soft wet thud as it is thrown down by the paddles. The outside of the clay mass will be softening and starting to be dragged by the paddles. Sometimes Ludt adds a bucket of sand here, just enough to give the ooey-gooey clay something to push and grind against. The force of the whole, big clay mass being thrown by the paddles softens the clods and begins to "persuade" them to open up. After about five minutes start dribbling in more water, and then keep dribbling until the clay mass starts to open up and loosen even more. Pretty soon the clay-lime is in a thick, creamy glop in the bottom of the mixer—and is about of the consistency of really good thick pudding. Then you are ready to add the rest of the ingredients of the mix. This method is best for rough base coats, as any rocks that were in the soil will remain in the mix.

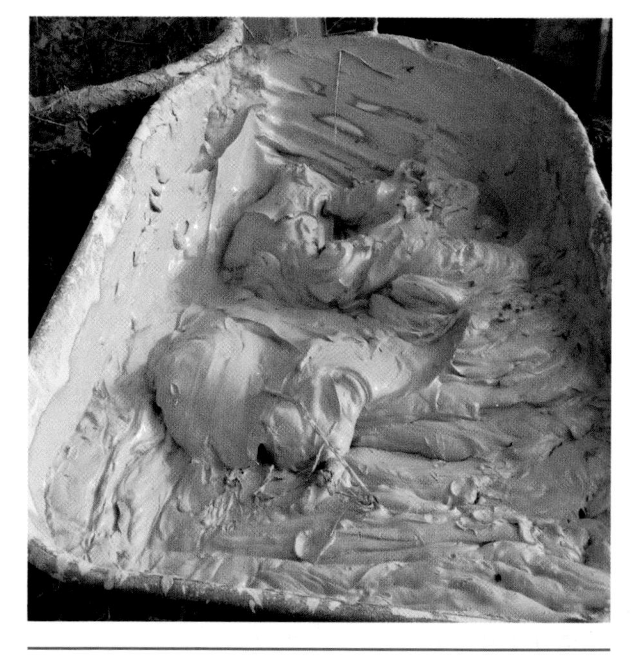

Lime added to clay adds plasticity and strength. PHOTO BY ACE MCARLETON.

Bagged Clay

We have started working much more with bagged clay in recent years. It is predictable and efficient, easy to order, store, and use. Its consistency as a product takes much of the "reinventing the wheel" out of working with different clay site soils. Specifically we have been using ball clay, a type of clay that exhibits strong bonding and resilience and low cracking. It comes in 50-pound bags from ceramic suppliers and can be added to the mortar mixer with water to be easily mixed into the consistency needed for plaster. The dryness of the material is also quite helpful in making a plaster that is consistent in moistness, which is good for application and for minimal cracking in the drying stage. The cost of the bagged clay can often be less than the cost to pay builders to process site soil.

Bagged Dry Lime

There are two ways to add Type S hydrated lime to your mix. The first way, which has rapidly become our favorite, is to mix the dry material into the plaster

Health Considerations

It is appropriate to take a moment to talk about self-health and protection while working with dry, powdered materials. Face masks should be worn to protect from the inhalant risk of dust with both clays and limes. Eye protection is also important. Limes also have the additional risk of being caustic, which can cause burning and irritation to skin, eyes, and mucous membranes with its alkaline nature. Gloves are recommended. If you are exposed, rinsing with white vinegar, a mild acid, is recommended to neutralize the alkalinity.

mix. The water that you add to the plaster becomes the water that the lime soaks in. This way, the soaking time (which for lime is a minimum of one day for it to gain activity) happens in the mix, which we have found yields a much more well-integrated and workable mix. It's akin to making a really good soup, curry, or sauce—the ingredients need to "blend" for a while so they can cease being separate ingredients and become something new: a sum of the parts.

The second way involves soaking the lime by itself in sealed containers. Usually we recommend using a 30-gallon trash can or similar receptacle, and adding first water to the bottom of it, then one bag of lime (carefully). The best way to add the powder without spreading the powder everywhere is to gently submerge one end of the bag in the water in the bottom, then reach your utility knife in and slice it open under the water. This way, all the powder is captured by the water and cannot become airborne.

The powdered lime will mix with the water and "fatten up" and settle to the bottom of the container in a thick, creamy paste (which is what we want for plastering). On top of this will be a layer of extremely clear water. This water is important for protecting the lime putty from reacting prematurely with the air; however, once it is time to access the putty, carefully bail out the water (trying not to mix it back into the putty) to remove the pure putty from the bottom. This is the best and really only way to store hydrated lime long-term, as lime left dry in bags will eventually react with moisture in the air and become inactive. Always try to source the freshest lime possible from your retailer, as quality suffers with age when lime is stored dry in bags.

Lime Putty

Hydrated lime can be acquired in putty form in buckets. Often the quality is higher than that of the partially hydrated, dried bagged lime. This is because lime putty is the result of quicklime being slaked. The bagged lime is this putty partially hydrated, then dried out and put in bags, where it immediately begins reacting with moisture in the air. Therefore the bucket of lime putty is purer and more active. This material is, however, more expensive and harder to source and ship. We have had quite good luck with the quality of bagged Type S hydrated lime for most applications, but for limewashes or applications where quality and strength are paramount, ordering lime putty is sometimes the right call.

Hydraulic lime, quarried where it geologically occurs, is also available for order through distributors in the United States. It tends to be quite expensive; we had one client who bought $19,000 worth of it for his project—whereas clay and hydrated lime would have been a fraction of the cost. There are some in the natural building world who would recommend hydraulic lime be put on bales in a three-coat system, regardless of climatic variations. Hydraulic lime, though definitely of high quality, is not the single answer for all circumstances, and in the Northeast, clay and hydrated lime make an affordable base coat with the best moisture management we know of.

STEP 8: APPLICATION

**On DVD, Chapter 17,
see PLASTERING APPLICATION TECHNIQUE**

There are many possible ways to apply plaster to a substrate. We will focus on plaster application by hand in this section. It is also possible, and at times optimal, to apply plaster to a wall using a pneumatic sprayer. Equipment like the Quikspray Carrousel has been used to good effect by many straw bale builders, especially for applying the thick clay base coat that can quickly tire someone working hard to key it into the straw substrate by hand. However, every approach has its pros and cons, and there are times when the noise and machinery of the tow-behind compressor and hopper/sprayer of the equipment feel overly cumbersome and industrial. And since we are covering all kinds of plastering here—fine finishes on drywall as well as bulk clay coats over straw bale—reviewing hand application is the most relevant and accessible for all situations.

Work Area Setup

Before applying plaster, take the time to set up the space carefully. Here is a space setup checklist:

- walkways clear from clutter to move people and material
- good lighting
- setups at good working height for plaster application; when troweling by hand, it is never comfortable or sustainable to work above your shoulders or below your waist, so have ladders, planks, or scaffolding at the ready—and able to move to adjust height as you work
- have the plaster itself within easy reach for the plasterer, and within easy refill location for the person delivering the plaster; we like to use tables, or boards, set up waist-high where we are working, which allows for us to "chop up" and work the plaster to integrate it before we pick it up and apply it (often we will have a vessel of water with us and will adjust the mix on the table to our preferences)
- tarps, tape, or other site or surrounding-materials protection, as covered in the substrate preparation section earlier
- if working with lime, evaluate the weather and make sure to hang tarps to protect new plaster from harsh sun or wind; burlap can be suspended from eaves and dampened with a hose daily to help create the necessary conditions for lime to optimally cure

Plastering with Hands

This is a truly enjoyable experience to try at least once. This method is best for clay plasters with little to no lime content, and best for rough coat work most commonly found on natural wall systems; pushing the material into the substrate with the heel of your hand is quite effective and satisfying. You can also try "harling," a term used to describe the process of taking a good handful of plaster and lobbing or pitching it underhand at the wall surface. Clumps of well-made plaster will stick to the slip-coated wall with ease; this is also sometimes used as a test to make sure there is adequate clay content in a plaster. Like with any method we have described, there are pros and cons to this approach. Applying with hands is great for people new to plastering, or who don't yet have the control necessary to use trowels. It is also more useful than trowels if the wall is quite lumpy.

Setting up the plasterers on scaffolding, with plaster at an ergonomic height and at an easy reach, greatly increases efficiency and comfort. PHOTO BY ACE MCARLETON.

Trowels and Tools

There are many different types of trowels and tools available for many different types of work. Here are the basics, at a glance:

Flat square/rectangular trowels:

- These are good for laying up and troweling on rough work.
- Check out the "tang" or rib of steel along the back of the trowel that the handle attaches to. Flat square/rectangular trowels typically have

An assortment of trowels will tackle any job. PHOTO BY JULIE KROUSE.

long tangs that run along the back side of the trowel all the way along the steel to provide strength and rigidity.

- These are good when paired with a pool float for two-handed finish-coat applications. The square trowel does the heavy lifting, adhesion, and spreading, while the pool float smoothes and integrates the surface to the desired look.

Pool floats:

- Oval shaped, they are good for when edges digging in is a bad thing, such as finish coats.
- They are more supple and flexible than the square flat trowels.
- The tang typically stops in the center of the back of the trowel and does not run all the way out to support the edges.
- They are great for smoothing.

Japanese trowels:

- These come in assorted shapes and sizes.
- Larger, hard steel ones are good for laying up.
- Flexible steel trowels are good for finishing, and for making curves without digging in.
- Small, specialty flexible trowels are good for niches and sculptural work.

Wood floats:

- They are generally rectangular.
- These come in assorted sizes.
- They are used to compact plasters after they are up, specifically lime plaster finish coats, to tighten pore spaces and create greater resistance to weather.
- They are also used to scour lime plaster, to create a rougher texture to the plaster to also create greater weather resistance.

Hawk:

- This tool is used in traditional plastering or in mudding and taping for drywall.
- It has a flat plane with a handle to stage plaster, while holding a trowel in the other hand.
- Using a hawk is ideal for hard-to-reach areas where you want to take a good amount of plaster with you.

Scratch tool:

- This tool is used for scratching gouges in the plaster while it's still wet.
- It provides key for the subsequent coat, which is essential to stability, strength, and performance of the plaster system.
- It should be used at a low angle to the wall to ensure deep, even grooves that minimally tear the surface of the plaster.

Most of these things are easily acquired at a building supply house in the cement finishing area. Japanese trowels are a special order: LanderLand (www.landerland.com) and Lee Valley (www.leevalley.com) are two good sources for Japanese plastering trowels and hawks.

Trowel Skills

Hand-troweled application of plasters is a whole other world of experience and is difficult to articulate. When we teach plaster classes, we have taken to asking the class to watch us plaster for a while because this seems to be the best way to learn (and is the way that we learned as well, by watching others' technique). There are so many variations of application, depending on body type, training, and even personality. Here are some basic guidelines to follow:

- Plaster must be pushed into the wall with force to ensure a good connection. Air pockets or loosely adhered areas will cause major failures.
- Set up your body so you can use good body mechanics to support delivering the force with a minimum of injury and stress. Use major muscle groups: your legs, back, and shoulders, rather than just your forearms or wrists.
- Work top to bottom, especially when doing finishes (so you don't drop on your work).
- Start at the edges to define the area you will work and the plane, and work into the fields.

After having tried many methods of hand application, we feel strongly that the two-trowel method is the best: best for the body of the plasterer, best for force, and best for efficiency. Our colleague David Ludt discovered it one day when working by himself and his dominant arm started hurting. He set himself up with a board at waist height, took up a trowel in his nondominant hand, and started challenging himself to use this arm just as much as his other. And so the two-trowel method was born (as we call it in our area, and now teach students from across the United States and around the world). The beauty of this method is that it focuses on ergonomics. The heavy mud is easily within reach, at a comfortable height, and it is easy to pick up. You can alternate between picking up and laying up with one arm and then trowel out and integrate with the other. In doing finish work, you can lay up with a more rigid square trowel, and then immediately work out the lines from the edges with a more flexible finish trowel in your other hand. It does depend on a good setup, and attention to moving your setup as you move along your wall can feel annoying—until you realize you are flying when you are comfortable, using your whole body, and having just what you need, where you need it. This is the method used by traditional plasterers in years past—another example of how the natural building renaissance is both rediscovering and updating the use of classical methods.

Other tips include:

- Put a good amount of plaster on, enough to cover an area in front of you that feels workable. Usually it's about 3½ feet by 3½ feet. Don't try to lay up just small amounts of plaster at a time and overwork them. A common beginner's mistake, this will, first, take an exceedingly long time and, second, yield poor results.

Two-handed troweling is efficient and saves the body from imbalanced strain. PHOTO BY JULIE KROUSE.

- Don't think you have to make it look perfect immediately—one key to efficient plastering is to get enough up on the wall so that you can move it around. Once it's up there, you can use your trowel and the force of your body to lean in and move the material, integrating it and adjusting for thickness.

- You can adjust for thickness—if you want more in some areas and less in others—by controlling your trowel angle and amount of pressure. A bigger angle of trowel to wall, with a good amount of force, will pull up and drag more material away. A smaller angle of trowel to wall will do the opposite. It will tend to leave more material where it is and to press it more firmly into the wall and compact it. Using the edge and pressure of your trowel in this way, you can quickly and efficiently integrate a field of plaster into a plane, shape, and thickness you are happy with.

- One of the most fun ways to use the edge and pressure of your trowel to create the shapes you want is with curves, such as on window reveals.

- If plastering above windows onto metal lath, or really anytime going over lath, make a special plaster mix with either no fiber or small enough fibers to not get caught on the face of the lath. The lath itself is acting as fiber support, and the long straws will just be an element to fight, and will prevent good key.

STEP 9: AFTERCARE

Clay plasters, to review, dry out mechanically. They are dry when the water has completely gone. They do not perform best if allowed to dry out too quickly, however. Therefore, ensure that your new walls are protected from direct sun and wind until they are fully set up.

It is important to note that lime-sand plaster put on as a finish coat on natural wall systems will also cure best if it is protected from wind, sun, and direct water for a minimum of a week to two weeks—ideally a month in damp conditions above 50°F. That means dampening the coat, after it is applied, for one to two weeks (so leave the staging up if it was used). If plastering on drywall, this much dampening is not necessary, but light misting for one to three days is recommended. While perhaps more of an application technique than aftercare, lime-sand finish coats benefit from being compacted after they are applied to the wall. This process, achieved using a wood float, is done after the plaster has set up enough to be firm but is still workable at the surface. The idea is to scour the wall with the float, applying moderate pressure using a circular pattern, thereby compacting the material on the wall. This serves to reduce pore space, which reduces moisture absorption from wetting, reduces cracking from void spaces, and improves the strength of the finish. The finish can then be hard-troweled with a pool float, if desired, although multiple applications of limewash will smooth out the finish.

We do not consider lime-sand plaster fully finished until it receives one to three coats of limewash. Limewash helps fill in the pores between the sand grains of the plaster and acts as the sacrificial coat to the weather. For more on limewashes and other paints, topcoats, and washes for natural plasters, see chapter 20.

This fresh lime-sand plaster is gray owing to the color of the sand; a limewash will brighten it to bright white. PHOTO BY ACE MCARLETON.

CHAPTER 18

Roofs for Natural Buildings

It could well be argued that the most important part of a building is the roof. After all, a structure with a roof but no walls is still a structure—be it a band shell, gazebo, or tropical bungalow. Walls without a roof, however, are simply a landscape feature. Roofs do more than provide definition to purpose and form, of course; their primary function—keeping out the weather—is of particular importance in climates where there is a great abundance of precipitation. Depending on the design, another key role of the roof assembly can be to help control the interior climate. In a warmer climate, this might mean keeping the sun from overheating a space, through reflective roofing materials or the targeted use of mass and insulation. In a cold climate, the demand is opposite—heat must be kept in; see chapter 7 for a discussion on the importance of a well-insulated roof in the function of the structure. It should be noted here, however, that these functions are roles of the "roof assembly," or "roofing system," not of the roof itself. The roof is the part of the structure one might see from the outside—in most cases—which sheds the rain. A roof assembly or system, on the other hand, includes not only the roof, but the underlayment the roof rests on, any ventilation components for the roof (more on that later), structural framing for the roof, and any insulation that is involved with this assembly. Insulation is not always a component of the roof assembly; for example, if a building has a cold attic with an insulated second-story ceiling, there would be no need for insulation in the roof assembly. In this chapter, we will look at both roofs and roof assemblies to get an idea of how these components can best be designed in natural buildings. We will begin with a broad look at roof-assembly strategies, and then move on to specific types of roofing to integrate with these assemblies.

Roof-Assembly Strategies

Roof assemblies can be classified in two ways: vented and unvented. The vent in question here is a channel of air that is directed immediately under the roofing—or in most cases, under the roof sheathing. This vent is located above, or on the "cold side," of the insulation, if insulation plays a role in the roof assembly. The vent must have a direct air inlet at the bottom of the roof—most commonly found at the soffit—as well as at the top of the assembly, either through the ridge or through gable-end vents. The purpose of this vent is to capture any heat that may escape from the interior of the building, migrating up through gaps, voids, or other defects in the insulation, and carry it out through the ridge or gable-end vents before it can warm the underside of the roof and roof sheathing. The vent also provides a pathway for moisture in the roof assembly to dry out.

As we learned in chapter 7, batt insulation placed between rafters—the most common form of roof insulation—gives opportunity for heat loss not only through the thermal bridging of the rafters themselves, but also through the inevitable flaws in installation, or gaps formed when green (wet) rafters dry and shrink away from the installed batts. Why is warming the underside of the roof a problem? The answer can be

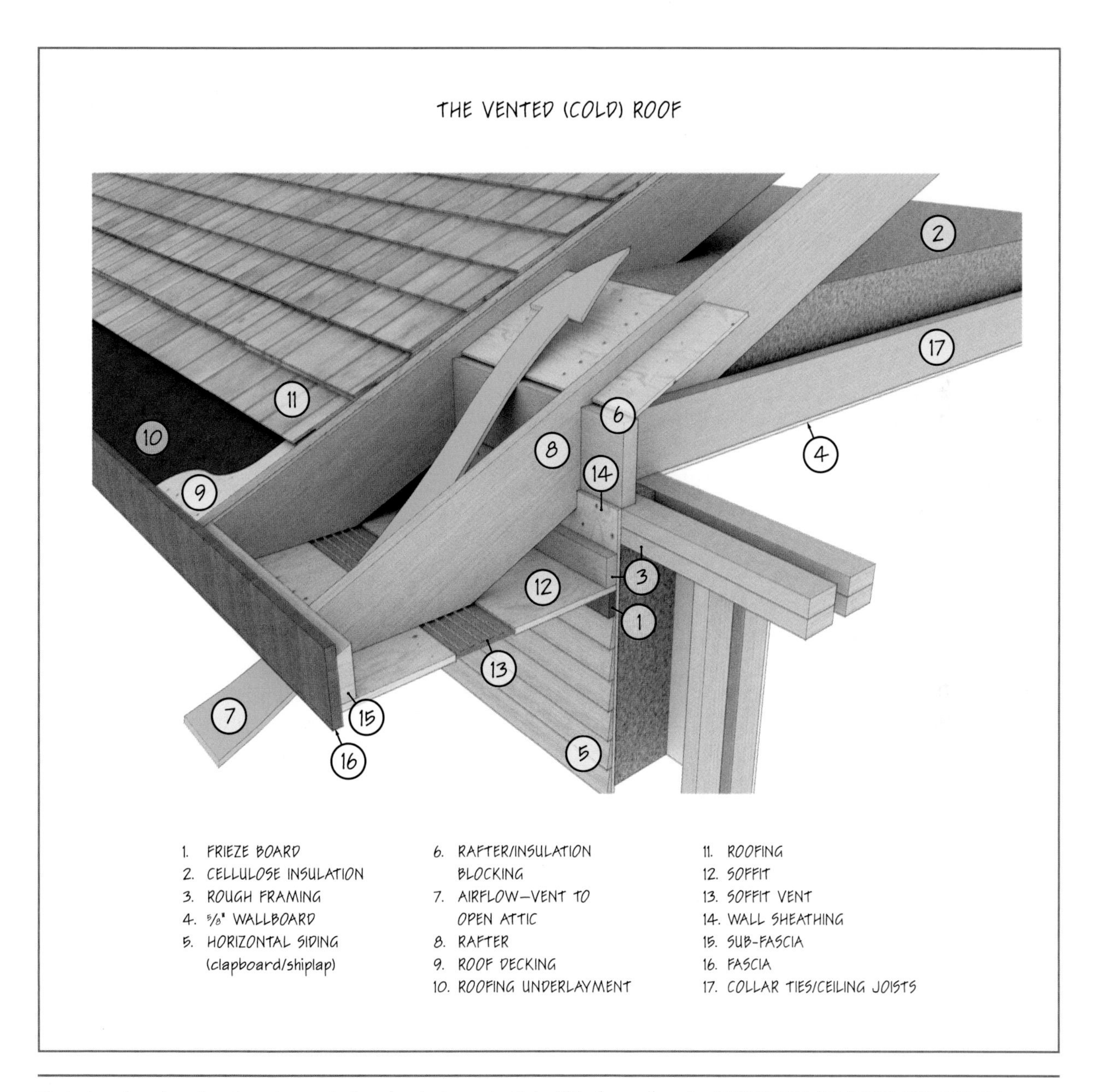

Vented roofs reduce ice damming and allow for drying potential within the roof cavity. ILLUSTRATION BY AARON WESTGATE/FREEFLOW STUDIOS.

found in the dangerous and damaging ice dams found along the eaves of many heated buildings throughout cold climates. They are dangerous for the massive icicles that form, frequently over walkways, parking areas, and entryways. They are damaging because of the drainage block they create, causing melting water to back up under shingles, flashing, or other cracks and crevices that may be found in the roof and create a moisture problem—or catastrophe—inside the building.

How does this happen? Snow that has built up on the roof will start to melt into liquid water on the surface of the warmed roof. This water then runs down the roof, only to refreeze once it reaches the unheated eaves of the building, forming icicles and ice dams. In the vented strategy, escaped heat from the interior is ducted safely away, keeping the snow from melting steadily from below. This is why vented roofs are frequently called "cold roofs."

Using the correct type of vent is a critical part of the design. Most commercially available vents are weak, thin pieces of Styrofoam that are stapled to the sides of rafters or underside of the sheathing. These are designed to be used with puffy batts of insulation, gently yet firmly placed by hand. These same vents get smashed, torn, compressed, collapsed, and otherwise deformed by the snaking of hoses and pressure of densely packed cellulose; remember, a proper dense-pack installation can have enough force to blow drywall off studs if it's not well attached. Every one of these deformities will compromise or outright disrupt the vent channel under the roof, turning your vented roof into an unvented roof. In this case, a better choice would be to use a more durable vent, such as one made of a more rigid and durable plastic.

One can also make vents on-site, using 1 × 2s for side nailers and plywood or similar solid material as the vent structure. Some builders will use 2-inch foam to get a little extra insulation out of the deal at the same time. Regardless, the vents must serve the purpose of keeping an uninterrupted air channel in place above the insulation, holding the insulation away from the sheathing by an inch or two.

An additional venting option is to "double-sheathe" the roof. In this case, the insulation fills the cavity all the way up to below the primary sheathing. Rather than place a vent in that cavity, 1 × 3 furring strips are laid vertically on top of the primary sheathing; these strips are forming the vent cavities and must be open to both the soffit and the ridge or gable vents. Secondary sheathing is then placed on top of these furring strips, and the roof is finished out from there (underlayment, flashing, roofing, etc.).

An unvented roof addresses the issue of ice damming by relying on the integrity of the thermal envelope to contain the heat within the building. One example of such a superinsulated roof is a SIP roof, in which foam-core panels are laid atop the roof framing, with the panel gaps sealed with spray-applied foam. This is commonly done over timber-frame structures with widely spaced timbered rafters, which may also be receiving SIP wall panels.

As we've already discussed, however, the disadvantages of foam are serious enough to steer us as natural builders elsewhere for insulative solutions. Double-rafter systems, either as rafter trusses or as cross-rafter framing, can create very deep cavities with a minimum of thermal bridging, allowing an installation of dense-blown cellulose to complete a very sound thermal envelope to help keep heat from leaking up to the roofing. One must be very, very confident with the insulation strategy—and its execution—in the roof assembly to rely upon an unvented roof. While venting a roof can add more expense and difficulty in construction, a failed unvented roof results in expense and difficulties many times greater in order of magnitude.

Our favorite strategy is to combine the above two approaches. It is foolish to design for insulation failure and heat loss and rely solely on a vent, however reliable it may be, when there are good options out there for a very efficient roof assembly. We have used both of the double-rafter assemblies mentioned above, and they can be built from relatively small-dimension, locally milled framing stock; we've even used reclaimed plywood for roof truss gussets. Dense-blown cellulose is our favorite roof insulation material; as mentioned in chapter 15, cellulose behaves very similarly to straw in its relationship to moisture while offering the important benefit of its flexible

Foam-panel roofs are frequently unvented, or "warm," roofs that rely on their high performance to avoid ice damming. PHOTO BY ACE MCARLETON.

form, which allows for a thorough and complete fill in the roof assembly. However, installing a roof vent above even such a solid insulation envelope as the ones described is cheap insurance. Though it can be difficult to vent some roofs—shed roofs coming off a vertical wall, for example, as in a lower clerestory roof or the roof below a wide shed dormer—whenever possible, a vented roof provides redundancy in a place where the margins of error are slim and the results of even a small error, over time, can create significant, widespread damage.

INNOVATIVE ROOF SYSTEMS

We have been steadily working on improving the roof design of our buildings. The standard roof with 12-inch rafters does not provide enough space for insulation if a superinsulated structure is the priority, and it does not answer the issue of thermal bridging from the rafters. (See chapter 7 for more information on the importance of insulating the roof.) A common improvement to this situation is to install a layer of 2-inch foam to the underside of the rafters, raising the R-value by 10 and isolating conductive heat losses. For a simple and low-cost roof, or a renovation situation, this can be an attractive option, but it involves the use of large amounts of foam, something we are trying hard to avoid. The next step up from that is the installation of a SIP roof (see beginning of this chapter), which is a very common approach for natural buildings in the Northeast owing to its speed of construction and high performance. However, we are now not only using even more foam to achieve the performance goal, but potentially creating a situation in which two fixed-form insulation planes (foam panels and straw bale walls) must meet, in a place where the pressures are greatest on convective heat loss in the building (again, see chapter 7). While this issue may be somewhat relieved using straw-clay or woodchip-clay walls, it is still a difficult transition to make. Accordingly, we have been working on different roof designs that utilize dense-pack blown cellulose as the insulation at a minimum thickness of 16 inches with no continuous framing through the building.

One approach is to build a "cross-framed" or "lattice-framed" roof, which has been popularized in our region by Josh Jackson of TimberHomes in Vershire, Vermont. We use rough-cut 2 × 8s as the primary rafters, as one would find in standard rafter framing. This dimension is plenty strong to carry all necessary roof loads for most designs, as well as support a generous eave-end cantilever over thick walls. Next, 2 × 8 "purlins" are run perpendicular to the rafters, or parallel to the ridge, at even spacing up the rafter, creating the lattice form. The purlins are installed primarily with 10-inch TimberLoks pilot-screwed through the junction between the two members, although toe-screws are occasionally used as well. The purlins give the strong advantage of cantilevering off the gable end to catch the overhang, a frequently challenging framing situation in bale-wrapped timber frames when upwards of 48 inches of cantilever needs to be achieved (or designed around!). The roof is then sheathed with long-length (12-foot minimum) 1 × 8 rough-cut boards, tying the purlins together and keeping them from rocking downhill. Metal strapping can be used running over the gable to further support gravity's pull on the purlins, especially on a steep-pitched roof, although we haven't found this to be necessary. This provides a nominal 16-inch-thick cavity (when using rough-cut 2 × 8s) and thermal bridging only at the cross-points

This lattice roof frame creates a 16-inch cavity using 8-inch-deep rafters and purlins with minimal thermal bridging. PHOTO BY ACE MCARLETON.

of the rafters and purlins. In practice, we find ourselves needing to put a fair amount of blocking between the purlins for stability during construction, which increases conductive losses. While this is a very successful roof in terms of achieving its design goals, it uses quite a lot of wood and is especially time-consuming to build when dormers—especially gable dormers—are involved.

Another approach we have used is to site-build mini roof trusses, using 2 × 6s for the upper chords, 2 × 4s for the lower chords, and reclaimed plywood gussets. With Ben Graham of Natural Design/Build of Plainfield, Vermont, we were able to design these "trusses" such that the upper chords were fully structural and did not rely on the entire unit for structural success; were the loads greater, we would need to either up-size the top chords or have the trusses professionally engineered and designed. This would have added significant cost to the project but may well have saved significant time (and therefore balanced out cost) were we to simply order custom-built trusses, especially for a larger project. This particular design was for a clerestory roof and required double birdsmouths between both upper and lower chords of the truss, which meant that it was difficult to achieve a consistent plane for sheathing, and a lot of shims under the strapping for the drywall ceiling—although improvements in the design and execution, which was done largely by students, could have alleviated this problem (as could having the trusses built to spec in a factory!). Despite those inaccuracies, however, we were able to design a clerestory roof, incorporating a double-stud clerestory wall, that was nearly free of thermal bridging, achieved near R-60 insulation value, and was built out of locally sawn lumber and reclaimed plywood; overall, it was a very satisfactory experience.

We plan on trying a variation on this theme: framing up standard rafters of the minimum-required dimension, enclosing the structure, and then, after

Double-rafter framing provides another opportunity for a deep-cavity roof built with small-diameter rafters with minimal thermal bridging. PHOTO BY ACE MCARLETON.

the walls are closed in, hanging a drop ceiling of sorts with lower rafters in parallel across the building. This would maintain the benefits of the systems discussed here already (deep cavity, minimal thermal bridging, local and accessible materials), while simplifying and speeding up initial roof construction and allowing the "lower rafters" to be hung to a string line, avoiding the need for strapping the ceiling before sheetrocking (or shimming laths, if lath-and-plastering). While it is safe to say that such an approach has been attempted before by a pioneering builder, refinements of this approach may prove to yield the next evolution of the innovative roof design.

On DVD, Chapter 18, see DOUBLE-RAFTER FRAMING

RAINWATER CATCHMENT AND ROOFS

Another consideration in designing the roof assembly is creating a catchment surface for rainwater harvesting. As water scarcity becomes a greater and greater threat to both ecological and human communities, reducing water consumption and improving the efficiency of water sourcing are pressing needs to be addressed by the way we design, build, and use our structures. Rainwater harvesting can be used to supply water for agricultural or landscaping uses, domestic non-potable uses, potable freshwater supply, or a combination of the three. It is important to identify what the intended use of the water will be, as this will affect the suitability of any given catchment surface—the roof—to maintain the appropriate level of contamination that may be sourced from the surface itself. While rainwater catchment is a controversial topic and may not be an option for you—it is prohibited in some states and permitted in others—administrative hurdles can be expected to lessen as issues of water scarcity gain more awareness. The topic of rainwater catchment is too involved to be covered here in full—see the reading list in appendix A for more—but the appropriateness for rainwater catchment of each of the roofing materials described in the chapter will be briefly addressed here.

Roofing Materials

There are many different types of roofing materials that can be used on top of both vented and unvented roof assemblies. The conventional approach found on most residences in America is the asphalt shingle: petrochemical in origin, currently cheap to make, easy to install, and relegated to the landfill after twenty to fifty years of service. There are both "cradle" and "grave" issues staring us in the face with this material, so we will examine a variety of different alternative roofing materials, organized by their material of origin—some of which may be more "natural" in their origin, and some more easily categorized as "green" building materials. Our conviction as natural builders is that natural building techniques must be integrated with green building technologies to achieve the goals of impact remediation and regenerative design; this is the core of the integrative design/build approach that we advocate. As we show in this section, roofing materials provide ample opportunity to explore this edge.

STONE

Stone roofs are some of the most resilient, durable, beautiful, natural, and, in the best of cases, local roofs available. They are also, perhaps unsurprisingly, some of the most expensive. Slate is the most ubiquitous stone roofing material in New England. Quarrying roofing-grade slates is an expensive process; slate deposits tend to be geo-specific, and this makes for a specialty, supply-limited product. There is also a lot of waste in cleaving slate shingles of just the right thickness, size, and shape to be used as roofing material. The labor in installation is another liability, as the art and skill of slating is one that has fallen to a select few craftspeople; again, it is a specialty supply-limited trade that commands a specialty rate. A strong roof assembly is necessary to support slate shingles. If one can ensure the structural stability and increased budget, a slate roof is an investment in beauty, quality, and ecological harmony that will pay back for many, many decades, if not centuries. Quarrying of slate generally does not require deep underground intensive mining and has been integrated in a scale-appropriate manner in many slate-rich communities for hundreds of years. To that effect, the geo-specific nature of the material greatly supports a connection to place and use of a localized, indigenous material. Slate roofs make acceptable catchment surfaces for rainwater harvesting.

Slate is a time-tested, beautiful, heavy, and expensive natural roofing option. PHOTO BY ACE MCARLETON.

WOOD

Cedar shakes are perhaps the most commonly found of wood roofing materials. Red cedar is particularly favorable, as it boasts a higher rot resistance than other cedars, although eastern white cedar has been used as a local equivalent. In contrast to slate roofs, the lightweight nature of a wood-shingled roof is a benefit to the building and the builder alike. Naturally occurring chemicals in the wood—thujaplicins, lignans, terpenes, and plicatic acid—provide cedar's rot resistance. Cedar splits more evenly and is more abundant than other rot-resistant species, such as black locust and tamarack, and is therefore found more frequently as roofing material. There is certainly an increased risk of fire from having a wooden roof; if you live in a fire-prone area, or one in which building codes specifically legislate against wooden roofs, this may not be a suitable option.

Longevity is another liability to the use of wood; care must be taken to provide a vent space below the shakes to allow them to "breathe" and to fully dry out after being saturated by a heavy rain. Even then, cedar shakes will not last as long as slate; depending on exposure and installation details, the life expectancy of cedar shingles ranges from fifteen to forty years. But given the rate of regeneration of cedar trees and the ease of their repurposing as biomass for heating or landscaping, the increased rate

Cedar shakes bear the mark of a forested region. PHOTO BY ACE MCARLETON.

of replacement can be justified if the labor variable of this equation can be satisfied. This justification, and the initial use of cedar shakes as a roofing material, is contingent upon the responsible and sustainable forest management and harvesting practices of the manufacturer. Please see chapter 13 for discussion of sustainable wood production. While untreated wood shakes are suitable for rainwater harvesting, especially for non-potable uses, treated wood should be avoided, especially for potable use.

EARTH

Given what we know of the properties of earth, and in particular clay, it can be understood why "raw" earthen roofs might be successful in a hot, dry climate, where the mass of earth can help buffer indoor temperature swings, and clay's relationship with water can be kept in balance for longevity with low-frequency high-volume moisture events. In a cold climate, however, where the effects of mass are less relevant and the prevalence of moisture is often much higher, "raw" earth becomes less appropriate. In some regions "cooked" earth may well be a viable alternative. By "cooked" we of course mean "fired," as in tile. Vitrified clay boasts significantly higher moisture repellency and can readily be formed into convenient forms for installation. Clay tiles, known for their prevalent and historical use in the countries of the Mediterranean, and later Mesoamerican and Caribbean regions as a result of Spanish colonialism, are now used worldwide, with supply in the United States provided by importers and domestic manufacturers alike.

While weight issues similar to those of slate can be a factor, there are some manufacturers offering lighter-weight versions of traditional clay tile products. There are many styles and colors available in clay tile; most common are the red-toned barrel-shaped tiles, although a relatively wide palette of earth-toned hues in a variety of designs (including barrel, S-shape, and flat roof tiles) is now offered by many manufacturers. Clay, as we now well know, is about as benign a substance as can be found in the building industry—dirt cheap, widely available, often a by-product of other mining or manufacturing operations, relatively nontoxic, and easy to reintegrate into the environment upon disposal. Cost is certainly higher compared to an asphalt shingle, and aesthetic considerations would need to be weighed if vernacular precedence was a component of the design program, as there is not a lot of historical use of clay tile in northern cold climates. That said, availability, track record, and design flexibility all point toward clay tiles as a favorable option for roofing. Like slate, earth tiles are a fine collection surface for rainwater harvesting.

PLANT

It may seem incongruous that plant-based roofing materials would be at all effective. However, certain applications of plants in roof assemblies have proven themselves in both historic and modern contexts. Thatch is perhaps the best example of a natural building approach that includes plants as a roofing material; it is regenerative, adaptable to a wide variety of climates, and low-tech, has a low carbon and energy footprint, returns easily to the earth, and is traditionally highly localized. Thatched roofs have been used for hundreds, if not thousands, of years, and the material used varies widely with the location in which it is applied. It was utilized in the colonial Northeast until stone and wood replaced it. In the tropics, palm leaves, sugarcane leaves, reeds, and native grasses are all frequently used; there, thatched roofs serve primarily as shade canopies, although water-repellent qualities can also be found with some species used in certain applications. Temperate thatch is of more relevance to us in moderate and cold climates; here, the water reed phragmites is favored for its water repellency and longevity, and the reed can be found widely throughout North America and Europe.

The basic procedure for thatching a house is to harvest sheaves of the plant species of choice and tie them in large bundles. These bundles are then laid in rows across horizontal purlins that have been affixed perpendicular to the rafters. The bundles are then cut

Thatching is a lost art in the United States, although it is still practiced regularly in many parts of the world. PHOTO BY BEN GRAHAM.

and spread across the roof, starting from the eaves and working up to the gables, secured with another run of strapping near the tops of the row of thatch, and sewn through to the purlins below. The next row of thatch overlaps the lower row, shingle-style, covering up the strapping and sewing. When done correctly and with proper maintenance, a temperate thatched roof can last upwards of sixty years when the best-quality materials are used, providing both insulation and adequate protection from rain and snow. The primary form of maintenance comes in replacing the cap over the ridge, which is the most vulnerable and wears the fastest, needing replacement every ten to fifteen years. Thatch has liabilities of fire susceptibility, vulnerability to insect damage, and occasional unecological harvesting practices.

Living roofs, or green roofs, have been gaining tremendous popularity among the green building community, particularly in urban commercial settings. Living roofs refer to roof systems that feature a planting of vegetation integrated directly into the roof assembly. There are many different takes on living-roof design, and many companies selling living-roof systems, each with its own proprietary materials and combination of features. In essence, though, a living roof consists of the following components: a roof structure, sheathing, a waterproof protective membrane, a root barrier to prevent failure of the membrane, a growing medium, and finally the plants themselves.

The roof structure may well need to be built to support a heavier-than-average dead load, depending on the weight of the other components of the system. While frequently found on low-pitched or flat roofs, there have been many successful installations on steeper-pitched roof structures, or even vertical surfaces ("living walls").

The waterproof membrane, on top of the sheathing, is the true "roofing material" in the system. It is this membrane that does the real work of keeping moisture out of the building; the plants and growing medium are in fact inclined to keep moisture around to help support the biology of the roof. If nothing else were present, this membrane would in and of itself constitute the roofing, perhaps with a protective layer of gravel. This membrane can be a roller- or spray-applied material, but it is more often an impervious solid sheet, such as EPDM (ethylene propylene diene monomer) or bituthene.

The root barrier is generally a synthetic material, similar to landscaping fabric. The growing medium is usually soil, but lightweight synthetic mediums are often used as well. A nonpetroleum material that can be used as a root barrier is metal roofing; if the roof is of a steep enough pitch, this roofing—when properly installed—can also be used as the roofing membrane, further reducing the use of petrochemical materials. In this case, a terraced application of the plants in their growing medium would be required.

Finally, the plants enter the picture (and may be the first time a "natural" building material is introduced to the roofing assembly, depending on the roof framing material). While many different plants are used for different systems, of particular favor are shallow-rooted, drought-resistant varieties, such as sedges. Plants with taproots or deep-rooting habits run the risk of interfering with the waterproof membrane below. Because of the relatively low water-retention properties of the growing medium (as compared to the topsoil of the earth's crust), even in moist climates the plants run a greater risk of drying out between rain events as the moisture leaves the system quickly.

Living roofs have many benefits, which explains why they have gained popularity in the last decade as green technologies have been embraced by the marketplace and industry. One of their greatest assets is their ability to mitigate storm-water runoff, helping to slow down the flow of water from buildings into overwhelmed sewer systems during times of intense rain. Another benefit comes from the added green space, in which the roof may be compensating for the reduction of green space that occurred during development of the building site. Associated advantages here include a carbon sink to sequester carbon dioxide out of the atmosphere, increased air quality, evaporative cooling through plant transpiration, habitat and oasis for migratory and nonmigratory birds, insects, and other creatures, and reduced temperatures in urban environments (if a large area of roof space is planted with living material).

This last asset addresses the "heat island effect" that refers to a large geographic area of a city being much hotter than the region in general, which in turn can dramatically affect the energy used by buildings throughout the city to keep the interior climates at a comfortable temperature. As living roofs cool the building through evapotranspiration, and as a collection of green roofs lower the temperatures of a heat island in part or whole, all the benefits of improved energy efficiency are realized, including reduced carbon emissions, resource depletion, and operating costs (U.S. EPA n.d.). Catchment from living roofs can also be used for landscaping and agricultural uses around the building, depending on the materials used in the system, although is not as well suited as other roofing systems for potable water catchment due to contamination from the various system components.

Living roofs are effective in managing water runoff, reducing heating of the building, providing green space and habitat, and helping a building integrate visually into its site. PHOTO BY ACE MCARLETON.

It should be noted that most if not all of these benefits are realized in urban or suburban environments. A building in a rural environment generally does not pose a threat to the local ecology through reduced green space, nor are storm-water issues generally a concern. Heat islands are certainly not an issue, and while cooling a building passively is always helpful, for those of us in cold climates this may not be nearly as much of a concern as keeping it warm. Given the high level of dependence on synthetic materials—again, some systems may feature the plants themselves as the only natural materials in the entire system—and high cost of installation, whether the benefits offered by this green technology are best realized in a given building application should be carefully evaluated.

MODERN MATERIALS

There are many other green roofing options available if none of the previously mentioned options quite fits your need or budget. While these modern, technologically developed products and systems are certainly more "green" than "natural" in isolation, when integrated into the structure they can well support the mission and program of the natural building. In that the most ubiquitous material for roofing is the asphalt shingle, as noted earlier, many of these products seek to improve upon or replace the shingle. There are now many shingles made from postindustrial or postconsumer waste, reducing the embodied energy and ecological footprint of their manufacturing while not straying too far from the aesthetics and logistics of installation. Old tires and milk jugs are two common waste products being repurposed as roofing materials; there are even plastic composite shingles styled to look like slates for a lightweight, long-lasting, easy-to-install, economical roof. Verify with the manufacturer to ensure suitability for intended use in a rainwater catchment system.

Bark Houses

Another plant-based material worth mentioning is bark shingles. The use of tree bark for building protection was well known to Native American communities, and the bark of the chestnut tree was particularly favored; bark houses enjoyed considerable popularity in the 1920s and 1930s, when poplar bark was also used. After the chestnut tree was largely wiped out by blight in the mid-1900s, use of bark for roofing material faded into obscurity. Today, manufacturers such as Highland Craftsmen are creating poplar bark shingles, under the trade name Bark House, for use as house siding (not as a roofing material). Bark shingles are treated primarily through heat, similar to charring wood for use below grade; the sugars in the bark, which are the primary foodstuffs of both macro- and microbiological activity, are destroyed, along with any existing biological life; without need for further treatment, it is claimed that bark shingles can last upwards of seventy-five years, although they are significantly more expensive than cedar shakes. While not suitable for a long-lasting roofing material, this product represents an interesting plant-based option for wall protection.

Building-integrated photovoltaics, or BIPV, are becoming a popular strategy for those concerned with the energy footprint of a structure or designing for net-zero energy rating. While people have been mounting solar panels on roofs for decades, today's owner can now choose from a selection of "solar shingles," combining the features of roofing product and electricity generator in one. Given the proper design of roof pitch, roof profile, and building siting, this can be an attractive option for a roofing shingle. Another popular form of roof-integrated PV is the laminated PV panel; designed to fit in the panels of a standing-seam metal roof (more on metal roofs below), these peel-and-stick PV modules simply roll out and adhere permanently to the roof and are wired through the ridge.

Metal roofs are a departure from the shingle approach yet offer many benefits as a roofing material. In fact, metal is our material of choice for the combination of function, flexibility of style, price, and performance. Screw-down metal roofs feature the use of neoprene-gasketed screws that simply screw sheets of recycled-content aluminum or steel onto sheathing or strapping. Although tricky to lay out and execute professionally, this is a simple procedure requiring little more than a cordless drill-driver, a predrill bit, and a tape measure. Because of their strength, in many cases full-coverage sheathing can be replaced with skip-sheathing, or even strapping; it is not uncommon, when the design allows, for us to simply strap rafters with rough-cut 1 × 4s placed 24 inches on center, saving a significant amount of both money and plywood. Metal roofing comes in a wide variety of stock colors—most manufacturers will even color-match—and all come in a selection of different rib profiles. The roofs are long lasting—generally on the order of thirty to fifty years or longer—and when removed can easily be recycled. Screw-down metal roofs are one of the most common roofs topping natural buildings in the Northeast, especially those of simpler design or needing to fit within a tighter budget.

Standing-seam roofing is a step up in quality from the screw-down version and is the other most common roof used on natural buildings in our region. With

On DVD, Chapter 18,
see SCREW-DOWN METAL ROOFING

its sleek appearance and hidden fasteners, standing-seam roofs offer not only a cleaner look but improved performance and reduced chance of leakage, as the fasteners are all hidden from rain and snow. Although there are some do-it-yourself standing-seam products available, for the most part this is a specialized procedure using specialty equipment and training, which all adds up to a more expensive product.

Another step up from the average standing-seam painted metal roof is a copper roof. Gleaming when newly installed, a copper roof weathers to a beautiful green patina as it oxidizes with age. Although the most expensive of metal-roofing options, it does make a beautiful roof.

With all metal roofing, it should be known that flashing around a perforation can be more difficult than with a shingled roof, and creating the fewest perforations—such as plumbing vent stacks or chimneys—as possible (a good idea regardless of the material) should be part of the design, as well as locating perforations to one side or the other of the ridge whenever possible to allow the ridge cap to flash over the "boot" used to penetrate the roof. While unpainted metal roofs should not be used for catchment, painted (epoxy) metal or copper roofs are fine for harvesting rainwater. Metal roofs also offer the advantage of shedding snow easily, helping to avoid the ice damning described in the beginning of this chapter.

CHAPTER 19

Floors for Natural Buildings

There are many ways to approach flooring in a natural building. In many cases, the floor is the last thing to be installed before occupancy. In others, it is one of the first, and is merely cleaned up before the furniture goes in. Some floors will receive very little wear at all, while others may see high levels of daily vehicular or pedestrian traffic. Some floors may need regular mopping, or even scrubbing; others will be asked to help contribute to the heating strategy of a building. Understanding the use pattern of a room, and the structure, is the key to developing an effective flooring solution. In this chapter, we will evaluate a series of different flooring options and explore in depth the construction of earthen floors.

In looking at the options for flooring solutions, keep in mind all of the standard criteria we have been considering throughout the book thus far: cost, ecological impact, performance, and aesthetics. Indoor air quality (IAQ) is a big consideration in choosing flooring material and floor-finishing options. In conventional construction, there is significant petrochemical and formaldehyde off-gassing into the structure from carpet materials, carpet adhesives, tile and flooring adhesives, plywood subfloors, laminate floor tiles and panels, and concrete sealers. These are harmful agents that will continue over time to have toxic and even carcinogenic effects on the human body, and it's important not to include them in the design of the building. Comfort is another consideration that is unique to a floor's use. In consideration of priorities, comfort can sometimes be lower on the list than durability and maintenance, which may be valued as higher priorities in a floor owing to the great potential for wear and tear and the inconvenience and expense of floor refinishing. For example, a concrete floor will wear better than a cork floor, whereas the cork floor will be far more comfortable underfoot, so the builder/owner must weigh these considerations when choosing the flooring material.

Also worth considering is how the floor relates to heating in the building. The primary attribute of a floor in heating will be its mass content, and whether it has the ability store and release heat slowly over a diurnal cycle. Floors are also in direct physical proximity to the elements most in need of heat in the building—the occupants—and are distributed throughout the building. There are two main ways of heating a floor: through the sun, and through distributed heat from a mechanical system, occasionally in the form of air, but more frequently in the form of liquid (water or a glycol mixture, depending on the design) or electricity.

In the former, with the floor as an element of passive solar design, a few conditions need be in place for the system to be effective. First, the floor needs to be of sufficient mass—a wood floor just isn't going to get the job done. This mass is most commonly found in the form of earth, concrete, stone, or tile, although these latter two run the risk of being too thin to hold sufficient mass, depending on the design. This brings us to the next point: the amount of mass is relevant as well. There are precise calculations that can be done to assess the appropriate amount of mass for solar heat storage. The mass should also be isolated thermally from the ground below, if on grade, or from a cold basement or crawl space. The sun also needs to hit the floor to be able to "charge" the mass, for it to store heat. That means that the floor must be directly within sight of the southern windows in the structure and must not be covered by built-ins, rugs, etc. That also means that there must be sufficient glazing on the south, and that the type of glazing used must be considered for its ability not only to keep heat in the building, but to allow heat, in the form of light, to enter the building (see chapter 7).

In the mechanical heating of a floor, known as radiant floor heating, it is most common to embed tubing into a massive floor and circulate hot water or glycol mixture through the tubing—in this case, the system is known specifically as hydronic floor heating, although electric-heated radiant floors are also common, particularly in smaller installations or where hydronic heat distribution is not practical. Again, appropriate design of insulation, floor thickness, and layout and control of the tubing loops must be considered. The floor is acting as a heat exchanger, pulling heat from the liquid circulating in the tubes and storing it, to be released to the air above the floor and the feet, back, or bellies of the occupants. The source of the heat is generally a boiler or furnace, although solar thermal systems can be used in part or whole to deliver this heat. Warm floor heating has many benefits, which we explore in chapter 21.

Earthen Floors

We will begin our exploration of floors with the most "natural" of options: earthen floors. We will take the time here to flesh out the application and installation process, as it is different from other techniques presented in this chapter, and similar to some of the other natural building techniques presented in this book, such as cob and plaster. Earthen floors have been under the feet of inhabitants of homes large and small, across the world and throughout the ages. Remarkably simple in their construction, earthen floors boast many benefits in regard to their light environmental impact, low skill requirements for installation, low cost of materials, stunning beauty, comfortable feel, and moderately durable performance. While easiest and most common to install directly on grade, they can also be installed over basements or at higher stories. From an environmental perspective, earthen floors have an incredibly low embodied energy, 80% to 90% less than concrete, its closest conventional-building counterpart. Construction waste from making an earthen floor can simply be reincorporated into the landscape. Additionally, the end-of-life impact of an earthen floor is very light (essentially, nontoxic earth fill).

Like other components of natural buildings, earthen floors lend the benefit of reduced cost when installed by owner-builders as a result of their inexpensive materials—the sealing oil generally being the single largest expense for the floor. Similarly to other earthen technologies, earth floors also feature high embodied labor. If this labor is to be hired out, this may prove to be a more expensive flooring system than others. A wide range of looks and feels can be achieved with earthen floors. Many different colored finishes are achievable by selecting a single or mixing multiple colored clays, by tinting the floor with pigments, or by applying a colored clay paint to the floor prior to sealing. Inlays, such as flagstone or tile, are possible when selected and applied appropriately. The texture is luxurious—often compared to leather once finished—and is much softer and more forgiving than concrete under foot (and dropped glass!). A higher-gloss finish may also be applied if desired. The longevity of the floor is well ensured by its hard, durable surface, which is resistant to most scratching and general wear.

As a massive floor, an earthen floor lends itself well to radiant-heating strategies, whether by embedded hydronic tubing, electric heat mats, or direct passive-solar radiant heating. All the same basic rules apply as to placement, testing, and detailing as they would for a concrete floor, as the mass level between earthen and concrete floors is similar.

The downsides to earthen floors are similar to those of other natural building systems. Finding skilled tradespeople to perform the work may be difficult in many areas. A certain knowledge, or willingness to experiment and learn, is required for

Earthen floors are beautiful, comfortable, and easy to maintain and have a light ecological footprint. PHOTO BY ACE MCARLETON.

the owner-builder to work with raw materials to produce a suitable product for installation. Drying times may slow down the construction process, as it can take quite a while for a floor system to fully dry out, especially if placed over a thick moist-applied earthen subfloor (as opposed to a concrete slab). And while quite durable, earthen floors are not as resilient as concrete or tile, and scratching and denting can—and eventually will—occur, particularly in high-traffic areas such as entryways and below desks and dining tables; earthen floors are particularly vulnerable to point loads, such as chair and table legs, stiletto heels, and dog nails. Fortunately, repairs are quite simple in the event of such incidents, and most damage is avoidable by appropriate selection of throw rugs and runners, chair and table coasters, and wheeled chair protection mats. Additionally, flagstone or tile may be installed in high-traffic areas to reduce early wear and maintenance needs.

Much of the durability of the floor relies on the quality of the subfloor. Just as a quality plaster job requires a sound substrate, so does a quality floor job. This means a very even, well-compacted, stable base that will resist deflection and movement; therefore, it is more difficult to apply an earthen floor over a wooden suspended floor frame over a basement, crawl space, or second story, due the increased structural support in the subfloor structure needed to resist movement that may translate stress cracks into the finish floor (note that this issue may be largely resolved by steel framing, which is common in many apartment buildings).

DESIGN AND PLANNING FOR EARTHEN FLOORS (PRE-APPLICATION)

There are many different approaches to earthen floor construction, adapted from a variety of different traditions. Our approach is to apply one 2½-inch layer, or "lift," of a base floor followed by one ½-inch layer of finish floor over a stable subfloor. The floor is installed first with wood floats, smoothed with steel trowels, and then hard-troweled, if desired (for the final lift only). Finally, the floor will be sealed with successive coats of oil and thinner, or sealer of choice.

One consideration, when planning for an earthen floor (and all floors, for that matter), is how it will intersect with walls, emerging posts, doorways/ entryways and thresholds, built-ins, appliances, and any other elements that may interface with the floor. An appropriate order of operations should be determined, and a finished height for the floor should be considered. How the process will fit into the construction schedule should also be considered, both so as not to interfere with, or be hindered by, other processes, and to avoid having to fight with the elements as little as possible. Setting up a mix station in the snow is a situation best avoided, and in that the material must be able to dry without freezing, not having to add the burden of keeping the structure heated may prove to be an important factor around which to plan the schedule (freezing while still wet will at best weaken the floor, and at worst it will cause cracking to the point of failure).

THE SUBFLOOR

There are two basic strategies for subfloors that lie below earthen floors: an earthen subfloor, or a concrete slab. Even if considering an earthen floor on a second story or above a basement, it is still advisable to lay a thin slab or earth subfloor prior to the finish floor. In this case, additional design requirements for the floor structure must be considered, including provisions to carry the increased weight load of the earthen floor, handle the heightened moisture level during construction, and reduce deflection and movement beyond what is acceptable for wood- or tile-floor installation. While the engineering can get complicated and is project-specific, such provisions may include increasing the size and spacing of floor joists, reducing joist spans, using stronger joist material such as steel or engineered lumber, allowing for a full-thickness subfloor, and including 'slip sheets' below the subfloor to isolate frame movement from the floor. However, these provisions are invariably costly and complicated and potentially difficult to employ in a retrofit scenario, and ultimately they may not be fully successful in reducing cracks relating to frame movement. The more important it is for the owner to have a permanently crack-free earth floor (a tall order in any scenario), the less desirable it is to attempt to place the floor on a suspended floor frame.

The benefit of installing a concrete-slab subfloor is its speed, its strength and durability (especially relevant if it will act as a construction floor), and the access to subcontractors familiar with the medium. The liabilities lie largely in the realm of cost (if not counting labor), causticity, rapid set time (which is a liability for those not familiar with doing "flatwork"), embodied energy, and toxic footprint. A standard 4-inch slab is more than adequate for a subfloor; well-designed and appropriately reinforced thin slabs can be poured as well.

An earthen subfloor is ideally 4 to 6 inches thick and made of what is commonly known around New England as "road base." This material is ¾ inch minus, meaning it is well graded from particle sizes ranging from ¾ inch down to fines; it also has enough clay content to hold the material together, but not so much as to cause the material to crack upon drying. This is the material that is commonly laid down as the base for dirt roads across the Northeast, and a comparable material should be available in most regions.

Prior to laying down the subfloor, in addition to addressing any framing requirements, it may be necessary to install a vapor barrier, or even insulation. If foam insulation is to be installed—for example, if the floor is to be installed directly on grade—extra care must be taken to ensure a very level and stable base onto which the foam is installed. The foam must

The subfloor should be installed at a minimum depth of 4 to 6 inches in lieu of a slab. PHOTO BY ACE MCARLETON.

Use a transit or levels to establish a datum line around the perimeter of the building to help keep the floor level and ensure the finish height will be accurate. PHOTO BY ACE MCARLETON.

be laid in such a fashion as to keep humps and dips out of the equation, as any movement and flexing in the foam may well translate up as cracking in the finish floor. This is not hard to do; it just takes a steady hand and eye, and perhaps the help of a level. A 1-inch sand bed below the foam layer can help create a level surface.

The approach to installing an earth subfloor is similar to that of the finish floor, which will be described in detail in this chapter. A level line marking the height of the subfloor should be struck; we will often strike an additional line ¼ inch above this first line to use as a reference once the true line is covered. The subfloor material should be moistened just well enough to stick, but not so much that it is sticky (i.e., sticks to feet or tampers). The material is raked evenly over the floor area, approximately 2 inches thick. This first "lift" is then compacted, with either a hand tamper (very laborious, but effective in small areas) or a plate compactor (quick and efficient, yet unwieldy, loud, and stinky, as it is usually gas-powered). Use an optical or laser level, or measure off the datum line and sweep with a 4-foot or 6-foot level, to ensure the subfloor's first lift is level, then continue with one or two more 2-inch lifts until the datum line is reached. The second, higher line may aid in visibility when compacting down to the final height.

At this point, whether using an earthen base or concrete slab, a stable, level base is created at the correct elevation to receive the designed finish floor. If in-floor radiant tubing is desired, this should be installed in the subfloor, generally in the second or third lift of an earthen floor or near the top of a concrete slab. Note that insulation must be placed below the subfloor if radiant tubing is installed, and it should be installed in any case if needed to maintain continuity of the thermal envelope. The tubing and any associated fastening should be completely buried by the subfloor, so make sure it is installed at an adequate depth so as not to interfere with finish-floor installation or raise the finish floor height above an acceptable level. Also be sure to pressure-test the tubing for leaks before continuing forward with the finish floor or pouring the concrete slab.

EARTHEN FLOOR WITH FROST-WALL FOUNDATION

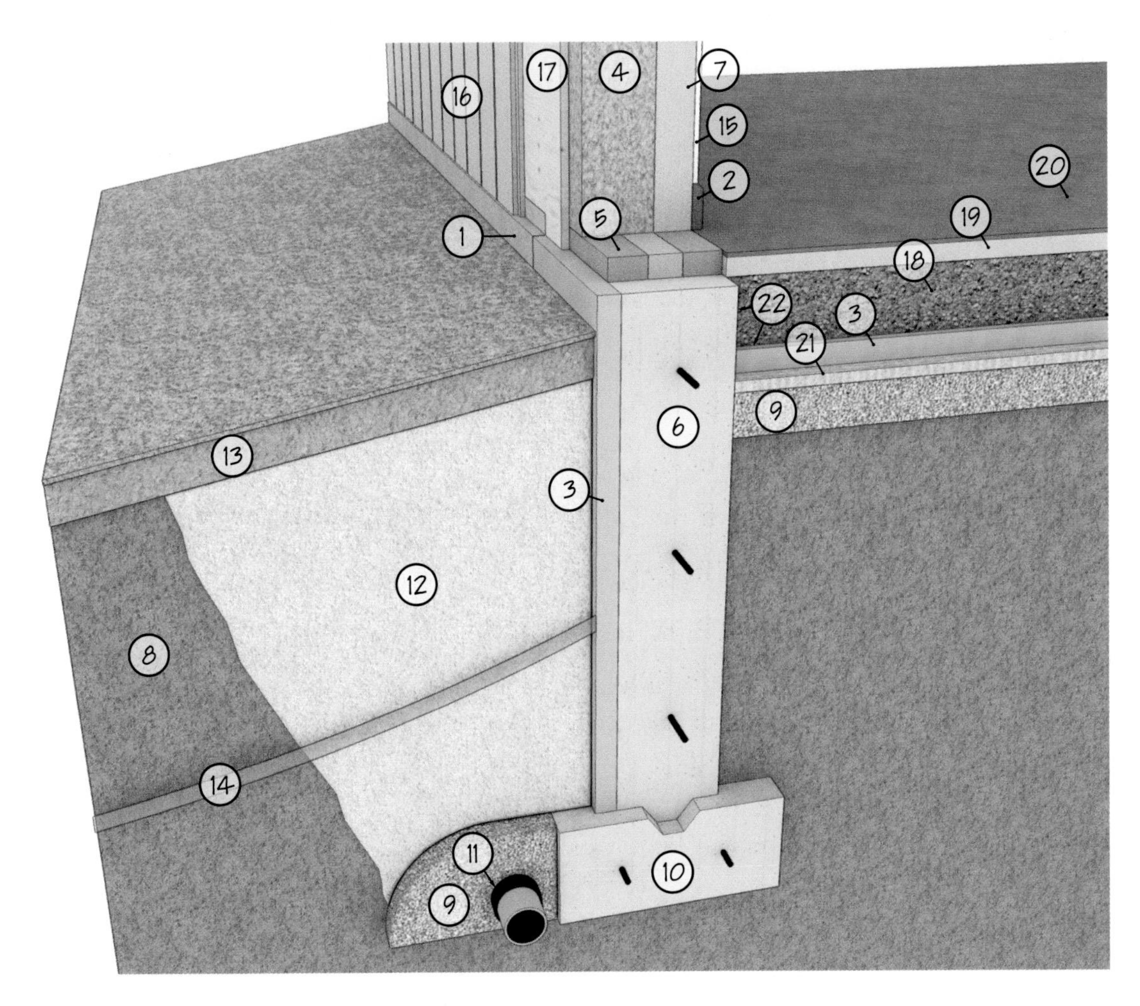

1. METAL FLASHING
2. BASEBOARD TRIM
3. 2" RIGID INSULATION
4. CELLULOSE INSULATION
5. ROT-RESISTANT SILL (moisture barrier not shown)
6. CONCRETE STEM WALL
7. ROUGH FRAMING (staggered-stud wall)
8. UNDISTURBED/COMPACTED NATIVE SOIL
9. WASHED CRUSHED STONE
10. FOOTING
11. 4" DRAINPIPE (wrapped in filter fabric)
12. POROUS BACKFILL (sand)
13. CLAY CAP
14. FROST DEPTH
15. ½" WALLBOARD
16. VERTICAL BOARD SIDING
17. WALL SHEATHING
18. 6"–8" ROAD BASE
19. 2½" ROUGH FLOOR
20. ½" FINISH FLOOR
21. 1" SAND BED
22. VAPOR BARRIER (not shown)

Section drawing of an earthen floor. ILLUSTRATION BY AARON WESTGATE/FREEFLOW STUDIOS.

SITE PREPARATION

To begin, it is best to start by organizing the site. Prepare the subfloor by sweeping away all loose debris, remove all tools and other obstructions, and tape and mask all sensitive areas (such as wood baseboard trim, if already installed); make sure that the removal of the tape will not interfere with the finish of the floor! Draw a level line 2½ inches above the subfloor to mark the base layer, and another ½ inch higher than that for the finish layer, around the perimeter of the floor and upon any protruding elements such as posts to aid in creating an accurate depth and establishing the finish-floor height. This can be done with a laser or optical level and a meter stick, or with a water level (essentially a clear, flexible, long plastic tube filled with water). Using a level is preferable to measuring off the slab, as that will only translate any inaccuracies in the slab pour to the finish-floor level. Make sure, if using a marker of any thickness, that the reference is to the top of the line, such that the line will ultimately be covered by the floor, as opposed to the floor running up to the underside of the line. You can either make a dashed line, spaced every so often along the wall, or a solid line, as long as it is accurate. This is not a necessary step, but it is highly recommended.

Outside, clear away unnecessary tools and materials, and establish and organize a convenient location to set up the mix station. This work area should allow enough space for material storage and mixing, as well as convenient access to the floor area and vehicle access (for delivering materials and equipment). The area should be reasonably level, well drained, and preferably directed out of high-traffic patterns (foot or motorized). If mixing by hand, a couple of options are available. Either set up an area on flat ground with large mortar tubs (or wheelbarrow, if preferred) with which long-handled hoes can be used for the mixing, or erect a table—easily done with sawhorses and a sheet of ¾-inch plywood—onto which mortar tubs can be set and short-handled mortar hoes can be used for the mixing. If mixing by foot, select an area large enough to accommodate a 6 × 12-foot tarp. If mixing by mortar mixer, select an area where it is easy to bring in and remove the mixer, where the noise and fumes of the mixer won't be unduly intrusive, and where a wheelbarrow can be securely placed below the mixer drum and wheeled away to the floor. Again, the mix station should be placed strategically in proximity to both materials and floor.

Once the site is prepared, bring in the materials. Sand— in either a pile or bags—should be protected both from moisture wicking up from below and from precipitation falling from above; generally, good tarping will keep the pile reasonably dry. Clay should also be stored nearby, although if bagged and stored for a long period of time, indoor or more securely covered storage may be desired. During mixing, however, space should be made available for clay to be stored close to the point of mixing. If lime is to be used, similar provisions should be made as with the bagged clay. Try to organize materials so that the fewest possible moves are necessary. For example, if a load of pit sand requiring screening is necessary, arrange the site so that the sand can be delivered easily by the truck, then sifted into a separate pile, which can be fed directly into the mix station, from which the mix can be delivered directly to the floor. Additional materials, such as straw, pigments, clay-paint materials, and oil and thinner, should be kept on hand and available for use when necessary, in a secure environment (particularly the straw, which is highly moisture-sensitive).

MIXING

Achieving a good mix is best done by feel, and the information in this section will help guide you as you begin to develop a relationship with the materials. The basic base mix will consist of approximately 1 part clay to 5 parts sand, with an additional ½ to 1 part straw or other fiber, such as horse manure. However, this ratio will vary widely depending on the types of clay or sand used. Stabilizers such as cow manure or lime may be desired to increase hardness of this mix, as discussed in chapter 17. It is imperative that test batches be made prior to installation of the floor to ensure a proper ratio is determined for the materials available; generally, a 3 × 3-foot wood form can be constructed for this purpose. Testing can be done to evaluate hardness, dusting, and cracking, as well as color, texture, and set time, if other additives are to be used. Upon drying, if a mix is dusting, it means there is not enough clay (or clay quality is poor) or that there is too much sand. If a mix is cracking, it means there is not enough sand, or too much clay. Note that a small amount of dusting (short of erosion) will be fixed by the sealing. Upon application, if a mix is not sticking well, it means there is not enough clay, or it is too dry; if a mix is too sticky, it means there is too much clay, or it is too wet. The finish mix is of a similar 1:5 clay-to-sand ratio for the base layer, although without the straw or fiber (unless desired for aesthetic purposes).

Water should be added "to taste," depending on the consistency of the clay and moistness of the sand; an appropriate consistency will be damp but not wet (when a handful is grasped, it should ball together nicely, without any water dripping or oozing from the material). A mix that is too wet will be easier to mix but will be more liable to crack, owing to the extra water leaving the matrix of the mix. Hydrated Type S lime may be added (in ratios of 1:3 to 1:1 lime to clay, depending on the nature of the clay) to help stabilize poor clay soils, enhance set time, and/or increase hardness of the floor (see the section "Lime Floors" later in this chapter for more information). Cement may also be used for this purpose, if preferred, but can lead to cracking or other problems if not executed with care.

It should be noted that the use of either lime or cement will greatly increase the pH of the mix, making it caustic enough to require the use of gloves, so avoid skin contact; both lime and cement are highly caustic in their undiluted forms, and particular care must be exercised when handling these materials—particularly when mixing—to avoid inhalation or contact with eyes or mucous membranes.

As mentioned earlier, chopped straw may also be added to the finish mix, primarily for aesthetic purposes, although the straw will also help control cracking. Small amounts are generally used; start with a ratio of 1:3 straw to clay by volume, and go from there. Color can be controlled either by selection of a colored clay or colored sand or by adding pigments (in powder or liquid form). Ensure adequate preparation of the pigment (for powder pigments, soak in water for a smoothie-thick consistency, and strain if necessary) and complete dispersion into the mix for best results. Pigment may also be applied after installation directly onto the wet floor, to create a wide variety of creative and artistic finishes. Creating sample boards to explore all of these possibilities—as well as to develop your technique—is well advised, before committing to the finish floor. Be sure to note carefully all ingredients and other pertinent variables when creating the samples. It is important to seal the samples as well, especially when evaluating for color, hardness, or dusting, as all of these variables and conditions will ultimately be affected by the sealing process (oil will darken, harden, and stabilize the floor).

Once the samples have been created, dried, and evaluated, go ahead and mix the full floor batches. To evaluate the amount of material you will need, simply calculate the square footage, multiply by the depth to get cubic footage, multiply by 30% for overrun and compaction, and divide by percentages of the mix for each ingredient. For example, to calculate a finish mix for a 10-foot-square room: 10 × 10 = 100 square feet, multiplied by 1⁄24 foot (to convert from feet to inches for one ½-inch lift) = 4.17 cubic feet of total floor, times 130% = 5.42 cubic feet. For a 1:5 clay-to-sand mix, multiply 5.42 by ⅙ to get clay (0.9 cubic feet of clay), and by ⅚ to get sand (4.52 cubic feet). Don't forget to order extra for samples, and remember, the construction waste can simply be raked back into the landscape.

It is often advisable to mix the bulk of the material a day in advance, as this helps to allow any un-hydrated pockets to fully hydrate before application; additional tempering with water before (or during) installation may be necessary. This may also prove advantageous if renting a mixer, especially for a smaller project, as the whole floor may be mixed in a day with the mixer, which can then be returned; store the mix in drums, tubs, or a temporary bin built of straw bales and lined with poly. Make sure to fully cover and protect the mix to avoid both drying out and additional moisture.

Directions for Hand Mixing

The most comfortable way to mix by hand, using mortar hoes, is to pull the material toward you; pushing the material is less effective and messier and will tire you out rapidly. Place an appropriate amount of dry clay and sand in the correct ratio into the tub, and mix materials together thoroughly until uniform (this you can do with your hands if you like). Push material to one end of the tub, flatten the pile and create a small well in the middle, and pour in a reserved amount of water. Proceed slowly at first when adding water—what seems like a small amount can quickly over-wet a mix, and it is much more difficult to remove excess water than to add more. Incorporate the water into the mix with small chopping motions, always pulling the mix toward you. Once all the material has been pulled to your end, walk around to the other side of the tub, or pass the hoe to your partner on the other side of the tub, or step to the side and work the other half of the mix, if your partner has a hoe as well and you are both working separate halves of the tub simultaneously (this is the fastest method, if the two of you work well together). Continue wetting needed areas of the pile, chopping and pulling, until the mix is uniformly wet and of the desired consistency. Make sure there are no dry or excessively wet pockets. If using wet-processed clay, simply add clay to sand measured to the appropriate ratio, and mix as described above, adding water only as necessary. If lime is to be used, add with dry clay, or with sand if using wet clay. If straw is to be used, mix the clay and sand to a slightly wetter consistency than desired, and then mix in straw.

Directions for Foot Mixing

Foot mixing for earthen floors is similar to mixing cob, as discussed in chapter 16. In the center of a fully intact 6 × 12-foot (or comparable) tarp, place dry ingredients to be mixed, and roll back and forth in the tarp by pulling one end of the tarp over the mix and flipping it in place on the tarp (a second person is very helpful here) until the mix is reasonably uniform. Reconsolidate the mix in the center of the tarp, add water (or wet-processed clay, if using), and mix, using the balls of the feet in a twisting motion; again, working with a partner can be very helpful. Add water as necessary, and occasionally fold the tarp over in the direction necessary to consolidate and/or invert the mix to aid the process. Again, straw should be added at the end after achieving a full mix between sand and clay. *Note*: this practice is not recommended if there is lime or rough stones in the mix (although duct tape can be wrapped around your feet to aid in protection, if need be).

Directions for Mixing with a Mortar Mixer

If using dry clay, put in one measure of sand and the clay. After engaging the paddles, immediately begin to mist the material to control excess clouding, taking care not to over-wet the mix. Continue to add the remaining sand, mixing to the desired consistency while adding water as needed. If using wet-processed clay, begin with sand and a bit of water in the mixer, then alternate adding measured amounts of clay, sand, and water until the desired ratio and consistency are achieved. Lime may be added with dry clay or measured in alternately with wet clay and sand. Straw can be incorporated at the end of the mix in both instances.

Once the base of your mix is complete, go ahead and introduce any desired coloring agents into the finish-floor layer. Ensure that they are added in correct proportion and mixed thoroughly to avoid streaking and undesired color variation throughout the mix. Color matching between batches will be difficult, so if a fully uniform color is required, extra care must be given to keeping measurements as accurate as possible—and measuring by weight, rather than by volume—and multiple batches will in turn need to be mixed together to ensure color consistency (not unlike working with multiple gallons of paint). If clay paint is to be applied, coloring the floor itself is unnecessary. Be sure a lime- or cement-stable pigment is used if lime or cement is in the mix, as not all pigments are suitable for use in a lime- or cement-rich environment (see chapter 20 for more information on using powdered pigments and making clay paint). Please note that most, if not all, milled clay contains silica, which can cause severe lung damage if one is exposed to prolonged inhalant exposure. Always wear a high-quality fine-particulate filter mask when working with milled clay, powdered pigments, lime, or sand, even when outdoors.

APPLICATION AND INSTALLATION

Once the mix has been made—either in its entirety in advance, or in batches as you go—it is time to install it on the subfloor. Begin by giving the slab or subfloor another good sweeping, removing any loose debris. Review the section "Site Preparation" (earlier in this chapter) and make sure all tasks are completed. If not already done, strike a finish-level line, as previously mentioned, around the perimeter of the floor area; this will give a reference point to true level around the edge of the floor. This step isn't critical, especially if you trust the level of the subfloor. If you choose to skip this step, set in pins (such as rebar or 4-inch nails) protruding 2½ inches above the subfloor and use the ½-inch depth-gauge sticks for the finish floor as the sole measure to guide the floor installation; the downside to this, however, is that any inaccuracies in the subfloor will be translated up to the finish floor. Striking a finish-level line should also be considered if you are trying to match the height of the floor to wall elements (such as trim); it may be more accurate to establish a level line for reference, rather than relying upon the relative height off the subfloor. This exercise will at the least show the level accuracy of the subfloor (this can also be roughly determined by checking across the subfloor in various places with a 6-foot level). Additionally, if there are any cracks or defects in the subfloor that haven't been fixed, now is the last chance. Ensure that the subfloor is as uniform, solid, and intact as possible; cracks and chunks can be filled/sealed with a variety of products found commonly in most hardware stores such as floor-leveling compound, in the case of a slab, or additional earthen subfloor mix.

Finally, get all of your application tools together and ready for use, as well as a supply of water. Strategize the best way to work within the building, working backward toward an exit so as not to box yourself into a corner. If windows are to be used in the ventilation strategy, this must also be considered, and they should be opened or closed accordingly. Similarly, if fans or heaters are to be hung or mounted to aid in drying, these should be set up ahead of time.

Bring wheelbarrow or bucket loads of mix into the room, and begin to dump them in the proximity of where you will be working, but not directly over the area you will be starting; don't bring in more than you plan on spreading at a given time. Like judging the mix, this is best done by experience. Space the gauge pins evenly across the floor in close enough proximity to ensure level application in the fields between where they are placed. Using the wood trowel, pack the material against the edge of the wall and work toward yourself between the pins. Begin by pressing in the material, keeping both hands on the wood float as you establish the correct depth; continue by floating back over the mix, making small sweeping arc motions, to help consolidate the lift, filling in any voids or cracks.

Wood floats are used to install the layers of the floor. PHOTO BY ACE MCARLETON.

The presence of empty gaps in the lift will lead directly to cracking, so it is this thorough packing and consolidating of the mix that will most aid in reducing cracking (assuming a good mix and sound subfloor, of course). However, do not try to get a beautiful, smooth finish at this point; your objective is only to get a consistent depth and a well-packed, gap-free application.

Use a screed board—either one of the gauge sticks, an older inexpensive level, a clean and true piece of wood, a piece of sound aluminum angle or other metal, or other straightedge—to help remove high spots and identify low spots in the floor, and check periodically throughout the installation to ensure a level surface—while deviations in the floor surface can be addressed by the finish layer, variations in thickness will either translate out through the finish floor or result in cracking due to uneven application of this final thin layer. Continue to work your way through the building until you can exit the room. A rougher finish should be left so that there is more surface area for the second lift to "key" into; roughing up the surface with the wood float or, even better, scratching with a scratch comb (found in masonry supply stores, and designed for the purpose), devil float (wood float with nails protruding from the face approximately ⅛ inch), or notched tile float will aid in this regard. Be sure to scratch the floor as you go, as it will not be possible to do so once the floor has dried.

Now begins the controlled drying process; even drying is very important, therefore direct sunlight is to be avoided, and ventilation should be applied evenly and consistently throughout the room (as opposed to a fan blowing directly against one part of the floor). Uneven drying can lead to cracking along the edges of differential drying zones. Make sure there is adequate ventilation to remove the moisture vapor being released by the floor, both to aid in drying and to reduce moisture buildup issues in the structure; this is of particular concern in winter applications, where there may already be a considerable amount of built-in construction moisture (e.g., from drying plaster or green timbers) and less-than-ideal drying or ventilation conditions. Because of the relatively dry nature of the mix and the fact that the subfloor will be absorbing a considerable amount of the moisture, drying potential shouldn't be too much of an issue, but drying will be greatly affected by seasonal conditions and should be considered before installing the floor. Allow complete drying of the floor if possible before proceeding to the next step.

Before applying the finish coat, apply a light misting of water—enough to moisten, but not to puddle or even come close to saturation. This can be done with a hand or backpack sprayer (a hose will most likely be too powerful, unless fitted with a fine-mist nozzle) over a "reachable" area before laying in that section of the floor.

To apply the finish coat, get out the gauge sticks, moisten them well, and place them on the floor parallel to each other—and to the direction you will be working. They can be as close as slightly less than a trowel's length apart to begin with, such that the trowel can run across them to determine the accurate depth between; as your skill and eye improve, they can be spaced wider apart.

By hand, take a bit of the mix and anchor in the sticks by applying the mix on either side of the sticks to keep the sticks from moving. Then spread more mix around between the sticks.

Once you've worked almost to the ends of the gauge sticks on both sides, the sticks can be removed; clean them off by wiping, and occasionally remoistening, before replacing, to avoid buildup.

The finish layer goes on thin—½-inch thickness maximum—to avoid cracking. PHOTO BY ACE MCARLETON.

In the spaces left by the sticks, use the wood float to angle in the edges, making 45-degree V-shaped edges where they once were square—this will strengthen the joint, reducing the chance of future cracking. Sprinkle in material and pack with the float, making sure not to add too much material at once—it is easier to add a little bit more than to work out a hump. In the event you get humping in the floor as you go, use a screed board and, holding it at an angle against the floor, slide it back and forth while pulling slowly yet firmly toward yourself, maintaining even and level pressure; this should pull off the offending lump while maintaining an accurate level with the material on either side. If there are any pits or depressions, simply sprinkle on a conservative amount of mix and compress well with the wood float until uniform and level.

When you have laid up as much as you can reach in one area, switch to the steel trowel (square, for edges and corners) or float (round, for fields), and do a quick but careful pass over your work to smooth out any remaining lumps or dips, work out any marks, and begin to smooth the surface. Hold the trowel with both hands (perhaps one hand on the handle and one placed on the trailing half of the trowel—see what feels comfortable and makes sense), making small arcs, keeping the trowel at a low angle and maintaining pressure about halfway between the handle and the trailing edge of the trowel, while ensuring that the leading edge does not catch against the material that has been laid down. Don't work the material too hard or too much, as this will only create more marks or disturb the integrity of the wet floor.

Shift yourself further along (you are ideally wearing knee pads or using a kneel pad), lay down the next set of gauge sticks, and continue as above. Periodically check the level of the floor with a 2-, 4-, or 6-foot level, depending on the space (the bigger the level, the more accurate the result); address any issues as early on in the process as possible, and remember that an accurate depth but an unlevel floor is most likely a result of inaccuracies in the subfloor. If you are not going to be continuing directly along the same edge you've created, make sure to taper it to a minimum 45-degree angle, similar to the edges of the spaces left by the gauge sticks (a lesser angle is better than a steeper angle); this increases the surface area between the two batches, minimizes the chances of gaps or voids, and reduces the chances of cracking along that edge.

Compacting the layer with a wood float creates a harder and more stable floor. PHOTO BY ACE MCARLETON.

If the design is for a rougher, less polished floor, then it is time at this point to leave the floor to dry and then seal. However, if a smoother, more polished look is desired, the next step is to go back over the floor with a steel trowel or float in a process called "hard troweling." The idea is to tighten and smooth the floor to a more polished consistency, although this will be limited by the coarseness of the material—the larger the aggregate particles, the harder it is to achieve a smooth finish. The trick to a successful hard troweling rests in the timing; one must wait until the floor has firmed up considerably—enough so that it can be walked upon without disturbing the floor and leaving dents—but is still damp enough that it can be worked and is responsive to the pressure of the trowel. Using both hands (generally one on the handle and one on the trailing half of the nose of the trowel), pass the trowel back over the floor, in broad sweeping arcs, at a low angle while maintaining a good amount of surface area contact between the trowel and the floor. Be sure not to catch the leading edge in the mix, or drag the trailing edge too deep into or across the floor. If you are leaving more marks than you are smoothing, you may need to wait a bit longer, or use a gentler touch. If it seems nothing is happening at all and the floor seems quite dry, you may have missed your window. You can use a sprayer to mist particularly dry areas in the floor and rework them; this may work, or may not. Be careful to keep your trowel clean so as not to mar the surface by dragging aggregate about, and be sure you are not staining the floor by any rust or other metallic buildup on the face of the trowel. Use kneeboards or broad, flat kneepads to protect both floor and knees, try not to dig in with your toes, and remoisten the boards or pads frequently to keep them from sticking to the floor. To get this timing right, you may need to come back and hard-trowel the first part of the floor before completing the finish floor installation.

Now the floor must be left to dry, as laid out earlier in this chapter. It is important that the floor fully dry before sealing; residual moisture left in the floor that gets sealed in will weaken the floor and perhaps even lead to delamination between layers of flooring material or of the surface sealer. The drying process can take anywhere from a few days to a few weeks to over a month, depending on the weather, ventilation, temperature, subfloor condition, and moisture content of the mix itself. It can be difficult to tell when the floor is completely dry; a general rule if thumb is to double the length of time it takes for the surface to fully dry out, but this is only a rule of thumb, and it depends on how the floor is drying. Some builders place a small amount of grass seed into their floor and wait for the sprouted grass to wilt to indicate that the floor is dry.

A final burnishing with a steel trowel brings a smooth finish to the floor. PHOTO BY ACE MCARLETON.

FINISHING

After your finish floor has been installed and hard-troweled (if desired) and has fully dried, you can proceed with the finishing. If you would like a different color than the floor as it stands, or wish to paint on designs or other artistic expressions, you can apply a clay-based paint, commonly known as alis (see chapter 20). This is an optional step, however, and not a standard part of earthen-floor construction. The benefits of using a paint as a finish are that less pigment is needed than to tint the whole floor, a cheaper, less-desirable (aesthetically speaking) clay can be used for the floor, and the pressure of matching batches of floor material is relieved, as is the need to decide on a finish color by the time you are ready to install the floor. The drawbacks are that you are adding an extra step, which will add more time in application and testing, and more cost in hired labor. Also, in the event there are scratches or defects in the surface of the floor, the color of the floor below will be exposed, highlighting the defect. Repair will also therefore involve touch-up painting as well as oiling, which we'll now discuss.

The final stage in the finishing process is sealing the floor. As mentioned earlier, there are a variety of different sealing options available. The most common is a penetrating oil sealer. Oiling is typically done in four or five coats, depending on the thickness of the floor, the nature of the subfloor, and the type of oil used. The timing on when to apply the sealing coats depends on a variety of factors; on the one hand, it is nice to have the floor fully sealed before using the space further, to avoid damage. On the other hand, it may be desirable to be able to continue with the construction process if the schedule calls for it, and by waiting a spell one can see if cracks develop in the floor—or damage due to continued construction—which can more easily be repaired before sealing. If you decide to wait a bit before sealing, this is fine as long as precautions are taken to avoid damage to the floor by moisture, wear, or impact. More information on types of oil to use in finishing can be found in chapter 20.

The sealing process begins by setting up the supplies: brushes, rags, pails, etc. You will want to set up an outdoor station for heating the oil, which improves penetration into the floor; an electric hotplate is ideal for fire prevention, but a well-supervised small gas camp stove will work as well. Make sure the burner is set up on a stable base, and use a large pot to avoid splattering oil coming in contact with the burner or flame. This is an inherently dangerous operation, so try to set up the station away from buildings or other fire-sensitive areas, and be sure to keep a fire extinguisher suitable for oil fires nearby (don't rely on water—water does not work to put out an oil fire!), or at least a large pile of sand with a couple of shovels.

Make sure you are using proper protective gear, including gloves, a respirator, and goggles, and wear long sleeves to avoid burns from splattering oil. The oil should be heated as warm as possible without boiling; don't push the envelope—better to pull it off the burner early than to start a fire. Transfer the oil to a pail to bring to the floor. Before applying the first coat, sweep the floor free of all loose debris. There is great debate as to whether the oil is first applied diluted, and successive coats work toward a pure oil coat, or whether the first coat is applied pure, with successive coats increasingly diluted. Our approach is to apply the first coat undiluted—pure oil. The floor will soak this coat up quite quickly; try to apply it as evenly as possible, and ensure there is no puddling on the surface. Use a 4-inch high-quality oil brush, or if you prefer (and the oil isn't too hot) rags or sponges. You can try pouring out the oil onto the floor and brushing or sponging in, but be cautious to avoid an uneven distribution.

Once you have worked over the entire surface of the floor, wait until it is dry before continuing with a second coat. You should be able to go back over with another full-strength oil coat. You will need to use much less material for this second coat, so continue your diligence to avoid puddling or pooling on the surface. Be sure to carefully observe how the floor is absorbing the oil; if it does not seem to be soaking in very well, either wait for the first coat to dry out a bit longer, or consider diluting the second coat. Allow to dry.

The third coat (or second coat, if desired) will be diluted, 75% oil to 25% thinner, such as citrus thinner, turpentine, or mineral spirits (thinner options are discussed in chapter 20). Again, watch to see how the floor takes this in; you may want to wait for more drying between coats, or the floor may tell you to just go ahead. Try two coats at 75:25; then move to 50:50, and apply one or two more coats at this strength, allowing the floor to dry between each coat (the drying time should be faster as the thinner concentration gets stronger). As you continue, the floor will be absorbing less and less oil as it reaches a higher saturation point; this is why we continue to heat the oil, and then add thinner to help encourage its absorption. Keep the thinner and an extra rag handy, in the event you do have pooling that cannot be absorbed, as you do not want the oil to dry on the

Earth floors can be created to look like tiles. PHOTO BY JONAH VITALE-WOLFF.

surface; you will be left with a tacky, sticky mess that is much harder to deal with after it has dried.

If a surface protection is desired in concert with the oil application, waxing is an option. Some wax products are designed to be applied separately, whereas others may be applied in emulsion with oil in the final coat. Linseed oil and beeswax is the standard for floor sealing; we recommend the use of "stand oil" or other oil products that reduce the inclusion of petrochemical and metallic drying agents (see chapter 20 for more information). There are companies, such as BioShield, that create a variety of low- or no-VOC surface sealers that range in gloss and strength; a hard sealer would be desirable for a higher-gloss finish. Note that surface sealing, while increasing durability and sheen, will also complicate touch-ups and repairs, depending on the type of finish you use. In all, the oiling process may take upwards of a few weeks, depending on the depth and nature of the floor, type of oil, oil and thinner concentrations, and drying conditions.

MAINTENANCE

A well-executed earthen floor is indeed quite durable, yet damage will undoubtedly occur at some point. The first step for a successful long-term maintenance program is to avoid undue wear and damage in the first place. The most fundamental approach is to protect the floor from point loads, as mentioned earlier, such as table and chair legs; furniture coasters or pads and protective mats will go a long way in this regard. Additionally, when moving furniture or other large items, be sure not to drag and scratch the floor; it's better to lift and place. The other primary source of wear comes through high traffic, particularly in areas involving footwear, moisture, and other high-impact elements. Appropriate selection of flooring given the area's use should be considered, of course—perhaps tile would be better in the mudroom, for example, if you are looking for less maintenance—but beyond that, throw rugs, runners, or other protective floor coverings will come in handy in reducing wear and impact. Flagstone or tile can be laid into the floor as well, although careful detailing must be practiced to ensure a successful application. It should be emphasized that since the weakest areas of the floor are the edges, the more edge put into the floor, the more opportunity for failure. Selective use of larger tiles is therefore more advisable than a large mosaic of shards and fragments, from a durability and maintenance standpoint.

All that said, maintenance is inevitable and begins with keeping the floor clean to avoid large particulates from grinding into the finish. Cleaning is as simple as sweeping and vacuuming (be sure to use a vacuum without a beater bar, often used for rugs), followed by an occasional damp mopping. Any mild detergent soap will do, although oil soaps such as Murphy's will help reinforce the oil-based finish for the floor. Go light on the soap and light on the water, and spot-clean trouble spots by hand.

After a period of time, you will need to refinish the floor, particularly where there has been lots of traffic or extra mopping. If you applied wax or another surface sealer, you will have to reference directions from the manufacture as per refinishing; however, if you are satisfied with the traditional oil finish, maintenance will be as simple as reapplying an oil or oil/thinner blend, as in the initial sealing. This can be done as a spot application, unlike a surface finish that generally must be stripped and refinished in its entirety; however, there will often be somewhat of a difference as the rest of the adjoining floor will undoubtedly have experienced a degree of wear as well, so consider it an opportunity to refresh the whole floor. Certainly, ensure that you reseal the floor before physical wear to the earthen material itself takes place. In the event that there is wear or damage to the body of the floor, this can be addressed by simply remoistening the floor in the surrounding area, troweling on a bit more floor mix, then resealing. Certain flaws, such as deep yet narrow cracks, may need to be opened up slightly to be able to pack in the patch material fully. Keeping on hand a portion of the dry floor mix may prove valuable as a future safeguard to ensure a proper match of look and consistency; if you have painted the floor, preserving some of the paint will also be helpful. It may be difficult to match the floor absolutely perfectly, but relative to concrete, wood, or tile, repairs are much easier, and relatively seamless to the unknowing eye.

Other Flooring Options

Earthen floors are only one of a number of different natural flooring options for a building. In the remainder of the chapter, we will examine other options in natural flooring.

LIME FLOORS

For the same reasons lime is so effective as a plaster finish, it is also favored as a floor finish. The differences between a lime floor and an earthen floor are similar to that between a lime- and earth-plaster finish: lime is more durable, finishes differently, and requires curing, rather than drying. While lime floors have yet to be fully rediscovered in the natural building community in the Northeast, they are rooted in thousands of years of history. Fascinating recipes were uncovered and published by Holmes and Wingate in their book *Building with Lime* (1997/2002), wherein they talk about Roman architect Vitruvius's various lime-floor recipes; we highly recommend reading the book and trying his recipes.

Jumping ahead to the present, you can find a smattering of lime floors in residential construction. Builder Aaron Dennis of Ithaca, New York, installed a lime floor in an earthship—a unique style of sustainable architecture—that ultimately received tile. Prior to tiling, however, Aaron confirmed that "it was noticeably more durable than its clay counterpart." Aaron's mix was similar to that of a sand-rich plaster: 1 part lime, 1 part chopped straw, and 4 parts sand, laid in two ½-inch layers over compacted road base. Builder Chris Magwood installed a lovely lime floor—on his exterior covered porch! It was laid over a slab and burnished in a similar fashion to *tadelakt*, and Chris avows: "It has been there for seven years and has handled snowy boots, dogs, a couple of skunk sprayings and general wear and tear. The polished surface was a bit slippery for snowy boots, but we used runner carpets to solve that problem."

STONE AND TILE FLOORS

Another great natural material that is abundant in most areas is stone, as has been noted previously in this book. Varying from region to region, there is a wide variety of different rough or finished stone products quarried locally to create a durable, low-maintenance, attractive, easy-to-install floor, which can show off the beauty of native geology and be supportive of regional quarrying industries. Here in the Northeast, we find ourselves with abundant slate, bluestone (a type of sandstone), granite, and marble, to be used as flooring, thresholds, countertops, hardscaping, and roofing. It is unfortunate that, given the current economies of scale, petroleum subsidies, and international trade policies, it is quite often less expensive to purchase imported slate from China or other far-flung regions from a national chain building-supply center than it is to purchase Vermont slate directly from the quarry. That said, regardless of where it is sourced, working with natural stone can be a very cost-effective and practical flooring solution, when weighing long-term performance and aesthetics in the equation. Be wary when purchasing local "discount" stone, however! Using a material with great discrepancies in thickness will at best greatly drive up installation costs as a result of needing to level out each piece individually, and at worst it will leave a permanently lumpy floor, which will gather dust and stub toes. It is worth the cost of a product that is consistent—at least in thickness, if not in length and width.

There are other tiling products that can serve a comparable role as stone in a building, capitalizing on similar benefits of aesthetics, durability, and ease of installation and trade support. Using materials such as tile begins to move us away from the traditional "natural building" approach and toward the "green building" approach, with a greater reliance on more sophisticated and centralized manufacturing. The

Concrete Floors

Concrete slab floors are very practical things for a variety of reasons. If you have a slab foundation, then the floor is done for you at the same shot, and you have a great construction floor until you are ready to finish with no additional trouble. There is abundant trade support for concrete slab installation, and the slab cures very quickly, keeping the project on schedule. You can bed in tubing for a radiant floor (see beginning of this chapter for more information) and use it to help create a highly efficient and comfortable heating system. It is incredibly durable and very easy to clean and maintain. You can stamp, texture, engrave, color, stain, inlay, and otherwise create nearly any visual effect you may desire. What's not to love?

As we discussed in chapter 12 and in the "Earthen Floors" section of this chapter, we have identified a number of issues with concrete that may cause us to look a bit more critically at its use as a floor in the natural home. It should be clear at this point that the embodied energy of Portland cement—the binder in concrete—is significant, in both its manufacturing and, to a lesser extent, its transportation to site (especially when compared to more local options). Any steps that can be taken to reduce the carbon footprint of our structures without sacrificing their function or performance should be taken, and reducing Portland cement usage is a key part of this strategy, as well as a way to reduce the toxic footprint of our materials. We offer a number of strategies in reducing Portland consumption in chapter 12, including the use of magnesium-based cements, incorporating supplementary cementing materials, or replacing Portland in part or whole with lime or, as explored in this chapter, clay.

Even more direct implications of using concrete as a floor are found in its effects on the inhabitants and users of the buildings in which the floors are laid. Spending a day standing on a concrete floor is a painful and exhausting experience; the hardness of the floor, while treasured in a manufacturing facility or garage, wreaks havoc on the human body, particularly on the spine and lower-body joints. And this says nothing of the dangers of impact, be it a dropped drinking glass or a small child's skull after tripping. Additionally, studies conducted by members of the Bau-Biologie movement suggest a possible link between direct exposure to Portland-based concrete and cellular function in the human body, in which communication between cells is disrupted and fatigue is exacerbated; these studies have purportedly led the German government to enact regulations limiting workers' exposure to concrete slabs in workplaces. Living on a concrete floor may not, ultimately, be in our best long-term interest (Swanson n.d.).

Building is nothing, however, if not a series of compromises. An obvious one would be to pour a slab (if called for as part of the foundation design) and take whatever steps are appropriate to reduce the amount of Portland in the mix. Then, at finishing time, lay down a finish floor of a different material; any of the flooring options presented in this chapter are suitable for installation over a slab. An earthen floor atop a superinsulated concrete slab may well be one of the more salient examples of the integrative approach we espouse in this book and in our practice.

benefits of many of these other tiling products, however, lie in contributing to a lighter footprint; for example, there are many glass tile products made largely from postconsumer recycled material, such as auto glass or glass bottles. Others focus on improved indoor air quality or lower toxic footprint; there are lots of options for carpet tiles and plastic floor tiles that feature recycled PVC or PVC-free ingredients, and low-VOC binders and resins. It is becoming easier for the consumer to select choices of conventional materials whose impact is a bit lighter than what could be found even five years ago.

Brick is another material worth considering. There are many reclaimed/salvaged brick products out there on the market, as well as bricks that are now being manufactured from recycled and reclaimed materials. Green Leaf Brick, made by the Red Tree Group of North Carolina, offers a 100% recycled-material brick, made of waste from ceramic plants and mining and steel operations, recycled glass, and incinerated sewage ash. This and other products are available in ½-inch and/or ¾-inch thicknesses for use as tile, making it a unique and functional option for a masonry floor.

WOOD FLOORS

A close second to an earthen floor for "most natural" would be a locally harvested, locally milled wood floor. The farther up in latitude one travels, the closer one is to the bounty of the northern hardwood forest, featuring some of the finest hardwoods in the world: sugar (or rock) maple, red maple, yellow and red birch, beech, cherry—the list goes on (see chapter 5 for more information). For a less expensive, easier to work, and more rustic (read: easier to dent and wear!) floor, spruce, pine, or even hemlock are available. Although not abundant, there are reliable mills that can produce beautiful, consistent V-groove tongue-and-groove flooring for a fraction of the price of what could be found in a retail store (see sidebar "2 × 6 Tongue-and-Groove V-Groove Spruce Flooring" on page 332). Given the importance of maintaining a working landscape that contributes to local economies—as well as the positive role that a scale-appropriate wood-products industry reliant on sustainably harvested trees can have on protecting and diversifying woodland ecology—carefully and intentionally selecting a wood flooring product can go a long way in making a positive impact both under your feet and in your region.

A wood floor is also easy to install; there is plenty of familiarity with the process, the tools are common (especially compared to those needed to work stone and tile), and trade labor is abundant. Different woods have different looks, and different finishes will give you the same range of color, sheen, and protection as earthen floors (see chapter 20 for more on products that can be used to finish wood). If you have a good resource for local spruce but would prefer a hardwood floor, or you don't have quite enough cherry to floor the whole room, then there is still ample opportunity to utilize native wood throughout the house in trim, built-ins and cabinets, interior windowsills, and floor treads. Meanwhile, if you are planning on purchasing new wood products outside of the local market, do your research in determining how the trees were grown and harvested. Relying on a trustworthy rating system, such as the Forest Stewardship Council (FSC) rating, can help make that selection process a bit easier. Green Mountain Woodworks of Oregon also sells a flooring material produced from the thinnings of national forests in the western United States; while wood from national forests is excluded from receiving FSC rating, this product is indeed supporting both healthy silvicultural practices and healthy logging-industry communities.

It is also possible now to find salvaged wood-flooring products, from old barn timbers and siding boards to old-growth southern yellow pine from torn-down manufacturing facilities below the Mason-Dixon line. While not necessarily local, and generally more expensive than local wood, this approach provides access to ancestral wood that is no longer available—they really don't grow them like they used to—and encourages the remanufacture of existing products, rather than reliance on the extraction of new resources. After all, the scale-appropriate nature of a local timber industry is a key part of what makes it a sustainable operation. Finally, the use of endangered tropical hardwood products for flooring is not an acceptable option, regardless of how "natural" and "regenerative" the source, because of the many forms of ecological damage caused by their use, including deforestation and desertification, loss of biodiversity, and increased global warming potential from the reduction of the forest as a carbon sink and the carbon emissions resulting from long-distance transport of the material.

CORK AND BAMBOO FLOORS

In chapter 5, we discussed the renewable resource cork, which is the exterior bark of the cork oak tree, and bamboo, a vigorous grass, and how both outpace wood in their propensity to regenerate and, in the case of cork, even create sustainable habitat through its commercial production rather than its depletion. In that regard, both cork and bamboo present themselves as strong candidates for ecologically sound building materials. Considering that virtually all buildings need floors, bamboo's ability to provide abundant flooring material without the significant ecological impacts of the kind frequently found with commercial-scale wood management (such as clear-cutting and the associated degradation of ecosystems) is highly relevant. On the other hand, depending on how heavily you are weighing the values of local sourcing and production, these materials may not seem as favorable, considering that most cork comes from the Iberian Peninsula, and most bamboo from Asia.

Pros and Cons of Cork Flooring

Cork is favored as a flooring material for many reasons. It is soft underfoot, and its cushioning is valuable in

2 × 6 Tongue-and-Groove V-Groove Spruce Flooring

One of our favorite flooring systems is 2 × 6 tongue-and-groove V-groove spruce flooring. We do a lot of work in timber-framed homes, and in homes that have interior timber-frame elements (for example, a second-story loft or second-floor system, within a stud-wall structure). The floor joists in these homes are often quite beautiful, sporting handsome hardwoods or custom taper detailing at the joinery. To hide these gorgeous works of art with drop ceilings would be a crime against craftspersonship, an assault on the aesthetic design of the building, and a waste of time and money. That said, slapping down a plywood subfloor sets up the painstaking (read: expensive) proposition of installing some sort of trim finish between the joists from below, lest the beauty of the frame be ruined by the exposed plywood.

The answer? Two-by-six tongue-and-groove V-groove spruce flooring. The "2 × 6" part ensures the boards are structural enough to span the 36 inches or more on-center spacing of the timbered floor joists. The "tongue-and-groove" milling provides for simple and secure installation, with no visible fasteners. The "V-groove" part not only aids in installation but is beautiful to look at from below and helps break up the flat plane of the "ceiling." And the "spruce" part—well, that can be a mixed bag. On the one hand, it is plentiful, easy to work, light, and best of all readily available from local sawmills; we work regularly with one mill that has been operated as a small family business for generations, and it supplies us with very high-quality flooring for a fraction of the price of what could be found at a flooring retailer. On the other hand, as a softwood, spruce is, well, soft. It dents much more easily than a hardwood, and that could be a problem in certain cases. In lower-traffic areas—which are frequently upstairs, atop those lovely joists—this may not be a problem.

Consider taking an "if you can't beat 'em, join 'em" approach, and allowing the floors to dent under the wear of children and houseguests. After all, a house is for living in and perhaps can be allowed to look the part. This kind of flooring also works well in situations where a subfloor is not desired. In that most subfloors are constructed of sheet goods that, more likely than not, contain an appreciable amount of formaldehyde as a binder in their formation, this system offers a way to save money and relieve compromises in indoor air quality.

There are some downsides, of course, in the approaches mentioned above and the product itself. For one, drop ceilings themselves can go a long way in reducing sound transmission from the first to the second floor. They also provide terrific wiring chases and places to hang flush-mount overhead light fixtures, and they can cover up second-floor plumbing for tubs and toilets. All of this must be considered when "spec'ing" an exposed-joist ceiling; you may find yourself installing a drop ceiling in certain areas anyway. Not having a subfloor can present problems, too. Over a basement, air is quick to travel through the many gaps that can form in the flooring if it is allowed to get wet before being finished (and which may well form even if treated with kid gloves), bringing with it moisture (in many cases) and causing drafts on the floor. From above, dust, spilled water, and even small Lego pieces can find their way down between gaps, knotholes, or other imperfections in the flooring.

About those gaps: unlike click-locking and pre-finished hardwood products, if you are buying locally milled spruce, you can expect some opening up between the floor boards. If this is going to be an issue, look elsewhere for a more stable product. But if you are able to embrace a bit of imperfection, this may be the floor for you.

Two-by-six tongue-and-groove spruce makes a fine floor. PHOTO BY KELLY GRIFFITH.

areas such as kitchens, where the user may be standing for extended periods of time. Cork is reasonably durable, especially once sealed. Cork's structure of air-entrained dead cells makes it not only waterproof but also incredibly lightweight and somewhat insulating, which helps improve acoustical performance (for example, as flooring on an open-joist floor system to help isolate the upstairs bedroom from the downstairs living room) as well as thermal performance (such as when laid over an unheated slab). Cork comes in a variety of different forms, most commonly as tiles, planks, or panels. These different systems may be floating, in which units lock together and "float" over the subfloor, or glue-down, in which an adhesive is used to install the units to the subfloor. Whereas some products may be pure cork, others are laminate systems that also include fiberboard cores. There are some cork products on the market that incorporate use of recycled material—specifically, sliced cork rounds or ground cork from wine stoppers, which of course helps offset consumption of virgin cork.

Sealing cork floors is necessary to keep out stains and excess moisture (especially in kitchens and bathrooms); many products come pre-finished, in a polyurethane or varnish, while others are unfinished.

One of the greatest drawbacks to cork comes from the impact on indoor air quality as a result of the manufacturing and installation process. First, in products containing fiberboard cores, there is a strong likelihood that the fiberboard is bound with formaldehyde-based resins (for more information on the dangers of formaldehyde, see chapter 2). Adhesives used to install the flooring to the subfloor could well be formaldehyde-based as well, or have other noxious VOCs that will release into the interior environment of the home, unless specific low-toxicity adhesives are sought out, and supported by the manufacturer for use with their product. Finally, the sealer can be a significant source of air pollution; if purchasing pre-finished flooring, carefully vet the potential impacts of the finish if this is a concern. Another option would be to select an unfinished product and find a sealant that will minimize the amount of airborne contaminants; see chapter 20 for more information on low- or nontoxic sealants.

Cork floors are beautiful, comfortable, and durable. PHOTO COURTESY OF WIKIMEDIA COMMONS.

The other consideration with cork is its cost—generally in the \$3–\$8/square foot range, not including installation or sealing; cost increases with finishes, locking systems, and fancier patterns.

Pros and Cons of Bamboo Flooring

Bamboo also offers many strong benefits as a flooring option (see chapter 5 for more information on bamboo). Although a grass, bamboo behaves similarly to, or better than, wood. It is stronger than concrete in compression, and comparable to steel in tension when used structurally; as a floor, it is generally comparable to, or harder than, hardwood flooring. This illuminates one of bamboo's strengths and weaknesses: while high-quality bamboo products are incredibly hard—harder than red oak—there is great variability in the marketplace. As the demand for bamboo has grown and more manufacturers have entered the field, ensuring a quality product has become more difficult; cutting bamboo too early—younger than three years—will yield a weaker product, and it is important to confirm that the manufacturer is reputable and selling a quality product if the hardness of the floor is valued.

Bamboo flooring comes in a variety of different options, similar to cork: tiles or strip flooring are available, both in locking, floating systems and in glue-down applications. There are different orientations of the bamboo, which will yield different appearances, and steam-treating the bamboo is done to caramelize the sugars in the plant to create a darker finish. While this steaming weakens the flooring slightly, a high-quality product will still be on par with, or stronger than, hardwood flooring. Like cork, some bamboo products are laminates that layer bamboo around a fiberboard core, whereas others are pure bamboo. There are dozens of different types of bamboo construction products available, from flooring to trim and molding to stair parts to sheet goods. Bamboo is available either pre-finished or unfinished. Again, as with cork, indoor air contamination is a concern in the laminates, the adhesive (if using an adhesive application system), and the finish—and diligence should be taken in sourcing a low-toxic product. Formaldehyde-free binders are beginning to show up in bamboo laminate products. The cost of bamboo flooring ranges widely, from $2–$8/square foot, not

Bamboo floors perform comparably to hardwood in feel and durability. PHOTO BY ACE MCARLETON.

The Green vs. Natural Divergence

Cork and bamboo beautifully exemplify the differences between green-building and natural-building philosophies, and they provide an opportunity to contemplate our choices through the lens of "integrative design."

As opposed to the relative scale, control, and impact of utilizing site soil, stone cleaved from a local hillside, or maples sawn in a mill down the road, these natural materials are fed through a global manufacturing and industrial pipeline that simply cannot be readily reproduced in most communities in North America. They are also potentially laden with chemicals that deplete the air quality of the buildings in which they are installed. A thorough analysis of the development of the material—from its sourcing and manufacturing to its transportation and chemical inputs—displays characteristics that distinguish it from those of natural building materials, yet are in keeping with many of the shared objectives between the natural and green building communities, highlighting their relevancy as part of an integrative building strategy. Therefore, this should not dissuade the use of cork or bamboo solely because we do not consider them to be "natural building materials." If, at the end of an honest and thorough evaluation of the project's needs and the available options, bamboo or cork rise to the top of the list, go forth with the knowledge that these are fine options indeed for the floor in your natural building.

including installation; pre-finished, locking, and low- or no-formaldehyde options will increase the cost.

Another potential ecological downside to bamboo comes from cultivation practices. While some bamboo forests are responsibly managed—and have been for hundreds of years—others are less so, and are displacing existing forestland for bamboo plantations and contributing to issues such as loss of biodiversity and erosion. FSC certification (see the "Wood Floors" section in this chapter) is beginning to be applied to bamboo products, confirming the integrity of the cultivation practices when evaluating different manufacturers, although, on balance, bamboo production tends to have a more positive environmental effect in regards to land-use patterns in comparison to other products (Malin and Boehland 2006).

CHAPTER 20

NATURAL FINISHES

The world of natural finishes is unique in contrast to many other forms of natural building. For one thing, finishes are often the most visible. Whereas your foundation, your insulation, possibly even your framing are all hidden from view once the building is complete, the coats of color on the wall are applied to be seen, and that lends a different energy to the process. For another, they are called "finishes" for a reason—you are nearly done! Applying paints, washes, and glazes generally signify that the project is close to completion, short of moving in the furniture. For a third, the techniques and recipes presented in this chapter represent what is arguably the most accessible form of natural building. Not everyone can incorporate a woodchip-clay wall or a timber frame into their project—especially if they own an existing structure. However, there is always a need for a fresh coat of paint on the walls, and these materials play as nicely with wallboard and wood trim as they do with earth and lime plasters. For the remodeler or more conventional builder looking to bring a look or element of natural materials into the project, working with natural finishes is a great way in, and represents a terrific bridge between the worlds of "natural" and "green" building. Finally, considering that more than 1.2 billion gallons of paint are used each year in the United States, a great opportunity exists to dramatically improve the impact of the built environment with relatively simple modifications in product selection and design.

We are referring to finishes as being brush-, roller-, rag-, or sponge-applied thin-surface films or penetrating coats that could be applied over plaster, wood, masonry, wallboard, or many other surfaces. Within this broad definition we include paints (thicker, more opaque films), washes (similar to paints, but thinner and less opaque/semitransparent), glazes (translucent, occasionally colored, pure-binder or binder-heavy films), and sealers (translucent finishes designed primarily to seal the substrate, whether as surface films or penetrating). For the sake of organization, we have kept our plaster finishes in chapter 17. In this chapter, we will first evaluate the criteria used in determining the right finish for a given project and analyze the different components of a paint. We will then take a closer look at recipes of different finishes, categorized by the binder ingredient, which we will explain later in this chapter. Finally, we will explore the process of coloring natural finishes, including the materials used to create color, basic color theory, and how to conduct tests to create different colors of natural finishes.

Choosing the Right Finish

There are many different reasons we might choose to use natural finishes in natural buildings, and knowing the rationales will help us select the right finish for the job.

Here are some attributes of finishes that may be desired:

Color. This may be the most common reason to apply a finish; adding color to trim, a wall surface, or a ceiling will completely transform a space and heavily influence the user's relationship with that space.

The beauty of natural finishes is unparalleled and cannot be replicated by synthetic industrialized products. PHOTO BY ACE MCARLETON.

Smoothness. This feature is especially evident in the case of earth and lime plasters; it may be preferable to use a finish of clay paint or limewash to cover over minor trowel marks, sand chatter, hairline cracks, or other "imperfections" in the surface.

Texture. On the other hand, it may be preferable to add a bit of texture to the wall. Adding some sand, finely chopped straw, or flecks of mica is a way to break up the monotony and static feel of a painted drywall-finished room; decorative painting techniques such as sponging, ragging, skip-rolling, brush patterning such as *strie* (adapted from the French meaning "striate" or "stripe") or crosshatching, and color layering with semitransparent washes and glazes is a very effective way of bringing vibrancy to a room.

Durability. Whether applied on the exterior over wood, plaster, or porous masonry finishes or on the interior in vulnerable areas such as on floors, baseboard trim, or wall surfaces behind sinks, applying a paint, wash, or sealant is, in many instances, a requisite part of the building system to prolong the life of the substrate and keep the building functional. It should be noted, however, that "durability" in the realm of natural building, for the most part, takes a different approach than that of conventional building. Whereas a synthetic paint aims to protect a surface by creating an impenetrable layer strengthened by the inclusion of petrochemicals, the goal of a natural finish is to act as a sacrificial coat that will wear slowly over time; the more slowly the coat wears, the more durable it is considered to be.

This is an important distinction, particularly with regard to how these finishes affect the building's response to moisture. For the most part, the ability of most of these sacrificial coats to both repel liquid moisture and allow vapor to permeate through them is highly prized in a cold, wet climate. The "impenetrable fortress" approach of synthetic paints is much more liable to trap any moisture that is able to migrate into the wall and may well cause the very problem it is intending to solve. At the least, it can be expected that the nonporous vapor-impermeable paint layer will blister and peel when applied over a porous surface such as a plaster and ultimately delaminate from the substrate as the effects of moisture and the freeze-thaw cycle take their toll.

Control of dusting. Even as experienced plasterers, we sometimes find our walls dusting. When this happens, it sure is nice having a few tools in the belt to control dusting. It is interesting to note that clay finishes, being negatively charged, are inherently repellent to dust, whereas common latex and acrylic petrochemical-based synthetic paints are statically charged and will attract dust.

As with many of the other building elements presented throughout this book, having a clear set of criteria with which to evaluate our options is important. For the most part, those criteria are familiar: cost, ecological impact, effect on the immediate environment (in particular, for finishes, indoor air quality, or IAQ), durability, ease of maintenance or reapplication, and aesthetics are the primary considerations that will guide our evaluations of these different materials.

The Anatomy of a Finish

Similar to how we defined plasters in chapter 17 as being composed of binders, aggregates, fibers, and other additives, so too can we classify the essential ingredients of paints and washes. These ingredients are binders, solvents, pigments, and additives.

Binders are an essential ingredient of every paint and wash. Similar to plasters, the binder is the glue that forms the substance of the film that remains once the paint is dry. It is also the substance in which the pigment particulates are suspended. In synthetic paints, the binder is usually acrylic or vinyl, both petrochemicals (see chapter 4 for more on how petrochemical binders are produced). In natural paints, binders include clay, lime, casein (milk protein), egg, plant oils (linseed, tung), flour pastes, and animal glues (hide or rabbit-skin sizing). In contrast to their nonrenewable petrochemical equivalents, natural binders' characteristics—the abundance of raw material, how easy they are to produce, the light footprint of their manufacturing, and the growth potential for small-scale agricultural and agroforestry industries as material sources (for nearly all binders except lime and clay)—point toward immediate or potential social and ecological benefits.

Solvents, also know as **carriers**, are also essential ingredients that are used to create a suitable consistency so that the paint or wash can be easily applied. After application, the solvent evaporates, leaving the binder-based film entrained with pigment and additives. As mentioned earlier, it is the evaporating solvent that is generally the source of the VOC content of paint. The primary solvents in synthetic paints are petroleum-derived mineral spirits or synthetic turpentine, water, or denatured alcohol. In natural paints, the solvents can be plant-based turpentine (distilled from pine tree resin) or thinner (from citrus fruits), water, or plant-based alcohol.

Pigments are frequently formulated into paints and washes, although not always. The pigments lend the color; they will be discussed in greater detail at the end of this chapter.

Additives of various sorts are added for different purposes, primarily in more heavily manufactured synthetic and "eco" paints that are water based, to compensate for restrictions by governmental regulators on solvent usage as a result of VOC emittance. These additives can include biocides, emulsifiers, stabilizers, antifoaming agents, and other materials to improve flow control, viscosity, surface tension, UV resistance, and other qualities. Additives are often used in natural paints as well, such as clove oil or borax as preservatives.

A Brief History of Paint

People have been grinding up minerals and plants and using them to paint walls, objects, and even their bodies for tens of thousands of years, since the Paleolithic era. In fact, many ancient cave and artifact paintings still exist, dating from the earliest in Australia at 50,000 years ago, to some of best known in Spain from 15,000 to 20,000 years ago, with other sites located in France, India, and elsewhere across the globe. These early artists created their paints through grinding lime and chalk (white), soot and charcoal (black), and the reds, browns, and yellows of natural earth clays and minerals, in hollows in the ground with bones and stones. These were combined with water, blood, and animal fat to create sophisticated and durable paints that were applied with twigs, feathers, fur, and hair, or sprayed through hollow bones and reeds (the first airbrushes).

As time went on, further discovery of boiling pine tree resin, harvesting beeswax, and using residues from the lac beetle all increased humans' ability to express themselves in artistic medium. A major leap forward was achieved by the Egyptians 2,000 to 5,000 years ago, who, through their unprecedented technological advancement, innovation, and discovery, brought to the world of artistry refined gums, resins, and waxes, lapis lazuli and azurite minerals (blue), processed lead ore (white and red), blues and greens through the treatment of copper plates with ammonia, vinegar, urine, and grape leaves (some of the first synthetic pigments), oils, mastics, eggs, milk and casein paints, and alcohol.

Developments in technology continued throughout the Greek, Phoenician, and Roman empires, leading up to the European Renaissance, which featured further innovation and application with heightened levels of craftspersonship and artistic mastery. Subsequent phases of extravagance and simplicity rose and fell throughout Europe and the New World in the ensuing centuries, all the while making use of these ancient materials and techniques. Then, things changed in the latter half of the nineteenth century, when entrepreneurial Americans began to mine and develop petroleum oil into a vast array of products, including paints. While the use of traditional materials was standard practice all the way up until the 1950s, in the last fifty to sixty years, with the rise of the petroleum age and the invention of vinyl and acrylic, synthetic paints have all but eliminated the common use of traditional paints. The durability of petroleum as a binder and the cheap cost of petroleum-based oils and thinners as products of the new oil economy secured the ubiquity of petroleum's use in modern paint formulation. Accordingly, traditional paints have been relegated to niche and specialty fields or to markets without access to the advantages provided by industrial manufacturing.

Why such a dramatic change? With the tremendous post–World War II construction boom and industrialization of building came the promise of maintenance-free, easy living, and paint was no exception. Pre-manufactured, replicable in color, easy to apply, more durable, and best of all very inexpensive, these new synthetic paints have ruled the market ever since. However, we have since realized that they do indeed come with a price, although not one paid at the register. Instead, the health effects of volatile organic compounds (VOCs) emanating from paint's evaporating solvents are real and well documented. Caution must be paid by the consumer who seeks out even the "low VOC" or "no VOC" paint, as labeling is not as strictly accurate as it could be.

A Sampling of Natural-Finish Formulations

As discussed in the sidebar "A Brief History of Paint" earlier in this chapter, there are many, many different formulations of natural finishes that have been used across the globe throughout time. Below, we will explore a few of the most common we use in our contemporary practice.

CLAY PAINT OR *ALIS*

On DVD, Chapter 20, see MIXING AND APPLYING CLAY PAINT (ALIS)

Simple clay-based paint finishes, or *alises*, have been used for centuries to protect and beautify earthen walls across the globe. In its most basic form, an *alis* is a mixture of fine clay, sand or mineral powder, and starch paste. The word comes from the Spanish word *alisar*, which means "to make smooth." In North America, the tradition of applying alis finishes is best known—and still actively practiced—in the desert Southwest, where the finish is used on both the interior and exterior of adobe walls. Traditionally, the women of Native American communities were the masters of clay plastering and painting, known as *enjarradoras* (plasterers), and would apply the finishes using untanned woolly sheepskins.

Benefits of Alis

Whether tinted with mineral pigments or colored by the natural clay binder, sparkling with mica or textured with straw flecks, the beauty of a decorative alis finish is unparalleled, and it is one of our favorite finishes over plastered walls. As a highly vapor-permeable finish, alises work in concert with a vapor-permeable wall system; being clay-based, they will also help buffer excessive vapor buildup in moist areas of the building and are inherently antistatic. The materials in alis are inexpensive and accessible, and working with it is a rewarding and gratifying experience, both during the process of formulation and for years to follow as its beauty radiates from the walls.

Where to Use

Alises can be applied over almost any substrate. As discussed above, they are particularly well suited over porous substrates and are a favorite finish over plastered walls. We have also had wonderful results when applying alises over painted and unpainted drywall, plywood, and even insulated metal doors. Drywall finished with conventional joint compound can be unpredictable, however. The gypsum component is very sensitive to water and the chemicals contained in the product may vary, and alis paint has been known to peel when applied over untreated joints. Therefore, it is best to prime drywall with a flat primer before applying an alis; we have also had favorable results priming the joints with a starch-paste glaze (see page 350). Smooth and porous walls can often be covered with two coats, whereas latex-painted or other nonporous walls will often take several coats for complete coverage (see sidebar in this chapter, "Working with a Nonporous Substrate").

Clay alis paints can be applied over a wide variety of substrates, helping bring natural finishes into many environments. PHOTO BY ACE MCARLETON.

Working with a Nonporous Substrate

Whereas nonporous synthetic paints are incompatible with porous substrates like plaster, natural paints and finishes often have trouble adhering to nonporous finishes, such as previously painted walls, especially those with higher-gloss finishes. There are a few different approaches that can be taken to help address this situation, as it is ultimately much easier—and healthier for the building—to get a porous paint to endure over a nonporous substrate than the other way around.

Evaluate the substrate. Is the surface chipping, peeling, flaking, or breaking? Are there cracks, gaps, or holes? Like plaster, your paint is only as good as the substrate. Remove any friable (unstable) material and patch larger holes or gaps with a high-quality wall spackle or joint compound. Note that different patch compounds have different chemical formulations, which may produce unpredictable results as a direct substrate for natural finishes. Unless you are interested in experimenting, it is always best to prime over patch compounds before painting.

Clean the substrate. Dust, dirt, grease, and oils will all disrupt the paint's ability to adhere to the wall. Vinyl-based paints often attract dust as a result of their chemical nature, and surfaces that are liable to be frequently touched, such as along a stairwell without a handrail, may accumulate skin oil. Kitchens are also notoriously greasy areas, depending on the type of cooking involved. Washing with a simple soap-and-water solution will often do the trick. Greasier or higher-gloss finishes can benefit from the use of trisodium phosphate (TSP) cleaners, followed by a thorough rinsing; they are more powerful than common soap and will help break down the gloss and open the pores of latex paint. Be cautious when using TSP, however; it is very caustic (a strong alkali) and can cause staining and damage to metal, glass, wood, ceramic tile, and grout. Additionally, the use of TSP has been severely curtailed since the 1960s, when it was recognized that large amounts of phosphorus contamination in watersheds was leading to uncontrolled algae blooms and damaging lake, river, and estuary ecological systems. Use this product judiciously and take care to keep it out of the watershed if at all possible. In some cases, a light sanding with a medium-grit sandpaper may be warranted to help prepare the surface or to reduce the need for more aggressive cleaning agents; do this before cleaning.

Prime the substrate. We have had great success in applying the clay alis discussed later in this chapter directly over the substrate. That said, sometimes we will use a wheat-paste glaze, also discussed later, as a transition coat to help adherence for an alis, if going over a particularly slippery paint. Other finishes, such as limewash, are much more difficult, and we generally will not use a limewash over a latex-painted wall. There are products available, however, that are designed to create a stable coat onto which limewash can be applied; one such product, Lime Prep Primer by the Earth Pigments Company, is specifically formulated for this purpose; alternatively, a thin plaster can be applied over the surface to ultimately receive a limewash. We have also had reliable success applying casein paints over a latex-paint substrate, although most manufacturers of these products recommend against this practice as being unreliable, and needing a porous substrate to guarantee adhesion. Certainly, if the goal is complete coverage over the paint, selecting the correct finish is important; you will receive much more opacity from a paint than a wash. If a washed look is desired, or if full opacity is desired and the base is either dark or inconsistent (different colors, lots of patches), then priming the substrate with a flat white or gray wall primer will be a helpful step in taking the pressure off the finish wash or paint to cover over the existing imperfections in the wall. We find it takes an extra coat or two to achieve complete opacity over a nonporous substrate.

Enhance the paint. It may be necessary to augment the paint itself by adding additional binding additives. In alis, we may increase the flour-paste content, as it is a very effective binder. In the case of casein paints or limewash, it is frequently recommended by manufacturers to add in a water-based acrylic binding agent. When purchasing such products, however, be sure to ensure that they are compatible with the paints you are using and that they are as nontoxic as possible. It should be noted that these additives may also affect the vapor permeability of the finish, although if being applied over an existing latex paint, one is to assume that this is not an issue.

Test test test! Always test a new paint over a tricky substrate! Whipping up a small batch of paint base and finding a discreet 1-foot-square area to sample on will either save you a world of trouble or give you the confidence you need to proceed without worry.

And, as always, if working on top of existing paint, it is possible substrates may contain lead. All necessary precautions must be followed to safeguard human health. For more on guidelines for safe working procedures with lead paint, see the U.S. EPA's website on lead paint listed in appendix A.

Alises have little resistance to moisture, meaning that they are not suitable for exterior use unless they are combined with oil or covered with a sealant (see the last section of this chapter). Similarly, they are not suitable for washing and wiping, unless similarly treated, so be cautious when applying in wet and greasy areas, such as bathrooms and kitchens. Another consideration is the starch-paste component; we have experienced mildew problems as a result of the inclusion of starch paste when applied in chronically moist areas, such as poorly ventilated basements and subterranean rooms. We have addressed this by including borax as a preservative (see the "Starch Paste" sidebar in this chapter); it may also be worth considering reformulating the paint without the use of starch paste, or using a hygienic finish, such as a limewash, instead.

Ingredients

We frequently use white kaolin clays—EPK is a favorite—for our paints, as they are light colored and receive color very easily. Kaolin clays are not as strong, however, as ball clays, and if a darker shade is tolerable, using a ball clay, such as OM-4, will make a stronger and more adhesive paint. Additionally, colored clays such as Newman's Red (red-orange), Red Art (red-purple), or C-Red (dark red) can be used for their own color, or in concert with pigments to create new colors (see chapter 4 for more about clay).

While there are many different recipes for clay alis, we prefer to formulate ours with a percentage of powdered aggregate to strengthen the mix and add body to the paint. Our favorite aggregate is a manufactured silica sand, which has the consistency of a fine powder. We have also used marble powder with great effect, although it is more expensive. Mica powder, discussed below, is a wonderful addition as a mineral powder. All of these minerals are light or white in color, helping to receive pigments more easily. If a darker shade is desired, granite powder can be used; we have a terrific source of waste granite powder from a nearby quarry and are able to make use of a waste product to make a high-quality finish when a gray mineral powder can be used.

The recipe can be adapted further from that point. If a slight texture is desired, or the paint is needed to fill in larger pore spaces or imperfections in the plaster finish, a fine-grit sand (40–60 grit) may be used in place of, or in concert with, a finer, powdered sand or mineral powder. In the final coat, the sand may be omitted altogether, if the paint is applied at a thinner consistency. Mica is commonly introduced into alis paints, and with good reason. Mica is a mineral closely related to clay, and as such it has a very flat crystalline structure, which makes the paint easy to apply and lends wonderful visual effect without necessarily changing the texture. Fine powdered mica used as part or whole of the mineral-powder component of the mix gives the paint a luminous glow, whereas flecks or even chips of mica suspended in the paint will brilliantly reflect light when burnished to remove covering clay. When adding mica, include the amount as a portion of the sand in the mix. Small quantities of finely milled straw or grasses can also be used for a different textural effect. Starch paste is a standard part of the recipe, although it can be removed or augmented with other binding additives, such as casein powder. Small quantities of oil can be emulsified into the paint to enhance durability or to darken the finish. Borax—a common, nontoxic, inert mineral used as a preservative and fire retardant—can be included as well, to help control mold during drying and mildew over time in damp locations.

Sealing

If you want to further protect them from moisture, or aid in their maintenance by making the paint washable, alises can be sealed with an oil glaze, casein glaze, a clear sodium- or potassium-silicate ("water-glass") sealant, or a wax product. It is always recommended to do a test before application, to ensure both favorable performance and aesthetics. The following is a list of sealants appropriate for use with alis.

Oils, while very effective at sealing the paint, will darken the finish to the point that the original color may be jeopardized. On the other hand, an oil glaze that has been diluted with citrus thinner can contain a much smaller quantity of oil and darken the color only a small amount. Excessive oiling will reduce vapor permeability.

Casein glazes are simple, inexpensive, and odorless and will not substantially affect the shade or gloss of the finish. They are less durable, however, and may not be strong enough to give the protection you need, and they will require more frequent reapplication.

Silicate-based sealants are highly protective and very durable, forming a chemical bond with the silicates in the paint. They are vapor-permeable while being very water resistant. Sodium silicate will darken the surface, whereas potassium silicate will not (or will to a much lesser degree). They are expensive, however, and are the product of a more intensive manufacturing process (see later in this chapter for more information).

Waxes, such as beeswax or pre-manufactured wax sealers, are similar to oils, although more concentrated in their effects of both protection and reduced vapor permeability. Multiple applications will build up more of a surface protection, which will also affect the gloss.

Storing and Reapplication

As with most natural paints using food ingredients, starch paints such as this alis formulation need to be used soon after they are made. They can be refrigerated for a short period of time (three to seven days) but the binding ability of the paint will suffer. The best way to store an alis for future use is to dry the material in pans in the sun, or at low temperature in the oven. Then break the dried paint into chunks and store them in an airtight bag or container, labeled with the location of application and details of the recipe (size and quantity of mica used, pigment type and quantity).

To touch up a wall, even years down the line, simply grind a chunk of dried paint in a coffee grinder, or grind thoroughly with a mortar and pestle, until it is the consistency of a very fine powder. Add enough water to reconstitute to the consistency of paint, and apply.

How to Make Alis

The process of making alis is quite simple. Clay—generally powdered, but occasionally slip if using a raw soil—is combined with sand or mineral powder and starch paste to the consistency of paint. We generally make our alises in 5-gallon buckets with the aid of an electric drill fitted with a paint basket or paddle-bit, although smaller batches can be made by hand with a whisk in a bucket, and larger batches in a mortar or cement mixer, or by hand with a hoe in a mortar pan. *Always use a fine particulate mask when working with powdered materials!* Inhalation of powdered silicate-based minerals—including clay and silica sand—exposes you to permanent respiratory damage, silicosis, and cancer; always protect yourself, and mix in a well-ventilated area.

Ingredients:

- 1 part bagged clay (kaolin, ball, or colored)
- ½–1 part sand and/or mineral powder (fine-grit sand, silica sand, powdered limestone, ground mica, granite powder, marble powder)
- 2 parts, approximately, diluted starch paste (see recipe on page 349)
- pigments, as desired

Instructions:

1. Prepare diluted starch paste and set aside.
2. Combine the clay and sand/mineral powders, mixing thoroughly until fully blended.
3. In a separate container, pour in three-quarters of the starch paste. Then, slowly add the powder while blending constantly to avoid lumps. Continue to add powder until the desired paintlike consistency (thick cream) is achieved. Add more diluted starch paste if necessary.
4. Add prepared pigment (see "Using Pigments" later in this chapter) and blend until color is uniform.

Coverage: 100–250 sq. ft./gal., depending on texture and porosity of wall surface.

How to Apply

These paints can be applied with sheep's wool, a brush, or a roller. If brushing, use an all-purpose medium-bristle brush; experiment to find the brush that works best for your application. Standard all-purpose rollers are acceptable as well.

It is always advisable to protect non-painted surfaces in a room. Lay down a drop cloth and tape it to the floor or baseboard; use 1½-inch to 2-inch, fourteen-day blue painter's tape to mask all edges of the painted field, especially over porous surfaces.

If applying over a porous surface such as plaster, mist the wall down before beginning to prepare for the paint, and then again just before painting to help control suction; this will make application easier and drying more even. Use a brush to apply, and work the paint well into the plaster by brushing in all directions; finish with even vertical strokes. Wait until fully dry (the color will be much lighter and will not rub off), then apply a second coat; repeat with a third coat if desired.

Making alis is simple, inexpensive, and best of all fun! PHOTO BY ACE MCARLETON.

Tips for Mixing

We like to prepare wet and dry ingredients separately—like baking!—and then combine into a third container. This way, you can continue to amend the mix as need be, which is especially helpful as you are getting familiar with the recipe and your materials. This strategy really comes in handy if you have a limited amount of material; if the quantity of dry ingredients is restricted by the amount of clay or mineral powder you have, and you pour it all into the starch-paste bucket only to find out it wasn't enough and you now have a bucket of muddy water, either your project will come to an abrupt halt, or you will have to change plans to apply an alis wash, rather than paint.

It is good protocol to add dry ingredients to wet ingredients. This serves a few purposes. For one, you are less likely to get a faceful of powder than if you pour liquid into a container of dry mix, especially if you do so at any speed. More importantly, you are far less likely to get lumps in the mix, and it is much easier to fully suspend or dissolve the dry materials when added slowly and continuously mixed. Finally, you greatly reduce the likelihood of dry pockets of material in the corners of the container, which can be quite cumbersome to release into the mix, especially if using a rounded whisk or paint basket.

If applying over a nonporous surface such as a latex-painted wall, follow wall preparation instructions in the sidebar "Working with a Nonporous Substrate." For best results when an even, monochromatic coat is desired, approach the wall from one end directly through to the other; do not leave a dry edge (for example, by cutting in all the edges in a wall and then returning to the fields), as this will lead to an uneven application and may cause the drying paint to "lift" as wet paint is then rolled over. Start by cutting in edges along the top, bottom, and sides of the wall in a neat and uniform fashion, using a trim roller or 2-inch brush; immediately switch to roll-applying the fields with a standard 7-inch roller, coming as close to the edges as possible. Continue across the wall until a uniform coat has been applied. As above, wait until fully dry to reapply.

The brushstrokes of the finish coat may be apparent, depending on the amount of burnishing and type of sealant coat. Burnishing will help integrate brush strokes and reveal mica flakes, chopped straw, or sand particles. To burnish, wait until the coat has partially dried but is still lightly damp, in the case of a porous substrate; for a nonporous substrate, allow the paint to dry completely. Using either a clean, well-wrung large yellow tile sponge or plastic trowel disk (such as the cut top of a plastic yogurt container), rub the surface of the wall in circular motions. Start lightly at first in one corner, and adjust the vigor and pattern of the burnishing until the correct balance of exposing the mica and rubbing out the brush strokes while not washing the paint off the wall has been achieved. Keep the sponge or plastic clean to avoid marring the finish.

Cleanup

Cleaning alis paints is a snap, particularly over nonporous surfaces. The easiest cleanup involves a quick wipe with a wrung sponge or rag while still wet; slight abrasion with soap and water will lift dried drips. Extra paint is simply clay, flour, and pigments and can be disposed of safely.

Traditionally spread on with sheepskin, clay alis can be applied on a wall like a conventional paint, by brush or roller. PHOTO BY ACE MCARLETON.

Casein paint, or "milk paint," has been used for hundreds of years in North America. PHOTO BY ACE MCARLETON.

CASEIN PAINTS, WASHES, AND GLAZES

On DVD, Chapter 20, see MAKING MILK CURDS (QUARK) AND FORMULATING A CASEIN PAINT

Casein is a milk protein isolated from fluid milk that produces a highly adhesive and quite durable binder. Casein paints, washes, and glazes have been used for centuries by everyone from peasant farmers rich in milk to Shaker craftsmen. Casein is a finicky paint and needs to be formulated carefully; the proportions of casein to binder must be kept in balance, or else cracking, peeling, or dusting may occur. Subtle adjustments to the basic formulas are often needed to obtain consistent results. Well-made casein paint that is fresh can last indefinitely on interior surfaces. Like all truly natural paints made from food products and without preservatives, it will spoil, but on the positive side, like most food products it can be safely composted, as can the cleaning water. The dried paint is not prone to fungal growth.

Benefits of Casein Paints

Casein is one of the most adhesive of natural binders; you have seen it mentioned throughout this chapter as an adhesion additive to various other paints and finishes. In combination with lime or borax, casein can make a very durable paint that is lightly washable on interior surfaces. As a translucent binder, casein is able to retain more of the hue of a pigment, allowing for richer colors. Perhaps best of all, casein applies like more-familiar synthetic paints, especially over nonporous surfaces; it is smooth and thick and provides great coverage. Making casein paints—especially from milk—is an exciting process, incorporating elements of folkloric tradition, creative joy, and artisanal craft, that produces a finish that is as practical as it is beautiful. Finally, being vapor permeable, casein paints are fully safe to use on vapor-permeable wall systems, such as those presented in this book.

TABLE 20.1. ALIS TROUBLESHOOTING

Problem	Cause	Solution
Paint is dusting onto hands, clothing	• too much sand/powder • too little clay • too little/too dilute starch paste	• reformulate mix accordingly and recoat • seal with starch-paste glaze sealer
Paint is cracking	• too much clay • too little sand/powder • paint applied too thick • porous substrate too dry before application	• reformulate mix accordingly and recoat • adjust wall preparation and application technique and recoat
Paint is too gritty	• sand is too rough • impurities in mineral powder • impurities in pigment	• reformulate mix with less coarse sand and more powder • examine powder and pigment for impurities, sift if necessary or source new material
Paint is uneven, mottled	• paint is still drying • paint was applied unevenly • substrate was improperly prepared	• wait for paint to dry • adjust application technique and recoat
Paint is peeling/not sticking to wall	• substrate was improperly prepared	• wash paint off wall, adjust wall preparation techniques, and reapply • confirm quality of paint formulation on sample board.

Notes: Use this table to help troubleshoot your alis, first by recognizing the symptom, then by identifying the cause, and finally by selecting a solution to address the problem.

Where to Use

The same condition of compatibility of porous surfaces with porous paints applies to casein paints as well. As a superior binder, however, casein makes it easier to achieve proper bonding over smoother surfaces and is frequently used as an additive to help other finishes achieve the same goal. Because of its biologically active and unstable organic composition, this paint is best used indoors and avoided in areas that have constant high humidity such as a basement, root cellar, or laundry area, unless significant amounts of lime or borax are in the mix. Because of its moderate durability, it can be used on surfaces that require light cleaning; it is not a fully washable paint and is not suitable for exterior use in wet climates.

Ingredients

As mentioned above, casein is a protein found in milk, and in its simplest form it can be most easily isolated by making quark—the curds made from curdling milk with an acid. Quark is enjoyable to make, but it is a process, and it is less concentrated and less consistent than using casein powder, which is a more concentrated and refined isolation of casein protein, in a convenient powdered form; it can be expected that a finish made from casein powder will be durable than that made from quark.

To make a full-bodied paint, fillers are needed to give the paint body. Chalk is the most common filler and adequate for most uses. In combination

Starch Paste

Starch paste is a natural glue, usually made from wheat or rice flour. We use starch paste for many different purposes in the world of natural building: as an additive to improve the adhesion of the mix and to control dusting, as a prime coat, and as a glaze. Use a white flour, as the germ found in a whole wheat flour adds no value and is inert filler. High-gluten bread flours are the best, although standard all-purpose white wheat flour works perfectly fine. Starch pastes have a very short shelf life. Make your starch paste shortly before you incorporate it into your mix and use it, or safely store your mix shortly after incorporating the starch paste (i.e., dry it, as described earlier in this chapter). It may be helpful to add a preservative, such as borax, to a formulation that includes starch paste to avoid mold or mildew, especially when applied in damp or wet areas.

How to Make Starch Paste

1. In a heavy-bottomed 6-quart pot, bring 1½ quarts water to a roiling boil on a full-strength burner; electric burners and camp stoves may not be able to maintain a boil.

2. Meanwhile, in a separate large pot, thoroughly blend 1 quart flour into 2 quarts cold water, ensuring there are no lumps. The mixture should be similar in consistency to a thick pancake batter.

3. When the water is boiling, slowly stir in the flour/water mixture into the boiling water, whisking continuously. Reduce the heat, while maintaining a boil; the boiling water will help keep lumps from forming and cause the paste to form faster. If the boil is lost, turn the heat up and simply continue to stir. Keep stirring, constantly scraping the bottom and sides, until the mixture thickens and turns translucent. After the mixture stops thickening and the color changes from milky white to translucent white, turn off the heat and remove from the burner. *Be careful that the paste does not burn!* Burnt paste will add lumps, darken the color, and affect the smell. Use a heavy-bottomed pot and stir continuously to prevent burning.

4. If there are lumps, strain through a fine colander or mesh, or multiple layers of cheesecloth. Be careful! The paste will be extremely hot. This paste can be used full strength as an additive to plaster.

Diluted Starch Paste for Alis Paint

To prepare a batch of starch paste for use in alis paint, dilute paste 50% with water; this should yield approximately 5–6 quarts of diluted starch paste, based on the above recipe. This will be the water and binder base for an alis paint.

Starch Primer/Glaze

A starch paste can be made into a primer or glaze, to prepare a wall surface to receive paint or to control dusting in an applied paint made without enough binder; it is translucent and matte and will not affect the look of the finish onto which it is applied. Although simple, inexpensive, safe, and effective, starch paste makes a relatively weak primer or glaze and is very vulnerable to moisture. It may not be strong enough to prime all wall surfaces, especially in chronically damp environments, and will not make a surface washable. A colleague of ours, Ryan Chivers in Colorado, always uses Elmer's glue as a base for plaster on wallboard as a more durable option in such cases where wheat paste may not be sufficient, such as in damp conditions.

To prepare a starch primer/glaze, make a batch of starch paste as above. Dilute, if necessary, with enough water to create a brush-able consistency—25% to 50%—and coat the wall evenly (if a primer) or work well into the substrate using even strokes (if a glaze).

with casein, it produces an excellent white and avoids the use of titanium white. Powdered marble also makes a good filler, as do any of the mineral powders identified in formulating alis paints. A small amount of kaolin clay increases paint stability. Silica powder will further improve the mix, and powdered mica can add a subtle sheen to the paint. Practically any inert substance can be used; the characteristics of the paint will vary according to what is added.

To increase the durability, permanence, and stability of casein paints, lime and borax are frequently added. With their alkaline nature, these materials react with the acidic proteins to create stronger finishes that are more water-repellent, while remaining vapor permeable. The increase in pH by their introduction also helps retard microbiological growth and decomposition.

Sealing

It is possible to seal the surface of casein paint so that it will be even more water-resistant and harder wearing. Casein-lime paint is more water-resistant than casein-borax paint, and an additional coating may not be necessary except in the case of areas that might receive a lot of wear or need frequent cleaning. Similar to those mentioned in the commentary for alis, an oil glaze, silicate glaze, or wax sealer could each be used, with varying results and effects. Wooden furniture that was painted with casein, such as that made by the Shakers, was traditionally sealed with linseed oil or waxes.

Storage and Reapplication

Casein paint should be used the same day that it is mixed, as its binding power will diminish and mold or spoilage can occur. However, the dry ingredients can be stored in dry conditions for a long time; there are many pre-manufactured casein paints that have a prolonged shelf life and simply require the addition of water. Plan on using or disposing of all you make in a day; disposal is quite safe, and as noted earlier, the paint can even be composted.

How to Make Casein Finishes

Because of the need for more precise formulation than with other recipes, casein paints are most often prepared by weighing the ingredients. A simple digital kitchen scale that will measure in both grams and ounces will be useful. It's also good to remember that these recipes typically need small adjustments of either water or filler to control consistency. Similar tools and scales of production can be used as for limewash or alis, with the addition of common kitchen appliances, mentioned below.

Quark (curds) are the principle binder for many of the following recipes; here are instructions to make quark (curds) from milk.

Ingredients:

- 1 gallon nonfat (skim) milk
- 1 pint lemon juice

Instructions:

1. If possible, leave milk to sour in a warm place for several days until fully curdled. Try to collect the curds right when they have formed; otherwise they will become tough and stringy. If you can't let the milk sour on its own, gently heat milk in a large pot, stirring frequently to keep it from burning and controlling the heat to keep from boiling. Pour lemon juice very slowly into the milk, and continue stirring until the milk curdles. Continue to stir for a minute more until the curdling is complete.
2. Strain the curds into cheesecloth. The use of lemon juice will make the curds more acidic, so if you've used lemon juice, thoroughly rinse the curds to remove most of it.
3. Hang the curds in a linen cloth or cheesecloth to let the excess whey drain off.
4. You now have quark, which can be used in the recipes that follow.

Making casein paint is just like making cheese; make enough, and you'll have dinner, too! PHOTO BY ACE MCARLETON.

Casein Wash

This wash can be used as a protective finish over clay plasters or paints, or as a primer coat, similar to starch-paste wash, over a substrate before applying the paint to increase bonding and coverage. It is durable, but not washable. The inclusion of borax will increase the adhesion and stability of this wash.

Ingredients:

- 8 parts quark, or 1 part casein powder
- 1 part borax powder, dissolved in 8 parts hot water (optional)
- prepared pigments (optional)

Instructions:

1. If using quark, blend with 4 parts water until smooth and lump free. If using casein powder, soak in 4 parts water for several hours, or overnight.
2. Blend in borax powder, if using.
3. Blend in pigments, if you desire a colored wash.
4. Add water as needed to achieve a washlike consistency.

5. This wash is generally best brush-applied in an even pattern. Two coats may be needed.

Casein Glaze

This formulation in undiluted form is the basis of traditional wood glues and can also be used as a base for the casein paints below when mixed with filler. The inclusion of lime will make the glaze more durable and adhesive. This is a terrific glaze for coating over finishes to make them not only more durable, but even lightly washable. Test first for results, as the oil present in this glaze may darken the finish.

Ingredients:

- 8 parts quark, or 1 part casein powder
- 2 parts linseed oil
- 1½ parts lime putty (optional)
- prepared pigments (optional)

Instructions:

1. If using quark, blend with 4 parts water until smooth and lump free. If using casein powder, soak in 4 parts water for several hours, or overnight.
2. If using lime putty, blend slowly and thoroughly into casein solution. Add water as necessary to keep the solution smooth, but do not over-thin.
3. Drizzle in oil very slowly while blending the casein solution.
4. Blend in pigments, if you desire a colored wash.
5. Add water as needed to achieve a glaze consistency.
6. This glaze is generally best brush-applied in an even pattern. Two coats may be needed.

The oil content may make this glaze take upwards of a few days to fully dry.

Casein Paint

Making the curds from nonfat milk is more troublesome than using casein powder, but it is less expensive and more accessible—and perhaps more enjoyable, if you enjoy the process! Casein powder is easier and more concentrated, and it tends to produce a more durable finish. The use of lime will create a substance called calcium caseinate, a more powerful binder than that which is created using borax, and therefore it's more appropriate for a water-resistant paint than a light wash.

Ingredients:

- 4 parts quark or ½ part casein powder, soaked in 2 parts water for a few hours or overnight
- 1 part lime putty
- 1 part oil (optional)
- 1 part water
- prepared pigments (optional)
- 2 parts filler (silica sand, chalk, granite, marble, mica, or other mineral powder)

Instructions:

1. Blend casein and lime putty thoroughly, adding a small amount of water if necessary to create a smooth, lump-free paste. We've found that one of the easiest ways to do this is in a blender, if working in small quantities.
2. If using oil, slowly drizzle into blending mixture until emulsified. Strain the mix when finished if any lumps remain, and wash equipment immediately, before the lime-casein dries.
3. Add the remaining water and blend until smooth.

4. Add pigments, if using; mix until color is fully uniform.

5. Mix filler into this base until the desired paint consistency is reached; as with the alis, achieve proper consistency by balancing dry ingredients and wet base. Add water if necessary.

Note: if desired, substitute ¾ part borax powder, prepared in hot water and cooled, in place of lime putty.

How to Apply

Casein paint, washes, and glaze do best applied in multiple thin coats; avoid the temptation to apply them in thick coats. It is helpful to apply a casein wash (described above) prior to painting, especially on porous surfaces. Allow each coat to dry before applying another. Apply as one would an alis.

Cleanup

As with the other water-based finishes in this chapter, casein paint cleans up easily with soap and water, although it is a more water-resistant paint and may take a bit more effort than for an alis. Be attentive to cleaning your mixing equipment quickly, as the undiluted casein-lime mixture is difficult to clean once dried.

LIMEWASH

Limewash is a classic mineral wash, perhaps the simplest to make of all the formulations presented in this chapter; it is simply lime mixed with water. The subtlety and nuances—and successes and failures—of limewash lie in the quality of the lime (hydrated or hydraulic lime, as opposed to agricultural lime), the application conditions and technique, and use conditions. As we discussed in chapter 17, "lime" as

TABLE 20.2. CASEIN PAINT TROUBLESHOOTING

Problem	Cause	Solution
Paint is dusting onto hands, clothing	• too much filler/pigment • too much water	• reformulate mix accordingly and recoat • seal with casein glaze or other sealer
Wash is crazing, cracking, or peeling/lifting	• too much casein-lime/borax base • paint is too thick • paint is applied too thickly	• reformulate mix accordingly and recoat • adjust wall preparation and application technique and recoat
Paint is uneven, mottled	• paint is still drying • paint was applied unevenly • substrate was improperly prepared	• wait for paint to dry • adjust application technique and recoat
Paint is peeling/not sticking to wall	• substrate was improperly prepared	• wash paint off wall, adjust wall preparation techniques, and reapply • confirm quality of paint formulation on sample board

Note: Use this table to help troubleshoot your casein finishes, first by recognizing the symptom, then by identifying the cause, and finally by selecting a solution to use to address the problem.

we are using in this application is calcium hydroxide, a material partway through a cycle that begins and ends as calcium carbonate, or limestone. Unlike clay paints, which simply dry by releasing their water carrier, limewashes chemically "cure," by reabsorbing carbon dioxide from the atmosphere and "carbonating" back into calcium carbonate. As for clay paints, this is a solvent-free, zero-VOC-emitting process. When we apply a limewash, we are essentially painting a very thin coat of limestone onto the wall. This practice has been done for thousands of years, wherever lime was used, as a sacrificial protective layer over masonry buildings, and is still in practice today, revitalized by both historic preservationists and natural builders.

Benefits of Limewash

Limewash is favored for its old-world, soft, matte appearance. Lime crystals are able to refract and reflect light in duplicate, creating a stunning visual effect known as a "dual refractive index" that manifests as a luminous surface glow not found in other finishes. Brilliantly white in color, unpigmented limewashes add terrific brightness to a room or building. In Mark Twain's *The Adventures of Tom Sawyer*, the protagonist was himself an early proponent of the pleasures of working with limewash, evidenced by his ability to convince a group of his peers to whitewash the fence in his stead. This exceptionally white base can make it difficult to achieve deeper, richer hues when pigmenting a limewash, however. Pigments of a strong tinting strength are needed, and in general lighter tints should be expected when coloring limewashes. This will be explored further later in this chapter.

The high alkalinity of limewash makes it an effective hygienic finish, ideal for use in persistently damp locations. In fact milk parlors and barn interiors were traditionally limewashed to sanitize the operations within and keep animals healthy and milk safe.

Perhaps of greatest benefit, however, is the role limewash plays in protecting walls. As the exterior finish of choice for vapor-permeable wall systems, limewash will not cause moisture to be trapped within the wall, as would an oil-based or synthetic paint. In that limewash is a wash and not a full-thickness paint, successive thin coats are applied; this also supports lime's carbonization process and keeps the material from cracking. This process of adding multiple thin layers helps fill in small cracks and pores in an exterior plaster coat, reducing liquid moisture penetration and serving as a protective barrier for the wall. Any absorbed moisture or interstitial condensation is able to safely evaporate to the exterior through the permeable finish. Limewash's porous nature also enables it to be applied onto a green coat of lime plaster, as it allows carbon dioxide to reach the carbonating plaster behind; this process will allow both the plaster and wash to carbonate together and strengthen its bond.

Where to Use

As mentioned throughout this chapter, porous finishes including limewash are best applied over porous substrates. See the sidebar "Working with a Nonporous Substrate" for more information on limewashing over painted surfaces. We frequently use limewash in both interior and exterior applications over plastered walls. However, without additives or sealers, limewash is not washable and should not be used in high-impact areas without further protection. Although it is ideal as a finish over lime plasters, we have also used limewash over earthen plasters, as well as cement and other porous masonry surfaces. As noted above, limewashes

Limewash is a traditional finish for exterior lime plasters. PHOTO BY JACOB DEVA RACUSIN.

are very well suited to damp areas such as basements and bathrooms, as they will not mold or mildew.

Ingredients

In that there is only one principle ingredient in limewash, the quality of lime used for the mix is paramount. Access to quality lime is particularly important when applying limewash as an exterior protective coat over a plaster finish, and it is worth it for us to source the highest-quality lime available for the wash. Especially considering how little lime putty is needed to create a limewash, it is worth the premium in cost. Hydraulic limes can be used as well, and they're particularly favored in perpetually damp conditions, although care should be taken that the correct hydraulicity is being used for your substrate; a weaker hydraulic lime, such as an NHL2, should be used over a stronger substrate.

To further support the longevity of this sacrificial layer, a number of additives can be entrained in limewash to prolong its life. Linseed oil, the most common of natural oil finishes, can be emulsified into limewash in small amounts; many recipes call for 2% to 5% oil of total volume. This will increase the durability and water repellency of the limewash. The fats of the oils react with the lime to create a lime soap, calcium stearate, that lends the desired characteristics; this is a same mechanism as the function of *tadelakt* (see chapter 17 for more information). Small amounts of oil will not dramatically affect vapor permeability. Casein powder or quark can be used for similar means and also acts as a binding additive when being applied over more difficult substrates. The acidic casein reacts with the basic limewash and forms a natural glue, calcium caseinate, which is a strong binder and the substance of casein paints, described earlier in this chapter.

Sealing

The same profile of sealants presented for alis can be used for limewash as well. Silicate paints are frequently used to prolong maintenance periods on the exterior. Oils are also very effective; as mentioned earlier in the alis section, the craft of using a black oil soap to seal a finely polished lime plaster, known as *tadelakt* (chapter 17), results in a finish that is durable enough to hold water for a bath.

Storage and Reapplication

Hydrated limewash can be stored indefinitely, as long as a layer of water is kept over the settled body of the wash. Simply keep an extra bucket of pigmented limewash, labeled by pigment type and application location, in a basement or closet to pull out when touch-ups are needed. Hydraulic limewashes cannot be stored long, as they can set up under water, meaning that they will set up in the bucket in which they are stored. Dry hydraulic lime powder can be stored, although it will deteriorate in quality with prolonged exposure to air, so keep it in an airtight container.

How to Make Limewash

Our experience is that limewash is most easily made in 5-gallon buckets with electric drills outfitted with paint whisks, but it can also be made by hand using whisks or in mortar mixers for varying quantities. *Proper safety protection is essential when working with lime!* In addition to the inhalation danger that exists with all powders, lime has the additional liability of being very caustic, and gloves and goggles should be worn at all times when working with lime (see chapter 17 for more information about lime hazards and safety). Making a basic limewash is fast, easy, and far less finicky than other recipes in this chapter.

Ingredients:

- 1–1½ gallons lime putty
- 3 gallons water
- pigments, as desired

- 2 tablespoons linseed oil (optional)
- 1¼ cups casein powder, or 10 cups curds (optional)

Instructions:

1. Pour lime putty into a 5-gallon bucket.
2. Slowly add approximately 3 gallons of water, stirring continuously, until the consistency of whole milk is reached (do not be tempted to stop at half-and-half!).
3. Mix in any prepared pigments, and blend thoroughly until color is completely uniform. Generally, limewash can hold up to 7% to 10% by weight pigment content before powdering or dusting; we find this translates to somewhere around 30% by volume, but it can vary.
4. If using an oil additive: heat oil gently, and slowly drizzle in as the wash is being blended. Mix regularly during application to avoid separation.
5. If using a casein additive:
    - For a casein powder, stir into a small amount of hot water, and let stand for two hours; then stir thoroughly into limewash, using additional water if necessary to achieve whole-milk consistency.
    - For milk curds, blend with measured amounts of water until smooth, creamy, and free of lumps, then stir thoroughly into limewash, using additional water if necessary to achieve whole-milk consistency.

Note: If using hydraulic lime or hydrated lime in powder form in place of lime putty:

- Add 3 gallons of water to a clean bucket.
- While blending, slowly sift in powder until consistency of whole milk is reached, being sure to fully incorporate all lumps.
- If using dry hydrated lime powder, allow to soak for at least one hour; one day is a basic minimum, and one week or more is preferable. If using hydraulic lime powder, continue immediately with the recipe; do not allow to soak, as hydraulic lime will begin to cure once hydrated.
- Continue recipe above at no. 3.

Note that whereas pure hydrated lime putty can be kept indefinitely, hydraulic limewash and limewash with casein or oil should be used immediately.

How to Apply

Being much thinner than a classic paint, a minimum of three coats of limewash are needed, especially over porous surfaces, to achiever proper coverage and uniformity. For freshly plastered walls, five or six coats are recommended to build up a strong protective barrier while filling in surface imperfections, especially on the exterior. Coats should be applied very thinly; multiple thin coats are far superior to fewer thicker coats, which are liable to crack or dust. Limewash is nearly transparent when first applied and doesn't develop opacity until it begins to cure, so be sure not to accidentally over-apply. Make sure to keep agitating the limewash during application to keep the lime in suspension, especially if oil is used; the last brush should be as thin as the first.

Limewash application follows the same general principles spelled out for alis concerning site preparation and integration into porous substrates. A short, thick-bristled brush, such as a stain brush, is frequently used to hold the thinner-consistency wash; be cautious of additional dripping. Moistening the substrate repeatedly and thoroughly is all the more important—particularly on exterior walls during drying conditions—as lime needs to cure in a damp state. Rapid suction from the substrate can cause premature drying and will cause damage to, or failure of, the finish. Wait twenty-four hours for curing between coats, and lightly mist coats before subsequent application.

The same instructions for nonporous substrates apply for limewashes as they do for alis; see the sidebar "Working with a Nonporous Substrate."

Cleanup

As with alis, similar protection and cleanup strategies apply. Limewash differs in two ways: first, lime has a much greater tendency to stain wood, especially cherry, and extra caution should be taken when limewashing near wood, particularly unfinished wood; second, limewash will carbonate onto stone very durably—a testament to its efficacy on the wall—if not quickly washed off. Thoroughly mask or pre-moisten stonework, and inspect and clean at more regular intervals. Mechanical abrasion or an acidic cleaning agent, such as muriatic acid, may be needed to remove cured limewash.

SILICATES

While silicate finishes stray somewhat from natural building products by virtue of their somewhat complex manufacturing process, they are highly compatible with natural building systems and perform remarkably well in concert with the paints and plasters presented in this book. Silicates were first discovered 125 years ago in Germany by chemist Adolf Wilhelm Keim; Keim went on to found an eponymous company that continues to be a leader in the manufacturing of silicate finishes, although other European and U.S.-based manufacturers are now producing similar products (see appendix A for more information). The binder for these finishes, called

TABLE 20.3. LIMEWASH TROUBLESHOOTING

Problem	Cause	Solution
Wash is dusting onto hands, clothing	• wash mixed too thickly • wash applied too thickly • too much pigment in wash	• reformulate mix accordingly and recoat • seal with sodium silicate or other sealer
Wash is crazing or cracking	• porous substrate is too dry before application • limewash is too thick	• reformulate mix accordingly and recoat • adjust wall preparation and application technique and recoat
Paint is uneven, mottled	• paint is still drying • paint was applied unevenly • substrate was improperly prepared	• wait for paint to dry • adjust application technique and recoat
Paint is peeling/not sticking to wall	• substrate was improperly prepared	• wash paint off wall, adjust wall-preparation techniques, and reapply • confirm quality of paint formulation on sample board

Note: Use this table to help troubleshoot your limewash, first by recognizing the symptom, then by identifying the cause, and finally by selecting a solution to use to address the problem.

Fresco

Fresco painting is the skilled art of painting with lime-stable pigments prepared with water or limewater directly onto actively curing plaster; as the plaster carbonates, the pigments are permanently, chemically fused into the plaster, leaving a durable and vivid finish. As a result of painting with pure pigment, richer colors and hues more true to the pigment's powdered appearance can be attained, something particularly favorable given lime's strong tinting power. Famous examples of this style of decorative painting include Michelangelo's Sistine Chapel ceiling and Diego Rivera's murals; they can also be found gracing the walls of many a 3,000-year-old Greek temple.

Diego Rivera is well known for his fresco murals. PHOTO COURTESY OF WIKIMEDIA COMMONS.

A great degree of skill and experience is required for fine and detailed fresco application. The plaster must be expertly prepared (a flat, smooth wall is not easily attained by novice plasterers). The timing must be judged correctly; there is a point during the curing of plaster called "green cure," when the plaster is leather-hard, resistant to impression from a brush or finger, yet still soft, damp, and early on in its curing phase. Diego Rivera once said that part of what fueled his creative process was the pressure behind needing to apply the paint before too much time had passed and the canvas had cured. Finally, there is no second-guessing with fresco painting; each brush mark is permanent. That said, simpler fresco painting and color application can be achieved with quality results by an attentive and moderately skilled builder (you don't have to want to be the next Michelangelo to try this technique).

"water glass," is made of either potassium or sodium silicate; quartz sand is combined with potash and water and heat- and pressure-processed to create a clear liquid that forms a chemical bond with silicate-based substrates.

Silicate finishes offer many benefits to natural builders. They are incredibly durable, with reports of some original applications still surviving over a hundred years later; general recommendation is for reapplication once every fifteen to twenty years! Their maintenance is very low, requiring occasional washing; reapplication can be executed without stripping. They are highly water-resistant, while still remaining very vapor permeable, making them ideal glazes for protecting exterior and high-impact interior clay- and lime-based plaster or paint surfaces. Application is simple, either by brush or sprayer. As their binding strength comes from a stable chemical reaction, silicates are solvent-free, no-VOC, nontoxic finishes that will not impact IAQ and for the most part contain no petroleum-based ingredients (some manufacturers offer silicate products with petroleum additives). Silicate finishes are noncombustible (unlike oils or synthetic paints), antimicrobial (a

Silanes and Siloxanes

One of the liabilities of exterior lime-based plasters is their propensity to absorb large amounts of water. Many builders—owner-builders and professionals alike—now use siloxane-based products to help seal their walls. *Siloxanes* are compounds of silicone and oxygen and are used in a wide variety of different applications, including sealers. These sealers—which are frequently mixed with more alkaline-tolerant *silanes*, also silicone compounds—serve the purpose of largely reducing liquid absorption, while leaving vapor permeability unaffected; this is a highly desirable characteristic for a plaster sealer, as many sealers close pore spaces in the plaster to the degree that vapor permeability is significantly reduced and greater moisture problems are potentially caused within the wall. In his report "Moisture Properties of Plaster and Stucco for Straw Bale Buildings," author John Straube concludes, based on extensive testing, that "siloxane appears to have no effect on the vapour permeance of . . . cement:lime stucco while almost eliminating absorption. The use of siloxane can be recommended based on these tests" (Straube 1999). A terrific example of the integrative approach to design and construction is the application of a silane or siloxane sealer—an industrial product, to be sure—over a natural plaster to prolong the life of the finish. We have not worked with siloxanes ourselves, but the precedence of use within the natural building community, coupled with data such as that presented in this study, leads us to believe siloxanes/silanes are a good tool in the chest of the modern plasterer.

result of their high pH), and colorfast when using mineral-based pigments. More expensive silicate formulations, such as lithium and colloidal silicates, are produced for the concrete finishing industry, although the subtle nuances of these improved finishes are generally lost in our applications and are therefore not worth the increased cost.

Again, these finishes are particularly well suited to paints and plasters, as well as stone and masonry substrates, because of the crystalline minerals contained in these substrates. Silicate finishes bind only with calcium salts (such as those found in limestone-parented materials, including lime plasters and cement), silica (such as clays and silica sands), ceramics (including brick and terra cotta), and certain metals, including iron. They are incompatible with wood, plastic, or synthetic paints, although some manufacturers sell priming products that can be used to bond silicate finishes to these substrates. Through a process called *silification* or *petrification*, the silicate finishes form integral, insoluble, permanent crystalline matrix bonds with the salts and silica in the substrate; this lends these finishes both their durability as well as their flexibility, allowing them to move with the substrate and eliminating the potential for delamination. Silicate finishes can be found or formulated in three forms:

Clear water-glass glazes provide superior protection to clay, lime, and stone surfaces without affecting the look of the finish, an attribute particularly favorable when you're looking to add protection to a wall—for example, instead of an exterior limewashing, or over an interior plaster in a kitchen—while maintaining the natural beauty of the existing finish. We have seen sodium silicate darken and reveal brush marks when applied over colored limewash, however, so be sure to test before committing to an entire wall. To prepare, the water glass may need to be diluted up to 50% with water; consult the product instructions and perform tests to determine what is best with your product. Apply with a brush, working well into the surface with uniform strokes, leaving a thin, translucent coat. Be sure to apply multiple thin coats, as too thick of an application can cause cracking or bubbling and will likely lend a milky, slightly opaque look to the finish. Two coats are generally recommended.

Colored water-glass glazes can be achieved in the same way as the clear glaze, but by adding in a desired amount of water-prepared alkali-stable pigment. The pigment will hold its hue very nicely, and this will produce a lovely layered effect to the color of the wall. Multiple coats may be necessary to build up full coloration. Adam Weismann and Katy Bryce recommend that "a more effective finish will be achieved if an undiluted coat is first applied onto the wall, before the pigmented coat. The pigmented coat should be covered by another undiluted sealant coat" (Weismann and Bryce 2008, 229).

Silicate paints are manufactured by a number of different companies and come in a very wide range of stock and custom colors. Some companies also produce specific formulations for certain applications, such as in high-traffic or high-moisture environments. By adding a filler such as mineral powder or chalk, along with mineral-based pigments, to the water-glass glazes mentioned above, you can create your own silicate paint. Be sure to allow any lime substrates to fully cure before application, as silicate paints will retard carbonation of the substrate.

OILS, WAXES, AND THINNERS

As we have seen throughout the water-based recipes presented above, plant-based oils are frequently used as an additive to enhance durability, water repellency, gloss, or hue of a finish. These oils also provide the same qualities when used as bases of their own paints, washes, and glazes. In fact, plant-oil-based paints were the standard household paints prior to the introduction of petroleum oil products. Today, we use oil finishes not only over wall surfaces, but as protection for wood in the home, whether a floor, trim, or timber frame. Unlike surface finishes, including paints, oil finishes are penetrating finishes that, through their slow drying time, are designed to seep into and fill the pores of the substrate, be it a wall surface or wood. This penetrating nature keeps oil finishes highly flexible, moving with the substrate and avoiding delamination and refinishing issues with surface finishes. An additional benefit is revealed during refinishing and touch-ups: as a flexible penetrating finish, oils can be spot-applied and will blend easily and evenly into the surrounding finish without a need to strip and refinish an entire surface. Oils will inherently darken a substrate and increase the gloss of a finish; oil application will enhance the grain pattern of wood.

Types of Oil

The most commonly used oil is boiled linseed; sourced primarily from pressed flaxseed, it is inexpensive and dries quickly. All linseed-oil products will yellow with time, and this can either be a detraction or lend a patina to the finish. The great downside to using boiled linseed, at least those found in your average hardware store, is that the oil is not actually "boiled" anymore (heat-treated and oxidized); conventional linseed oil is now cut with high-VOC petroleum-solvent-based or metallic-based drying agents, which are added to the oil, replacing the slower, healthier, traditional way of refining the oil. The damaging effects of these driers has been discussed at the beginning of this chapter. Fortunately, there are other options:

- *Raw linseed oil* is less toxic than commercially available "boiled" oil as there are fewer solvents and driers introduced to the oil; that said, it will take significantly longer to dry and may not dry tack free, even when cut with a thinner; testing is needed if this approach is to be used. There are quality, even organic, linseed oils available from better manufacturers.

- *Solvent-free, low-VOC boiled linseed oil* is available through more progressive manufacturers and offers the convenience of a boiled linseed oil without the toxic loading. This will be a more expensive proposition, however, and will involve more planning to ensure that you have enough, as these are not products commonly found in hardware stores and will often involve sourcing through mail-order companies (see appendix A for sources of low-toxicity, prepared linseed oils).

- *Stand oil,* also known as sun-thickened oil, is the most traditional approach. It's more effective than raw linseed oil and cheaper than specialty prepared linseed oil products. Commercially prepared stand oils may be described as "polymerized oils," which describes the chemical change in the oil as a result of heat treatment. Tried and True is one such common and commercially available product. Another is Landark, which was originally formulated to finish timber frames; Landark's recipe also includes pine rosin and natural wax. To make stand oil:

1. Pour raw linseed oil into a large dark-colored tub (like a black mortar pan), 1–2 inches thick.

2. Cover with glass or Plexiglas, making sure to leave a small vent space between the glass and the tub to allow for vapor release (prop up with a piece of wood or scrap foam).

3. Place in the sun for a period of a few days to a few weeks, depending on the weather; ensure that the oil remains completely dry and free of contaminants.

4. Decant into an airtight container. The result is a material that dries similarly to conventional boiled linseed oil, but without the drying agents. Stand oil can also be purchased, but as with solvent-free boiled linseed oil it is harder to find and more expensive.

- *Tung oil* is another that may be used successfully; tung oil is particularly attractive over wood, offering many of the same properties of a boiled linseed oil without the long-term yellowing effects. Different manufacturers offer different products with different ingredient lists, and all the same considerations of off-gassing, drying time, cost, and access apply.

- *Walnut oil* is favored for its color and luster; while expensive, this may be an attractive oil for certain wood installations in the home.

Waxes

As mentioned earlier, waxes are akin to oils in their properties and effects on a finish; you might consider waxes a highly concentrated oil. Waxes are particularly favored when a surface-protection finish is needed, and they are frequently used in concert with oil for a combination penetrating-and-surface-protection system. As with oils, spot application and refinishing can be achieved with waxes, even with the surface protection it offers.

Classic natural waxes are primarily beeswax and carnauba wax, often found mixed together to take advantage of beeswax's flexibility and carnauba's strength. Petroleum-based waxes are often used today in modern formulations, in part or in replacement of natural waxes. Many oil/wax blends are available off the shelf, and a thorough examination of the ingredients list should point toward products that are safer to use and more benign in their formulation. To begin experimenting with making your own wax finish, follow these instructions:

1. Melt beeswax in a double boiler until liquid in form.

2. Thoroughly stir in 2 parts boiled linseed (or comparable) oil to 1 part melted beeswax; this will allow the mix to remain viscous and able to be easily applied at a cooler temperature.

3. If necessary, a natural thinner (see below) can be mixed in to further improve viscosity.

4. Apply with a brush and burnish with a clean cloth, or rub in with a rag.

Thinners

Thinners are solvents, used with oils to cut a finish to decrease its viscosity. Turpentine is a common option; many people also use mineral spirits. Again, like boiled linseed oil, both are inexpensive and easy to find, but much more toxic; a preferable option from both a human health and ecological health perspective would be to use a plant-based thinner, such as a citrus-based thinner or true pine-resin turpentine. As noted earlier in the chapter, even natural thinners release VOCs and should be used with care. Like the specialty oils, natural thinners are harder to find and more expensive, but worth the investment.

Making Oil Finishes

Working with oil finishes is a straightforward and intuitive process, involving using a few materials to create a range of different products:

- Use straight oil, perhaps thinned with a natural solvent to create a glaze or wash, as a clear sealant and to darken a surface; brush-apply and burnish with a cloth, or use cloth to rub oil into the surface.

- Float pigments into an oil glaze to stain a finish. For example, floating umber and red-oxide pigments into an oil applied over pine will color the wood to look like mahogany. Working with pigments is explored in depth later in this chapter.

- Mix in a mineral-based filler such as whiting, chalk, or mineral powders to create a full-bodied oil paint. Brush-apply with an oil brush.

When working with any of these applications, it is important to consider a few basic tips:

- Heating the oil gently will greatly aid in penetration. Be very careful when heating oils, as they can easily catch fire! Use a non-flame heat source when possible, such as an electric burner, and always keep the heat low, heating the oil slowly. Gently stir the oil, never filling the pot more than half full. Attend the oil at all times, and remove from heat before it starts smoking. Finally, always keep a fire extinguisher rated for use on oil fires handy at all times.

- Applying an oil finish to wood may raise the grain. Sand lightly with a fine-grit sandpaper to remove the grain and wipe clean with a tack cloth between applying coats.

- Never throw away a bundled rag that has soaked up any oil or thinner! The drying of oils and thinners is exothermic, and we know directly of multiple projects that have suffered fire damage—from minor to extensive—as a result of careless disposal of oily cloths and rags. The safest option for disposal is to burn the rags in a controlled environment, such as a woodstove. Another acceptable practice is to spread out the rags, so that the heat is able to easily dissipate, in a well-ventilated area (preferably outdoors) on a noncombustible surface.

OTHER NATURAL SEALERS

There are currently a few products on the market that make use of non-resinous plant-based materials as the bases of clear wood sealers. A number of these are soy based; they use distillates from the oil pressed from soybeans to create a solvent that is formulated with resins and other materials to create a protective finish. Another promising sealant is a whey-based product manufactured by Vermont Natural Coatings. Whey, a milk protein that is a by-product of the cheese-processing industry, is used as a replacement for petroleum-based ingredients in a line of floor, furniture, and exterior wood-finishing products. Their line of products is low-VOC and quick drying; in both look and application, our experience with these finishes leads us to recommend them as a good water-based polyurethane replacement.

Color and Natural Paints

A fundamental role of paints and finishes is to bring color to a surface. As we identified earlier, the mechanism for coloring a finish is pigment. The world of pigments is vast and ancient. When you use an iron oxide to tint the color of your limewash, you are engaging in a process whose roots can be traced back to a time before the written word. Pigments can be classified in a couple of different ways: organic pigments are those that are sourced from carbon-based materials, such as plants or insects, whereas inorganic pigments are sourced from minerals and metals. Another way to categorize them is natural pigments, whose sources are from naturally occurring elements with minimal or no human intervention, and synthetic pigments, which are produced through human ingenuity, frequently to replicate or improve upon characteristics found in nature.

For the most part, we favor the use of natural, inorganic pigments. Organic pigments, while beautiful, are for the most part very unstable, particularly in the presence of ultraviolet light; we've seen plant-based pigments fade and distort even in dark hallways illuminated only with intermittent fluorescent light. As natural builders, we certainly favor the use of natural pigments over synthetic pigments when possible. The beauty of natural pigments is unparalleled. The crystalline structure of these mineral pigments is able to refract and reflect light to our eye, lending a depth and vibrancy to the walls they grace that is in no small part responsible for why naturally finished walls are so pleasing. There are instances, however, when synthetic pigments may be the only option, or even preferable to natural pigments for their qualities (for example, a greater tinting strength or unique color, or lower embodied energy when compared to titanium dioxide, one of the highest embodied-energy materials in paints).

In the world of inorganic pigments, there are a number of different terms whose definitions vary depending on the source. Oxides are derived from oxide minerals, such as iron or magnesium; these pigments are often heat-treated to derive a variety of colors. Ochres are derived from limonite, an iron-oxide ore, in combination with clay and other silicates. Ochres are essentially used in their natural state directly from the earth and appear as a range of yellow, red, orange, brown, and even purple hues. Siennas (yellows/browns) and umbers (browns) refer to oxides that range from "light" to "burnt" (richer and darker), achieved through heat treatment. Spinel colors (green, blue, purple) are sourced from vibrant minerals that are rarer, such as lapis lazuli, and are therefore more expensive than oxides; often, synthetic colors are generated to replicate spinel colors for a lesser cost. Ultramarines (blue, red, purple) are aluminum sodium silicates (clay) that have been heat-treated, sometimes with sulfur.

Just as with solvents, simply because a pigment is "natural" does not inherently make it "nontoxic." Many greens and blues, in particular, are made from naturally occurring carcinogenic heavy metals such as cadmium and chromium; only purchase pigments from reputable sources that can confirm that their pigments are free from these dangerous metals.

FORMULATING WITH PIGMENTS

Knowing how the pigment will behave is just as important as knowing where it is sourced. If you are to be using pigments in a limewash, for example, you will need to ensure that the pigments are "lime-fast" or "lime-stable," that is, they will remain true to their original color in the high-alkaline environment of lime. Most inorganic pigments will be lime-fast, although many pigments available in catalogs are in fact blends of multiple pigments, some of which

Pigments come in a wide range of colors and can be combined to create an infinite palette of options. PHOTO BY KELLY GRIFFITH.

may be organic or synthetic and potentially liable to distort in color in the presence of lime.

Another significant distinction between pigment types is their tinting strength. Some greens, such as "bohemian" or "earth" green, which are simply green clays, will be very weak in their tinting strength—meaning a lot of pigment will be needed to alter the color of the mix. On the other hand, many black pigments are quite strong in their tinting strength, and "a little dab will do ya," especially if the intention is to simply drop the shade a notch or two. This is particularly relevant when trying to achieve a certain degree of saturation with a pigment of weak tinting strength.

Depending on the binder, you will only be able to add so much pigment before the binding capability is overwhelmed and the paint starts "dusting," meaning it will literally dust off the wall onto hands or clothing. We find this especially with limewash, when a deeper or richer color is desired. Because lime makes for a brilliantly white wash, it takes quite a lot of pigment to bring a stronger saturation to the color; if pushed too far, no matter how well cured the

Changing the concentration of pigment in the mix will affect the tint of the color. PHOTO BY ACE MCARLETON.

limewash is, the unbound pigment will end up on the shoulders of unwitting occupants and visitors unless otherwise sealed. It is interesting to note that it is the saturation of pigment in the paint that determines in large part the gloss of a paint. A matte finish has a very high proportion of pigment—70% pigment to 30% binder—whereas a high-gloss finish will be the reverse. This, of course, also depends on the nature of the binder.

The final consideration, although generally more of a minor one in the world of wall painting (as opposed to studio painting), is the absorption of different pigments. Some—particularly those that are clay based—will thicken a paint or wash readily, whereas others will have a lesser effect on the viscosity. This will be relevant, especially if larger quantities of pigment are used, but in general this necessitates simple adjustments in the ratios as you are mixing the paint and should have no major bearing on the process or product.

USING PIGMENTS

To prepare dry pigments for use in your paint, it is best to start by soaking them in the appropriate solvent for your medium. For water-based paints, use water, or for more stubborn pigments that resist going into suspension, use alcohol, such as vodka or grain alcohol. Soaking overnight is recommended to ensure complete dispersion of the pigment in the solvent. For oil paints, use the same oil as you are using in the base. The "base" is any of the uncolored paint recipes presented earlier to which the pigments are to be added, such as clay paint, limewash, casein paint, or linseed oil.

Start by mixing equal volumes of pigment and solvent, adjusting as necessary to make a smooth, pourable paste; the goal is to completely suspend the pigment in the solvent, noting that, as inorganic minerals, the pigments will not dissolve. Care should be taken to work out all lumps of dry material in the preparation. Finally, thoroughly mix the desired amount of the prepared pigment into the paint base until the color is consistent. The purpose of this process is to integrate the pigment evenly throughout the entire body of the paint base, and to ensure that the pigment is completely suspended in solvent. Dry clumps of pigment will lead to streaking or "explosions" when painted onto the wall. While this can be a beautiful effect, it is best if it is executed intentionally, and not by surprise. Similarly, incomplete integration of the prepared pigment will lead to swirling or mottling—a potentially beautiful discovery, or a potential disaster.

Depending on the source of the pigments, you may find it necessary to strain the prepared pigment before use in paint, or to crush it further using an artist's muller or common mortar and pestle. This is generally not an issue with artist-grade pigments, but if you are using cement or mortar pigments, you may find particulates that are large enough to add unwanted texture to the paint. If this is the case and you opt to strain rather than grind, use extra solvent in preparing the pigments and pass the mixture through a fine-mesh strainer or multiple layers of cheesecloth; be sure to factor in the increased solvent content of the prepared pigment in the formulation of the paint.

SAFETY

It is important to note a few safety measures in the handling and disposing of pigments. As with all powdered materials, inhalation is the greatest danger, and care should be taken to protect yourself while working with powdered pigments before they are bound in solution. At a minimum, a tight-fitting fine-particulate mask should be worn, and the working environment should be well ventilated. As mentioned briefly earlier, it is common for some pigments to contain carcinogenic heavy metals, such as cadmium and chromium. Although noted before, it bears repeating: when sourcing pigments, find nontoxic alternatives to these dangerous products. Finally, be responsible in disposing of pigments; as inorganic minerals, pigments will not biodegrade, and care should be taken to keep pigment residues and waste out of aquifers and waterways.

Terms of Color

It is helpful to understand the meanings of terms used in discussing color, as a common language has been developed over the centuries by artists to explain the subtleties and nuances of color. Here is a glossary of the most common terms used when discussing color:

- *Colors* are created by the different wavelengths of light, which the eye receives and the brain interprets. Colors are initially divided into *primaries*, blue, yellow, and red, and these primaries can then be mixed to create *secondaries*: blue + yellow = green; yellow + red = orange; red + blue = purple. Further combinations of secondaries give us *tertiaries.*

- *Hue* refers to the properties of a particular color. For example, when mixing two different colors in various proportions, a range of different hues are being created

- *Shade* speaks to the change in a hue when black is added. Shading makes a hue darker.

- *Tint* speaks to the change in a hue when white is added. Tinting makes a hue lighter.

- *Tone* refers to the change in hue when gray is added. Toning makes a hue less rich, but not necessarily lighter or darker; it varies the intensity of the hue.

- *Saturation* is related to *shade*, *tint*, and *tone*; saturation describes the amount of hue present. The greater the saturation, the richer the color.

- *Temperature* is a value system placed on different colors and how they feel to us, generally described at *hot*, *warm*, *cool*, and *cold*. Red is a hot color, yellow is warm, green is cool, and blue is cold.

- *Gloss* is technically a reference not to color, but to the reflective nature of a surface. A high-gloss surface is highly reflective of light and will look shiny; these surfaces are frequently high in binder and low in pigment. A matte surface is the opposite, with low light reflectivity and a high percentage of pigment relative to binder.

There are many other ways to describe different color attributes, such as *chromaticity*, *luminance,* and *value*, but the terms listed above will get you well on your way toward having an informed and productive conversation about color.

Source: Art and Architecture Thesaurus Online, Getty Research Institute. http://www.getty.edu/research/tools/vocabularies/aat (accessed January 30, 2012).

SAMPLING PIGMENTS

Most references on using natural pigments will simply tell you that it can be difficult to predict how a pigment will behave, so be sure to make sample boards first; they may perhaps offer a tip or two. From experience, we can tell you confidently that working with pigments to create a color that is the same color as was shown to you by the client in a picture clipped out from a magazine, or replicable from the sample board to the finish product on the wall, is the single most difficult part of the process of making your own paints. The standard protocol in the world of synthetic commercial paints is to hand the client (or your spouse, or yourself!) a large book of color chips, with interesting names like "tapioca sunset" or "burning desert," with instructions to "choose one." Alternatively, a small sliver of the living room wall is deposited into a plastic baggie, if matching an existing paint, which is brought to the paint store, where it is fed into a spectrophotometer, that magic machine that analyzes the spectrum of hues that make up your sample. Specific instructions are given to the clerk as to how much of which colors are to be mixed into their "neutral base" and set into the paint shaker while you go look for the other things on your list. All that is left to do is pay at the register and get painting!

Not so for us do-it-yourself paint manufacturers. Instead, we must develop a good relationship with the pigments to understand what their capabilities are, and this comes through experience. To begin gaining that experience, we start with sampling, but when faced with a large pallet of pigments and endless combinations, none of which look the same mixed into the paint as they did in the jar, utilizing a methodology for understanding how to work with pigments is very important. To test the potential range of hues and tones of a pigment in a base, we perform what we call a "color study." This can be done with one, two, or even three pigments. Let's walk through the steps for a two-color study, in which we will look for two characteristics: change in hue, by varying the proportions of the two pigments; and change in saturation, as more pigment is added to the base.

On DVD, Chapter 20, see CONDUCTING A COLOR STUDY WITH DRY PIGMENTS

Here are the materials you will need:

- a prepared sample board, approximately 2 × 3 feet
- a fine-tipped Sharpie or ball-point pen
- ruler or straightedge
- 7 plastic cups
- 7 plastic spoons
- 5 1- or 2-inch chip brushes (wood-handled disposable bristle brushes)
- 2 clean pails
- a clean rag
- 2 pigments
- base paint/wash
- hair dryer (optional)

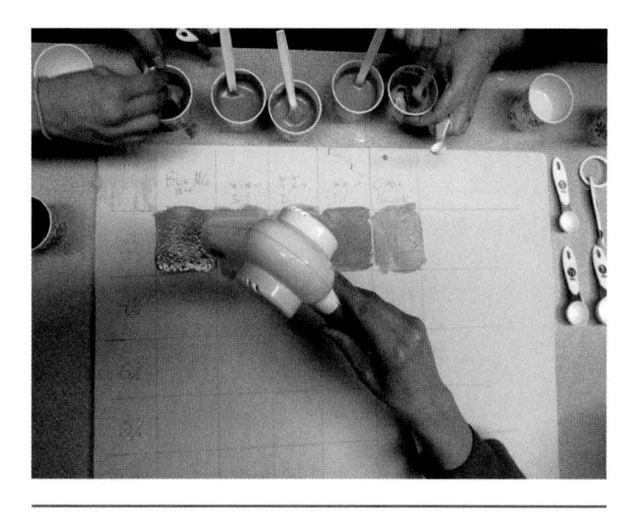

Laying out the sample board. PHOTO BY ACE MCARLETON.

Start with an Accurate Board!

It is important that your sample board be the same color and texture as your in situ substrate. If you are going over primed drywall, use a piece of drywall primed with the same color and gloss as the wall. If you are going over an existing paint, try to replicate that surface as closely as possible. This is especially the case with washes, which are semitransparent and will be influenced by the color beneath.

Lay Out the Board

To begin laying out the board, you need to decide how many columns across you will need to explore the two colors, making sure you leave enough room for each sample (2–3 inches by 2–3 inches is plenty, although you can go bigger if you have enough room). We will start with one color to the left of the board (pigment A), and the other to the far right (pigment B). We usually have three or four varying concentrations in between, for a total of five or six sample columns. We will also need an additional column to the far left, to label the saturation levels of each row. The ratios will vary depending on the tinting strength of the pigments. For two comparable strengths, the ratios should be 1:2, 1:1, and 2:1. You may find it easier to work in percentages, in which case you will notate 75/25, 50/50, 25/75; regardless the system, always maintain consistency in the order in which you write the ratio. If you are trying to shade a weak-tinting yellow (to the far left) with a strong-tinting black (to the far right), your ratios of yellow:black may read 6:1, 4:1, 2:1, and you may choose to eliminate the pure black column altogether.

Working your way vertically down the rows, we start with 5% of pigment, then add in increments of 5% until we reach 30% (at which point we may be in danger of overwhelming the binder, depending on the base). What you now have is a neatly labeled grid, with a range of pigment (hue) ratios across the x-axis at the top that label the columns, and a range of saturation percentages running down the y-axis on the left that label the rows. If you are looking for more of a pastel color, consider starting with a lower percentage, such as 2%.

Prepare the Materials, Base, and Pigments

Set out a small plastic cup above each pigment ratio column. Set another on each side of the board, to prepare pigments A and B. Set a larger cup or small container off to the side for the base. Place a spoon and a small 1- or 2-inch chip brush next to each cup, and keep handy a set of measuring spoons and a pail of clean water for rinsing, with a clean rag for drying or wiping up spills. Fill the small container with your base.

Fill each of the cups flanking the board halfway with the dry pigment. Add in small amounts of water and stir continuously, trying to break up any lumps, until the mix is soupy; you want something thin enough to pour easily, yet still quite thick—think "milkshake." This is the prepared pigment you will use to fill each of the cups along the top of the pigment sample columns. Next, fill each of these cups with 10 teaspoons of base. Finally, in the upper-left square where the top row labels and left column labels intersect, paint and label a sample of the base using a small chip brush. Be sure to paint the base on thickly enough to achieve opacity, unless a wash is desired.

Mix and Apply the First Row of Samples

Starting with the left-hand pigment, pigment A, add ½ teaspoon into the first (left-hand) column; you now have a 5% saturation of that pigment. Working with the 2:1, 1:1, and 1:2 ratios, continue across the row, adding the correct amount of pigment, *remembering that the total pigment of both colors added must equal ½ teaspoon!* Therefore, we would add ⅓ teaspoon to the 2:1 column (2:1 = 3 parts total, ⅔ of which is pigment A; ⅔ of ½ teaspoon = ⅓ teaspoon); in practice, add a shy ½ teaspoon. Add ¼ teaspoon to

With a bit of organization, a clear study of the effects of blending two pigments can be conducted to produce a range of colors. PHOTO BY ACE MCARLETON.

the middle column (1:1), and 1⁄6 teaspoon (or a shy 1⁄4 teaspoon) to the fourth column. Wash your spoons, and then add pigment B in reverse: 1⁄2 teaspoon to the fifth column, and working to the left with a shy 1⁄2, a 1⁄4, and a shy 1⁄4 teaspoon. Now, mix each cup with a spoon until well blended (don't mix spoons unless you clean first!). Using each chip brush, paint on a sample of each paint in its designated spot in the first row, and clean and pat dry the brushes. As before, be sure to paint on the paint until it is opaque, unless you are specifically testing color for a semitransparent wash, in which case paint the sample as you would on the substrate.

Continue to Complete the Board

Repeat this process for each of the remaining five columns. Note that the saturation percentages are very approximate; you are continuing to add in more pigment, and therefore increasing the volume, at each stage, skewing the percentage. At the same time, you are also removing volume with each sample application. As this is simply a study, we are not overly concerned with the accuracy of the percentages and ratios. If you are in a rush and are not concerned with the integrity of the samples, go ahead and dry the board with a hair dryer. You will notice that the colors are dramatically lighter when dry; never judge a color until it has completely dried! Force-drying the board in this fashion will likely cause the samples to crack,

Accuracy of the Color Study

It is important to understand that this "study" is simply that—a focused observation of the behavior patterns of a pigment (or pigments) in varying concentrations. It is premature at this point to formulate directly for a specific color, although working in that direction is obviously helpful. Multiple rounds of color studies may be needed to hone in on the color of choice; at that point, however, the sample must be replicated in volume. It is highly unlikely that your sampling process will have been accurate enough to give you a replicable formula for 5 gallons of paint; rather, it will get you very close, at which point your final color samples can be executed, in either partial or full scale, depending on the complexity of the formula and the volume of paint needed. Always save your sample boards, as they will be helpful to reference in case there is a question as to the accuracy of the finished product; it is all too frequent for a client not to recognize a color once it has jumped from a 1-inch chip to a full room. Also, as you are developing familiarity with your pigments, having old color studies to reference can help put you further down the road toward that next final color.

dust, or delaminate, but if you are not intending on keeping the board, it may be worth the trade-off in time, as the color will still be accurate.

When the board is dry, the final step is to bring the board into the room you will be painting, under the predominant lighting conditions (e.g., direct sunlight, overhead light, etc.), to evaluate the color. Especially given the natural pigment's propensity to play with light, it is very important to view the samples in the conditions in which they will be replicated. You may find that none of the samples work, and you need to go in a different direction; that's fine, you have learned what doesn't work, which is a very valuable part of the process. You may find that certain areas need further study (e.g., trying ratios from 1:1, 1:1.5, 1:2) or to go further off one side of the chart (e.g., 1:3, 1:4, 1:5). You may find something you like, in which case you can go ahead and proceed to replicating the sample in volume. Regardless, you now have a procedure in place to evaluate any set of pigments in any base to bring you toward selection of a final color.

CHAPTER 21

Mechanicals and Utilities for Natural Buildings

The modern home is no longer just a shelter from the elements, nor even a shell to be heated in times of cold or cooled in times of heat. The conveniences of industrialization have gifted us—or saddled us, depending on your perspective—with a host of mechanical and utility systems that must be designed carefully in any building. In fact, these services drive fundamental elements of the design program and make for some of the most complicated and expensive rooms in our homes (bathrooms and kitchens). We will take a look in this chapter at all the other major utilities and services you are likely to encounter in a modern natural building, except for telephone and datacom service, and evaluate options for future designs.

Water

We are currently living in a world in which nearly 3 billion people (approximately 40 percent of the earth's population) live in water-scarce conditions. Millions of women and children worldwide spend several hours daily simply collecting water and transporting it back to their homes. This water, like much of it across the world, may be dirty; 2.5 billion people worldwide lack access to improved sanitation—the hygienic separation of human excreta from water supplies. Accordingly, lack of sanitation is considered to be the leading cause of infection globally. Considering that less than 1% of the water on the planet is accessible for human consumption, this makes the phenomenon of indoor plumbing quite miraculous indeed (SIWI n.d.). The goal then—in light of both the scarcity and the importance of safe water supply and treatment—is to identify ways to maximize freshwater treatment and conservation, as well as to provide appropriate wastewater management.

Shifting climate patterns as a result of global warming leading to increased drought and flooding are expected to exacerbate issues of freshwater supply and safe water treatment in the decades to come. Therefore, it is critical we start now to address these and other issues relating to the use of water in our homes.

PLUMBING (DISTRIBUTION)

Once the source has been brought to the building, the plumbing in the structure handles distribution of freshwater and collection of wastewater. There are few options at this point for "natural building" approaches to plumbing. Rather, we can examine a few basic ideas about ecology and resource awareness:

Materials selection. Freshwater distribution was most commonly run through copper piping, until recent developments made cross-linked polyethylene (PEX) plastic tubing widely available. While PEX tubing allows for far easier installation (crimping flexible tubing, rather than soldering rigid pipe), and therefore more flexible design ("home-run" distribution from zoned manifolds is a frequent feature of PEX-run plumbing systems), many—including the authors of this book—harbor concerns about plastic compounds leaching into the water source, despite industry assurances to the contrary. Copper piping, on the other hand, risks contamination from copper, lead, and flux, particularly in the presence of acidic water. The available evidence suggests that PEX may be safer for private well systems, while copper is better for chlorinated municipal systems (Riversong 2010, 1). For now, design logistics and client proclivities will be the determining factors of preference. For wastewater drains and vents, ABS black plastic pipe is frequently used as a replacement for PVC, which is highly toxic in its production and disposal as a result of its chemical formulation and therefore is more ecologically harmful than other plastics (see chapter 2). Both are recognized for use under the Uniform Plumbing Code.

Efficiency. The heart of sustainable water management comes from conservation and efficiency measures incorporated into the structure's design. Options for improved efficiency measures abound, beginning with lifestyle habits that include actions such as turning off the faucet while brushing teeth, taking shorter showers, and planting low-water lawns and gardens. Concentrating all plumbing fixtures spatially in the building will reduce pipe and tubing runs and will allow for short runs from the hot water heater, leading to reduced heat losses in distribution. Insulating hot water pipes will further reduce heat loss (see sidebar "Heating Domestic Hot Water (DHW)" later in this chapter). Finally, installing low-flow showerheads, water-conserving washing machines, and low-flush or dual-flush toilets will all substantially reduce water consumption over the life of the building's use.

Access and service. Good plumbing design centers on access to key features of the systems, allowing easy service for maintenance or extension. Exposed plumbing runs in basement ceilings or plumbing chases in bathrooms are great places to start; this highlights a liability of a slab-on-grade, in which the plumbing is frequently encased in concrete, making servicing and adaptation of the system very difficult. Venting is important to design in early, to ensure compatibility of the venting layout with the floor plan and structural system. Perhaps most important, *keep plumbing supply and waste runs out of exterior walls at all costs!* There is no need to tempt fate—keep the plumbing away from the cold, and from the moisture-sensitive and essential insulation and structure components of the building. There is rarely a good exception to this rule!

WASTEWATER TREATMENT

Wastewater is often broken into two categories: graywater and blackwater. *Graywater* refers to water that is less likely to have biological contamination, such as lavatory or bathwater; *blackwater* refers to contaminated water, often defined as toilet and kitchen-sink water, although different places will have different definitions. The state of Vermont, for example, considers all effluent except from toilets to be graywater but requires a septic tank and an engineered graywater-treatment system for its treatment, even if separated from blackwater. If you are building in a region with wastewater requirements, you must inform yourself of the existing statutes as early on in the process as possible.

A number of different strategies can be employed to reduce the volume of effluent needing to be treated, and to utilize the valuable nutrient load in wastewater as a resource, rather than shouldering it as a disposal burden. In addition to conservation measures mentioned earlier, separating graywater from lavatories, clothes washers, and bathtubs from toilet blackwater allows this high volume of water to be treated with simpler means, and potentially reused for non-potable water uses such as cooling or landscape irrigation, further reducing potable water consumption. This can easily be designed into the plumbing layout, with separate graywater and blackwater plumbing. There are a variety of products on the marketplace that manage graywater, by filtering, disinfecting, and rerouting it into toilet tanks or drip irrigation systems.

Composting toilets of a variety of designs, from ultrasimple sawdust bucket toilets to larger engineered systems designed to handle commercial volume, can be used to drastically reduce blackwater effluent. The collected solids, once fully broken down by bacteria, become a valuable nutrient source for landscape plantings.

Composting toilets can range from simple sawdust bucket systems to industrial-capacity manufactured units. PHOTO COURTESY OF WIKIMEDIA COMMONS.

Natural wetlands are the most effective water-treatment systems imaginable. PHOTO COURTESY OF WIKIMEDIA COMMONS.

Treatment options also vary widely, depending on scale, cost, and climate. Many of the more successful ecological wastewater treatment systems involve re-creating the highly effective natural systems that treat wastewater in macro-ecological systems: wetlands. The principle is simple: the high concentrations of biological activity, from bacteria to plants to invertebrate and vertebrate aquatic life, remove toxins and nutrients from the wastewater, and the plants help reduce effluent volume through evapotranspiration (pulling water up through their roots and evaporating the liquid through their leaves).

Constructed wetlands—wetland ecologies that are built into the landscape to treat a building's effluent—are effective for sites that can support a larger installation, although they must be very carefully designed to ensure that they are efficiently serving their intended goal of treating wastewater, and not simply supporting semiaquatic growth. In cold climates requiring year-round active treatment or on smaller sites, technologies termed "living machines" model similar systems in a highly controlled and efficient design, involving passing the effluent through a series of sand filters, treatment tanks (each populated with a carefully selected community of flora, fauna, and microbiological organisms), and in some designs a constructed wetlands at the end of the process.

Heat from wastewater can also be reclaimed. There are a number of products available for residential and commercial systems alike—as well as simple DIY setups—that route outgoing hot wastewater through a heat exchanger, preheating cold water piped into a water heater.

Energy: Electricity

Energy consumption during the lifetime operation of a home accounts for the vast majority of the carbon output of that building (Malin 2008). In addition to carbon and other greenhouse gas emissions, energy production and consumption have additional major negative impacts, including (1) the ecological devastation caused by mountaintop removal practices for coal mining; (2) underground aquifer contamination from hydrofracturing practices involved in natural gas extraction; (3) human displacement and cultural destruction from widespread flooding occurring when industrial-scale hydroelectric dams are built; (4) the future legacy of a radioactive nuclear waste disposal problem that will linger for millennia, not to mention the potential for incidental radiation contamination when man-made or natural disasters strike in the vicinity of a nuclear power plant, as was witnessed in the Japanese earthquake of 2011; and (5) the damage caused to sensitive ecosystems, such as remote ridgelines, through wind-power development. As such, the importance of reducing energy consumption in the home is critical to achieving program goals of ecological sensitivity. We will be evaluating energy consumption in its two major forms in a building—electricity and heat—from source to distribution. Please see chapter 2 for discussion of the ecological impact of the built environment, and chapter 7 for information concerning the physics of heat as it relates to buildings.

SUSTAINABLE SOURCES

Many utilities offer electricity generated from renewable sources as a part of their portfolio. For some it is an indistinguishable part of the whole; for others, a premium may be charged if you choose to have your energy sourced from renewables (note that they don't segregate the electrons; this is first and foremost an economic distinction, which at least allows you to voice financial support for more sustainable generation sources).

Another power option is to create your own electricity. Photovoltaic (PV) panels, wind turbines, and micro-hydroelectric turbines are three technologies that are sophisticated and affordable enough to be effectively implemented as a home-scale power generation system. There are many incentives available from municipal, utility, state, and federal sources to support purchase of these technologies, and most states allow "grid intertie" technologies, in which the on-site renewable-energy generation can be fed into the grid network during times of abundance, and electricity can be drawn from the grid during times of low or no production.

The decision to live off-grid is one that a small but enthusiastic segment of the North American population makes by choice or necessity. The consumer is the producer and is wholly responsible for all power used on the premises. This may simply mean running a gasoline generator, but more often than not it involves the harvesting of electricity from the sun, wind, waterways, or biomass from the forest. When one is required to produce every watt of electricity one uses, one's attention to consumption rates and generation potential is greatly heightened. Whether born of practicality, environmental ethics, or a combination of the two, an off-grid lifestyle—and it is indeed a lifestyle—creates a far more informed and conservative electrical consumer. It also requires discipline, investment (financial and time), tinkering, and flexibility. There is no utility ready to come out at all hours of the night to get the lights back on; as the power producer, the homeowner *is* the utility, and must be prepared to deal with maintenance and repair of the system.

With every passing year, there are more developments in power production technologies and renewable energy sources, from more efficient inverters to home-scale methane digesters. Contact your local utility and state energy authority to get more information on what options and incentives are available to you in seeking clean electricity to power your project.

CONSERVATION

It is not enough to simply replace one energy source for another, without exploring the amount of energy we consume to power our homes, our businesses, and our lives. Fundamentally, the issue we face is one of consumption rates, rather than simply one of energy sourcing. To be effective in reaching the goal of creating structures that do not contribute excessively to the proliferation of greenhouse gases in the atmosphere, we must understand that the most efficient—and affordable—watt is not one generated by coal, solar power, or any other source; it is the one that is never created.

Electrical conservation measures come in many forms, and a combination of strategies is required to create an effective low-energy building. Accessible and accurate meters are used to enhance our awareness of consumption. The monthly power bill, the inverter or battery meter, and even plug-in meters to evaluate power consumption of individual appliances all help us learn more about our consumption patterns. System operation can be improved through the use of well-designed controls. More sophisticated electrical designs feature intuitively located switches that will power off a circuit at the end of a day. A power strip can also control plug-in appliances. Replacing standard switches with timers for lights and fans reduces incidents of wasted energy when these are accidentally left on; motion sensors are an even more automated method of control for lighting. Integrating efficiency measures into the design such as daylighting and the use of energy-efficient appliances will return long-term benefits in building performance. Finally, a personal commitment to energy efficiency will ensure success of the efficiency strategy. This largely involves following through with efficiency practices over time. It may start by turning the lights off when you leave the room; it may lead to getting active in energy-related issues in your community. The potential is limitless!

WIRING

Wires are most commonly jacketed in a PVC sheath that is treated with halogenated flame retardants (HFRs), both of which are explored in depth in chapter 2 with regard to their ecological and human-health dangers. Lead is often used as a plasticizer, and other carcinogenic heavy metals are frequently used as pigments for coloring wiring jackets. In fact, the millions of miles of abandoned wires throughout the built environment will continue to present a long-term occupant safety issue due to the documented production of lead dust that can travel from within walls and plenums to occupied parts of the building. To reduce the inclusion of toxic wiring into the building, here are few guidelines:

- Use PVC-free, halogen-free, heavy-metal-free wiring. While more difficult to find and more expensive, there are a few such wiring products on the shelf today; as demand increases, so will product selection. Using metal-jacketed (MC or BX) cable will also avoid this problem, although it is more expensive to use as well.
- Keep wiring runs short. Shorter runs mean less wire; design electrical runs for efficiency.
- Keep wiring runs accessible. Accessible wires are easy to remove, reroute, extend, and modify.
- Don't over-wire. While code will often dictate much of the electrical specification, be judicious with additional wiring. You probably don't need outlets every 4 feet.
- Minimize wiring in exterior walls. The more interruptions in exterior walls, the more interruptions in insulation and air barriers. Try to minimize the services in the shell whenever possible.

Another significant yet controversial danger of indoor wiring comes from electromagnetic field (EMF) radiation. There is great controversy and debate within the scientific community as to the true level of danger posed by EMF radiation from wiring, as well as the radio frequency waves from Wi-Fi, cellular, GPS, and other wireless transmissions. Findings of a host of academics, researchers, doctors, biologists, physicists, and other scientists have found conclusive links between EMF radiation and cellular disruption, leading to a variety of illness including neurological disorders, cancer, immune-system disruption, and more (Riversong 2011; *Environmental Building News* 1998).

Given the relative lack of testing and data on the subject, and the potential danger in both scope and impact, we advocate for invoking the precautionary principle (introduced in chapter 2) when considering EMF exposure. Simple steps can be taken to reduce exposure to harmful radiation:

- Install circuit-interrupting wall switches that shut off bedroom current at night.
- Avoid sleeping near electric motors, such as plug-in alarm clocks, and avoid electric blankets.
- Keep wiring runs, service and circuit panels, and large-load appliances away from heavy use areas.
- Avoid separating hot and neutral wires, whose magnetic fields neutralize each other when in close proximity.
- Use three-strand wire when wiring three-way switches.
- Be judicious in the use of wireless technologies in the home and keep transmitters away from high-use areas.
- Use low-radiation computer monitors and televisions and keep a safe distance when viewing.
- Use 12-volt DC self-generated current and wiring.

NATURAL BUILDING AND ELECTRICITY

As with plumbing, the best design strategy for wiring natural wall systems is to keep the services out of the walls. Unlike with plumbing, it is generally necessary that some electrical installations, such as switches, will be located in natural wall systems. Therefore, appropriate design practices should be implemented. To reduce in-wall wiring, consider running wires through basement ceilings, where they can be accessible in the future, and stub wires up into wall cavities only when needed to service specific boxes. Another option when dealing with a slab-on-grade is to build a small pony wall—which may already be required for straw bale construction for wall protection—or a built-up chase in which to run your wires. Interior walls are also great places to run wires.

Remember that if you are working with an electrician, he or she may very well have strong feelings about running wires in nonconventional wall systems. Make sure you start a conversation with your electrician early on to design and implement systems with which he or she will feel comfortable. Understanding the code that governs your design is equally important, and make sure that any alternative wiring strategies are all code compliant, first and foremost.

When installing wiring in exterior walls, there are three main concerns: safety, function, and integrity of the thermal envelope. Safety concerns are specific to appropriate wiring runs, wire selection, and box and appliance installation; this is where consulting your electrician and your code will be very important. Function has to do with boxes that don't shift and move when you plug in appliances, and that can be mounted in convenient locations, often specified by code. It also relates to future adaptation and access to wiring—best achieved by keeping the wiring out of the walls in the first place or using wiring conduit when possible. Integrity of the envelope must be considered when thinking about displacing and disrupting insulation, as well as perforation of interior air barriers. Using sealed electrical boxes, such as those designed for airtight-drywall approach (ADA), or taking the time to seal conventional boxes or install airtight liners is well worth the effort, as is caulking between boxes and rough-coat plaster or plastering over an integral flange on the box.

Specific detailing for electrical installation in straw bale walls can be found in chapter 14. Installation into stud walls insulated with woodchip-clay or straw-clay follows standard practice as with conventional insulation; the difference is that the electrical inspector may require the use of UF (moisture-resistant) wiring or conduit owing to the wetness of the wall during installation, and care must be taken when tamping the insulation into place.

Wiring in mass walls is different, and techniques vary; consult the resources in appendix A for more information.

Energy: Heat

The other major use of energy in a building is space heating and water heating. As with electrical energy, there are similarities in sources, ecological impact, and the importance of conservation. As heating systems can be quite varied, we will explore the benefits and drawbacks of different combustion and distribution options in designing the optimal heating strategy for any given project.

RENEWABLE SOURCES: SOLAR, BIO-BASED

The obvious winners among sources (not considering appliance efficiencies or distribution losses) are those that are readily renewable and that can be managed for long-term production without enacting irreparable ecological or social damage. The most common is firewood. The management of firewood, as a low-value wood product is often easily integrated into other woodland management practices, making it an important value-added product in the forest products industry. High-yield management practices such as coppicing can efficiently produce large volumes of small-diameter fuelwood on a relatively small land base.

Coppicing is a technique for increasing the biomass output of a woodlot, producing trees that are very effective for fuelwood. PHOTO BY MARK KRAWCZYK.

Wood can be high maintenance and messy (in the area of the woodstove or wood furnace). Indeed, people who burn wood like to say that it heats us at least three times: when we fell and buck it, when we split and stack it, and when we burn it. Wood can also be very inefficient and polluting, depending on the source and combustion appliance. This creates issues of both atmospheric pollution and indoor air quality (IAQ). Wet wood will make for cooler and slower burns that are incomplete and release more harmful gases and particulates. Dry hardwood, on the other hand, burned quickly at a high temperature—preferably in an appliance that achieves secondary combustion of released gases—makes for an efficient and clean fuel source. Wood, particularly hardwood scraps, can be manufactured in regional-scale plants into pellets for use in space heaters or furnaces/boilers, achieving a cleaner burn using a secondary-value regional product while largely eliminating the mess, inconvenience, and work of cordwood. Note that these units require electricity to run, and the fuel is often sold in plastic packaging, increasing its ecological footprint.

Solar heat is the cleanest of all energy sources: in utilizing it we are making use of a readily abundant low-intensity energy source that requires no additional extraction or combustion to make the heat energy available. The most common form of solar heating—beyond the significantly effective passive-solar-heating strategies discussed in chapter 10—is solar hot water, which most commonly involves water or antifreeze circulating via an electronically controlled pump in a loop that heats the fluid up in collector panels (often roof- or ground-mounted) and dumps the heat through a heat exchanger into a supply tank for domestic hot water or space heating. A benefit of solar hot water, relative to solar electricity, is that it represents a more efficient use of the sun's energy; solar thermal is able to convert, on average, 60% to 70% of the sun's energy into heat, compared to conversion efficiencies of 15% to 20% for photovoltaic panels. Solar-thermal panels also perform much more efficiently than solar-electric panels in dappled or partially shaded environments.

In the cold and cloudy climate in New England, solar hot water (SHW) is best used for domestic hot water (DHW) loads, especially in smaller or more efficient buildings. To use SHW for space heating, it would most frequently have to be integrated into a central boiler, requiring a heating system that may be more expensive and complicated than necessary. Additionally, as SHW's output is reduced during the primary heating season, it becomes a less efficient option for space heating than for the smaller load of DHW. This is not necessarily the case in cold and sunny regions, like many parts of the United States and Canadian Rockies.

Solar hot air (SHA) or thermo-siphon air panels (TAP) operate under similar principles—air is heated in collector panels and circulated into the building—although use and distribution are a bit different. SHA is an open-loop system, in which fresh air enters the unit and is heated before entering the building. SHA heaters can be used either as in-room space heaters or as preheaters for ventilation systems.

HEAT PUMPS

Heat pumps use electricity in combination with a heat sink/store (they work very much like refrigerators) to either heat or cool a space. These pumps use the air (air-source heat pump) or ground or groundwater (ground-source heat pump) as a heat sink into which excess heat can be dumped for cooling purposes, or as

Two roof-mounted solar hot water panels are next to rows of PV panels on this Vermont residence. PHOTO BY WILL WHITE.

a heat source from which heat can be collected. Heat pumps are most effective in moderate climates, where both heating and cooling are required.

COMBUSTION AND DISTRIBUTION

There are two primary approaches to heating a residence or small commercial building: space and central. "Space heating" uses a heater that is designed to heat a room or given area, with limited or no distribution system. "Central heating" uses a heater that creates a lot of heat, which is then distributed to different regions throughout a building. Both approaches have advantages and disadvantages, which will be explored below. In general, if one is building a small, tight, and well-insulated structure, the additional capacity, cost, and complexity of a central heating system is unnecessary. That said, if a single controlled heater is desired to supply heat to different regions throughout a structure, a central heater will be the best approach. An exception to these two approaches can be found in Passive Houses, in which super-efficiency of the building's shell allows for the elimination of a heating system, utilizing the building's ventilation system to distribute the heat.

Heating-system design can be complicated, and slight nuances in the specifications can make big

Biofuels and a Truly Sustainable Energy Future

There has been a sharp increase in heating from bio-based fuel sources in the United States for environmental, political, and financial reasons. Some of these fuels are actually quite traditional and simple, such as wood or corn cobs; some regions of the world still burn peat or dried animal dung. Others are more complicated including diesel fuel oil made from plants such as soy and ethanol made from plants like corn.

While technological developments that allow us the ability to grow our own fuel source are very exciting when compared to the dangers and impacts of extractive energy sources, we must be careful not to replace one problem with another. The industrialization of agriculture that would be necessary to produce fuel crops on the scale required to replace extractive sources—without accompanying and much-increased conservation measures—presents a host of other vulnerabilities such as dependence on petroleum to power the agricultural operations and impacts including uncontrolled release of genetically modified organisms (GMOs), water contamination from pesticide and fertilizer use, and erosion from poor land-management practices. Indeed, some forms of biofuel such as ethanol from corn and biodiesel from soy or sunflower oil actually consume more energy than they produce (Lang 2005, 30). Furthermore, we quickly wade into dangerous waters once land devoted to food production is planted for energy production.

On the other hand, there are opportunities to utilize currently unused land to grow perennial plants for biofuel use, and in some locations there is enough forest cover to provide a significant fuel source, even when utilizing sustainable harvesting practices. It is clear that a very complicated, yet valuable, conversation concerning energy and food pricing, sovereignty, and equitable distribution is necessary, which challenges us to evaluate these technologies in a much larger context. We must be careful in the selection of appropriate technology, and ensure that technological advances become but a part of a larger solution to safely fulfill our energy needs.

differences in comfort, performance, and efficiency. In the sections that follow, we discuss the many options available for more efficient heating technologies to help inform an integrated, holistic design. Bringing in qualified support to model building heat loads, assess heating-system efficiencies, and analyze potential options may be warranted for more complicated or higher-performance buildings.

Finally, in this analysis, we must keep in mind the primary goal of heating a structure: the comfort of the occupants. Human comfort in a building is affected by five primary factors: thermal radiation (heat radiating to a person), air temperature, humidity, air movement, and the thermal conductivity of surfaces in contact with a person. Increasing thermal radiation, air temperature, relative humidity, and the temperature of the floor, while decreasing air movement, will all increase the perception of comfort. Other factors, such as air temperature stratification and radiant asymmetry, are also important to consider owing to their negative influence on human thermal comfort.

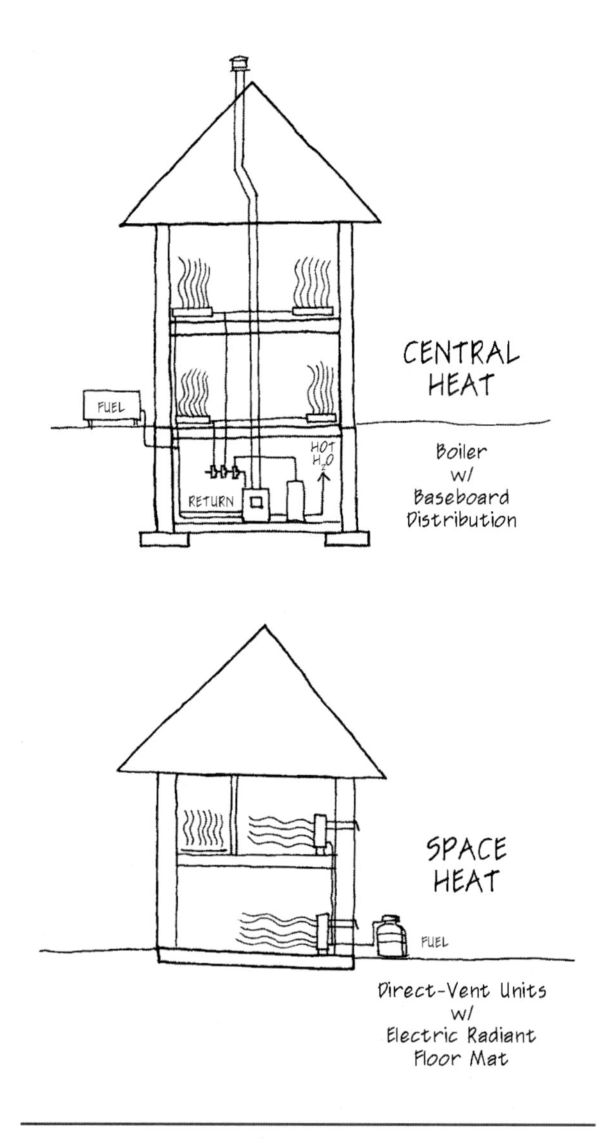

Space and central heating are two different approaches toward heating a building. ILLUSTRATION BY BEN GRAHAM.

Space Heating: Wood (Cordwood or Pellet)

Heating with wood is a lifestyle choice, as discussed earlier. The best way to fully experience "third heating" with wood is in a space heater, located within the living space of a building. In small-to-moderate efficiently built buildings, this space heater may be the only heater needed, or it may be a primary heater with additional sources required for the coldest winter nights or to supplement areas where the heat does not reach. Fans—mounted on heaters, from ceilings, or in interior doorjambs—can help direct air heated by wood heaters to isolated rooms in a building, as can wall and floor grates. In general, however, distribution of heat from any space heater is limited, and this should be considered carefully in system design as it relates to floor plan. Also worth considering from an ecological standpoint is the significant air pollution that can be emitted by wood-burning heaters. There are a number of factors that contribute to how dirty wood smoke may be, including wood species, moisture content, combustion temperature, circulation and secondary combustion of flue gases, and particulars of combustion appliances and venting. While we will look here at more efficient heaters, it is ultimately the responsibility of the operator to make sure all the variables are accounted for in reducing air pollution and protecting IAQ when burning wood.

Fireplaces are the most primal form of wood heat, but they are also the least efficient heaters, owing to the rapid exfiltration of heated air and subsequent draw of frequently unconditioned air into the house. Rumford fireplaces feature a few design improvements, including efficient firebox shape and flue throat to temper drafting, that improve the overall efficiency of a fireplace. The inclusion of dampers at the top of the flue will also help control convective heat losses when the fire isn't burning. A Rumford fireplace we constructed featuring a damper at the top of the chimney proved to be nearly airtight when we subjected that building to a blower-door test. Note that many building or energy codes require fireplaces to be outfitted with gasketed doors for both safety and efficiency.

Woodstoves are the most common wood-powered space heaters, for their ease of installation and use, convenience, affordability, and modularity (they can be moved, replaced, and modified). Not all woodstoves are created equal, and many experienced wood burners have tremendous loyalty to one design or manufacturer over another (for example, side-loading fireboxes versus top-loaders). The better woodstoves will have a few important features, beyond those of personal preference:

A Rumford fireplace is the most efficient design for a fireplace. PHOTO BY SARAH BRODEUR.

- tight construction, with well-fitting gaskets in any doors and no leaks between components of the stove
- an exhaust pathway that encourages secondary combustion of flue gases (more on this below)
- stones or other masonry components mounted onto or built into the firebox to retain heat
- direct-combustion air supply from the outdoors into the firebox, to reduce drafts in the building and avoid the potential for "back-drafting," the failure of effective flue gas exhaust from the building

Pellet stoves are another option for woodstove technology. Wood pellets, which are made of ground sawdust and agricultural waste, are compressed through heat and used as fuel in pellet stoves, which use electricity to power their automatic feeders and fans for both combustion air and distribution into the room. Compared to cordwood, pellets burn cleaner (owing to their low moisture content and high burn temperatures) and are lower maintenance, easier to regulate temperature, and less messy, but they are significantly more expensive for both the fuel and the appliance and will not run in the event of a power outage.

A big step up in efficiency, as well as in cost and permanency, is the masonry heater. Using the heat retention properties of the massive stone, brick, block, or other masonry materials of which they are built, these heaters accomplish two things very well. First, the circuitous pathway of the flue gases from the firebox to the chimney encourage their complete combustion, reducing particulates and other pollution for cleaner combustion, and extracting more heat from the fuel (this is in comparison to fireplaces, in which the gases go straight up the chimney from the fire). Second, the large amount of surface contact between the hot fire and gases and mass of the heater ensures that the majority of the heat is kept in the heater within the building envelope, rather than being sent out of the chimney. Combined, these make for fantastically efficient and comfortable heaters. A client of ours uses only 1–1.5 cords of wood in his masonry heater to keep his 1,300-square-foot Vermont home warm, with no other space heat source in the building. Masonry heaters are significantly more expensive, however, and they require careful planning in the design phase; improvements in the foundation, modification in the framing, and allowances in the floor plan all are considerations that must be taken into account when designing for a masonry heater.

Rocket stoves were developed in the 1980s to reduce heating and cooking-fuel consumption and to bring clean combustion technologies to developing regions suffering from poor indoor air quality as a result of open-pit-fire cooking in homes. These heaters and cookers encourage clean combustion by using simple principles in combustion dynamics known as the three Ts: time, temperature, and turbulence. By slowing down the exfiltration of flue gases (time), encouraging a very hot burn (temperature), and encouraging mild turbulence to mix flue gases with oxygen and help slow exhaust, a more complete combustion is achieved. These principles are the same as those used for a masonry heater; the hallmark of the rocket stove is a simple and accessible technology that is easily adaptable in scale—small rocket stoves can be built in the corner of a room from clay and

Masonry heaters are among the most efficient wood-heating options available. PHOTO BY WILLIAM DAVENPORT, TURTLEROCK MASONRY HEAT.

sand, for example. The great flexibility in their design makes them appropriate for inclusion in a wide variety of different heating-system designs.

Space Heating: Gas/Electric

For those not living in a forested region and not interested in the mess and inconvenience associated with wood, and who desire greater control and lower maintenance, a range of different efficient gas (natural gas or propane) and electric space heaters are available. One of the greatest benefits of a thermally efficient structure is the potential for significantly downsizing the heating system, allowing for one or two efficient and well-placed space heaters to handle all heating loads of the building. Between the low purchase price and simple direct-vent exhaust options, this is one of the best options for efficiency-minded designers on a budget.

The simplest and least expensive options for space heaters come from wall- or floor-mounted electric heaters, which come in a variety of styles, from baseboard units to panels to fan-driven heaters. Perhaps the best example of electric space heating we have seen is electric heating mats that are installed beneath tile; these have the benefit of providing heat to a bathroom, for example, which might otherwise be isolated from a space heater in another part of the building, while also heating a body of mass—albeit a small one—for retained, slow-release heat providing comfort underfoot.

The benefits of heating with electricity are the lack of combustion and associated fire and air-quality safety issues (or rather, the combustion is outsourced to the electrical generation facility), the removal of venting and shell penetration from the system, and the consolidation of energy sources (presuming electricity service to the building). Heating with electricity, however, is inherently inefficient compared to on-site combustion technologies as discussed at the beginning of this section, and it will likely cost more to run depending on current utility rates; off-grid buildings will most likely not have the option to heat with electricity.

Gas-fired space heaters are also very inexpensive and highly modular (they can be placed in a variety of locations and come in a variety of styles and models, including sealed-combustion fireplaces and visible flame burners). The more efficient models have annual fuel utilization efficiency (AFUE) ratings (the amount of energy in the fuel that is converted to heat output from the appliance) of 85% or higher. While chimney-vent options (through-roof) are available, direct-vent (through-wall) options are even more favorable, reducing venting runs and keeping shell perforations closer to the neutral pressure axis of the building, located at the volumetric center of the building (roof penetrations are more likely to result in convective heat loss in the building).

While "vent-free" or "unvented" heaters are able to achieve higher efficiency ratings (over 90%), the probable compromise to indoor air quality is not worth the efficiency savings. These heaters, as their names imply, vent the exhaust from their combustion directly into the building. We do not promote the use of these heaters for regularly occupied spaces, and they are illegal in many jurisdictions.

Space Heating: Air-Source Heat Pumps

As described earlier, air-source heat pumps (ASHP) work by circulating refrigerant through a loop, where it is alternately compressed and expanded to induce phase change from gas to liquid, allowing it to absorb and release heat from one source to another. The most effective and efficient ASHPs available for use as space heaters are "mini-splits." The "mini" refers to their relatively small size—frequently ductless units operating in the room requiring conditioning—while the "split" refers to the dual-component nature of the system, having an indoor air handler unit and an outdoor compressor unit (unlike a packaged window-mount air conditioner, which operates as a single unit).

As heating sources, these non-combustion units offer the air quality and fire safety of other electric heaters, with efficiencies comparable to those of high-efficiency direct-vent gas heaters. Their greatest strength may lie in their dual role of heater and cooler, indicating use in moderate climates for which they are best suited. We were able to find one product—the Mitsubishi Mr. Slim M-Series—that is able to retain most of its heat output (30,000 BTUs for a 35,000 BTU-rated unit) at -15°F. For those living north of the mid-Atlantic and central midwestern states additional heat supply will be necessary, while the demand for air-conditioning will be reduced. The other major downside of mini-split ASHPs is their cost. While they may prove to be quite cost-effective as a replacement for, or component of, more complex mechanical systems, for simple, small, and/or efficient structures they may not provide a quick economic payback relative to simpler space-heating and air-conditioning options.

CENTRAL HEATING: BOILERS AND FURNACES

For buildings that cannot rely solely on space heating for comfort, we can look to the conventional heating systems of most heating-climate North American homes: centralized, distributed heating. The two most common forms of this heating system are the furnace and the boiler. Furnaces heat air, which is then distributed in ductwork throughout the building. Boilers heat water, which circulates through baseboards, panels, or hydronic tubing. Fuel sources for these units can vary widely, including gas (NG and propane), fuel oil, coal, cordwood, wood pellet, or other biomass (such as corn). Multi-fuel options are available as original manufactures or add-ons; for example, a multi-fuel boiler may use wood with an oil backup for when the wood fire is not maintained, or a wood combustion unit may be added on to an existing oil-fired furnace.

Furnaces are simple in their design; hot air is simply ducted out to registers throughout the building, with one or more return ducts. One of their greatest attributes is the ability to use this same ductwork to distribute heat, chilled air, and fresh-air ventilation. There are significant inefficiencies involved with ductwork, however, and the design and installation of the distribution network must be done carefully to ensure that losses are minimized. Many people find such "forced hot air" heating systems to be uncomfortable, due to their turbulence and the dust and pollutant circulation from dirty vents and unchanged air filters (although this should not be an issue for properly installed and well-maintained systems); noise from the furnace also circulates through the ductwork, which can be a nuisance as well.

Best-quality boilers also feature sealed combustion (meaning supply air, the combustion chamber itself, and combustion exhaust are isolated from the indoor environment) and similar AFUE ratings. Direct venting is an option for many boilers. Condensing boilers are able to capture more energy from combustion gases; the resulting lower flue-gas temperatures result in water condensing from the gas, which increases heat output by further capturing the heat of vaporization as it changes back to liquid form. Boilers offer the additional benefit of being able to supply domestic hot water, eliminating the need for a stand-alone hot water heater.

In addition to indoor boilers, there are many outdoor wood boiler (OWB) options on the market. Originally, OWBs were very polluting and inefficient, even attracting nuisance bans in certain communities. Newer, more efficient OWBs achieve efficiencies comparable to those of indoor wood boilers, achieving very high burn temperatures and secondary flue-gas combustion for much cleaner emissions. These units have the benefits of keeping both fuel and combustion outside of the building envelope, enhancing convenience, indoor air quality, and safety, and the ability to pipe heat to more than one building, creating greater efficiencies of scale.

Outdoor wood boilers provide an opportunity to centrally heat multiple buildings and keep the mess and combustion safely outdoors. PHOTO BY KELLY GRIFFITH.

Boilers circulate hot water to baseboard units, or occasionally radiant wall panels that can be more space efficient than baseboard units as distribution points. One of the more popular distribution systems is in-floor hydronic tubes that circulate heated liquid, warming up the floor (see chapter 19 for more discussion on in-floor heating). These radiant floor heaters are beloved by many for their comfort, as they warm you by direct contact and minimize air turbulence and temperature stratification. Efficiency is another hallmark of radiant floor heating because room air temperature can be lower for the same level of comfort, requiring less heat input because the heat is kept closer to the occupants, which is particularly significant in high-ceiling or cathedral-ceiling spaces. Quiet operation and no obstructions to furnishings (such as with baseboard heaters and heat registers) are additional benefits, though care must be taken not to overly insulate a radiant floor with heavy carpets or rugs.

As wonderful as these systems can be, they tend not to play out so favorably in tighter, well-insulated buildings, for a few reasons. The investment in a well-built, high-quality shell offers the opportunity to utilize a simple, small, and inexpensive heating system. A radiant floor system that requires sophisticated and energy-consuming controls and circulating pumps with a centralized, distributed heating system may be over-designed and overengineered—and overpriced, at $10,000 or more—for a building that might only require a few hundred dollars of heat a year, which could be easily provided by one or two space heaters. The result is a dramatic reduction in the savings that could be realized from the investment in the building envelope. Even if this savings is unimportant to the homeowner, system-design issues related to correct sizing of the boiler and programming of the controls, as well as heat moderation issues, have been shown to arise in high-performance buildings needing only a small amount of heat. Irregular solar gain patterns may additionally load on top of the massive heat of a radiant floor, which cannot be easily "turned off," exacerbating the potential of overheating. Heat losses through under-insulated slabs-on-grade are another concern. Careful cost-benefit analysis and

Heating Domestic Hot Water (DHW)

For buildings running a boiler, DHW heating is efficiently achieved through combination heaters or indirect DHW tanks powered off the boiler. For buildings without a boiler, a stand-alone domestic hot water heater is required. This load is significant—it's typically the second-largest energy load in a residence, after space heating, unless in a climate where air-conditioning is desired. We discussed many options for making plumbing systems more efficient earlier in the chapter; here we'll look at options for the water heaters themselves. There are two basic approaches to heating DHW: storage and demand. In a storage heater, a large insulated tank keeps a volume of water (generally 30–50 gallons) heated at a set temperature, ready to use.

Storage units readily accommodate variable-heat input sources such as solar, wood, and heat pump and are low maintenance, but they suffer standby losses. In a demand heater (often called "on-demand" or "tankless"), the water is flash-heated instantaneously as the controls in the heater recognize a load. There are advantages and disadvantages to both.

Tankless heaters can be more efficient, but they are expensive and more finicky and need to be sized carefully to ensure delivery of the right quantity of sufficiently heated water quickly. Most units require a minimum flow rate before they turn on and so can be problematic with low-flow fixtures. Specific units must be selected for use with preheat systems.

At the end of the day, the correct choice must be vetted against the particulars of the given design. Regardless of which choice is made, remember that the heater should be considered as a part of a whole system including plumbing, heating, distribution, and use.

holistic-systems evaluation should be conducted before committing to a radiant-floor system in a high-efficiency building.

CENTRAL HEATING: RENEWABLE SOURCES (HEAT PUMPS, SOLAR)

In addition to the renewable biomass fuel sources used to power boilers and furnaces, solar-thermal systems can also be used as renewable sources for centralized heating systems, and heat pumps can take advantage of atmospheric conditions to improve efficiency. The space heating offered by the ASHP can also be configured to work as a centralized system, with a single compressor unit servicing multiple indoor units. GSHPs, while costly to install, are used in concert with boilers as preheaters, reducing the heating load on the boiler to bring the circulating fluid up to heating temperature.

Solar heating can be integrated into the building in a couple of ways. Solar hot air (SHA) units while able to be used in spot applications as space heaters (frequently installed outside operable windows), are more efficiently used as preheaters for furnaces or as heaters for ventilation systems (noting that additional heat sources may be required, depending on the climate). Solar hot water (SHW) often works in a similar fashion, most frequently serving to preheat boiler or domestic hot water (DHW) supply. As a closed-loop system, water or antifreeze circulates through a loop with the assistance of circulating pumps and controls.

TABLE 21.1. COMPARISON OF HEATING OPTIONS

Heater	Fuel types	Space or central?	Cost—initial ($–$$$)	Cost—operating ($–$$$)	Efficiency (*–***)	Design flexibility (*–***)	IAQ (*–***)	Comfort
Fireplace	wood	space	$$–$$$	$$	*	*	*	*_**
Woodstove	wood	space	$–$$	$–$$	*_**	**	*_**	**_***
Masonry heater	wood	space	$$$	$	***	*	**	**_***
Rocket stove	wood	space	$–$$	$	**_***	**	**	**_***
Electric heater	electricity	space	$	$$–$$$	*	***	***	**_***
Gas heater	gas	space	$	$$–$$$	**	**	**_***	**_***
Furnace	multi	central	$$–$$$	$$–$$$	**	*_**	*_***	*_**
Boiler	multi	central	$$–$$$	$$–$$$	**_***	*	*_***	**_***
Solar heat	sun (requires electricity)	both	$$–$$$	$	***	**	***	**_***
Heat pump	ground/air (requires electricity)	both	$$–$$$	$–$$	**_***	**	***	**_***

Note: There are many different considerations to weigh in evaluating a heating system. Use this chart to help compare the pros and cons of different heating systems and select the system most appropriate for your design.

The sun shines on solar hot water panels, heating up the coolant in the loop running through the panels; this heat is then transferred to the boiler or DHW heater through a heat exchanger. There are different types of solar hot water panels, varying in complexity of design, efficiency, and cost; some simple heaters are open-loop, heating potable water for direct use. Solar heating is much more efficient that solar electricity in terms of solar energy transfer. The challenge for solar heating in cold climates, however, lies in the fact that solar heat potential plummets when the demand is greatest, in the winter. For this reason, it is generally cost-prohibitive to use SHW or SHA systems as primary heat sources in cold climates. However, they can work well as auxiliary systems to reduce fossil fuel consumption of, for example, a boiler powering a radiant floor. They are perhaps best suited to supply the smaller, year-round demand for DHW when a stand-alone DHW heater is required (see "Heating Domestic Hot Water (DBH)" sidebar).

Ventilation

Ventilation systems are critical for both human occupants and the buildings themselves. For occupants, ventilation safely exhausts stale air—particularly important for structures with features that compromise indoor air quality, such as off-gassing chemicals in paint or furniture, or combustion appliances such as woodstoves—and supplies fresh air that may be conditioned by filtration, humidity, or temperature. For buildings, ventilation removes moisture that builds up within the house before it is able to migrate into wall and roof cavities; some ventilation systems are also able to recover heat and moisture during the process, reducing heating loads on the building.

It can be very difficult to rely on normal building leakage alone to control the amount, location, and timing of airflow into the building to ensure adequate ventilation. Leaks in the upper half of a building during heating conditions (the reverse if during cooling conditions) are particularly troublesome because, rather than sending cold, dry air into the building, they actually allow warm moist air to exfiltrate into the wall or ceiling cavities, depositing moisture as well as allowing heat loss. While a certain degree of leakiness is allowable, it should never be designed for as part of a ventilation strategy.

Presuming that your building is not providing ventilation in an uncontrolled fashion, you must take responsibility for designing appropriate ventilation in your building. A good ventilation system must include the following functions:

- exhaust stale and moisture-laden air, particularly from high-moisture areas such as bathrooms
- supply fresh air, particularly to high-use areas such as offices and bedrooms, conditioned if necessary/desired
- operate routinely throughout the day and the year despite outdoor climatic conditions
- run efficiently, using as little power to operate as necessary and losing as little heat as possible
- be responsive to pressure dynamics in a structure, to ensure that air is entering and exiting the building where and when it should

There are a number of different approaches to handling ventilation, from very simple to more complicated. As in the other mechanical systems we have looked at in this chapter, there are advantages and disadvantages to each different approach, which we will tease out as we look at how to achieve the goal of an appropriately ventilated building. Note that for the sake of this section, we are referring to air exchange within the building as a whole, and not specifically to direct air supply to combustion appliances (although this will be referred to where relevant).

WHOLE-HOUSE MECHANICAL VENTILATION

On the whole-house end of the ventilation spectrum, we find mechanical systems that are able to very accurately control the amount and location of airflow into and out of the building, and in many cases help condition the air by filtration, temperature, and humidity. In some cases, whole-house ventilation units can be connected to the ductwork of a heating system such as a forced-hot-air furnace, or with a centralized air-conditioning system. In many cases, the ventilation ductwork will be stand-alone for its own purpose, with fans that will control the supply and exhaust through the ductwork. Ventilation supply ducts are brought to bedrooms and offices, while returns pull air from bathrooms and kitchens. These systems are generally "balanced systems," meaning they exhaust the same amount of air they take in.

The most efficient ventilation systems are heat recovery ventilators (HRVs) and energy recovery ventilators (ERVs). These units integrate a heat exchanger into the fan and ventilation hub assembly, forcing heat exchange from warm stale exhaust air to cold fresh incoming air in the winter, and in reverse during the summer, reducing heating and cooling loads. ERVs also feature a component that facilitates moisture transfer, minimizing change in indoor relative humidity as a result of ventilation, and recovering the latent heat of vaporization. Efficiencies for HRVs and ERVs are likely to pay for the energy it takes to run the units. Many of these units are highly programmable, allowing for great control over building pressure, flow rates, power consumption, wireless switching, and even CO_2 monitoring and response. Advanced filtration, including HEPA filters, is also available for many units, as are humidistatic control features. One unit that has been certified for use in Passive House construction features an option for in-line electric heating of supply air. The technology will only continue to develop and improve as the market embraces these ventilation systems. As with heating systems, the integrity of the ductwork is exceedingly important to ensure efficient operation

HRVs and ERVs provide balanced, targeted whole-house ventilation while reducing heat loss. PHOTO BY KELLY GRIFFITH.

of the ventilation system; the ventilation unit is only one part of the whole system.

While these units give us the control, reliability (while the power is on), and automation missing from more passive ventilation systems, they also come with the liabilities of any mechanized system. Cost is significant, for one; not only are the units themselves expensive, but the cost of well-installed ductwork is an additional expense as well—installed cost of a high-end ERV can easily be ten to twenty times that of a quality exhaust-only fan, discussed next in the "Hybrid Ventilation" section. Designing ductwork into an open floor plan is, in and of itself, an obstacle, although new snap-fit flexible smooth-wall "home-run" ducting makes this easier to accomplish. The reliance on electricity, electronic controls, motors, and moving fan parts are all steps away from passive survivability and do indeed require successful long-term operation of all these moving parts to ensure viability of the system. As with heating systems, while centralized distributed ventilation may be required for more complicated or segregated floor plans, they may be less relevant for smaller, open buildings. It is important to note that their efficient use relies upon a relatively tight building; too much uncontrolled air leakage in a building will lead to a reduction in their efficiency, and these units are not recommended for leakier buildings.

HYBRID VENTILATION

A middle ground of sorts does exist in ventilation between fully passive and fully mechanical strategies. Exhaust-only fans—ductless or short-run ducted fans that blow air out of the assembly—are placed in high-moisture load areas, such as in bathrooms and kitchens (use fans rated for damp locations), or wherever else in a building stale-air exhaust is required, such as in a laundry room. The best of these fans are incredibly efficient in their electricity consumption (up to 10 cubic feet per minute per watt for some models), and very quiet—some as low as a barely audible 0.3 sones. Again, features abound, from motion sensors to lights to variable speed control.

One particularly helpful feature for use as part of a hybrid ventilation system is a dual-speed option that allows for either incidental high-speed operation (for example, during a shower) or lower-speed, lower-power operation for continuous use as part of a whole-house ventilation system. Timer-controlled units further enhance control over the system and can be wired remotely from the bathroom. In fact, programmable timers are a necessary component of exhaust-only-fan-powered ventilation systems to ensure that the system runs as frequently as necessary. As Robert Riversong points out, "It can be argued that the most effective exhaust-only ventilation system is one that runs continuously, maintaining an interior negative pressure and a constant low-level mix of fresh air. There is a trade-off, however, since a constant low-level flow rate may not remove concentrated moisture (in a shower room) quickly enough to prevent potential problems. For that reason, I prefer the mix of occupant-controlled high-speed intermittent exhaust and longer-term timed exhaust" (Riversong 2011). A quality bath-exhaust fan can also be separately controlled by a 24-hour programmable timer to allow it to operate intermittently throughout the day to provide sufficient air for whole-house exchange.

The use of exhaust-only fans depressurizes the building, which induces a draft to supply fresh air. This is where the "passive" part of the hybrid system comes into play, using passive air inlets for fresh-air supply. Passive air inlets are exterior-mounted units that allow air to enter the building in targeted points. Some are more complex, integrating temperature, moisture, or occupancy sensors to control when the vent is open or closed; others are quite simple, with manually operated controls to govern flow rate. These units take the pressure off window layout and can be designed to readily predict the amount of airflow into a building and direct it appropriately. When combined with exhaust-only fans, a well-controlled and accurately designed whole-house ventilation system can be achieved for very little cost, with only a small number of high-efficiency ductless or short-ducted fans required to operate the system. In this strategy, there is a compromise being made in that we are sacrificing the heat-recovery potential of an HRV or ERV in favor of a more passive system.

Passive vent units allow the controlled inlet of fresh air, which can be coupled with exhaust-only fans for an inexpensive and efficient ventilation system. PHOTO BY ACE MCARLETON.

Air-Conditioning

Air conditioners can be one of the highest energy loads for a building. We approach efficient air-conditioning system design the way we approach efficient heating system design: rather than looking at the appliance first, we look at all other conditions that can be put into place to require the smallest and simplest appliance possible, and then we ensure that the appliance is high performance and well suited to the design.

COOLING STRATEGIES

Heat enters a building through four primary ways: solar radiation through windows and roofs, conduction through the building envelope, convection from warm air leaks, and internal sources such as appliances. Referring back to the "Energy—Heat" section of this chapter, in which we identify the five major factors for comfort, we see that air circulation is a factor that, if increased, cools us. There is much we can do to encourage airflow within buildings, including capturing breezes through operable windows oriented correctly to prevailing wind vectors, and taking advantage of the principles of stack-effect convection outlined in chapter 7. For cooling purposes, it is worth noting that there are two strategies for incorporating passive ventilation: continuous ventilation or "night flush" ventilation, in which cooler nighttime air is invited into the building to reduce temperatures, followed by restriction of daytime ventilation to keep hot air out.

This latter strategy follows a principle of mass cooling, discussed in chapter 7, in which a lot of mass is built into a structure to absorb higher amounts of heat during the daytime, controlling temperature spikes, and the heat is released back into the cooler nighttime environment. Interior mass works well with night-flush ventilation strategies. Exterior mass walls can also work well under similar principles, keeping the heat from moving through the wall until nighttime, when it can be released back to the

How Tight Do We Build?

Ventilation systems obviously have a relationship with the tightness of a building. In the worst form of this relationship, leaks in a building's envelope are used as ventilation inlets and outlets. In its best form, the envelope is very tight, and a ventilation system operates to help support the interior climate created by a tight envelope. What is the appropriate tightness of a building? We posed this question to John Straube of the University of Waterloo, a leading building scientist.

Based on extensive field experience in Canada and the United States, Dr. Straube observed that houses with 1.5–3.0 ACH50 seem to have very few problems with air-leakage condensation. If they do, it is limited to small focused areas of the enclosure. Once envelope tightness gets to this level, one needs some form of controlled, often mechanically driven ventilation to avoid moisture problems. Anecdotally, he finds that at the lower end of this range (1.5 ACH50) small houses with ASHRAE 62 levels of ventilation (an industry standard of acceptable ventilation rates) begin to have higher, and potentially too high, wintertime relative humidity levels. The energy savings gained by dropping much below 1.5 ACH50 become not cost effective, but for homes with extreme insulation levels (such as Passive House) targeting airtightness under 1.0 ACH50 can be justified when trying to meet demanding performance goals. However, even at levels under 3.0 ACH50, depressurization of a home due to the use of exhaust-air appliances (clothes dryer, bathroom fans, range hood) can cause fireplaces, woodstoves, and other unsealed combustion appliances like gas water heaters or furnaces to back-draft carbon monoxide and other pollutants, so all combustion appliances should be direct-vent, sealed combustion. Achieving tightness standards of 1.5–3.0 ACH50 is therefore recommended.

Ventilation Safety: Non-Sealed Combustion Appliances

Robert Riversong of Riversong HouseWright offers this important safety consideration for designing a safe ventilation system for a home: "It's important to consider makeup-air requirements for non-sealed appliances such as clothes dryers and woodstoves. I prefer to place a dryer in an air-sealed (weather-stripped door) laundry/utility room with a dedicated 6-inch fresh-air inlet with ducting that drops 3 feet before exiting to prevent back drafting. Any woodstove that is not direct-coupled to outside combustion air should have a close-coupled makeup-air inlet located as near as possible to the stove's air inlet and at the same level or lower. If a house is heated primarily with a woodstove, the stove's flue can serve as a continuous nonelectric exhaust 'fan' with makeup inlets located in the usual living and sleeping places. Just be careful not to overwhelm the chimney's stack effect with too many exhaust fans running simultaneously, particularly when the stove door is open or the stove is 'idling'" (Riversong 2011).

outdoors, or by which point the thermal lag (the time it takes for the heat to move through the body of mass) will allow for easier control of the heat than during the daytime. It is therefore unsurprising that earthen building forms are so popular in desert areas, such as the southwestern United States, where diurnal temperature extremes rise steeply above and sharply below the average indoor comfort temperature of 70°F. This effect is lost when nighttime temperatures are higher than the indoor comfort temperature.

Strategies to keep heat from entering the building are also important to consider. Shading from landscape features such as trees is very effective; deciduous trees will even allow desired wintertime sun to hit the building, while shielding summer sun. High levels of insulation are as effective at keeping heat out of a structure as they are in keeping heat in; all the same principles of insulation strategy are in effect when trying to keep a building from overheating. Reflective barriers, such as a metal roof painted white, or with a paint containing highly reflective pigments, will go a long way toward keeping heat from entering the attic, as will radiant foil barriers underneath the attic framing. Living roofs also help keep buildings cool, as the plants' evapotranspiration process, combined with their shading effect, helps lower temperatures. Avoiding over-glazing on the east and west walls of the building will also help keep additional heat gain at bay, as will using windows with lower solar-heat-gain coefficients (SHGC)—the metric of how much solar heat can transmit through the window. Refer to climate-specific guidelines for your region. Overhangs, awnings, and other building features will help control shading, as well.

Spot-ventilating areas of intense heat production, such as kitchens, or insulating other sources such as hot water tanks and pipes, will help keep internal heat sources to a minimum. Considering that humidity increases discomfort in a hot environment, removing moisture sources from the building is another helpful strategy.

AIR-CONDITIONING APPLIANCES

Although heat pumps were introduced in the "Energy—Heat" section above, it is helpful to recognize here their effectiveness as air conditioners as well. In fact, most central and window-mount air-conditioning units utilize the same basic compressive cooling technology; the heat pump allows the cycle to flow in reverse, offering the advantage of providing heat as well, where desired in moderate climates. The efficiencies of these units are improving all the time, reducing the energy required for their operation. While the refrigerant used in most heat pumps and compression coolers is generally harmful—most notably for ozone depletion—improvements in reducing this liability are being made in next-generation refrigerants. Improvements are also being made to the appliances themselves, including their being able to retain the refrigerant and preventing it from leaking into the environment. For ducted coolers, the same principles of distribution efficiency as described in the "Ventilation" section apply here as well; in many cases, a central air conditioner can be combined with a furnace to provide cold air through the same ductwork.

The other common style of cooling for residential and small commercial buildings is evaporative cooling. An evaporative cooler takes advantage of the heat of vaporization to cool air. In these systems, warm, dry air circulates past a wet membrane, evaporating the water and transferring the heat of the air into the latent heat of the water vapor. The cooler air is then blown by a fan into a room or through ductwork. These units are simple, inexpensive, and efficient and do not require the use of refrigerant to operate. They are, however, limited to climates with hot dry air, as hot moist air has limited capacity to absorb excess vapor from the membrane.

CONCLUSION

If we have done our job well in writing this book, there should be at least as many new questions as answers, for it is through the process of opening wide the context in which we as natural builders evaluate how to build well that the vast complexity involved in making good decisions concerning the buildings we construct can be recognized. Indeed, it is this very approach—a commitment to a complex process viewed with a broad perspective—that distinguishes natural building from much of green building or other design and construction modalities.

Though we've covered much ground in this book, touching on topics as diverse as sociology, geology, botany, building science, and ecological and human health—not to mention the nuts and bolts of natural design and construction practices—an exploration of any of these topics quickly shows us how much we still have to learn. In fact, it is the presentation of these questions, and the research, debate, and arguments that follow, that will compel the natural building movement to continue its evolution. The practice of natural building promotes a deep focus on, and commitment to, the process of design and construction, the experience of living in a building, and the relationships formed as a result of this intentional process. As long as we continue to ask good questions, we will continue to move toward good answers, and we will grow as designers and builders.

A Structure Not in Isolation

A question we took up in chapter 1 while looking at the history of natural building now must be revisited as we look forward to the future: "What is the relationship between the building and the community beyond its immediate site?" We have looked at various levels of ecological impact, climatological response, and social context of the building. How those relate to a building in the broader context of local and regional community, however, is in many ways every bit as important as the nature of the building itself in realizing the goals of the design program.

One can build a permaculture paradise, as described in chapter 10, and blend the house beautifully into the landscape. With this effort, however, we've only just begun to broaden the context of a building's relationship to its place. The relationship between our buildings and our communities cannot be an afterthought or even an incidental consideration. For our goals as natural builders to be truly realized in the broadest, most meaningful, and most effective context possible, so they might truly address the myriad ecological and social problems we currently face, they must be implemented on a community-wide scale and connected with other ongoing work with social and ecological justice. It is getting easier with each passing year to build a "sustainable" building in a "sustainable" landscape—these are things that can be accomplished and managed with time and money and assisted by technological innovations. But without a truly sustainable community in which time and money are not scarce resources to which only a privileged few have access, the potential is necessarily limited.

As we have discussed throughout this book, natural building has a strong legacy of seeking to address not only the ecological transgressions of the built environment, but its social transgressions as well. As natural builders, we are uniquely positioned to be aware of the role that the contemporary practice of architecture and the mechanics of the construction industry often play in disempowering and disenfranchising segments of our population.

On Social Justice in the Built Environment: An Architect's View

BY CAREY CLOUSE, CROOKEDWORKS ARCHITECTURE, WWW.CROOKEDWORKS.COM, AND YESTERMORROW DESIGN/BUILD SCHOOL LEAD INSTRUCTOR FOR SEMESTERS IN SUSTAINABLE AND INTEGRATIVE DESIGN/BUILD

As architects and builders look for opportunities to diversify their professional practice, their work should extend beyond the traditional barriers of race and class. The architecture profession's relatively recent affection for "green building" has made it a movement that highlights physical factors without necessarily addressing social and cultural needs. Moreover, this sustainability agenda disproportionately targets and serves privileged classes. To broaden the scope of design practice today is to make it more accessible, more inclusive, and more open to different ways of doing and being.

This opening up—particularly toward social justice in the built world—could help the design profession become more relevant today. Historically, public-interest design work has been a fringe movement within an otherwise established field. The dearth of community-driven architectural design, at least in the traditional definition of the practice, accounts for a growing disjunction that involves class, access, and the responsibility of designers to provide services to all kinds of clients.

Moreover, social justice is an important component of environmental stewardship. As environmental health relies on intact social and cultural centers, natural building can act as a link between architectural service and healthy communities. Similarly, natural building techniques could help level the playing field, providing all people with healthy and efficient living spaces.

In considering a more expansive view of sustainability, architects and designers have the opportunity to integrate critical social-justice components in the built world. By making the connection between sustainable design practice and social justice—and in recognizing that the two issues are inextricably linked—designers have the potential to become a bigger force for change in the world.

We are able to see large patterns of social inequity in which many in the United States are heavily pressured by cultural indoctrination, development patterns, and economic imperatives to chase a goal of home ownership beyond their means, frequently at the expense of living for decades under the burden of debt, requiring extensive investments of time and energy into the workforce to keep financially solvent (the word "mortgage" is from the Old French, with *mort*, meaning "dead," and *gage*, meaning "pledge" or "debt," so the literal translation is therefore "death pledge"). As was clearly evidenced by the United States financial crisis of 2008—triggered in large part by the implosion of a housing bubble inflated by unregulated securities investments of high-risk mortgages offered to millions of clients ill-positioned to afford them—this issue does not stay quietly in the working class of American society, but rather affects all but the very wealthiest. The burden of this debt, combined with other exacerbating factors such as rising cost of living, stagnant wage and employment levels, increased educational costs and associated student loan debt, and credit card debt, robs us of leisure time, as many are forced to work additional hours or jobs to cover their expenses. In many cultures this leisure time is used not purely for escapist recreation, but for cultural activities in the fields of art, sport, philosophy, and political discourse; our inability as a society to engage fully in this level of cultural involvement keeps us in an oppressed state of disempowerment.

And what is received in return for this death pledge? Might we at least as a society be provided with a quality built environment in which to conduct our lives? The answer, of course, is no. Rather, we are left with housing stock that is destroying our ecology, consuming vast amounts of energy and other resources, and toxic to its inhabitants, and from whose service and operation we are largely disconnected and removed, often unable to substantially repair or modify the building without significant investment in tools and knowledge. As we have seen in the 2008 collapse of the U.S. housing market, the investment in a place to live cannot even be expected to hold its value within the period of maturation of the death pledge, leaving homelessness, ruined credit, and financial destitution as the payoff for millions.

While this is the pattern for many of those who are able to seek the goal of home ownership, for the millions more for whom simply finding a home is a goal, never mind becoming a homeowner, an even grimmer pattern of housing insecurity is manifest in dangerous and toxic living conditions and communities struggling under the weight of racism, economic pressure, and violence. Even the patterns for housing security provided or supported by government institutions for those recovering from natural disaster are fraught with inequity and damage to the health of the human body and spirit, as is clearly evidenced from the ongoing recovery effort in New Orleans, Louisiana, in the wake of Hurricane Katrina in 2005.

If we truly want to create structures that are socially and ecologically responsible, we have to conceive of them—and build them—so that the buildings themselves, as well as the people who live and work in them, relate to, and enhance, their surrounding communities. To do this is to look toward the future—the future of ecological design, of green building, of social change—because it is on the community scale that much of the impact we seek to have with our buildings can be realized. Having looked long and hard throughout this book at the building and the site itself, we conclude by looking beyond the building to the community and society in which it is built.

The Places We Live: Challenges and Opportunities in Cities, Towns, Suburbs, and Rural Areas

Cities, suburbs, small towns, rural areas—all of these types of communities have within them differing challenges and opportunities to promote ecological and social well-being.

In terms of limiting ecological impact, urban communities have the potential to address many of the resource-intensive consumption challenges faced by rural and suburban communities by virtue of their population density. High-density populations are well positioned to extend the utility of a service or the potential of a resource to many more people than in lower-density populations. Transportation, utilities including water (freshwater supply and wastewater treatment) and energy, places of commerce and service, institutional and governmental access—all are increased in scale, consolidated in location, and accessed by larger populations, offering the potential for a far more efficient social model. In fact, two recent reports support the claim that urban populations have lower per-capita energy and carbon footprints. A 2009 study conducted by the London-based International Institute for Environment and Development (IIED) analyzed the greenhouse gas emissions of residents of eleven different cities on four continents. London's per capita emissions were nearly half of the U.K. average, with the rural northeast of that country identified as having the highest per capita emissions. New Yorkers' emissions were less than one-third of the average American's. Transportation and housing density were noted as being the largest two factors governing the lower emission levels (Vaughn 2009). Similar conclusions were drawn in a 2008 report conducted in the United States by the Brookings Institute, in which a study of one hundred American metropolitan areas revealed significantly lower per capita emissions than nationwide averages in each city (transportation and electricity consumption were primary factors identified in this study), despite being host to two-thirds of the nation's population and three-quarters of its economic activity (Brown et al. 2008, 4–7).

Merely living in a metropolitan area does not inherently create a lower carbon footprint, however—the potential of urban efficiencies is not always realized. In a study of two Finnish metro areas comprising eleven cities of both dense urban core and less-dense suburban population centers, the authors analyzed per capita carbon footprint not based purely on regional output, as was the mechanism for the IIED and Brookings Institute studies, but based on a hybrid life-cycle analysis (LCA) model in which consumption patterns were taken into account as well, tracing carbon loads of goods and services beyond the confines of the metro area to their regions of origin. The results given from this assessment model showed that urban residents in fact had much higher carbon footprints, as their heavier consumption patterns outweighed the efficiency benefits of their region (perhaps predictably in direct proportion to their income) (Heinonen and Junnila 2011). In both the IIED and Brookings Institute studies, it was further shown that there were wide variations in per capita emissions that related closely with the development patterns of the city. Tokyo and Seoul, for example, had comparably lower emission rates than less-affluent cities, challenging the allegation that affluent developed populations inherently have higher per capita emission rates. In the United States, east-central and eastern metro areas had much higher emission rates than western metro areas, with no obvious distinction between areas in heating and cooling climates. Urban consolidation, public-transportation options, policy initiatives targeting efficiency improvements, and progressive planning policies were all cited as primary factors for decreased emission rates among the lowest-rated metro areas.

In order for metropolitan areas to be able to reach the lowest potential emission scenarios, much more needs to be done; the IIED report indicates that all but the lowest two emission rates found in the study were still too high to curb climate change, despite being lower than national averages. Recommendations from the Brookings Institute report include promoting enhanced public transportation and freight delivery options and regional planning, as well as federal and market policies and initiatives driven toward increasing the efficiency of existing housing and commercial stock. The conclusions of the Finnish study strongly imply that efficiencies realized by urban density are quickly outweighed by consumption patterns—and that when consumed goods and services are subjected to LCA, the reliance of urban residents on regional and global industry dramatically reduces their efficiencies—pointing to the importance of greater self-reliance in metropolitan areas, including the development of sustainable manufacturing, agricultural, and energy-production sectors within the metro area or surrounding region.

All the emphasis on resource consolidation for the population-dense cities is not to suggest that living in a lower-density environment such as a rural or suburban area or town is inherently unsustainable. Rather, it is to identify that we must pursue what it means to create communities by exploring the problems we currently face and making evaluations on a community-wide scale, instead of as a series of neighboring parcels pursuing piecemeal permacultural installations. Transportation is one of the biggest issues, considering that the transportation sector is responsible for 33% of the U.S. emissions of carbon dioxide. Certainly fuel-efficient vehicles will help address this problem, but technology alone cannot save us. Carpooling and the extension of mass public-transit systems into rural communities is a step to be taken in both the grassroots-cultural and public-sector arenas. Creating land-based and home-based economic opportunities, alleviating the need to commute far distances for employment, is perhaps one of the most powerful approaches to take—and one of the most complicated and difficult to institute.

Whether through the promotion of small-scale agricultural or other land-based enterprises, the extension of high-speed Internet connectivity to outlying rural communities that would enable e-commerce and telecommuting, or the development of vibrant rural cultures that can support small rural economies—perhaps based on the development of regional-scale agricultural or manufacturing-sector industries, as was the hallmark of rural life in many parts of the United States up until the middle of the last century—there are many opportunities to create or re-create, as the case may be in many regions, a sustainable rural society.

As for the suburbs, we find a development pattern that features many of the liabilities of both rural and urban communities: there is complete reliance on the automobile for functionality; they are located far from the urban centers upon which they depend for culture, commerce, employment, and many services; they are culturally fragmenting, divisive, and exclusive; and they are frequently sited on once-productive agricultural land that supported the nearby urban center.

Whether in a new rural community or with the redevelopment of an existing urban or suburban neighborhood, there are common approaches in creating a sustainable community design supportive of the same ethics and values exemplified in the practice of natural building. Jeff Schoellkopf, an architect and planner with the Design Group of Warren, Vermont, which is experienced in community-scale design, offers these insights: "I like to think of communities as socially and culturally generative. We like to design buildings and plan communities that grow from the patterns and traditions of their region and landscape, but express themselves as a new generation. We like to think of buildings as grandchildren of the community, not replications, but bearing family resemblances and of a new generation.

"Good new or rejuvenated neighborhoods and community design present many opportunities. Clustering or joining building saves land and works against sprawl. The benefits can be to agriculture, forests, the visual landscape, and nature. This can also make for more walkable and bikable access to work, school, play, etc., and makes more concentrated access to public transportation possible. It can also take advantage of shared utilities and infrastructure—walks, paths, water and wastewater systems, power generation and distribution, gardens, stormwater systems, fields, playgrounds, and other common amenities. Natural and local building provides a deeper richness and ecological integrity in communities."

We cannot attempt here to provide in entirety solutions to the social dilemmas presented by the contemporary built environment. However, this should not deter us from the challenge. Throughout this book we have shown applications taken from the world of natural building that actively address issues of health, affordability, skills access, performance, ecological sensitivity, and community building and development. These advances are in no way unique to the world of natural building; housing nonprofits, community action organizations, and many members of the green building industry have all been working to address these same issues in a proactive and meaningful way, with great success. With action comes empowerment, and with empowerment comes change.

This social change is taking many forms, with the common theme behind them all a growing awareness of the importance of learning how to live and work with each other in different, often closer ways. As the pattern of individually owned single-family residences proves to be in many ways flawed, other patterns of more communal or integrated living provide some solutions. Cohousing communities, intentional communities, and community land trusts continue to be relevant options as we learn how to live more closely with each other. The Transition Town movement has spread across the country, galvanizing and strengthening communities on a grassroots level to proactively create a post-carbon world. In Buffalo, New York, a strong and growing reuse movement exists, led by ReUse Action, that counters the effects of racism, sexism, and crushing economic realities by creating employment for local people through dismantling crumbling infrastructure ("green demolition") and acquiring "waste" products and remaindered goods in order to build new, more ecological and just shelter. Many other solutions in many diverse places are happening, with more being developed every day.

ReUse Action in Buffalo, New York, is one of many organizations working to address social and ecological issues relating to the built environment, in their case through job training, deconstruction services, and building material reuse. PHOTO COURTESY OF REUSE ACTION.

In 2008, for the first time in history, more people on the planet were living in urban environments than rural environments—the effects of the largest wave of urban growth ever (UNFPA 2007). This means that looking to creating social and ecological vibrancy in cities, towns, suburbs, and rural areas is essential. For just as it is necessary to expand the context beyond the structure to the site, and beyond the site to the community, so too must we look at how our communities relate to and support each other. The patterns of globalized manufacturing and agricultural production and distribution systems are also proving to be flawed in many ways, and regional-based economies—once the norm—are resurging, supportive of creating an environment that can foster cultural change. It is necessary to continue to intentionally develop regional rural/urban connections, such as the establishment of regionwide foodsheds that maximize the potential of agricultural production areas to support larger regional population centers. It is not a question of whether rural, urban, or those societies in between will prove to be more sustainable; we need all of them, and all of them need to be mutually supportive of long-term goals of sustainability.

Call to Action

How do we realize the transition to a new cultural paradigm that embraces the building of vital communities, politics, and economies in relationship with the natural world in a thriving, interrelated whole? To define a way forward, we must look at the problems that are keeping us from obtaining this goal. This is a process of enormous scope and scale, but this should not permit us to avoid the effort. There is a principle in Judaism called *tikkun olam*, translated as "the renewing of the world." As with many principles in Judaism, many different meanings and interpretations of this concept have developed over time; we read it here as the contemporary call to social and ecological justice. In a famous parable, a distraught scholar seeks the council of his rabbi, asking, "What is the point of attempting this impossible task? Why should we follow the directive of a mandate beyond our ability to achieve?" The rabbi replies: "Simply because the task is not ours to complete does not mean that it is not ours to undertake."

We must remember that communities are built of a collection of individuals. Ultimately, the power of creating the changes necessary for our survival lies not with society as a whole, but with a society of individuals, each of whom must work, together, to maintain personal commitments toward acting and living in service to something greater than themselves. Such commitment is supported by the strength of the relationships we form—with each other, with our ecology, within our economy, with our food, with our homes, relationships that are built as we strive to break old patterns of individualism and isolation.

Natural building is about relationships. We choose to work with natural materials not just because they are "natural," but also because their use is the logical conclusion of a process in which we seek to develop and sustain as many relationships and connections as possible within the context of the development of a building. This process of natural building acts as a web, connecting us back to "place" and all those who help make that place. Even as we face unprecedented ecological and social challenges as a species—challenges that must be met on the scale of a society—we find ourselves, a collection of individuals, reaffirming our commitment to creating change through the power of relationships, one building at a time.

Asking good questions helps lead to good answers. We face an incredibly challenging question: in light of the transition we are undergoing as a species into a post-oil, post-carbon world, how can we adapt to create habitat that is safe, accessible, and beautiful? Current and future generations do not have the luxury, or the limitation, of remaining blissfully ignorant of the impact of our actions in developing our built world. This burden of consciousness, heavy as it may seem, provides us with the question, and the opportunity to come up with an answer. In this book, we have attempted to illuminate opportunities to provide answers, not through prescriptive formulas for "good building," but instead through principles developed around the case study presented by our northeastern region that can be applied anywhere, on any scale. There is no singular answer to this question. No industry, government, or movement will be able to provide the answer alone. It is an answer we each must provide ourselves, as builders, designers, planners, engineers, code officials, and owners, in response to our own ethics, values, beliefs, and truth. At the same time, it is a question that we, as natural builders, are answering together.

APPENDICES

A. Suggested Further Reading

Introduction

Bookchin, Murray. 1997. *The Murray Bookchin Reader*. London: Cassell.

Davis, Mike. 1999. "Fortress Los Angeles: The Militarization of Public Space." In *Variations on a Theme Park: The New American City and the End of Public Space*, ed. M. Sorkin, 154–80. New York: Hill and Wang.

Chapter 1. A Brief History of Natural Building in the Northeast

Garvin, James L. 2001. *A Building History of Northern New England*. Hanover and London: University Press of New England.

Chapter 2. Ecology

Baker-Laporte, Paula, Erica Elliott, and John Banta. 2008. *Prescriptions for a Healthy House: A Practical Guide for Architects, Builders & Homeowners*. 3rd ed. Gabriola Island, BC, Canada: New Society Publishers.

Meadows, Donella, Jorgen Randers, and Dennis Meadows. 2004. *Limits to Growth: The 30-Year Update*. 3rd ed. White River Junction, VT: Chelsea Green Publishing.

Moore, Charles, and Cassandra Phillips. 2011. *Plastic Ocean: How a Sea Captain's Chance Discovery Launched a Determined Quest to Save the Oceans*. New York: Avery Publishers.

Chapter 3. Beyond the Building: Siting and Landscape Design

Jacke, Dave, with Eric Toensmeier. 2005. *Edible Forest Gardens*. 2 vols. White River Junction, VT: Chelsea Green Publishing.

Kellogg, Scott, and Stacy Pettigrew. 2008. *Toolbox for Sustainable City Living: A Do-It-Ourselves Guide*. Cambridge, MA: South End Press.

Mollison, Bill, and Reny Mia Slay. 1994. *Introduction to Permaculture*. 2nd ed. Tyalgum, NSW, Australia: Tagari Publications.

Thallon, Rob, and Stan Jones. 2003. *Graphic Guide to Site Construction*. Newtown, CT: Taunton Press.

Chapter 4. Soil and Stone: Geology and Mineralogy

United States Department of Agriculture Natural Resources Conservation Service (NRCS). n.d. Web Soil Survey. http://websoilsurvey.nrcs.usda.gov/app/HomePage.htm.

Chapter 5. Flora and Fauna

Wessels, Thomas. 1997. *Reading the Forested Landscape: A Natural History of New England.* Woodstock, VT: The Countryman Press.

Chapter 6. Structure and Natural Building

Allen, Edward. 1980. *How Buildings Work: The Natural Order of Architecture.* New York: Oxford University Press.

American Wood Council. n.d. "Maximum Span Calculator for Wood Joists & Rafters. http://www.awc.org/calculators/span/calc/timbercalcstyle.asp (accessed September 17, 2011).

King, Bruce. 1996. *Buildings of Earth and Straw.* Sausalito, CA: Ecological Design Press.

Chapter 7. Heat and Natural Building

International Passive House Association (iPHA), Darmstadt, Germany. www.passivehouse-international.org.

Passive House Institute U.S. (PHIUS), 110 S. Race Street, Suite 202, Urbana, Illinois 61801. www.passivehouse.us.

Chapter 8. Moisture and Natural Building

Building Science Corporation, 30 Forest Street, Somerville, MA 02143. www.buildingscience.com.

Chapter 9. Fire, Insects, and Acoustics

National Fire Protection Association (NFPA). 2012. "2011 National Electric Code *and* NFPA 220: Standard on Types of Building Construction." Quincy, MA. www.nfpa.org.

Chapter 10. Design

Alexander, Christopher. 1979. *The Timeless Way of Building.* New York: Oxford University Press.

Alexander, Christopher, Sara Ishikawa, and Murray Silverstein. 1977. *A Pattern Language.* New York: Oxford University Press.

Brand, Stewart. 1995. *How Buildings Learn: What Happens After They're Built.* London: Penguin Books.

Energy Efficiency and Renewable Energy Clearinghouse. 2001. "Passive Solar Design for the Home." National Renewable Energy Laboratory. http://www.nrel.gov/docs/fy01osti/27954.pdf.

International Institute for Bau-Biologie and Ecology, Clearwater, FL. http://buildingbiology.net/.

International Living Future Institute's Living Building Challenge. http://ilbi.org/lbc.

Kellert, Stephen R., Judith Heerwagen, and Martin Mador. 2008. *Biophilic Design: The Theory, Science, and Practice of Bringing Buildings to Life.* Hoboken, NJ: John Wiley and Sons.

Mazria, Edward. 1979. *The Passive Solar Energy Book: A Complete Guide to Passive Solar Home, Greenhouse and Building Design.* Emmaus, PA: Rodale Press.

Melby, Pete, and Tom Cathcart. 2002: *Regenerative Design Techniques.* Hoboken, NJ: John Wiley and Sons.

Pearson, David. 2005. *In Search of Natural Architecture.* New York: Abbeville Press.

Savory, Allan, and Jody Butterfield. 1998. *Holistic Management: A New Framework for Decision Making.* Washington, DC: Island Press.

United States Green Building Council (USGBC) LEED Program. http://www.usgbc.org/DisplayPage.aspx?CMSPageID=1988.

Chapter 11. Before Construction

Chiras, Dan. 2004. *The New Ecological Home: A Complete Guide to Green Building Options.* White River Junction, VT: Chelsea Green Publishing.

Clark, Sam. 1996. *Independent Builder: Designing & Building a House Your Own Way.* White River Junction, VT: Chelsea Green Publishing.

Development Center for Appropriate Technology (DCAT), Tucson, AZ. www.dcat.net.

Journal of Light Construction. Hanley Wood, LLC, Williston, VT. www.jlconline.com.

Kennedy, Joseph, Michael Smith, and Catherine Wanek, eds. 2001. *The Art of Natural Building: Design, Construction, Resources.* Gabriola Island, BC, Canada: New Society Publishers.

Lstiburek, Joe. 1998–2005. *Builder's Guide to Climates.* Newtown, CT: Taunton Press.

Roy, Rob. 2008. *Mortgage-Free: Innovative Strategies for Debt-Free Home Ownership.* 2nd ed. White River Junction, VT: Chelsea Green Press.

Snell, Clark. 2004. *The Good House Book.* New York: Lark Books.

Chapter 12. Foundations for Natural Buildings

Ching, Francis D. K., and Cassandra Adams. 2001. *Building Construction Illustrated.* 3rd ed. Vol. 2. New York: John Wiley and Sons.

National Association of Home Builders. 1998. "Energy-Efficient Resource-Efficient Frost-Protected Shallow Foundations." http://www.nahb.org/assets/docs/publication/Energy-efficient-frost-protected-shallow-foundations_1211200244041PM.pdf.

Chapter 13. Framing for Natural Buildings

Chappell, Steve. 1998. *A Timber Framer's Workshop: Joinery and Design Essentials for Building Traditional Timber Frames.* Brownfield, ME: Fox Maple Press, Inc.

Sobon, Jack, and Roger Schroeder. 1984. *Timber Frame Construction: All About Post and Beam Building.* North Adams, MA: Storey Publishing.

Chapter 14. Insulative Wall Systems: Straw Bale

King, Bruce. 2006. *Design of Straw Bale Buildings: The State of the Art.* San Rafael, CA: Green Building Press.

Lacinski, Paul, and Michel Bergeron. 2000. *Serious Straw Bale: A Guide to Construction in All Climates.* White River Junction, VT: Chelsea Green Publishing.

Magwood, Chris, Peter Mack, and Tina Therrien. 2005. *More Straw Bale Building: A Complete Guide to Designing and Building with Straw. Mother Earth News* Book for Wiser Living. Gabriola Island, BC, Canada: New Society Publishers.

Magwood, Chris, Chris Walker. 2005. *Straw Bale Details: A Manual for Designers and Builders.* Gabriola Island, BC, Canada: New Society Publishers.

Chapter 15. Insulative Wall Systems: Straw-Clay, Woodchip-Clay, and Cellulose

Andersen, F. 2002. "Light-Clay: An Introduction to German Clay Building Techniques." In Kennedy, J.F., Wanek, C., and Smith, M. *The Art of Natural Building: Design, Construction, Resources*, ed. J. F. Kennedy, C. Wanek, and M. Smith. Gabriola Island, BC, Canada: New Society Publishers.

U.S. Forest Products Laboratory and Design Coalition. 2004. "Straw-Clay Thermal Testing." http://www.designcoalition.org/articles/Lansing-LHJ/research/Ktesting2.htm.

Chapter 16. Natural Mass-Wall Systems: Earth and Stone

McRaven, Charles. 1989. *Building with Stone*. North Adams, MA: Storey Publishing.

Minke, Gernot. 2006. *Building with Earth: Design and Technology of a Sustainable Architecture*. Basel, Switzerland and Berlin, Germany: Birkhäuser.

Snow, Dan. 2001. *In the Company of Stone:The Art of the Stone Wall*. New York: Artisan.

Steele, James. 1997. *Architecture for People: The Complete Works of Hassan Fathy*. New York: Watson-Guptill.

Walker, Peter, Rowland Keable, Joe Martin, and Vasilios Maniatidis. 2010. *Rammed Earth: Design and Construction Guidelines*. Watford, UK: IHS-BRE Press.

Chapter 17. Natural Plastering

Hida Tool. Source for Japanese plastering trowels. www.hidatool.com.

Holmes, Stafford, and Michael Wingate. 1997/2002. *Building with Lime: A Practical Introduction*. Rugby, UK: ITDG Publishing.

Lacinski, Paul, and Michel Bergeron. 2000. *Serious Straw Bale: A Guide to Construction in All Climates*. White River Junction, VT: Chelsea Green Publishing.

LanderLand. Source for traditional Japanese plaster tools. www.landerland.com.

Magwood, Chris, Peter Mack, and Tina Therrien. 2005. *More Straw Bale Building: A Complete Guide to Designing and Building with Straw*. Gabriola Island, BC, Canada: New Society Publishers.

Weismann, Adam, and Katy Bryce. 2008. *Using Natural Finishes*. Totnes, Devon, UK: Green Books.

Chapter 18. Roofs for Natural Buildings

Gould, John, and Erik Nissen-Peterson. 2002. *Rainwater Catchment Systems for Domestic Supply: Design, Construction, and Implementation*. London: ITDG Publishing.

Jenkins, Joe. 2003. *The Slate Roof Bible: Understanding, Installing and Restoring the World's Finest Roof.* 2nd ed. Grove City, CA: Joseph Jenkins, Inc.

Kinkade-Levario, Heather. 2007. *Design for Water: Rainwater Harvesting, Stormwater Catchment, and Alternate Water Reuse*. Gabriola Island, BC, Canada: New Society Publishers.

Letts, Joh, and James Moir. 1999. *Thatch: Thatching in England 1790–1940*. 5th ed. Leeds, UK: Maney Publishing.

Snodgrass, Edmund, Nigel Dunnett, Dusty Gedge, and John Little. 2011. *Small Green Roofs: Low-Tech Options for Greener Living*. Portland, OR: Timber Press.

Weiler, Susan, and Katrin Scholz-Barth. 2009. *Green Roof Systems: A Guide to the Planning, Design and Construction of Building over Structure*. Hoboken, NJ: John Wiley and Sons.

Chapter 19. Floors for Natural Buildings

Environmental Building News. BuildingGreen, LLC, 122 Birge Street, Brattleboro, VT 05301. www.buildinggreen.com.

Journal of Light Construction. Hanley Wood, LLC, Williston, VT. www.jlconline.com.

Chapter 20. Natural Finishes

American Formulating and Manufacturing (AFM; maker of Safecoat environmental finishing products). www.jlconline.com.

BioShield Paint Company, Santa Fe, NM. www.bioshield.com.

Crews, Carole. 2010. *Clay Culture: Plasters, Paints and Preservation*. Ranchos de Taos, NM: Gourmet Adobe Press.

Eco-House (mineral finishes), Fredericton, New Brunswick, Canada. www.eco-house.com.

KEIM (German mineral finishes), distributed in United States by KEIM Mineral Coatings of America, Charlotte, NC. www.keim.com.

Livos (German oil finishes), distributed in United States by Ocean Pacific International, Inc., Portland, OR. www.livos.us.

OSMO (German oil finishes), distributed in United States by OSMO North America, Seattle, WA. www.osmona.com.

Silicote USA (mineral finishes), Bluffton, OH, and Grass Valley, CA. www.silicote.com.

Sloan, Annie. 2001. *Paint Alchemy: Recipes for Making and Adapting Your Own Paint for Home Decorating*. New York: Collins and Brown.

U.S. EPA. n.d. "Lead in Paint, Dust, and Soil." http://www.epa.gov/lead.

Weismann, Adam, and Katy Bryce. 2008. *Using Natural Finishes*. Totnes, Devon, UK: Green Books.

Chapter 21. Mechanicals and Utilities for Natural Buildings

Home Power magazine, Ashland, OR. www.homepower.com.

Mr. Slim M-Series Mini-Split Heat Pump, Mitsubishi Electric & Electronics USA, Suwanee, GA. www.mitsubishipro.com.

United States Department of Energy, Office of Energy Efficiency and Renewable Energy (EERE), Washington, DC. www.eere.energy.gov.

Conclusion

Berry, Wendell. 1996. *The Unsettling of America: Culture & Agriculture*. Rev. ed. San Francisco: Sierra Club Books.

Hopkins, Rob. 2008. *The Transition Handbook: From Oil Dependency to Local Resilience*. White River Junction, VT: Chelsea Green Publishing.

Kunstler, James Howard. 1994. *The Geography of Nowhere*. New York: Free Press.

ReUse Action, Buffalo, NY. www.reuseaction.com.

Roy, Rob. 2008. *Mortgage-Free: Innovative Strategies for Debt-Free Home Ownership*. 2nd ed. White River Junction, VT: Chelsea Green Publishing.

Transition US (United States hub of the International Transition Network, www.transitionnetwork.org), Petaluma, CA. www.transitionus.org.

B. Selected Details

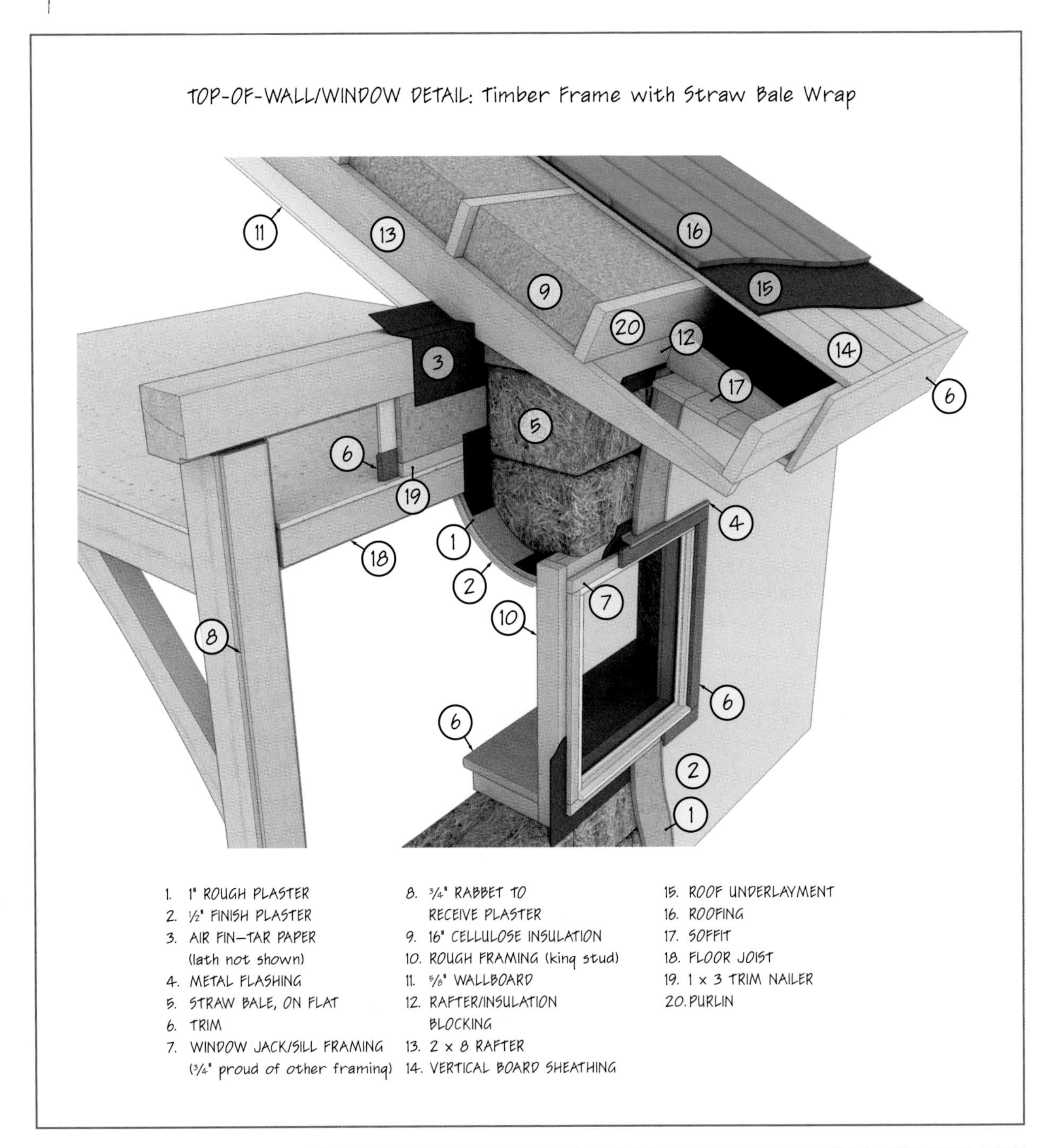

Top-of-wall detail featuring interior timber frame, straw bale (on flat) wall, lattice-framed and cellulose-insulated roof, and top of window.
ILLUSTRATION BY AARON WESTGATE/FREEFLOW STUDIOS.

BOTTOM-OF-WALL/WINDOW DETAIL:
Timber Frame with Straw Bale Wrap and Basement Foundation

1. 1" ROUGH PLASTER
2. ½" FINISH PLASTER
3. AIR FIN—TAR PAPER
4. METAL FLASHING
5. STRAW BALE (on flat)
6. TRIM
7. WINDOW JACK/SILL FRAMING (¾" proud of other framing)
8. 2 x 4, FLAT TOE-UP FRAMING
9. 2" RIGID INSULATION
10. ⅜" RABBET TO RECEIVE PLASTER
11. CELLULOSE INSULATION
12. ROT-RESISTANT SILL PLATE
13. SILL SEAL (two layers min.)
14. ICF STEM WALL
15. ROUGH FRAMING
16. TIMBER
17. STUCCO

Bottom-of-wall featuring interior timber frame partially shown, straw bale (on flat) wall, framed floor deck over ICF basement foundation, and bottom of window. ILLUSTRATION BY AARON WESTGATE/FREEFLOW STUDIOS.

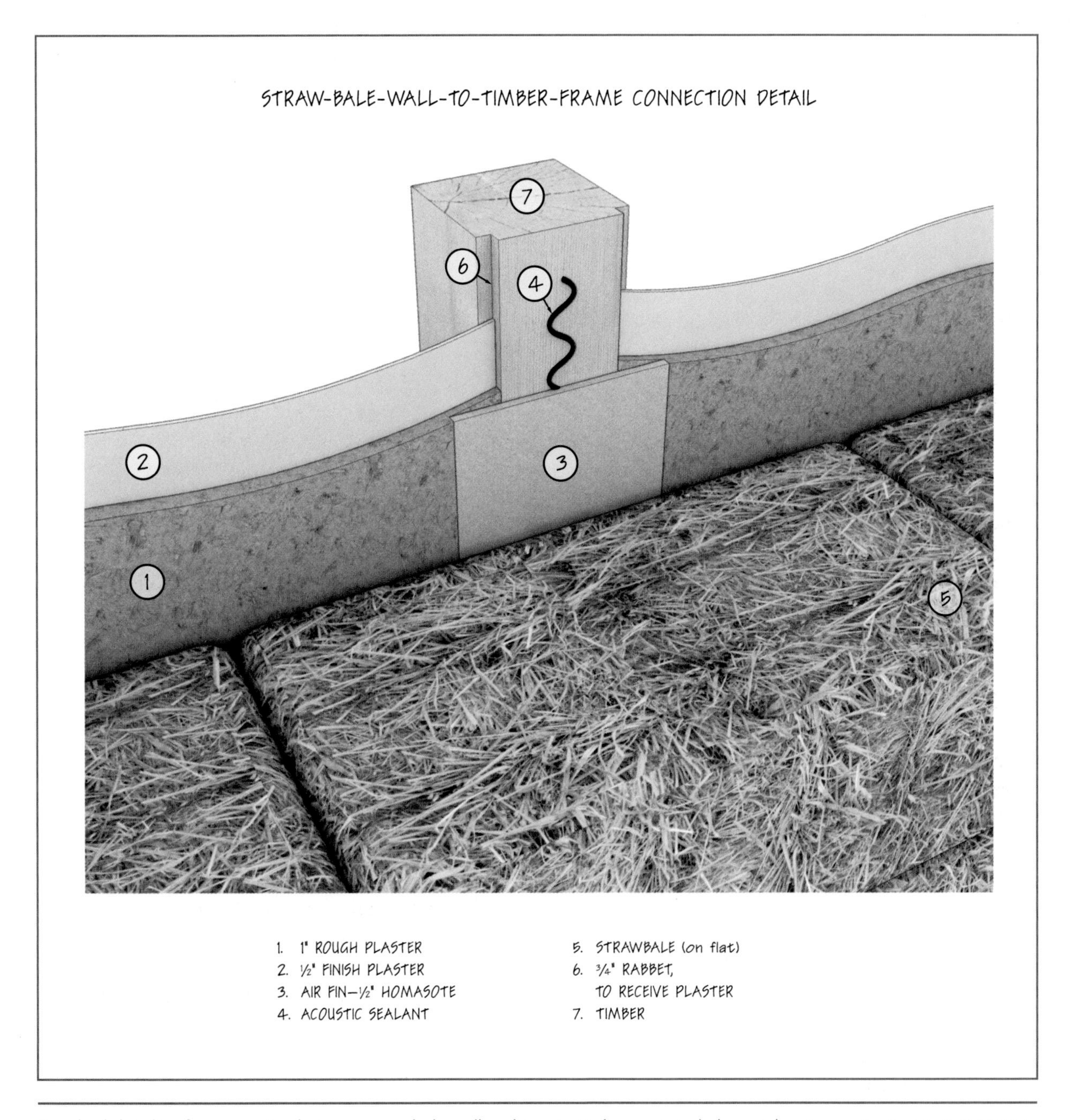

Standard detailing for connection between straw bale wall and interior timber post, including air-barrier continuity. ILLUSTRATION BY AARON WESTGATE/FREEFLOW STUDIOS.

C. Plaster Recipes

Note: Plastering is an art; all measurements are approximate. Always make samples!

Straw Bale Plaster System

Base Coat ("Lime-Stabilized Clay Plaster")

1 part clay
3–4 parts sand
¼ part slaked hydrated lime
1 part chopped straw
1 part manure (optional)

Brown Coat

Same as base coat, but with more sand and lime and less straw

Finish Coat ("Lime-Sand Plaster")

1 part hydrated lime
2½–3 parts sand (mortar)

Note: All plaster coats must be allowed to dry thoroughly before applying next coat. Always moisten walls before applying next plaster coat.

Other Base-Coat Plasters

Clay-Straw Plaster (or Light-Clay-Straw)

1 part clay (must stabilize with lime if following with lime coat)
1½–2 parts chopped straw

Clay-Sand Plaster

1 part clay
2–3 parts sand
1 part chopped straw

Manure Plaster (Litema)

1–2 parts clay
2–4 parts sand
4–6 parts manure (cow, horse)

Note: Horse manure is effective for fiber; cow manure is effective as a binder and to stabilize poor soil. This plaster reacts with clay to improve set time and water repellency.

Finish Clay Plaster

1 part bagged dry clay (OM-4 or EPK)
2–3 parts fine sand
¼–½ part wheat paste or casein
½ part soaked, blended cellulose, or manure, or fine straw

Finish Lime Plaster

1 part hydraulic lime, dry bagged
2–3 parts limestone sand

Note: This plaster must be applied very thin—⅛ inch.

D. Results Summary of Energy Performance of Northeast Straw-Bale Buildings Research

THERMAL CHARACTERISTICS OF SELECTED STRAW BALE HOMES—AS TESTED BETWEEN 3/10/11 AND 4/5/2011 BY BUILDING PERFORMANCE SERVICES LLC FOR NEW FRAMEWORKS NATURAL BUILDING, LLC

Town	Roof style	Ceiling style	Wall framing	Foundation type	Air-fin type	Air-fin quality	Heat source	Ventilation
Brookfield, VT	SIPs	cathedral; drywall finish	timber frame	slab-on-grade; 2 in. PIC edge insulation	Masonite	high	masonry heater w/in-slab radiant loop	Venmar HRV, concentric in/out
Barnet, VT (addition)*	N/A	cathedral; drywall/ paneling finish	timber frame	full basement + slab-on-grade	paper and lath	medium	woodstove	exhaust-only fan
Barnet, VT (total)*	N/A	cathedral; drywall/ paneling finish	timber frame	full basement + slab-on-grade + crawl	paper and lath	low (cabin)/ medium	woodstove	exhaust-only fan
Newbury, VT	16½ in. built-up rafters with ¼ in. foam thermal break, dense-pack cellulose	cathedral; drywall finish	timber frame	slab-on-grade + root cellar	½ in. drywall	high	woodstove	Venmar HRV
Middlesex, VT**	lattice-frame, 16 in. dense-pack cellulose, 2 shed dormers (south), 1 gable dormer (north)	cathedral; ADA**** finish	timber frame	full basement—R-20/28 Durisol block	paper and lath/ Homasote	high	propane-fired in-slab radiant, wood fireplace	HRV
Warren, VT	clerestory, 16 in. double-rafters and 12 in. clerestory wall with dense-pack cellulose	cathedral; ADA**** finish	timber frame	slab-on-grade, 2 in. foam insulation	paper and lath/ Homasote	high	woodstove, DV propane space-heated in-slab radiant	none
Granville, NY	site-built foam panels 12 in. foam thickness; clerestory salt-box	cathedral; T&G paneling finish	timber frame	walk-out two-side basement, 2 in. foam insulation (incomplete)	paper and lath/ Masonite	high	wood- and propane-fired (duel-fuel) in-slab radiant	bath exhaust fan exhausts to basement
Clinton, NY***	trusses, dense-pack cellulose	cathedral; T&G paneling finish	post-and-beam frame	rubble-trench; earth floor/ slab-on-grade with Durisol stem wall, 2 in. foam insulation	paper and lath/ Homasote	high	woodstove	none

Town	Built	WX	WX temps	BDT (CFM50)	Heated square feet	Volume cubic feet	Ext. surface square feet	Stories/ exposure	ACH50	ACH-NAT	CFM50/ext. square feet
Brookfield, VT	2009	rainy	68 inside 32 outside	557	1272	13360	3748	2/normal	2.50	0.18	0.15
Barnet, VT	2008	snow/sleet; breezy	58 inside 31 outside	2139	2096	27329	5829	2.5/exposed	4.70	0.40	0.37
Barnet, VT	1998	snow/sleet; breezy	58 inside 31 outside	3015	2692	32117	7705	2.5/exposed	5.63	0.49	0.39
Newbury, VT	2011	cloudy	58 inside 36 outside	767	1537	13400	4539	1.5/exposed	3.43	0.25	0.17
Middlesex, VT**	2010	cloudy	66 inside 34 outside	850	2448	19844	6506	2.5/exposed	2.57	0.20	0.13
Warren, VT	2009	mostly cloudy	62 inside 43 outside	552	456	5028	2535	1.5/low	6.59	0.36	0.22
Granville, NY	2010	rainy; breezy	65 inside 40 outside	2057	2931	33221	7091	3/exposed	3.72	0.34	0.29
Clinton, NY***	2010	snow/sleet; breezy	68 inside 35 outside	1756	1190	8921	2888	1.5/exposed	11.81	1.02	0.61

* (Addition) refers to large primary residence and breezeway; (total) includes original small cabin, attached

** Tested by Matt Sargent of Vermont Energy Investment Corporation, Burlington, VT

*** Tested by Richard Robinson, Advanced Energy Systems of New York, Utica, NY

**** Airtight drywall approach serving as ceiling air barrier

Notes:

- Brookfield: Masonry heater is only heat source
- Barnet (addition): Primary heat loss through cupola with no air barrier, breezeway connections to structure, occasional plaster edge detailing
- Barnet (total): Cabin featured no plaster-edge air barriers
- Newbury: Still in construction, nearing completion; chimney-to-ceiling connection and incidental plaster edge and dormer edge air sealing to be completed before finish
- Middlesex: Primary heat loss through chimney-to-ceiling connection, incidental plaster edge detail; ADA and majority of plaster edges tight
- Warren: Primary heat loss through chimney-to-ceiling connection, salvaged windows, clerestory window-to-frame (not sealed)
- Granville: Primary heat loss through ceiling tongue-and-groove paneling (runs through wall), eave-end rafter blocking, ceiling-to-frame connection, clerestory wall edges and windows, basement windows and doors (some windows missing, temp. plastic, no weatherstripping); plaster edges very tight
- Clinton: Primary heat loss through ceiling and knee walls (no air barrier), chimney-to-ceiling connection, through-wall ventilation port; plaster edges very tight
- Vermont Energy Star Homes (VESH)-rated 2009 building average value: 3.2 ACH50

E. Sample Budget

BUDGET ANALYSIS FOR BROOKFIELD RESIDENCE, 2009

Foundation	$14,922.86*
Timber frame	$15,634.85
Straw walls (to finish)	$34,367.50
Mechanicals	$10,758.46
Masonry heater	$13,351.80
Windows and doors	$13,623.69
Roof and ceiling	$22,987.11
Finishing and appliances	$7,195.65
Well and septic	$9,528.48
RE systems	$27,068.09
Tools and supplies	$1,404.21
Miscellaneous	$1,099.49
Total	**$171,942.19**
Approx. cost/square foot	**$132.26**

PROJECT DETAILS: BROOKFIELD RESIDENCE, 2009

Heated square footage:	1300 square feet
Foundation type:	slab-on-grade
Wall type:	straw bale, plastered finish on both sides
Roof type:	SIPs, standing seam
Heating system:	masonry heater, propane on-demand water
Renewable energy system:	solar hot water, solar electric (stand-alone)
Ventilation:	HRV
ACH50:	2.5
CFM50/ ext. square footage:	0.15
Built by:	owner-designed, owner-contracted, owner-built with subcontractors

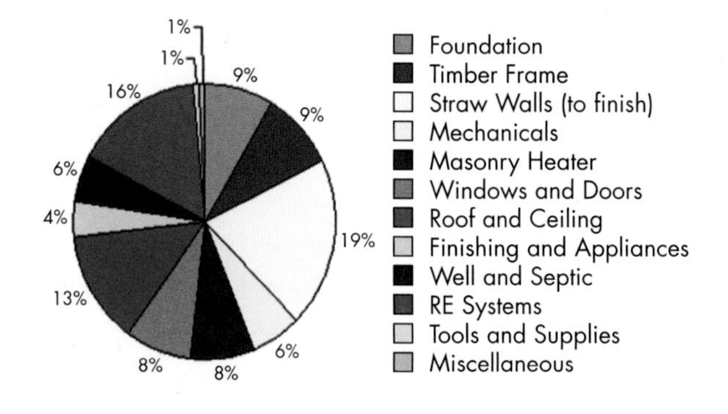

REFERENCES

Introduction

Bookchin, Murray. 1997. *The Murray Bookchin Reader*. New York: Continuum International Publishing Group.

Ware, J., J. Spengler, L. Neas, J. Samet, G. Wagner, D. Coultas, H. Ozkaynak, and M. Schwab. 1993. "Respiratory and Irritant Health Effects of Ambient Volatile Organic Compounds: The Kanawha County Health Study." *Am J Epidemiol* 137 (12): 1287–301.

Chapter 1. A Brief History of Natural Building in the Northeast

Beauregard, Robert A. 2006. *When America Became Suburban*. Minneapolis: University of Minnesota Press.

Bookchin, Murray. 1997. *The Murray Bookchin Reader*. New York: Continuum International Publishing Group.

City Repair. http://cityrepair.org.

Eberlein, Harold Donaldson. 1915. *The Architecture of Colonial America*. Boston: Little, Brown, and Company.

Garvin, James L. 2001. *A Building History of Northern New England*. Hanover and London: University Press of New England.

Gelernter, Mark. 1999. *A History of American Architecture: Buildings in Their Cultural and Technological Context*. Hanover and London: University Press of New England.

Hagelstein, Isabella. 1977. *A Primer on New England's Colonial Architecture*. Boston: Herman Publishing, Inc.

Kimball, Fiske. 1922/1950. *Domestic Architecture of the American Colonies and of the Early Republic*. New York: Dover Publications, Inc.

McGinn, Anne Platt. 2000. "Why Poison Ourselves? A Precautionary Approach to Synthetic Chemicals." Worldwatch Paper 153, Worldwatch Institute, Washington, DC. http://www.worldwatch.org/system/files/EWP153.pdf.

Upton, Dell, ed. 1986. *America's Architectural Roots: Ethnic Groups That Built America*. Washington, DC: The Preservation Press National Trust for Historic Preservation.

Watson, J. F. 1857. *Annals of Philadelphia and Pennsylvania*. 2nd ed. Vol 2. Philidelphia County.

Wiseman, Frederick M. 2001. *The Voice of the Dawn: An Autohistory of the Abenaki Nation*. Hanover, NH: University Press of New England.

Chapter 2. Ecology

Biello, David. 2008. "Green Buildings May Be Cheapest Way to Slow Global Warming." *Scientific American*, March 17. http://www.scientificamerican.com/article.cfm?id=green-buildings-may-be-cheapest-way-to-slow-global-warming.

Carmody, John, and Wayne Trusty. 2005. "Life Cycle Assessment Tools." *Implications* 5, no. 3. InformeDesign, University of Minnesota. http://www.informedesign.org/_news/mar_v05r-p.pdf.

Dobson, Stephen. 2000. "Continuity of Tradition: New Earth Building." Keynote address at Terra 2000: 8th International Conference on the Study And Conservation of Earthen Architecture, Torquay, England. http://members.optusnet.com.au/rogergarlick/Terra2000.doc.

Edwards, Lynn, and Julia Lawless. 2002. *The Natural Paint Book*. Emmaus, PA: Rodale Organic Living Books.

Ehrlich, Brent. 2010. "Reducing Environmental Impacts of Cement and Concrete." *Environmental Building News*, September 1. http://www.buildinggreen.com/auth/article.cfm/2010/8/30/Reducing-Environmental-Impacts-of-Cement-and-Concrete/.

Emrath, Paul, and Helen Fei Liu. 2007. "Residential Greenhouse Gas Emissions." On the website of the National Association of Home Builders, Washington, DC. http://www.nahb.org/generic.aspx?genericContentID=75563.

Environmental Building News. 2008. "All About Formaldehyde." Back Page Primer, July. http://www.buildinggreen.com/auth/article.cfm/2008/7/29/All-About-Formaldehyde/.

———. 2009a. "Nanomaterials May Not Lower Net Environmental Impacts." Newsbrief, February 1. http://www.buildinggreen.com/auth/article.cfm/2009/1/29/Nanomaterials-May-Not-Lower-Net-Environmental-Impacts/.

———. 2009b. "Brominated Flame Retardants." Back Page Primer, June 1. http://www.buildinggreen.com/auth/article.cfm/2009/5/29/Brominated-Flame-Retardants/.

———. 2010. "The Precautionary Principle." Back Page Primer, September 1. http://www.buildinggreen.com/auth/article.cfm/2010/8/30/The-Precautionary-Principle/.

———. 2011. "Embodied Carbon: Measuring How Building Materials Affect Climate." April 1. http://www.buildinggreen.com/auth/article.cfm/2011/3/30/Embodied-Carbon-Measuring-How-Building-Materials-Affect-Climate/.

Gemenne, Francois, and Shawn Shen. 2008. "Tuvalu's Environmental Migration to New Zealand." International Conference on Environment, Forced Migration, & Social Vulnerability," Bonn, Germany, October 9–11. http://www.efmsv2008.org/file/A6+Shen.

Gertsakis, John, and Helen Lewis. 2001. *Design + Environment: A Global Guide to Designing Greener Goods*. Sheffield, UK: Greenleaf Publishers.

Hammond, Geoff, and Craig Jones. 2011. *Inventory of Carbon and Energy (ICE),* ver. 2.0. Sustainable Energy Research Team (SERT). Bath, UK: University of Bath. http://www.bath.ac.uk/mech-eng/sert/.

Healthy Building Network. 2006a. "PVC Facts." January 11. http://www.healthybuilding.net/pvc/facts.html.

———. 2006b. "PVC-Free Alternatives." March 31. http://www.healthybuilding.net/pvc/PVCFreeAlts.html.

———. 2006c. "Screening the Toxics out of Building Materials: Interior Flooring and Finishes." September 14. http://www.healthybuilding.net/healthcare/Screening-the-Toxics.pdf.

———. 2008a. "Alternative Resin Binders for Particleboard, Medium Density Fiberboard (MDF), and Wheatboard." May. http://www.healthybuilding.net/healthcare/2008-04-10_alt_resin_binders_particleboard.pdf.

———. 2008b. "Batt Insulation: No Added Formaldehyde Alternatives." May. http://www.healthybuilding.net/healthcare/2008-04_fiberglass_batt_insulation_updated.pdf.

———. 2008c. "No Added Urea Formaldehyde Casework Substrates for Use in Health Care." May. http://www.healthybuilding.net/healthcare/2008-05_ProductComp_NoAddUrea_updated.pdf.

———. 2008d. "Toxic Chemicals in Building Materials." In conjunction with Kaiser Permanente. May. http://www.healthybuilding.net/healthcare/Toxic%20Chemicals%20in%20Building%20Materials.pdf.

———. 2011. "Formaldehyde and Wood." http://www.healthybuilding.net/formaldehyde/index.html.

Intergovernmental Panel on Climate Change. 1996. *The Science of Climate Change*. Contribution of working group 1 to the second assessment report, UNEP and WMO. Cambridge: Cambridge University Press.

Ketel, Hermen J. 2004. "Global Warming and Human Migration." In *Climate Change, Human Systems, and Policy, ed. Antoaneta Yotova, in Encyclopedia of Life Support Systems (EOLSS),* developed under the auspices of UNESCO. Oxford, UK: EOLSS Publishers. http://www.eolss.net.

Kostigen, Thomas M. 2008. "The World's Largest Dump: The Great Pacific Garbage Patch." Discover, July 10. http://discovermagazine.com/2008/jul/10-the-worlds-largest-dump/article_view?b_start:int=2&-C=.

Learn, Scott. 2011. "Reports of Pacific Ocean's Plastic Patch Being Texas-Sized Are Grossly Exaggerated, Oregon State University Professor Says." *The Oregonian*, January 4. http://www.oregonlive.com/environment/index.ssf/2011/01/reports_of_pacific_oceans_plas_1.html.

Lent, Tom. 2003. "Toxic Data Bias and the Challenges of Using LCA in the Design Community." Presented at GreenBuild 2003, Pittsburgh, PA. Healthy Building Network. http://www.healthybuilding.net/pvc/Toxic_Data_Bias_2003.html.

———. 2009. "Formaldehyde Emissions from Fiberglass Insulation with Phenol Formaldehyde Binder." Healthy Building Network, January 16. http://www.healthybuilding.net/healthcare/Fiberglass-insulation-formaldehyde-emissions-090116.pdf.

Malin, Nadav. 2008. "Counting Carbon: Understanding Carbon Footprints of Buildings." *Environmental Building News*, July 1. http://www.buildinggreen.com/auth/article.cfm/2008/6/27/Counting-Carbon-Understanding-Carbon-Footprints-of-Buildings/#sidebar-1.

Malin, Nadav, and Alex Wilson. 1994. "Should We Phase Out PVC?" *Environmental Building News*, January 1. http://www.buildinggreen.com/auth/article.cfm/1994/1/1/Should-We-Phase-Out-PVC/.

McGinn, Anne Platt. 2000. "Why Poison Ourselves? A Precautionary Approach to Synthetic Chemicals." Worldwatch Paper 153, Worldwatch Institute, Washington DC. http://www.worldwatch.org/system/files/EWP153.pdf.

Melton, Paula, and Alex Wilson. 2011. "Air Renew Wallboard Absorbs Formaldehyde from Indoor Air." *Environmental Building News*, March 1. http://www.buildinggreen.com/auth/article.cfm/2011/2/28/Air-Renew-Wallboard-Absorbs-Formaldehyde-from-Indoor-Air/.

Milne, Geoffe, and Chris Reardon. 2005. "Embodied Energy." *Sustainable Home Technical Manual 3*, Commonwealth of Australia. http://www.buildwise.org/library/energy/sustainable/green-technical-guide/fs31.pdf.

Roberts, Tristan. 2008. "Is Nano a No-No?" *Environmental Building News*, March 1. http://www.buildinggreen.com/auth/article.cfm/2008/3/1/Is-Nano-a-No-No-Nanotechnology-Advances-into-Buildings/.

Rogers, Stephanie. 2009. "EPA Cracks Down on Mercury Pollution from Cement Kilns." *Earth First*. http://earthfirst.com/epa-cracks-down-on-mercury-pollution-from-cement-kilns/.

Rossi, Mark. 2004. "Reaching the Limits of Quantitative Life Cycle Assessment." Clean Production Action, June. http://www.healthybuilding.net/pvc/CPA_EC_LCA_Critique.html.

Sell, Nancy Jean. 1992. *Industrial Pollution Control: Issues and Techniques*. New York: John Wiley & Sons.

Shapley, Dan. 2009. "Cement Plant Mercury—U.S. Will Finally Crack Down on Toxic Air Pollution from Cement Plants." The Daily Green (blog), January 20. http://www.thedailygreen.com/environmental-news/latest/mercury-cement-47012002.

Symbiotic Engineering. 2007. "Embodied Energy Considerations in Existing LEED Credits." September; Boulder, CO. www.symbiotic-engineering.com.

United States Department of Health and Human Services. 2007. "An Update and Revision of ATSDR's February 2007 Health Consultation: Formaldehyde Sampling of FEMA Temporary-Housing Trailers, Baton Rouge, Louisiana." October. http://www.atsdr.cdc.gov/substances/formaldehyde/pdfs/revised_formaldehyde_report_1007.pdf.

———. 2011. "Report on Carcinogens. 12th ed." National Toxicology Program. http://ntp.niehs.nih.gov/ntp/roc/twelfth/roc12.pdf.

United States Environmental Protection Agency. n.d. "An Introduction to Indoor Air Quality: Volatile Organic Compounds (VOCs)." http://www.epa.gov/iaq/voc.html (accessed March 22, 2011).

———. 2010. "Code of Federal Regulations, 40: Chapter 1, Subchapter C, Part 51, Subpart F, 51100." July. http://cfr.vlex.com/vid/51-100-definitions-19784887#ixzz1DwTNkz9j.

Weisman, Alan. 2007. "Polymers Are Forever." *Orion*, May–June. http://www.orionmagazine.org/index.php/articles/article/270/.

Weismann, Adam, and Katy Bryce. 2008. *Using Natural Finishes*. Totnes, Devon, UK: Green Books.

Wilson, Alex. 2004. "Flame Retardants Under Fire." *Environmental Building News*, June 1. http://www.buildinggreen.com/auth/article.cfm/2004/6/1/Flame-Retardants-Under-Fire/.

———. 2009. "Polystyrene Insulation: Does It Belong in a Green Building?" *Environmental Building News*, August 1. http://www.buildinggreen.com/auth/article.cfm/2009/7/30/Polystyrene-Insulation-Does-It-Belong-in-a-Green-Building/.

———. 2010. "Avoiding the Global Warming Impact of Insulation." *Environmental Building News*, June 1. http://www.buildinggreen.com/auth/article.cfm/2010/6/1/Avoiding-the-Global-Warming-Impact-of-Insulation/.

Wilson, Alex, and Nadav Malin. 1995. "Establishing Priorities with Green Building." *Environmental Building News*, September 1. http://www.buildinggreen.com/auth/article.cfm/1995/9/1/Establishing-Priorities-with-Green-Building/#sidebar-1.

Wilson, Alex, and Peter Yost. 2007. "Plastics in Construction: Performance and Affordability at What Cost?" *Environmental Building News*, July 1. http://www.buildinggreen.com/auth/article.cfm/2001/7/1/Plastics-in-Construction-Performance-and-Affordability-at-What-Cost/?.

World Nuclear Association. 2011. "Energy Analysis of Power Systems," January. http://www.world-nuclear.org/info/inf11.html.

Chapter 3. Beyond the Building: Siting and Landscape Design

Hemenway, Toby. 2001. *Gaia's Garden: A Guide to Home-Scale Permaculture*. White River Junction, VT: Chelsea Green Publishing.

Jenkins, Joseph. 1999. *The Humanure Handbook: A Guide to Composting Human Manure*. Grove City, CA: Jenkins Publishing.

Mollison, Bill. 1988. *Permaculture: A Designers' Manual*. Tyalgum, NSW, Australia: Tagari Publications.

Chapter 4. Soil and Stone: Geology and Mineralogy

Anthoni, J. Floor. 2000. "Soil Geology." www.seafriends.org.nz/enviro/soil/geosoil.htm.

The Ball Clay Heritage Society. 2003. *The Ball Clays of Devon and Dorset*. Cornish Hillside Publications.

Bryant, William Logan. 1995. *Dirt: The Ecstatic Skin of the Earth*. New York: Riverhead Books.

Cairns-Smith, A. G., and H. Hartman, eds. 1986. Clay *Minerals and the Origin of Life*. Cambridge: Cambridge University Press.

The Clay Minerals Society. http://www.clays.org/index.html.

Ferrario, J. B., C. J. Byrne, and D. H. Cleverly. 2000. "2,3,7,8-Dibenzo-*P*-Dioxins in Mined Clay Products from the United States: Evidence for Possible Natural Origin." *Environ Sci Technol* 34:4524–32.

Ferrario, J. B., and C. J. Byrne. 2002. "Dibenzo-*p*-Dioxins in the Environment from Ceramics and Pottery Produced from Ball Clay Mined in the United States." *Chemosphere* 46:1297-1301.

Ferrario, J. B., C. J. Byrne, and J. Schaum. 2007. "Concentrations of Polychlorinated Dibenzo-*P*-Dioxins in Processed Ball Clay from the United States." *Chemosphere* 67:1816–21.

Franzblau, A., E. Hedgeman, Q. Chen, S.-Y. Lee, P. Adriaens, A. Demond, et al. 2008. "Case Report: Human Exposure to Dioxins from Clay." *Environmental Health Perspectives* 116:238–42.

Hubka, Thomas C. 1987. *Big House, Little House, Back House, Barn: The Connected Farm Buildings of New England*. Hanover and London: University Press of New England.

Industrial Minerals Association—North America. 2007. "Ball Clay." http://www.ima-na.org/about_industrial_minerals/ball_clay.asp.

Natural Resources Conservation Service, United States Department of Agriculture. n.d. "Soil Survey." http://websoilsurvey.nrcs.usda.gov/app/HomePage.htm.

Raymo, C., and M. E. Raymo. 2001. *Written in Stone: A Geological History of the Northeastern United States*. Chester, CT: The Globe Pequot Press.

Chapter 5. Flora and Fauna

Adler, I., and R. Adler. 1966. *Tree Products*. New York: The John Day Company.

Boulger, G. S. 1908. *Wood: A Manual of the Natural History and Industrial Applications of the Timbers of Commerce*. London: Edward Arnold.

Chapman, G. P., and W. E. Peat. 1992. *An Introduction to the Grasses (including Bamboos and Cereals)*. Wallingford, UK: CAB International, and Melksham, UK: Redwood Press Ltd.

Commonwealth Economic Committee, Annual Review. 1966. *Vegetable Oils and Oilseeds: A Review of Production, Trade, Utilization and Prices Relating to Groundnuts, Cottonseed, Linseed, Soya Beans, Coconut and Oil Palm Products, Olive Oil and Other Oilseeds and Oils*. London: Her Majesty's Stationery Office.

Good Shepherd Wool Company. n.d. "R-Values of Wool Insulation." http://www.goodshepherdwool.com/Rvalue.htm (accessed May 2011).

Holmes, Stafford, and Michael Wingate. 1997/2002. *Building with Lime: A Practical Iintroduction*. Rugby, UK: ITDG Publishing, 171–78.

Lewis, D., ed. 1961. *Digestive Physiology and Nutrition of the Ruminant*. Proceedings of the University of Nottingham Seventh Easter School in Agricultural Science, 1960. London: Butterworths.

McDowell, R. E. 1977. *Ruminant Products*. Morrilton, AR: Winrock International Livestock Research and Training Center.

RoofingProducts.com. n.d. "Biomass Roofing for All Needs." http://www.roofing-products.com/biomass-roofs/biomass-roof-info/biomass-roofing-for-all-needs/ (accessed May 2011).

Weismann, Adam, and Katy Bryce. 2008. *Using Natural Finishes*. Totnes, Devon, UK: Green Books.

Wessels, Thomas. 1997. *Reading the Forested Landscape: A Natural History of New England*. Woodstock, VT: The Countryman Press.

Chapter 6. Structure and Natural Building

Allen, Edward. 1980. *How Buildings Work: The Natural Order of Architecture*. New York: Oxford University Press.

Ching, Francis D. K., and Cassandra Adams. 2001. *Building Construction Illustrated*. 3rd ed. Vol. 2, 8–30. New York: John Wiley and Sons.

Donovan, Darcey. 2009. "Seismic Performance of Innovative Straw Bale Wall Systems." PAKSBAB, November 5. http://nees.unr.edu/projects/straw_bale_house/PAKSBAB_Test Handout_11_5_09.pdf.

King, Bruce. 1996. *Buildings of Earth and Straw*. Sausalito, CA: Ecological Design Press.

———. 2006. *Design of Straw Bale Buildings: The State of the Art*. San Rafael, CA: Green Building Press.

Lstiburek, Joe. 2009. "BSI-023: Wood Is Good . . . But Strange." Building Science Insights. Building Science Corporation, July 31. http://www.buildingscience.com/documents/insights/bsi-023-wood-is-good-but-strange.

Network for Earthquake Engineering Simulation, University of Nevada at Reno. n.d. "Straw Bale Construction: A Solution for Seismic Resistant Housing in Developing Countries." http://nees.unr.edu/projects/straw_bale_house/NEES_Highlights2009_PAKSBAB.pdf (accessed March 6, 2011).

Southern Pine Council. 2010. "Span Tables." http://www.southernpine.com/span-tables.asp.

———. 2010. "Using These Tables." http://www.southernpine.com/downloadpdf.asp?filename=using_these_tables.pdf.

Vardy, Stephen, and Colin MacDougall. 2005. "Compressive Testing of Plastered Straw Bales." *Journal of Green Building*. Glen Allen, VA: College Publishing. Archived online at http://www.osbbc.ca/wordpress/wp-content/uploads/2009/02/final_submission_journal_of_green_building.pdf.

Chapter 7. Heat and Natural Building

Air Barrier Association of America. n.d. "About Air Barriers: Materials, Components, Assemblies, & Systems." http://www.airbarrier.org/about/materials_e.php (accessed April 7, 2011).

Allen, Edward. 1980. *How Buildings Work: The Natural Order of Architecture*. New York: Oxford University Press.

Altes, Tristan Korthals. 2008. "Making Air Barriers That Work: Why and How to Tighten Up Buildings." *Environmental Building News*, June 1. http://www.buildinggreen.com/auth/article.cfm/2008/5/29/Making-Air-Barriers-that-Work-Why-and-How-to-Tighten-Up-Buildings/?.

Forest Products Laboratory. 2004. "Engineering Report of Light Clay Specimens." http://www.designcoalition.org/articles/Lansing-LHJ/research/FPLreport.pdf.

Hoberecht, Mark. 2011. "Our Innovative Natural Building Systems." HarvestBuild Associates, Inc. http://www.harvestbuild.com/innovation.html.

International Green Construction Code (proposed). 2011. "Chapter 5: Materials Resource Conservation and Efficiency"; "Section 508: Straw-Clay." November. http://www.iccsafe.org/cs/IGCC/Documents/PublicComments0810/07-Chapter-05.pdf.

King, Bruce. 2006. *Design of Straw Bale Buildings: The State of the Art*. San Rafael, CA: Green Building Press, pp. 185–94.

Kosney, Jan. 1999. "New Values for High Mass Walls." *Home Energy*, September–October. http://www.homeenergy.org/archive/hem.dis.anl.gov/eehem/99/990912.html.

———. 2004. "A New Whole Wall R-Value Calculator: An Integral Part of the Interactive Internet-Based Building Envelope Materials Database for Whole-Building Energy Simulation Programs." Oak Ridge National Laboratory, August. http://www.ornl.gov/sci/roofs+walls/NewRValue.pdf.

Lstiburek, Joe. 2006. *Air Barriers*. Research Report 0403, Building Science Press. http://www.buildingscience.com/documents/reports/rr-0403-air-barriers.

Oak Ridge National Laboratory. 2004. "Hotbox Test R-Value Database: Straw Bale Wall Technology." http://www.ornl.gov/sci/roofs+walls/AWT/HotboxTest/Hybrid/StrawBale/index.htm.

Oak Ridge National Labs and Polish Academy of Sciences. 2001. "Thermal Mass—Energy Savings Potential in Residential Buildings."http://www.ornl.gov/sci/roofs+walls/research/detailed_papers/thermal/dynamic.html.

Riversong, Robert. 2010. "Hygro-Thermal Engineering." Course handouts, Yestermorrow Design/Build School, February 20–21.

Straube, John. 2009a. "The Passive House (Passivehaus) Standard—A Comparison to Other Low-Energy Cold-Climate Houses." BSI-025, Building Science Corporation, September 7. http://www.buildingscience.com/documents/insights/bsi-025-the-passivhaus-passive-house-standard.

———. 2009b. *Thermal Metrics for High Performance Enclosure Walls: The Limitations of R-Value*. Research Report 0901, Building Science Press. http://www.buildingscience.com/documents/reports/rr-0901-thermal-metrics-high-performance-walls-limitations-r-value.

———. 2010. "How Heat Moves through Homes." Building Science Podcast, Green Building Advisor, April 12. http://www.greenbuildingadvisor.com/blogs/dept/building-science/how-heat-moves-through-homes-building-science-podcast.

Wilson, Alex. 2009. "How Much Insulation Is Enough." Green Building Advisor, July 15. http://www.greenbuildingadvisor.com/blogs/dept/energy-solutions/how-much-insulation-needed.

Chapter 8. Moisture and Natural Building

Canada Mortgage and Housing Corporation (CMHC). 2007. "Straw Bale House Moisture Research." Technical Series 00-102, revised June 2. http://www.cmhc-schl.gc.ca/odpub/pdf/62573.pdf?lang=en.

National Fiber. n.d. "Cellulose Insulation, Moisture, and Vapor Barriers." National Fiber, Belchertown, MA. http://www.nationalfiber.com/docs/CelluloseInsulationMoistureAndVaporBarriers0909.pdf (accessed March 24, 2011).

King, Bruce. 2006. *Design of Straw Bale Buildings: The State of the Art.* San Rafael, CA: Green Building Press.

Lstiburek, Joe. 1999–2011. "Joe's Top Ten Rules of Wood Durability." Building Science Corporation. http://www.joelstiburek.com/topten/wood.htm.

———. 2004. *Insulations, Sheathings, and Vapor Retarders.* Research Report 0412, Building Science Corporation, November. http://www.buildingscience.com/documents/reports/rr-0412-insulations-sheathings-and-vapor-retarders.

———. 2009a. "Capillarity—Small Sacrifices." BSI-011, Building Science Corporation, January 27. http://www.buildingscience.com/documents/insights/bsi-011-capillarity-small-sacrifices/files/bsi-011_small_sacrifices.pdf.

———. 2009b. "Thermodynamics: It's Not Rocket Science." BSI-021, Building Science Corporation, June 8. http://www.buildingscience.com/documents/insights/bsi-021-thermodynamics-its-not-rocket-science?topic=doctypes/insights.

Mold & Bacteria Consulting Laboratories. 2005. "Factors That Affect the Growth of Moulds." September 14. http://www.moldbacteriaconsulting.com/2005/09/factors-that-affect-growth-of-moulds.html.

Riversong, Robert. 2009. *Permeability and Permeance of Lime:Sand Plasters.* Privately commissioned report, HouseWright Construction, Warren, VT.

———. 2010. "Hygro-Thermal Engineering." Course handouts, Yestermorrow Design/Build School, February 20–21.

Southern Pine Council. n.d. "Using Southern Pine—Special Topics: Mold, Moisture, & Lumber." Southern Forest Products Council. http://www.southernpine.com/using-southern-pine_special-topics_mold.asp (accessed March 24, 2011).

Straube, John. 1999. "Moisture Properties of Plaster and Stucco for Strawbale Buildings." University of Waterloo. Archived at http://www.homegrownhome.co.uk/pdfs/Straube_Moisture_Tests.pdf.

———. 2009. "Building Science for Straw Bale Buildings." *Building Science Digest* 112, January 30. http://www.buildingscience.com/documents/digests/bsd-112-building-science-for-strawbale-buildings/files/bsd-112_strawbale_performance.pdf.

United States Environmental Protection Agency. n.d. "Indoor Water Use in the United States." http://www.epa.gov/WaterSense/pubs/indoor.html (accessed March 30, 2011).

Water.org. n.d. "Water Facts." http://water.org/learn-about-the-water-crisis/facts/ (accessed March 30, 2011).

Chapter 9. Fire, Insects, and Acoustics

Allen, Edward. 1980. *How Buildings Work: The Natural Order of Architecture.* New York: Oxford University Press.

Dobson, Stephen. 2000. "Continuity of Tradition: New Earth Building." Keynote address at Terra 2000: 8th International Conference on the Study and Conservation of Earthen Architecture, Torquay, England. http://members.optusnet.com.au/rogergarlick/Terra2000.doc.

Ecological Building Network. 2007a. "1-Hour Fire Resistance Test of a Nonloadbearing Straw Bale Wall." Prepared for Ecological Building Network of San Rafael, CA, by Intertek Testing Services of Elmendorf, TX, revised July 9. http://www.ecobuildnetwork.org/images/stories/ebnet_pdfs/Non-Bearing_Clay_Wall.pdf.

———. 2007b. "2-Hour Fire Resistance Test of a Nonloadbearing Wheat Straw Bale Wall." Prepared for Ecological Building Network of San Rafael, CA, by Intertek Testing Services of Elmendorf, TX, revised July 9. http://www.ecobuildnetwork.org/images/stories/ebnet_pdfs/Cement_Stucco_Wall.pdf.

Fisette, Paul. 2007. "Controlling Termites and Carpenter Ants. Building and Construction Technology Program." University of Massachusettes–Amherest, November 28. http://bct.eco.umass.edu/publications/by-title/controlling-termites-and-carpenter-ants/.

King, Bruce. 2006. *Design of Straw Bale Buildings: The State of the Art.* San Rafael, CA: Green Building Press, pp. 195–201.

Marsh, Andrew. 1999. "2.4 Noise Transmission." Online course on acoustics, University of Western Australia. http://www.kemt.fei.tuke.sk/Predmety/KEMT320_EA/_web/Online_Course_on_Acoustics/transmission.html.

Shaw, Neil. 2002. "Architectural Acoustics." Menlo Scientific Acoustics, Inc., Topanga, CA. First Pan-American/Iberian Meeting on Acoustics, Cancun, Mexico, December 2–6. http://www.acoustics.org/archacouintrotoot.pdf.

Silex Innovations, Inc. 2002. "Sound Attenuation." http://www.silex.com/pdfs/Sound%20Attenuation.pdf.

Chapter 10. Design

American Feng Shui Institute, San Gabriel, CA. http://www.amfengshui.com/.

Eliason, Mike. 2011. "A Passivehaus Rebuttal: In Defense of the Standard. Green Building Advisor." April 1. http://www.greenbuildingadvisor.com/blogs/dept/green-building-blog/passivhaus-rebuttal-defense-standard?utm_source=email&utm_medium=eletter&utm_content=20110406-passivhaus-shakedown-garage-fumes&utm_campaign=green-building-advisor.

Energy Efficiency and Renewable Energy Clearinghouse. 2001. "Passive Solar Design for the Home." National Renewable Energy Laboratory, February. http://www.nrel.gov/docs/fy01osti/27954.pdf.

HarvestBuild Associates, Inc. n.d. "Relative Worth of Construction Materials." http://www.harvestbuild.com/baumaterials.html (accessed May 3, 2011).

———. n.d. "Top-10 Design Guidelines." http://www.harvestbuild.com/bauguidelines.html (accessed May 3, 2011).

Heerwagen, Judith, and Betty Hase. 2001. "Building Biophilia: Connecting People to Nature in Building Design." *Environmental Design and Construction*, March–April, 30–36.

Holliday, Martin. 2011. "Are Passivhaus Requirements Logical or Arbitrary?" Green Building Advisor, April 1. http://www.greenbuildingadvisor.com/blogs/dept/musings/are-passivhaus-requirements-logical-or-arbitrary?utm_source=email&utm_medium=eletter&utm_content=20110406-passivhaus-shakedown-garage-fumes&utm_campaign=green-building-advisor.

Malin, Nadav. 2004. "Integrated Design." *Environmental Building News*, November 1. http://www.buildinggreen.com/auth/article.cfm/2004/11/1/Integrated-Design/.

———. 2006. "LEED: A Look at the Rating System That's Changing the Way America Builds." *Environmental Building News*, June 1. http://www.buildinggreen.com/auth/article.cfm/2000/6/1/LEED-A-Look-at-the-Rating-System-That-s-Changing-the-Way-America-Builds/.

Malin, Nadav, and Jessica Boehland. 2003. "Spotlight on LEED." *Environmental Building News*, December 1. http://www.buildinggreen.com/auth/article.cfm/2003/12/1/Spotlight-on-LEED/.

Mang, Pamela. 2001. "Regenerative Design: Sustainable Design's Coming Revolution." DesignIntelligence, July 1. http://www.di.net/articles/archive/2043/.

National Renewable Energy Laboratory. n.d. "Energy Analysis and Tools." http://www.nrel.gov/buildings/energy_analysis.html (accessed May 2, 2011).

New Mexico Solar Energy Association. 1998. "Passive Solar Design." Albuquerque, NM. http://www.nmsea.org/Passive_Solar/Passive_Solar_Design.htm.

Wendt, Allyson. 2009. "The Living Building Challenge: Can It Really Change the World?" *Environmental Building News*, June 1. http://www.buildinggreen.com/auth/article.cfm/2009/5/29/The-Living-Building-Challenge-Can-It-Really Change-the-World/.

———. 2010: "First Buildings Certified under Living Building Challenge." *Environmental Building News*, October 12. http://www.buildinggreen.com/auth/article.cfm/2010/10/12/First-Projects-Certified-Under-Living-Building-Challenge-ILBI/.

Wendt, Allison, and Nadav Malin. 2010. "Integrated Design Meets the Real World." *Environmental Building News*, May 1. http://www.buildinggreen.com/auth/article.cfm/2010/5/1/Integrated-Design-Meets-the-Real-World/?.

Wilson, Alex. 2006. "Biophilia in Practice: Buildings That Connect People with Nature." *Environmental Building News*, July 9. http://www.buildinggreen.com/auth/article.cfm/2006/7/9/Biophilia-in-Practice-Buildings-that-Connect-People-with-Nature/.

Chapter 11. Before Construction

Connell, John. 1998. *Homing Instinct: Using Your Lifestyle to Design and Build Your Home.* New York: McGraw-Hill Professional.

Magwood, Chris, Peter Mack, and Tina Therrien. 2005. *More Straw Bale Building: A Complete Guide to Designing and Building with Straw.* Gabriola Island, BC: New Society Publishers.

McHenry, Paul Graham Jr. 1984. *Adobe and Rammed Earth Buildings: Design and Construction.* Tucson: University of Arizona Press.

Savitz, Andrew W. 2006. *The Triple Bottom Line.* San Francisco, CA: Jossey-Bass.

Chapter 12. Foundations for Natural Buildings

EcoSmart Concrete. 2004–2008. "The Facts: SCM Basics." http://www.ecosmartconcrete.com/facts_scmbasics.cfm.

Headwaters Resources. 2005. "Proportioning Fly Ash Concrete Mixes." March. http://www.flyash.com/data/upimages/press/TB.4%20Proportioning%20Fly%20Ash%20Concrete%20Mixes.pdf.

Koko, Sigi. 2003. "Rubble Trench Foundations: A Brief Overview." *Building Safety Journal*, May, 32–33. http://www.dcat.net/resources/bsj_may03feature.pdf.

National Association of Home Builders. 1998. "Energy-Efficient Resource-Efficient Frost-Protected Shallow Foundations." http://www.nahb.org/assets/docs/publication/Energy-efficient-frost-protected-shallow-foundations_1211200244041PM.pdf.

National Association of Home Builders Research Center. n.d. "Autoclaved Aerated Concrete." http://www.toolbase.org/Technology-Inventory/Foundations/autoclaved-aerated-concrete (accessed November 13, 2010).

Oak Ridge National Labs and Polish Academy of Sciences. 2001. "Thermal Mass—Energy Savings Potential in Residential Buildings." http://www.ornl.gov/sci/roofs+walls/research/detailed_papers/thermal/dynamic.html.

Swanson, George. n.d. "Magnesium Oxide, Magnesium Chloride, and Phosphate-Based Cements." *Building Biology Based New Building Protocol.* http://www.greenhomebuilding.com/pdf/MgO-GENERAL.pdf (accessed September 2011).

Velonis, Elias. 1983. "Rubble-Trench Foundations." *Fine Homebuilding* December–January, 66–68. http://digioia.com/wp-content/uploads/2010/09/rubble_trench_foundations.pdf.

Chapter 13. Framing for Natural Buildings

Goodman, Joshua. 2008. "Bamboo Is Catching On Among Green Architects." Associated Press, February 15. http://poorbuthappy.com/colombia/post/bamboo-is-catching-on-among-green-architects/.

Lamboo Architectural and Structural Bamboo. n.d. "Lamboo Structure Structural Applications." http://www.lamboo.us/images/stories/brochures/structural%20brochure%20portrait.pdf (accessed January 24, 2011).

Malin, Nadav. 1994. "Steel or Wood Framing: Which Way Should We Go?" *Environmental Building News,* July–August. Archived at http://www.buildinggreen.com/auth/article.cfm/1994/7/1/Steel-or-Wood-Framing-Which-Way-Should-We-Go/?.

Nakasone, Sarah. 2003. "Bamboo: An Alternative Movement." *Illumin* 5 (4). http://illumin.usc.edu/article.php?articleID=114&page=1.

Partnership for Advanced Technology in Housing (PATH)/National Association of Home Builders (NAHB). n.d. "Advanced Framing Techniques." *ToolBase: TechSpecs.* http://www.toolbase.org/pdf/techinv/oveadvancedframingtechniques_techspec.pdf (accessed December 3, 2010).

Chapter 14. Insulative Wall Systems: Straw Bale

King, Bruce. 2006. *Design of Straw Bale Buildings: The State of the Art.* San Rafael, CA: Green Building Press.

Lacinski, Paul, and Michel Bergeron. 2000. *Serious Straw Bale: A Guide to Construction in All Climates.* White River Junction, VT: Chelsea Green Publishing.

Magwood, Chris, Peter Mack, and Tina Therrien. 2005. *More Straw Bale Building: A Complete Guide to Designing and Building with Straw.* Gabriola Island, BC: New Society Publishers.

Chapter 15. Insulative Wall Systems: Straw-Clay, Woodchip-Clay, and Cellulose

Andersen, F. 2002. "Light-Clay: An Introduction to German Clay Building Techniques." In *The Art of Natural Building: Design, Construction, Resources,* ed. J. F. Kennedy, C. Wanek, and M. Smith. Gabriola Island, BC: New Society Publishers.

Canadian Mortgage and Housing Corporation. 2005. "Initial Material Characterization of Straw Light Clay." Research Highlight Technical Series 05–109, June.

The Canelo Project. http://www.caneloproject.com/.

International Green Construction Code (proposed). 2011. "Section 508, Straw-Clay Code," 5/4/10.

National Fiber. http://www.nationalfiber.com/index.htm.

U.S. Forest Products Laboratory and Design Coalition. 2004. "Straw-Clay Thermal Testing." http://www.designcoalition.org/articles/Lansing-LHJ/research/Ktesting2.htm.

Chapter 16. Natural Mass-Wall Systems: Earth and Stone

Arch-Net. n.d. "Geltaftan Process." http://archnet.org/library/sites/one-site.jsp?site_id=26 (accessed January 23, 2011).

California Institute of Earth Art and Architecture. n.d. "What Is Superadobe?" http://calearth.org/building-designs/what-is-superadobe.html (accessed January 23, 2011).

Chiras, Dan. 2000. *The Natural House: A Complete Guide to Healthy, Energy-Efficient, Environmental Homes.* White River Junction, VT: Chelsea Green Publishing.

Goodvin, Christina, Gord Baird, and Ann Baird. 2011. *Cob Home Performance Report.* August 26. Vancouver Island, BC: Eco-Sense.ca. http://ecosenseliving.files.wordpress.com/2011/09/science-research-report_sept_1.pdf

Kiffmeyer, Doni and Kaki. 2004. *Earthbag Building.* Gabriola Island, BC: New Society Publishers.

McHenry, Paul G. 1989. *Adobe and Rammed Earth Construction: Design and Construction.* Tucson: University of Arizona Press.

———. 1998. *The Adobe Story.* Albuquerque: University of New Mexico Press.

Nearing, Helen and Scott. 1970. *Living the Good Life: How to Live Sanely and Simply in a Troubled World.* New York: Schocken Books.

Roy, Rob. 1992. *The Complete Book of Cordwood Masonry.* New York: Sterling Publishing.

Van Lengen, Johan. 2008. *The Barefoot Architect.* Bolinas, CA: Shelter Publications.

Weismann, Adam, and Katy Bryce. 2006. *Building with Cob.* Totnes, Devon, UK: Green Books Ltd.

Chapter 17. Natural Plastering

The Canelo Project. http://www.caneloproject.com/.

Chivers, Ryan. 2011. "Lectures and Conversation from Spring 2011." Natural Building Certificate.

Francis, Harry. 2011. "Conversation on Straw Bale Social Club." Listserv on lime-stabilization of clay plaster. http://groups.yahoo.com/group/SB-r-us/message/14236.

Guelberth, Cedar Rose, and Dan Chiras. 2002. *The Natural Plaster Book: Earth, Lime, and Gypsum Plasters for Natural Homes.* Natural Building Series. Gabriola Island, BC: New Society Publishers.

Holmes, Stafford, and Michael Wingate. 1997/2002. *Building with Lime: A Practical Iintroduction.* Rugby, UK: ITDG Publishing, 171–78.

Lacinski, Paul, and Michel Bergeron. 2000. *Serious Straw Bale: A Guide to Construction in All Climates.* White River Junction, VT: Chelsea Green Publishing.

King, Bruce. 2006. *Design of Straw Bale Buildings: The State of the Art.* San Rafael, CA: Green Building Press.

Mitchell, David S. 2007. "Historic Scotland Inform Guide: The Use of Lime and Cement in Traditional Buildings." Technical Conservation Research and Education Group, Edinburgh, UK. www.historic-scotland.gov.uk.

Weismann, Adam, and Katy Bryce. 2006. *Building with Cob.* Totnes, Devon, UK: Green Books Ltd.

Weismann, Adam, and Katy Bryce. 2008. *Using Natural Finishes.* Totnes, Devon, UK: Green Books.

Chapter 18. Roofs for Natural Buildings

Bark House. Highland Craftsmen, Inc., Spruce Pine, NC. http://www.barkhouse.com/.

Gould, John, and Erik Nissen-Peterson. 2002. *Rainwater Catchment Systems for Domestic Supply: Design, Construction, and Implementation.* London: ITDG Publishing.

Houston, James, and John N. Fugelso. 2008. "Fabricating and Installing Side-Lap Roof Shingles in Eastern Pennsylvania." *APT Bulletin* 39, no. 1: 33–41. http://www.jstor.org/stable/25433936.

Johnson, Kirk. 2009. "It Is Now Legal to Catch a Raindrop in Colorado." *New York Times,* June 28. http://www.nytimes.com/2009/06/29/us/29rain.html?em.

Maunder, Elwood R., Betty E. Mitson, Barbara D. Holman, Virgil G. Peterson, Paul R. Smith, and Charles Plant. 1975. "Red Cedar Shingles & Shakes: Unique Wood Industry." *Journal of Forest History* 19, no. 2 (April): 56–71. http://www.jstor.org/stable/3983235.

Mendez, Carolina B., J. Brandon Klenzendorf, Brigit R. Afshar, Mark T. Simmons, Michael E. Barrett, Kerry A. Kinney, and Mary Jo Kirisits. 2010. "The Effect of Roofing Material on the Quality of Harvested Rainwater." *Water Research,* Texas Water Development Board, January. http://www.twdb.state.tx.us/innovativewater/rainwater/projects/rainquality/Final_Report_013110.pdf.

Stirling, Rod. 2010. "Residual Extractives in Western Red Cedar Shakes and Shingles after Long-Term Field Testing." *Forest Products Journal* 60, no. 4 (June): 353.

U.S. EPA. n.d. "Reducing Urban Heat Islands: Compendium of Strategies." http://www.epa.gov/heatisland/resources/pdf/GreenRoofsCompendium.pdf (accessed January 30, 2012).

Chapter 19. Floors for Natural Buildings

Ashish, Shukla, G. N. Tiwari, and M. S. Sodha. 2009. "Embodied Energy Analysis of Adobe House." Renewable Energy 34: 755–61. http://www.sciencedirect.com.ezproxy.uvm.edu/science?_ob=MImg&_imagekey=B6V4S-4SV12MV-6-1&_cdi=5766&_user=1563816&_pii=S0960148108001729&_origin=search&_zone=rslt_list_item&_coverDate=03%2F31%2F2009&_sk=999659996&wchp=dGLzVzz-zSkWb&md5=43207eabbfb9f0aa1c12bbc4310d641b&ie=/sdarticle.pdf.

Environmental Building News. 2000. "Restoring Forests and Making Flooring." http://www.buildinggreen.com/auth/article.cfm/2000/4/1/Restoring-Forests-and-Making-Flooring/.

Holmes, Stafford, and Michael Wingate. 1997/2002. *Building with Lime: A Practical Introduction.* Rugby, UK: ITDG Publishing, 171–78.

Malin, Nadav, and Jessica Boehland. 2006. "Bamboo in Construction: Is the Grass Always Greener?" *Environmental Building News,* March. http://www.buildinggreen.com/auth/article.cfm/2006/3/1/Bamboo-in-Construction-Is-the-Grass-Always-Greener/.

McHenry, Paul G. 1989. *Adobe and Rammed Earth Construction: Design and Construction.* Tucson: University of Arizona Press.

Swanson, George. n.d. "Magnesium Oxide, Magnesium Chloride, and Phosphate-Based Cements." *Building Biology Based New Building Protocol.* http://www.greenhomebuilding.com/pdf/MgO-GENERAL.pdf (accessed September 2011).

Chapter 20. Natural Finishes

Bikales, Norbert M, ed. 1971. *Adhesion and Bonding.* New York: John Wiley & Sons, Inc.

Caliwel paint. http://www.alistagen.com/products.htm.

The Earth Pigments Company. n.d. "Coating Existing Latex Paints and Primers with Natural or Traditional Pigmented Finishes." http://www.earthpigments.com/coating-existing-latex-paint.cfm (accessed December 27, 2010).

Edwards, Lynn, and Julia Lawless. 2002. *The Natural Paint Book.* Emmaus, PA: Rodale Organic Living Books.

Ehrlich, Brent. 2008. "Whey-Based Floor and Furniture Finishes." *Environmental Building News,* June. http://www.buildinggreen.com/auth/article.cfm/2008/5/29/Whey-Based-Floor-and-Furniture-Finishes/.

Malin, Nadav. 1999. "Paint the Room Green." *Environmental Building News,* February. http://www.buildinggreen.com/auth/article.cfm/1999/2/1/Paint-the-Room-Green/.

Martens, Charles R. 1974. *Technology of Paints, Varnishes, and Lacquers.* New York: Robert E. Krieger Publishing Co.

Penn, Theodore Zuk. 1984. "Decorative and Protective Finishes, 1750–1850: Materials, Process, and Craft." *Bulletin of the Association for Preservation Technology* 16, no. 1: 3–46. http://www.jstor.org/stable/1493913.

SAP Design Guild. 2003. "Color Glossary." http://www.sapdesignguild.org/resources/glossary_color/.

Straube, John. 1999. "Moisture Properties of Plaster and Stucco for Straw Bale Buildings." University of Waterloo. Archived at http://www.homegrownhome.co.uk/pdfs/Straube_Moisture_Tests.pdf.

Weismann, Adam, and Katy Bryce. 2008. *Using Natural Finishes.* Totnes, Devon, UK: Green Books.

Wilson, Alex. 2003. "Mineral Silicate Paints Tops for Durability." *Environmental Building News,* October. http://www.buildinggreen.com/auth/article.cfm/2003/10/1/Mineral-Silicate-Paints-Tops-for-Durability/.

Zinsser Co., Inc. 2003. "The Story of Shellac." http://www.naturalhandyman.com/iip/author/zinsser/shellac.html.

Chapter 21. Mechanicals and Utilities for Natural Buildings

Allen, Edward. 1980. *How Buildings Work: The Natural Order of Architecture.* New York: Oxford University Press.

Bryden, Mark, Dean Still, Peter Scott, Geoff Hoffa, Damon Ogle, Rob Bailis, and Ken Goyer. 2011. "Design Principles for Wood Burning Cook Stoves." Cottage Grove, OR: Aprovecho Research Center. http://www.rocketstove.org/images/stories/design-principles-for-wood-burning-cook-stoves.pdf.

BuildingGreen CSI Divisions. n.d. CSI section 23 52 01: "Residential and Small Commercial Heating Boilers." http://www.buildinggreen.com/auth/productsByCsiSection.cfm?csiMF2004ID=6997 (accessed May 25, 2011).

———. n.d. CSI section 23 52 03: "Cordwood, Pellet, and Multi-Fuel Boilers." http://www.buildinggreen.com/auth/productsByCsiSection.cfm?csiMF2004ID=7000 (accessed May 25, 2011).

———. n.d. CSI Section 23 54 00: "Furnaces." http://www.buildinggreen.com/auth/productsByCsiSection.cfm?csiMF2004ID=3741&show=all#articles (accessed May 25, 2011).

Environmental Building News. 1998. "EMF May Be Carcinogenic After All." October 1. http://www.buildinggreen.com/auth/article.cfm/1998/10/1/EMF-May-Be-Carcinogenic-After-All/?.

———. 2008. "Solar Reflectance Index and Cool Roofs." Back Page Primer, January 1. http://www.buildinggreen.com/auth/article.cfm/2008/1/1/Solar-Reflectance-Index-and-Cool-Roofs/.

Lang, Susan S. 2005. "Cornell Ecologist's Study Finds That Producing Ethanol and Biodiesel from Corn and Other Crops Is Not Worth the Energy." Cornell University News Service, July 5. http://www.news.cornell.edu/stories/july05/ethanol.toocostly.ssl.html.

Malin, Nadav. 2008. "Counting Carbon: Understanding Carbon Footprints of Buildings." *Environmental Building News,* July 1. http://www.buildinggreen.com/auth/article.cfm/2008/6/27/Counting-Carbon-Understanding-Carbon-Footprints-of-Buildings/?.

Malin, Nadav, and Alex Wilson. 2000. "Ground-Source Heat Pumps: Are They Green?" *Environmental Building News,* July 1. http://www.buildinggreen.com/auth/article.cfm/2000/7/1/Ground-Source-Heat-Pumps-Are-They-Green/?.

Massachusettes Executive Office of Environmental Affairs. 2006. "Massachusetts State Water Conservation Standards." http://www.mass.gov/Eoeea/docs/eea/water/water_conservation_standards.pdf.

Riversong, Robert. 2010. "Pex vs. Copper." Self-published white paper, Riversong HouseWright, Warren, VT.

———. 2011. "Electro-Pollution May Be the Greatest Threat to the Biosphere." Blog posting at Transition Vermont, March 23. http://transitionvermont.ning.com/profiles/blogs/electro pollution-may-be-the.

Roberts, Tristan. 2008. "Ductless Mini-Splits and Their Kin: The Revolution in Variable-Refrigerant-Flow Air Conditioning." *Environmental Building News,* August 1. http://www.buildinggreen.com/auth/article.cfm/2008/7/29/Ductless-Mini-Splits-and-Their-Kin-The-Revolution-in-Variable-Refrigerant-Flow-Air-Conditioning/?.

Sierra Club, Iowa Chapter. n.d. "Ecological Wastewater Treatment." Des Moines, IA. http://iowa.sierraclub.org/06WQS/Ecological WastewaterTreatment.pdf (accessed May 22, 2011).

Solomon, Nancy B. 2007. "Go with the Flow." *GreenSource,* July 1, BuildingGreen.com. http://www.buildinggreen.com/auth/article.cfm/2007/7/1/Go-With-the-Flow/.

Stockholm International Water Institute (SIWI). n.d. "Statistics." http://www.siwi.org/sa/node.asp?node=159 (accessed May 22, 2011).

Union of Concerned Scientists. 2010. "How It Works: Water for Natural Gas." October 5. http://www.ucsusa.org/clean_energy/technology_and_impacts/energy_technologies/water-energy-electricity-natural-gas.html.

United States Department of Energy. n.d. "Ventilation Preheating. Energy Savers." http://www.energysavers.gov/your_home/space_heating_cooling/index.cfm/mytopic=12510 (accessed May 24, 2011).

United States Energy Information Administration. 2009. "What Are the Major Sources and Users of Energy in the United States?" October 28. http://www.eia.doe.gov/energy_in_brief/major_energy_sources_and_users.cfm.

———. 2011a. "Net Generation by Energy Source: Total (All Sectors)." April 14. http://www.eia.gov/electricity/monthly/index.cfm#two.

———. 2011b. "Renewable and Alternative Fuels." March 16. http://www.eia.gov/renewable/.

———. 2011c. "Table 1.2 Primary Energy Production by Source, 1949–2009." http://www.eia.gov/totalenergy/data/annual/txt/ptb0102.html.

United States Environmental Protection Agency. 1980. "Design Manual: Onsite Wastewater Treatment and Disposal Systems." October. http://www.epa.gov/owm/septic/pubs/septic_1980_osdm_all.pdf.

Water.org. n.d. "Water Facts." http://water.org/learn-about-the-water-crisis/facts/#environment (accessed May 22, 2011).

Wilson, Alex. 1992. "Pellet Stoves: Wood Burning That's Better for the Environment." *Environmental Building News,* September 1. http://www.buildinggreen.com/auth/article.cfm/1992/9/1/Pellet-Stoves-Wood-Burning-That-s-Better-for-the-Environment/.

———. 1994a. "Building Design and EMF." *Environmental Building News,* March 1. http://www.buildinggreen.com/auth/article.cfm/1994/3/1/Building-Design-and-EMF/?.

———. 1994b. "Keeping the Heat Out: Cooling Load Avoidance Strategies." *Environmental Building News,* May 1. http://www.buildinggreen.com/auth/article.cfm/1994/5/1/Keeping-the-Heat-Out-Cooling-Load-Avoidance-Strategies/.

———. 2002. "Radiant-Floor Heating: When It Does—and Doesn't—Make Sense." *Environmental Building News,* January 1. http://www.buildinggreen.com/auth/article.cfm/2002/1/1/Radiant-Floor-Heating-When-It-Does-and-Doesn-t-Make-Sense/.

———. 2004. "Wire and Cable: Untangling Complex Environmental Issues." *Environmental Building News,* March 1. http://www.buildinggreen.com/auth/article.cfm/2004/3/1/Wire-and-Cable-Untangling-Complex-Environmental-Issues/?.

———. 2008. "Alternative Water Sources: Supply Side Solutions for Green Buildings." *Environmental Building News*, May 1. http://www.buildinggreen.com/auth/article.cfm/2008/4/29/Alternative-Water-Sources-Supply-Side-Solutions-for-Green-Buildings/.

———. 2011. "Water Heating: A Look at the Options." *Environmental Building News*, October 2. http://www.buildinggreen.com/auth/article.cfm/2002/10/1/Water-Heating-A-Look-at-the-Options/.

Wilson, Alex, and Nadav Malin. 1997. "Ecological Wastewater Treatment." *Environmental Building News*, July 1. http://www.buildinggreen.com/auth/article.cfm/1996/7/1/Ecological-Wastewater-Treatment/?.

Wilson, Alex, and Mary Rickel Pelletier. 2001. "Green Roofs: Using Roofs for More Than Keeping Dry." *Environmental Building News*, November 1. http://www.buildinggreen.com/auth/article.cfm/2001/11/1/Green-Roofs-Using-Roofs-for-More-Than-Keeping-Dry/.

Conclusion

Brown, Marilyn A., Frank Southworth, and Andrea Sarzynski. 2008. "Shrinking the Carbon Footprint of Metropolitan America." Metropolitan Policy Program, the Brookings Institute, Washington, DC, May 4–7. http://www.brookings.edu/reports/2008/~/media/Files/rc/papers/2008/05_carbon_footprint_sarzynski/carbonfootprint_brief.pdf.

Clendaniel, Morgan. 2011. "City Dwellers Have Huge Carbon Footprints, Because They Buy So Much." Fast Company, New York, April 25. http://www.fastcompany.com/1749726/city-dwellers-have-huge-carbon-footprints-because-they-buy-so-much.

Co-Housing Association of the United States. 2008. "Cohousing Neighborhoods." Brochure. http://www.cohousing.org/docs/WhatIsCohousingTriFold.pdf (accessed May 28, 2011).

Heinonen, Jukka, and Seppo Junnila. 2011. "Implications of Urban Structure on Carbon Consumption in Metropolitan Areas." *Environ. Res Lett* 6. http://iopscience.iop.org/1748-9326/6/1/014018/fulltext.

United Nations Population Fund (UNFPA). 2007. "Urbanization: A Majority in Cities." May. http://www.unfpa.org/pds/urbanization.htm.

Vaughn, Adam. 2009. "City Dwellers Have Smaller Carbon Footprints, Study Finds." *Guardian* (London), March 23. http://www.guardian.co.uk/environment/2009/mar/23/city-dwellers-smaller-carbon-footprints.

INDEX

THE YESTERMORROW DESIGN/BUILD LIBRARY

Chelsea Green Publishing and the Yestermorrow Design/Build School have partnered to produce a series of books giving professional builders, designers, homeowners, and DIYers state-of-the-art information and guidance on using sustainable design strategies, natural building methods and materials, renewable energy sources, and environmentally progressive construction techniques. Accompanied by instructional DVDs, books in the series strive to be accessible to readers of all skill levels.

About Yestermorrow

Yestermorrow Design/Build School trains students in the technical arts of design and craftsmanship as a single integrated process. Through experiential learning in the design studio and workshop and with an emphasis on designing in partnership with the natural world, students develop critical thinking skills and technical know-how, expand perception and perspective, and deepen their creative assets for living and building resourcefully through the work of their own hands. With more than 150 class offerings in sustainable design, construction, woodworking, and architectural craft, the students who come to Yestermorrow are DIYers and professionals, women and men, undergraduates and lifelong learners, hobbyists and those seeking a career change. Immersion-style classes are offered as one-day to two-week-long workshops, six-week to eleven-week certification programs, and accredited semester programs. Courses and programs are taught by top architects, builders, and craftspeople from around the country. Our classes provide significant hands-on experience, and several projects each year are built in collaboration with community partners to address community needs. By uniting the processes of design and building and exploring the relationships between the natural and built environments, Yestermorrow empowers students to create intentional and inspired buildings and communities that enhance our world.

www.yestermorrow.org

About Chelsea Green

For twenty-six years, Chelsea Green has been the publishing leader for books on the politics and practice of sustainable living. We are a founding member of the Green Press Initiative and have been printing books on recycled paper since 1985, when our first list of books appeared. We lead the industry both in terms of content—foundational books on renewable energy, green building, organic agriculture, eco-cuisine, and ethical business—and in terms of environmental practice, printing 95% of our books on recycled paper with a minimum 30% post-consumer waste and aiming for 100% whenever possible. This approach is a perfect example of what is called a "triple bottom line" practice, one that benefits people, planet, and profit, and the emerging new model for sustainable business in the twenty-first century.

www.chelseagreen.com

About the Authors

Jacob Deva Racusin, left, and Ace McArleton

Jacob Deva Racusin and Ace McArleton co-own New Frameworks Natural Building, LLC, a company offering services in green remodeling, new construction, consultation, and education featuring natural building technologies. Instructors at educational organizations including Yestermorrow Design/Build School, Jacob and Ace are involved with the Natural Building Certificate Program, which Jacob helped develop and has served as program director. They are also codevelopers and instructors of the graduate-level Professional Semester of Integrative Design and Construction at Yestermorrow. Their company is a member of the Northeast Sustainable Energy Association, Vermont Builders for Social Responsibility, and the Vermont chapter of the U.S. Green Building Council. Jacob is a Certified Passive House Consultant and a member of Seven Generations Natural Builders, which offers training in natural building technologies around the world. Ace, a former union tilesetter, has developed curricula and instructed at the Institute for Social Ecology. Jacob and Ace are members of Natural Builders Northeast and the Timber Framers Guild. Through their work as builders, consultants, and educators, Jacob and Ace are able to merge their passions for fine craft, ecological stewardship, relationship to place, and social justice.